De Gruyter Series in Nonlinear Analysis and Applications 20

John R. Graef
Johnny Henderson
Abdelghani Ouahab

Impulsive Differential Inclusions

A Fixed Point Approach

De Gruyter

Mathematics Subject Classification 2010: 34A60, 34A37, 34K09, 34K45, 34B37

Authors
John R. Graef
Department of Mathematics
University of Tennessee at Chattanooga
Chattanooga, TN 37403-2598
USA
john-graef@utc.edu

Johnny Henderson
Department of Mathematics
Baylor University
Waco, Texas 76798-7328
USA
johnny_henderson@baylor.edu

Abdelghani Ouahab
Department of Mathematics
University of Sidi Bel Abbes
BP 89 2000 Sidi Bel Abbes
Algeria
agh_ouahab@yahoo.fr

ISBN 978-3-11-029361-6
e-ISBN 978-3-11-029531-3
Set-ISBN 978-3-11-029532-0
ISSN 0941-813X

Library of Congress Cataloging-in-Publication Data
A CIP catalog record for this book has been applied for at the Library of Congress.

Bibliographic information published by the Deutsche Nationalbibliothek
The Deutsche Nationalbibliothek lists this publication in the Deutsche Nationalbibliografie; detailed bibliographic data are available in the Internet at http://dnb.dnb.de.

Typesetting: PTP-Berlin Protago-TEX-Production GmbH, www.ptp-berlin.de
Printing and binding: Hubert & Co. GmbH & Co. KG, Göttingen
⊗ Printed on acid-free paper

Printed in Germany

www.degruyter.com

John R. Graef: *To my wife, Frances.*

Johnny Henderson: *To my wife, Darlene.*

Abdelghani Ouahab: *To my parents, wife Zohra, my chidren Hemza, Fatima, Zohra and sisters, brothers and all the members of my family.*

Contents

Notations xi

1 Introduction and Motivations 1

1.1 Introduction 1

1.2 Motivational Models 8

1.2.1 Kruger–Thiemer Model 8

1.2.2 Lotka–Volterra Model 8

1.2.3 Pulse Vaccination Model 9

1.2.4 Management Model 9

1.2.5 Some Examples in Economics and Biomathematics 10

2 Preliminaries 11

2.1 Some Definitions 11

2.2 Some Properties in Fréchet Spaces 12

2.3 Some Properties of Set-valued Maps 13

2.3.1 Hausdorff Metric Topology 15

2.3.2 Vietoris Topology 18

2.3.3 Continuity Concepts and Their Relations 20

2.3.4 Selection Functions and Selection Theorems 28

2.3.5 Hausdorff Continuity 30

2.3.6 Measurable Multifunctions 32

2.3.7 Decomposable Selection 35

2.4 Fixed Point Theorems 36

2.5 Measures of Noncompactness: MNC 37

2.6 Semigroups 40

2.6.1 C_0-semigroups 40

2.6.2 Integrated Semigroups 42

2.6.3 Examples 44

2.7 Extrapolation Spaces 45

3 FDEs with Infinite Delay 47

3.1 First Order FDEs 47

3.1.1 Examples of Phase Spaces 48

3.1.2 Existence and Uniqueness on Compact Intervals . . . 50
3.1.3 An Example . . . 57
3.2 FDEs with Multiple Delays . . . 58
3.2.1 Existence and Uniqueness Result on a Compact Interval . . . 58
3.2.2 Global Existence and Uniqueness Result . . . 65
3.3 Stability . . . 66
3.3.1 Stability Result . . . 67
3.4 Second Order Impulsive FDEs . . . 69
3.4.1 Existence and Uniqueness Results . . . 71
3.5 Global Existence and Uniqueness Result . . . 76
3.5.1 Uniqueness Result . . . 77
3.5.2 Example . . . 82
3.5.3 Stability . . . 83
4 Boundary Value Problems on Infinite Intervals 86
4.1 Introduction . . . 86
4.1.1 Existence Result . . . 87
4.1.2 Uniqueness Result . . . 92
4.1.3 Example . . . 96
5 Differential Inclusions 98
5.1 Introduction . . . 98
5.1.1 Filippov's Theorem . . . 98
5.1.2 Relaxation Theorem . . . 111
5.2 Functional Differential Inclusions . . . 113
5.2.1 Filippov's Theorem for FDIs . . . 114
5.2.2 Some Properties of Solution Sets . . . 123
5.3 Upper Semicontinuity without Convexity . . . 125
5.3.1 Nonconvex Theorem and Upper Semicontinuity . . . 126
5.3.2 An Application . . . 130
5.4 Inclusions with Dissipative Right Hand Side . . . 131
5.4.1 Existence and Uniqueness Result . . . 131
5.5 Directionally Continuous Selection and IDIs . . . 136
5.5.1 Directional Continuity . . . 136
6 Differential Inclusions with Infinite Delay 140
6.1 Existence Results . . . 140
6.2 Boundary Differential Inclusions . . . 150

7 Impulsive FDEs with Variable Times 154
7.1 Introduction 154
7.1.1 Existence Results 154
7.1.2 Neutral Functional Differential Equations 155
7.2 Impulsive Hyperbolic Differential Inclusions with Infinite Delay 156
7.3 Existence Results 157
7.3.1 Phase Spaces 157
7.3.2 The Nonconvex Case 168
8 Neutral Differential Inclusions 171
8.1 Filippov's Theorem 171
8.2 The Relaxed Problem 182
8.2.1 Existence and Compactness Result: an MNC Approach 189
9 Topology and Geometry of Solution Sets 199
9.1 Background in Geometric Topology 199
9.2 Aronszajn Type Results 201
9.2.1 Solution Sets for Impulsive Differential Equations 206
9.3 Solution Sets of Differential Inclusions 208
9.4 σ-selectionable Multivalued Maps 208
9.4.1 Contractible and R_δ-contractible 212
9.4.2 R_δ-sets 218
9.5 Impulsive DIs on Proximate Retracts 219
9.5.1 Viable Solution 220
9.6 Periodic Problems 226
9.6.1 Poincaré Translation Operator 226
9.6.2 Existence Result 227
9.7 Solution Set for Nonconvex Case 231
9.7.1 Continuous Selection and AR of Solution Sets 232
9.8 The Terminal Problem 245
9.8.1 Existence and Solution Set 245
10 Impulsive Semilinear Differential Inclusions 254
10.1 Nondensely Defined Operators 254
10.2 Integral Solutions 255
10.3 Exact Controllability 267
10.3.1 Controllability of Impulsive FDIs 267
10.3.2 Controllability of Impulsive Neutral FDIs 276

10.4 Controllability in Extrapolation Spaces 282
10.5 Second Order Impulsive Semilinear FDIs 290
10.5.1 Mild Solutions 291
10.5.2 Filippov's Theorem 292
10.5.3 Filippov–Wazewski's Theorem 303

11 Selected Topics 306

11.1 Stochastic Differential Equations 306
11.1.1 Itô Integral 307
11.1.2 Definition of a Mild Solution 308
11.1.3 Existence and Uniqueness 311
11.1.4 Global Existence and Uniqueness 321
11.2 Impulsive Sweeping Processes 327
11.2.1 Preliminaries in Nonsmooth Analysis 327
11.2.2 Uniqueness Result 328
11.3 Integral Inclusions of Volterra Type in Banach Spaces 331
11.3.1 Resolvent Family 332
11.3.2 Existence results 334
11.3.3 The Convex Case: an MNC Approach 339
11.3.4 The Nonconvex Case 342
11.4 Filippov's Theorem 346
11.4.1 Filippov's Theorem on a Bounded Interval 346
11.5 The Relaxed Problem 351

Appendix 357

A.1 Čech Homology Functor with Compact Carriers 357
A.2 The Bochner Integral 359
A.3 Absolutely Continuous Functions 361
A.4 Compactness Criteria in $C([a,b],E)$, $C_b([0,\infty),E)$, and $PC([a,b],E)$ 363
A.5 Weak-compactness in L^1 365
A.6 Proper Maps and Vector Fields 367
A.7 Fundamental Theorems in Functional Analysis 367

Bibliography 369

Index 399

Notations

- BVPs: Boundary value problems.
- DIs: Differential Inclusions.
- FDIs: Functional differential inclusions.
- IDIs: Impulsive differential inclusions.
- IFDIs: Impulsive functional differential inclusions.
- IVPs: Initial value problems.
- $\mathbb{N}$ set of positive natural numbers.
- $\mathbb{Q}$ set of rational numbers.
- $\mathbb{R}$ set of real numbers.
- $d(x, A) = \inf\{d(x, y) : y \in A\}$.
- $\mathcal{O}_\epsilon = \{x \in X : d(x, A) < \epsilon\}$.
- $\overline{A} = clA$ closure of the set A.
- $B(x_0, r)$ open ball with radius r centred at x_0.
- $\overline{co}A$ closure of the convex hull of the set A.
- AR absolute retract.
- ANR neighbourhood retract.
- $\chi(A)$ Hausdorff measure of noncompactness of the set A.

Chapter 1

Introduction and Motivations

1.1 Introduction

Differential equations with impulses were considered for the first time in the 1960's by Milman and Myshkis [375,376]. After a period of active research, primarily in Eastern Europe during 1960–1970, early studies culminated with the monograph by Halanay and Wexler [266].

The dynamics of many evolving processes are subject to abrupt changes, such as shocks, harvesting, and natural disasters. These phenomena involve short-term perturbations from continuous and smooth dynamics, whose duration is negligible in comparison with the duration of an entire evolution. In models involving such perturbations, it is natural to assume these perturbations act instantaneously or in the form of "impulses." As a consequence, impulsive differential equations have been developed in modeling impulsive problems in physics, population dynamics, ecology, biotechnology, industrial robotics, pharmacokinetics, optimal control, and so forth. Again, associated with this development, a theory of impulsive differential equations has been given extensive attention. Works recognized as landmark contributions include the books [39, 40, 68, 335, 435] and the papers [9, 125, 126, 199, 209, 210, 356, 414, 473]. There are also many different studies in biology and medicine for which impulsive differential equations provide good models; see, for instance, [10, 327, 328] and the references therein.

In recent years, many examples of differential equations with impulses with fixed moments have flourished in several contexts. In the periodic treatment of some diseases, impulses correspond to administration of a drug treatment or a missing product. In environmental sciences, impulses model seasonal changes of the water level of artificial reservoirs. The theory and applications addressing such problems have heavily involved functional differential equations as well as impulsive functional differential equations. Recently, extensions to functional differential equations with impulsive effects with fixed moments have been done by Benchohra et al. [67, 70, 71] and Ouahab [397], with the aid of the nonlinear alternative and Schauder's theorem, as well as by Yujun and Erxin [495] and Yujun [494] by using coincidence degree theory. For other results concerning functional differential equations, we refer the interested reader to the monographs of Azbelez et al [38], Erbe, Qingai and Zhang [193], Hale and Lunel [268], and Henderson [273].

There is a great variety of motivations that led mathematicians, studying dynamical systems having velocities uniquely determined by the state of the system, but loosely upon it, to replace differential equations

$$y' = f(y)$$

by differential inclusions

$$y' \in F(y).$$

A system of differential inequalities

$$y_i' \leq f^i(y_1, \ldots, y_n), \quad i = 1, \ldots, n,$$

can also be considered as a differential inclusion. If an implicit differential equation

$$f(y, y') = 0$$

is given, then we can put $F(y) = \{v : f(y, v) = 0\}$ to reduce it to a differential inclusion. Differential inclusions are used to study ordinary differential equations with an inaccurately known right-hand side.

As an example, consider the differential equation with discontinuous right-hand side,

$$y' = 1 - 2\,\mathrm{sgn}\, y,$$

where

$$\mathrm{sgn}\, y = \begin{cases} +1, & \text{if } y > 0, \\ 0, & \text{if } y = 0, \\ -1, & \text{if } y < 0. \end{cases}$$

The classical solution of above problem is defined by

$$y(t) = \begin{cases} 3t + c_1, & \text{if } y < 0, \\ -t + c_2, & \text{if } y > 0. \end{cases}$$

As t increases, the classical solution tends to the line $y = 0$, but it cannot be continued along this line, since the map $y(t) = 0$ so obtained does not satisfy the equation in the usual sense (namely, $y'(t) = 0$, while the right-hand side has the value $1 - 2\,\mathrm{sgn}\, 0 = 1$). Hence, there are no classical solutions of initial value problems starting with $y(0) = 0$. Therefore, a generalization of the concept of solutions is required.

To formulate the notion of a solution to an initial value problem with a discontinuous right-hand side, we restated the problem as a differential inclusion,

$$y'(t) \in F(y(t)), \quad \text{a.e. } t \in [0, \infty), \ \ y(0) = y_0,$$

where $F : \mathbb{R}^n \to \mathcal{P}(\mathbb{R}^n)$ is a vector set-valued map into the set of all subsets of $\mathbb{R}^n$ that can be defined in several ways. The simplest convex definition of F is obtained by the so-called Filippov regularization [197],

$$F(y) = \bigcap_{\epsilon > 0} \overline{\mathrm{conv}(f(\{y \in \mathbb{R}^n : \|y\| \leq \epsilon\} \backslash M))},$$

where $F(y)$ is the convex hull of f, $conv$ is the convex hull, M is a null set (i.e., $\mu(M) = 0$, where μ denotes the Lebesgue measure in $\mathbb{R}^n$) and ϵ is the radius of the ball centered at y.

One of the most important examples of differential inclusions comes from control theory. Consider a control system

$$y'(t) = f(y,u), \ u \in U,$$

where u is a control parameter. It appears that the control system and the differential inclusions

$$y' \in f(y,U) = \bigcup_{u \in U} f(y,u)$$

have the same trajectories. If the set of controls depends on y, that is, $U = U(y)$, then we obtain the differential inclusion

$$y' \in F(y, U(y)).$$

The equivalence between a control system and the corresponding differential inclusion is the central idea used to prove existence theorems in optimal control theory.

Since the dynamics of economics, sociology, and biology in macrosystems is multivalued, differential inclusions serve as natural models in macrosystems with hysteresis.

A differential inclusion is a generalization of the notion of an ordinary differential equation. Therefore, all problems considered for differential equations, that is, existence of solutions, continuation of solutions, dependence on initial conditions and parameters, are present in the theory of differential inclusions. Since a differential inclusion usually has many solutions starting at a given point, new issues appear, such as investigation of topological properties of the set of solutions, selection of solutions with given properties, evaluation of the reachability sets, etc. To solve the above problems, special mathematical techniques were developed.

As a consequence, differential inclusions have been the subject of an intensive study of many researchers in the recent decades; see, for example, the monographs [34,35,112,230,301,311,445,465] and the papers of Bressan and Colombo [105,106], Colombo et al. [154,155], Fryszkowsy and Górniewicz [214], Kyritsi et al. [331], *etc.*

As for more specialized problems, during the last ten years, impulsive ordinary differential inclusions and functional differential inclusions with different conditions have attracted the attention of many mathematicians. At present, the foundations of the general theory of such kinds of problems are already laid and many of them are investigated in detail; see [55,61,74,146,194,241–243,280] and the references therein.

Some of this work is devoted to the existence and stability of solutions for different classes of initial value problems for impulsive differential equations and inclusions with fixed and variable moments. Yet, this monograph addresses a variety of side issues that arise from its simpler beginnings.

We now give an overview of the book's topical arrangement. In Chapter 2, we introduce notations, definitions, lemmas, and fixed point theorems that are used throughout.

In the first section of Chapter 3, we consider the initial value problem of impulsive functional differential equations with fixed moments,

$$y'(t) = f(t, y_t), \quad \text{a.e. } t \in J := [0, b],\ t \neq t_k,\ k = 1, \ldots, m, \tag{1.1}$$

$$y(t_k^+) - y(t_k^-) = I_k(y(t_k^-)), \quad k = 1, \ldots, m, \tag{1.2}$$

$$y(t) = \phi(t), \quad t \in (-\infty, 0], \tag{1.3}$$

where $f : J \times B \longrightarrow \mathbb{R}^n$ is a given function satisfying some assumptions that are specified later, $\phi \in B$ where B is a co-called *phase space*, $0 = t_0 < t_1 < \ldots < t_m < t_{m+1} = b$, and $I_k \in C(\mathbb{R}^n, \mathbb{R}^n)$, $k = 1, 2, \ldots, m$.

In the second section of Chapter 3, we give local and global existence and uniqueness results for impulsive functional differential equations with infinite delay and multiple delays.

In the first subsection, we consider the problem,

$$y'(t) = f(t, y_t) + \sum_{i=1}^{n_*} y(t - T_i), \quad \text{a.e. } t \in J := [0, b] \backslash \{t_1, t_2, \ldots, t_m\}, \tag{1.4}$$

$$y(t_k^+) - y(t_k^-) = I_k(y(t_k^-)), \quad k = 1, \ldots, m, \tag{1.5}$$

$$y(t) = \phi(t), \quad t \in (-\infty, 0], \tag{1.6}$$

where $n_* \in \{1, 2, \ldots\}$, and f, I_k, B are as in problem (1.1)–(1.3). In the second subsection, we give sufficient conditions for global existence and uniqueness and stability results for impulsive functional differential equations with infinite delay and multiple delays. More precisely, we will consider the impulsive differential equation,

$$y'(t) = f(t, y_t) + \sum_{i=1}^{n_*} y(t - T_i), \quad \text{a.e. } t \in J := [0, \infty) \backslash \{t_1, t_2, \ldots, \}, \tag{1.7}$$

$$y(t_k^+) - y(t_k^-) = I_k(y(t_k^-)), \quad k = 1, \ldots, \tag{1.8}$$

$$y(t) = \phi(t), \quad t \in (-\infty, 0]. \tag{1.9}$$

In the last sections of Chapter 3, we consider existence, global existence and uniqueness, and stability results for

$$y''(t) = f(t, y_t), \quad \text{a.e. } t \in J := [0, b],\ t \neq t_k,\ k = 1, \ldots, m, \tag{1.10}$$

$$y(t_k^+) - y(t_k^-) = I_k(y(t_k^-)), \quad t = t_k,\ k = 1, \ldots, m, \tag{1.11}$$

$$y'(t_k^+) - y'(t_k^-) = \overline{I}_k(y(t_k^-)), \quad t = t_k,\ k = 1, \ldots, m, \tag{1.12}$$

$$y(t) = \phi(t), \quad t \in (-\infty, 0],\ y'(0) = \eta, \tag{1.13}$$

where $\eta \in \mathbb{R}^n$, f, B, are as in problem (1.7)–(1.9), and $I_k, \overline{I}_k \in C(\mathbb{R}^n, \mathbb{R}^n)$, $k = 1, 2, \ldots, m$, and for the problem

$$y''(t) = f(t, y_t), \quad \text{a.e. } t \in J := [0, \infty),\ t \neq t_k,\ k = 1, \ldots, \tag{1.14}$$

$$y(t_k^+) - y(t_k^-) = I_k(y(t_k^-)), \quad t = t_k,\ k = 1, \ldots, \tag{1.15}$$

$$y'(t_k^+) - y'(t_k^-) = \overline{I}_k(y(t_k^-)), \quad t = t_k,\ k = 1, \ldots, \tag{1.16}$$

$$y(t) = \phi(t), \quad t \in (-\infty, 0],\ y'(0) = \eta, \tag{1.17}$$

where $\eta \in \mathbb{R}^n$, f, B, are as in problem (1.7)–(1.9), and $I_k, \overline{I}_k \in C(\mathbb{R}^n, \mathbb{R}^n)$, $k = 1, 2, \ldots$.

In Chapter 4, we present existence theory for initial and boundary value problems for impulsive functional differential equation. First, in Subsection 4.1, we study the global existence of the first order impulsive boundary value problem,

$$y'(t) = f(t, y_t), \quad \text{a.e. } t \in J := [0, \infty),\ t \neq t_k,\ k = 1, \ldots, \tag{1.18}$$

$$y(t_k^+) - y(t_k^-) = I_k(y(t_k^-)), \quad t = t_k,\ k = 1, \ldots, \tag{1.19}$$

$$Ay(t) - y_\infty = \phi(t), \quad t \in (-\infty, 0], \tag{1.20}$$

where f, B are as in problem (1.4)–(1.6), $I_k \in C(\mathbb{R}^n, \mathbb{R}^n)$, $k \in \mathbb{N}$, $\lim_{t\to\infty} y(t) = y_\infty$, $A > 1$, and $\phi \in B$.

Chapter 5 is primarily concerned with impulsive differential equations and inclusions on bounded intervals. The main object of this chapter is to prove a Filippov theorem and a Filippov–Wazewski theorem for impulsive differential inclusions. More precisely, we consider the following problems:

- Existence of solutions.
- Compactness of the solutions set.
- Impulsive form of Filippov's Theorem.
- Relaxation problems.
- Upper semicontinuity without convexity.

In addition, impulsive differential inclusions also are considered on noncompact intervals.

In Chapter 6, we give sufficient conditions for existence of solutions of first order impulsive functional differential inclusions with infinite delay,

$$y'(t) \in F(t, y_t), \quad \text{a.e. } t \in J =: [0, b] \backslash \{t_1, \ldots, t_m\}, \tag{1.21}$$

$$y(t_k^+) - y(t_k) = I_k(y(t_k^-)), \quad k = 1, \ldots, m, \tag{1.22}$$

$$y(t) = \phi(t), \quad t \in (-\infty, 0], \tag{1.23}$$

where I_k, ϕ, B are as in problem (1.4)–(1.6), and $F : J \times B \to \mathcal{P}(\mathbb{R}^n)$ is a multivalued map. In this chapter, we will give three theorems: one for the convex case and two for the nonconvex case. In the last subsection, we consider the boundary value problem,

$$y'(t) \in F(t, y_t), \qquad \text{a.e. } t \in J := [0, \infty),\ t \neq t_k,\ k = 1, \dots, \tag{1.24}$$
$$y(t_k^+) - y(t_k^-) = I_k(y(t_k^-)), \qquad t = t_k,\ k = 1, \dots, \tag{1.25}$$
$$Ay(t) - y_\infty = \phi(t), \qquad t \in (-\infty, 0], \tag{1.26}$$

where $I_k, B, \phi, A, y_\infty, F$ are as in (1.4)–(1.6), (1.18)–(1.20), and (1.21)–(1.23).

Chapter 7 extends the theory of some of the previous chapters to functional differential equations and functional differential inclusions under impulses for which the impulse effects vary with time. In particular, Chapter 7 is devoted to the relatively less developed area of impulsive differential equations with *variable* time impulses. Most of the results in this chapter deal with Benchohra et al. works [67, 72] for a system of impulsive functional differential equations with finite delay and variable impulse times; namely, first for the problem

$$y'(t) = f(t, y_t), \qquad \text{a.e. } t \in [0, b], t \neq \tau_k(y(t)), k = 1, \dots, m,$$
$$y(t^+) = I_k(y(t)), t = \tau_k(y(t)), \qquad k = 1, \dots, m,$$
$$y(t) = \phi(t) \in B, \qquad t \in (-\infty, 0],$$

B is a phase space, $f : [0, b] \times B \to \mathbb{R}^n$, $\tau_k : \mathbb{R}^n \to \mathbb{R}$ and $I_k : \mathbb{R}^n \to \mathbb{R}^n$, $k = 1, \dots, m$. This is followed by results for a system of impulsive neutral functional differential equations with infinite delay.

The results of Chapter 8 are somewhat in the spirit of the first two sections of Chapter 5 in that they are devoted to a Filippov's Theorem and to a relaxed problem in the context of neutral differential inclusions.

Chapter 9 is primarily concerned with topological and geometrical structures of solutions for impulsive functional differential equations and inclusions and impulsive differential inclusions on approximate retracts. More precisely, we will consider the following problems:

- Aronszajn Type Results.
- Contractible sets and R_δ-sets.
- Viable Solutions.
- Periodic problems.
- Solution sets of the terminal problem.

Chapter 10 extends results of previous chapters on semilinear problems to semilinear functional differential inclusions and functional differential operators that are nondensely defined on a Banach space; questions addressed include exact controllability of impulsive semilinear functional differential inclusions, controllability in extrapolation spaces, and Filippov and Filippov–Wazewski Theorems for second order impulsive semilinear functional differential inclusions.

Chapter 11 is a brief chapter dealing with impulsive stochastic differential equations, impulsive sweeping process and integral inclusions of Volterra type in Banach spaces. More precisely in Section 11.2, we will consider the problem,

$$d[y(t) - Ay(t)] = f(t, y(t))dt + g(t, y(t))dW(t), \quad t \in J := [0, b]\backslash\{t_1, \ldots, t_m\}, \tag{1.27}$$

$$y(t_k^-) - y(t_k) = I_k(y(t_k)), \qquad t \neq t_k,\ k = 1, \ldots, m, \tag{1.28}$$

$$y(0) = y_0 \tag{1.29}$$

where H a real separable Hilbert space, $f : J \times H \to H$ is a given function, $A : H \to H$ generates a C_0-semigroup $\{T(t) : t \geq 0\}$ on H, $t_0 < \cdots < t_m < b$, $I_k \in C(H, H)$, $W(t)$ is a Hilbert space Q-Wiener process, and ϕ is a suitable initial random function independent of $W(t)$.

Section 11.3 deals with the impulsive sweeping process problem,

$$-y'(t) \in N_{K(t)}(y(t)), \quad \text{a.e.} \quad t \in J := [0, \infty), \tag{1.30}$$

$$y(0) \in K(0), \tag{1.31}$$

where $K(t)$ is a convex time dependent set, and $N_{K(t)}(y(t))$ is the normal cone to $K(t)$ at $y(t)$.

Section 11.4 is devoted to semilinear integral inclusions of Volterra type,

$$y(t) \in \int_0^t a(t-s)[Ay(s) + F(s, y(s))]ds, \quad \text{a.e.} \quad t \in J := [0, b], \tag{1.32}$$

where $a \in L^1([0, b], \mathbb{R})$, $A : D(A) \subset E \to E$ is the generator of an integral resolvent family defined on a complex Banach space E, and $F : [0, b] \times E \to \mathcal{P}(E)$ is a multivalued map.

The final chapter is really an appendix containing a summary of a number of important and useful results in functional analysis that are used throughout the book.

Acknowledgments. The authors wish to thank all of their many colleagues who have contributed to this important area of research and a special thanks to Mouffak Benchohra who was Ouahab's doctoral thesis adviser and a research collaborator of Graef and Henderson. We also extend our special and sincere thanks to our editor Anja Möbius at Walter de Gruyter in Berlin (Germany) for accepting publication of this monograph in the "De Gruyter Series in Nonlinear Analysis and Applications."

1.2 Motivational Models

In this section, we present some common models in which impulsive constructions arise.

1.2.1 Kruger–Thiemer Model

The Kruger–Thiemer [327,328] model deals with adjusting the distribution of medicines absorbed orally in the gastro-intestinal system of the human body. If we denote by $x(t)$ and $y(t)$ the respective quantities of medicines at time t in the gastro-intestinal system and in the second compartment called "apparent volume of distribution" (the one that distributes them in the blood and in the muscles), we obtain the system

$$x'(t) = -k_1 x(t), \tag{1.33}$$

$$y'(t) = -k_2 y(t) + k_1 x(t), \tag{1.34}$$

where k_1 and k_2 are characteristic constants. If we assume that at the moments $0 < t_1 < t_2 \cdots t_m < b$, the medicines are absorbed in quantities $\delta_0, \delta_1, \ldots, \delta_m$, we also have

$$x'(t_k^+) = x(t_k^-) - \delta_k, \quad k = 1, \ldots, m, \tag{1.35}$$

$$y(t_k^+) = y(t_k^-), \quad k = 1, \ldots, m, \tag{1.36}$$

$$x(0) = 0, \quad y(0) = \delta_0. \tag{1.37}$$

The problem is then to minimize the function

$$f(\delta) = \frac{1}{2} \sum_{k=0}^{k=m} \delta_k^2$$

in order to obtain the desired therapeutic effect; that is, the quantity of medicines in the second compartment cannot descend below a certain level. The system (1.33)–(1.37) present in this model enters into the framework of impulsive differential equations systems at fixed moments.

1.2.2 Lotka–Volterra Model

The Lotka–Volterra model for population growth, with impulses at fixed times, is represented by a system such as,

$$x_i' = x_i(a_i + \sum_{j=1}^{n} b_{ij} x_j), \quad t \neq t_k, \ i = 1, \ldots, n, \tag{1.38}$$

$$x(t_k^+) - x(t_k^-) = I(t_k, x(t_k^-)), \quad t = t_k, \ k = 1, \ldots, \tag{1.39}$$

$$x(t_0^+) = x_0 > 0, \tag{1.40}$$

where $0 = t_0 < t_1 < \cdots t_k < \cdots$, $\lim_{k\to\infty} t_k = \infty$, and $x + I(t_k, x)) > 0$ for $(t_k, x) \in \mathbb{R} \times \Omega$. The constants $a_i \in \mathbb{R}$ represent the natural intrinsic growth rates of the ith species, if resources were unlimited and interspecies effects were neglected, while the constants $b_{ij} \in \mathbb{R}$ represent the growth inhibiting or enhancing effect that space j has on species i; for more detail, see Ballinger and Liu [44].

1.2.3 Pulse Vaccination Model

An application of impulsive functional differential equations with multiple delay arises in the study of pulse vaccination strategies. In [221], the authors considered the model

$$\begin{cases} S'(t) = b - bS(t) - \dfrac{\beta S(t)I(t)}{1+\alpha S(t)} + \gamma I(t-\tau)e^{-b\tau}, \\ E'(t) = \displaystyle\int_{t-\omega}^{t} \frac{\beta S(u)I(u)}{N(u)} e^{-b(t-u)}du, \\ I'(t) = \dfrac{\beta e^{-b\omega} S(t-\omega)I(t-\omega)}{1+\alpha S(t-\omega)} - (b+\omega)I(t), \\ R'(t) = \displaystyle\int_{t-\omega}^{t} \gamma I(u)e^{-b(t-u)}du, \\ S(t_k^+) = (1-\theta)S(t_k^-), & t_k = kT, k \in \mathbb{N}, \\ E(t_k^+) = E(t_k^-), & t_k = kT, k \in \mathbb{N}, \\ I(t_k^+) = I(t_k^-), & t_k = kT, k \in \mathbb{N}, \\ R(t_k^+) = R(t_k^-) + \theta S(t_k^-), & t_k = kT, k \in \mathbb{N}, \end{cases} \tag{1.41}$$

where $\mathbb{N} = \{0, 1, 2, \dots, \}$, $N(t) = S(t) + E(t) + I(t) = 1$, for all $t \geq 0$, and

- S denotes the susceptibles,
- I denotes the infectives,
- R denotes the removed group,
- E denotes the exposed but not yet infectious.

1.2.4 Management Model

A good reference for state-dependent impulsive models of integrated pest management (IPM) strategies and their dynamic consequences is Tang and Cheke [460]. IPM involves combining biological, mechanical, and chemical tactics to reduce pest numbers to tolerable levels after a pest population has reached its economic threshold (ET). The complete expression of an orbitally asymptotically stable periodic solution involves the model whose maximum value is no larger than a given (ET) presented. The existence of such a solution implies that pests can be controlled at or below their ET levels. In [460], the following predator-prey model is presented concerning (IMP) strategies:

$$\frac{dx(t)}{dt} = g(x(t))x(t) - h(x(t), y(t))y(t) + \epsilon f_1(x(t), y(t), \epsilon),\ x \neq ET, \quad (1.42)$$

$$\frac{dy(t)}{dt} = \gamma h(x(t), y(t))y(t) - dy(t) + \epsilon f_2(x(t), y(t), \epsilon),\ x \neq ET, \quad (1.43)$$

$$\Delta x(t) = -px(t) + \epsilon I_1(x(t), y(t), \epsilon),\ x = ET, \quad (1.44)$$

$$\Delta y(t) = \tau + \epsilon I_2(x(t), y(t), \epsilon),\ x = ET, \quad (1.45)$$

$$x(0^+) = x_0 < ET,\ y(0^+) = y_0, \quad (1.46)$$

where x and y are the population abundances of prey (or host) and predator (or parasitoid), respectively, $g(x)$ is the per captia net rate of increase, $h(x, y)$ is the per capita functional response of the predator, d is the per capita death rate of the predator population, γ is the conversion efficiency of the prey to predator, $0 \leq p < 1$ is the reduction proportion of the pest density by killing or trapping once the number of pests reaches ET, and $\tau \geq 0$ is the number of natural enemies released at this time. The functions f_1, f_2, I_1, I_2 can be considered as the effects of exogenous variables on the system (1.42)–(1.46).

1.2.5 Some Examples in Economics and Biomathematics

Here, we simply list titles of some additional motivational models.

1. Multiple-phase dynamics in economics; see, for instance Day [165].
2. Stock management in production theory; see Bensoussan and Lions [85–87].
3. Viability theory for implementing the extreme version of the "inertia principle;" see Aubin [32, 33].
4. Propagation of the nervous influx along axons of neurons triggering spikes in neurosciences and biological neuron networks (networks: instead of the continuous time Hodgkin–Huxley type of systems of differential equations inspired by the propagation of electrical currents which are the subjects of abundant literature). See the pioneering work by Hodgkin and Huxley [293] (see the "Integrate-and Fire" models in [108, 109, 440]).
5. For issues dealing with "qualitative physics" in Artificial Intelligence and/or "comparative statistics" in economics see, for example, Aubin [33] and Dordan [170].

We hope this monograph is timely and will fill the vacuum in the literature on the existence theory of differential, difference, and integral equations over infinite intervals. We also hope that it will stimulate further research and development in this important area.

Chapter 2

Preliminaries

In this chapter, we introduce notations, definitions, lemmas, and fixed point theorems that are used throughout this monograph. These include some necessary results for Fréchet spaces along with a number of topological and analytical properties of set-valued mappings, followed by some fixed point results and measure of noncompactness results in those contexts. The latter part of the chapter is devoted to material that will be used in semigroup settings along with some material on extrapolation spaces.

Let $J := [a,b]$ be an interval of $\mathbb{R}$ and let $(E, |\cdot|)$ be a real Banach space. We let $C([a,b],E)$ be the Banach space of all continuous functions from $[a,b]$ into E with the norm

$$\|y\|_\infty = \sup\{|y(t)| : a \le t \le b\},$$

and let $L^1([a,b],E)$ denote the Banach space of measurable functions that are Bochner integrable. A function $y : [a,b] \longrightarrow E$ is Bochner integrable if and only if $|y|$ is Lebesgue integrable. For properties of the Bochner integral, see for instance, Yosida [492]. We also let $L^1(J,E)$ denote the Banach space of functions $y : J \longrightarrow E$ which are Bochner integrable and normed by

$$\|y\|_{L^1} = \int_a^b |y(t)|dt,$$

and let $AC^i([a,b],E)$ be the space of i-times differentiable functions $y : (a,b) \to E$, whose i^{th} derivative, $y^{(i)}$, is absolutely continuous.

2.1 Some Definitions

Definition 2.1. A map $f : [a,b] \times E \longrightarrow E$ is Carathéodory if

(i) $t \longmapsto f(t,y)$ is measurable for all $y \in E$, and
(ii) $y \longmapsto f(t,y)$ is continuous for almost each $t \in [a,b]$.

If, in addition,

(iii) for each $q > 0$, there exists $h_q \in L^1([a,b],\mathbb{R}_+)$ such that

$$|f(t,y)| \le h_q(t) \quad \text{for all } |y| \le q \quad \text{and almost each } t \in [a,b],$$

then we say that the map is L^1-Carathéodory.

We remark here that conditions (i) and (ii) imply, for $t \in [0,b]$, that $f(t,u(t))$ is measurable for any measurable and almost everywhere finite function $u(t)$. This is a result of Carathédory; see Krasonsel'skii [325]. Also, (iii) implies that $f(t,u(t))$ is L^1-Carathédory.

Definition 2.2. A map f is said to be compact if the image of each bounded set is relatively compact. The map f is said to be completely continuous if it is continuous and compact.

2.2 Some Properties in Fréchet Spaces

For more details on the following notions we refer to [208]. Let X be a Fréchet space with a family of seminorms $\{\|\cdot\|_n,\ n \in \mathbb{N}\}$. We say that $Y \subset X$ is *bounded* if for every $n \in \mathbb{N}$, there exists $M_n > 0$ such that

$$\|y\|_n \le M_n \quad \text{for all} \quad y \in Y.$$

To a set X, we associate a sequence of Banach spaces $\{(X^n, \|\cdot\|_n)\}$ as follows. For every $n \in \mathbb{N}$, we consider the equivalence relation $\sim_n$ defined by $x \sim_n y$ if and only if $\|x - y\|_n = 0$ for all $x, y \in X$. We denote $X^n = (X/\sim_n, \|\cdot\|)$ the quotient space, the completion of X^n with respect to $\|\cdot\|_n$. To every $Y \subset X$, we associate a sequence $\{Y^n\}$ of subsets $Y^n \subset X^n$ as follows. For every $x \in X$, we denote by $[x]_n$ the equivalence class of x of subsets X^n, and we define $Y^n = \{[x]_n : x \in Y\}$. We denote by $\overline{Y}^n$, $\text{int}_n(Y^n)$ and $\partial_n Y^n$, respectively, the closure, the interior and the boundary of Y^n with respect to $\|\cdot\|$ in X^n. We assume that the family of seminorms $\{\|\cdot\|_n\}$ satisfy

$$\|x\|_1 \le \|x\|_2 \le \|x\|_3 \le \cdots \quad \text{for every} \quad x \in X.$$

Definition 2.3. A function $f : X \to X$ is said to be a contraction if for each $n \in \mathbb{N}$, there exists $k_n \in (0,1)$ such that

$$\|f(x) - f(y)\|_n \le k_n \|x - y\|_n \text{ for all } x, y \in X.$$

Theorem 2.4 (Nonlinear alternative, [208]). *Let X be a Fréchet space and $Y \subset X$ a closed subset in X, and let $N : Y \to X$ be a contraction such that $N(Y)$ is bounded. Then one of the following statements holds:*

(C1) *N has a unique fixed point;*

(C2) *There exists $\lambda \in [0,1)$, $n \in \mathbb{N}$, and $x \in \partial_n Y^n$ such that $\|x - \lambda N(x)\|_n = 0$.*

2.3 Some Properties of Set-valued Maps

Let (X, d) be a metric space and Y be a subset of X. We denote:

- $\mathcal{P}(X) = \{Y \subset X : Y \neq \emptyset\}$ and
- $\mathcal{P}_p(X) = \{Y \in \mathcal{P}(X) : Y$ has the property "p"$\}$, where p could be: cl = closed, b = bounded, cp = compact, cv = convex, etc.

Thus:

- $\mathcal{P}_{cl}(X) = \{Y \in \mathcal{P}(X) : Y \text{ closed}\}$,
- $\mathcal{P}_b(X) = \{Y \in \mathcal{P}(X) : Y \text{ bounded}\}$,
- $\mathcal{P}_{cp}(X) = \{Y \in \mathcal{P}(X) : Y \text{ compact}\}$,
- $\mathcal{P}_{cv}(X) = \{Y \in \mathcal{P}(X) : Y \text{ convex}\}$, where X is a normed space,
- $\mathcal{P}_{cv,cp}(X) = \mathcal{P}_{cv}(X) \cap \mathcal{P}_{cp}(X)$ where X is a normed space, etc.

Definition 2.5. A multivalued function (or a multivalued operator, multivalued map, or multimap) from X into Y is a correspondence which associates to each element $x \in X$ a subset $F(x)$ of Y. We will denote this correspondence by the symbol: $F : X \to \mathcal{P}(Y)$. We define:

- the effective domain $\mathrm{Dom} F = \{x \in X : F(x) \neq \emptyset\}$.
- the graph $\mathrm{Gra} F = \{(x, y) \in X \times Y : y \in F(x)\}$.
- the range $F(X) = \bigcup_{x \in X} F(x)$.
- the image of the set $A \in \mathcal{P}(X)$: $F(A) = \bigcup_{x \in A} F(x)$.
- the inverse image of the set $B \in \mathcal{P}(Y)$: $F^{-}(B) = \{x \in X : F(x) \cap B \neq \emptyset\}$.
- the strict inverse image of the set $B \in \mathcal{P}(Y)$: $F^{+}(B) = \{x \in \mathrm{Dom} F : F(x) \subset B\}$.
- the inverse multivalued operator, denoted by $F^{-1} : Y \to \mathcal{P}(X)$, is defined by

  $$F^{-1}(y) = \{x \in X : \ y \in F(x)\}.$$

  The set $F^{-1}(y)$ is called the fiber of F at the point y.
- Let $F, G : X \to \mathcal{P}(Y)$ be multifunctions. Then

  $$(F \cup G)(x) = F(x) \cup G(x), \text{ and } (F \cap G)(x) = F(x) \cap G(x).$$

Also, if $F : X \to \mathcal{P}(Y)$ and $G : Y \to \mathcal{P}(Z)$, then the composition $(G \circ F)(\cdot)$ is defined by $(G \circ F)(x) = \cup_{y \in F(x)} G(y)$. Finally, if $F, G : X \to \mathcal{P}(Y)$, then the product $(F \times G)(\cdot)$ is defined by $(F \times G)(x) = F(x) \times G(x)$.

Proposition 2.6. *The following properties hold.*

- *If* $F, G : X \to \mathcal{P}(Y)$ *and* $A \subseteq Y$, *then*

$$(F \cup G)^-(A) = F^-(A) \cup G^-(A), \ (F \cup G)^+(A) = F^+(A) \cup G^+(A)$$

and

$$(F \cap G)^-(A) \subseteq F^-(A) \cap G^-(A), \ F^+(A) \cap G^+(A) \subseteq (F \cap G)^+(A).$$

- *If* $F : X \to \mathcal{P}(Y)$, $G : Y \to \mathcal{P}(Z)$, *and* $A \subseteq Z$, *then*

$$(G \circ F)^-(A) = F^-(G^-(A)), \text{ and } (G \circ F)^+(A) = F^+(G(A)).$$

- *If* $F : X \to \mathcal{P}(Y)$ *and* $A_i, A \subseteq Y$, $i \in I$, *then*

$$X \backslash F^-(A) = F^+(Y \backslash A), \ X \backslash F^+(A) = F^-(Y \backslash A),$$

$$F^-\Big(\bigcup_{i\in I} A_i\Big) = \bigcup_{i\in I} F^-(A_i), \ F^-\Big(\bigcap_{i\in I} A_i\Big) \subseteq \bigcap_{i\in I} F^-(A_i),$$

and

$$\bigcup_{i\in I} F^+(A_i) \subseteq F^+\Big(\bigcup_{i\in I} A_i\Big), \ \bigcap_{i\in I} F^+(A_i) \subseteq F\Big(\bigcap_{i\in I} A_i\Big).$$

- *If* $F : X \to \mathcal{P}(Y)$ *and* $G : X \to P(Z)$, $A \subseteq Y$, *and* $B \subseteq Z$, *then*

$$(F \times G)^+(A \times B) = F^+(A) \cap G^+(B),$$
$$(F \times G)^-(A \times B) = F^-(A) \cap G^-(B).$$

This is also true for arbitrary products.

Definition 2.7. A multimap $F : X \to \mathcal{P}(Y)$ is convex (closed) valued if $F(x)$ is convex (closed) for all $x \in X$. We say that F is bounded on bounded sets if $F(B) = \cup_{x\in B} F(x)$ is bounded in Y for all $B \in \mathcal{P}_b(X)$

$$(\text{i.e., } \sup_{x\in B}\{\sup\{|y| : y \in F(x)\}\} < \infty).$$

The set $\Gamma_F \subset X \times Y$, defined by

$$\Gamma_F = \{(x, y) : x \in X, y \in F(x)\}$$

is called the graph of F. We say that F is has a closed graph, if Γ_F is closed in $X \times Y$.

2.3.1 Hausdorff Metric Topology

The Haudorff metric is defined on a metric space and is used to quantify the distance between subsets of the given metric space. Let (X, d) be a metric space. In what follows, given $x \in X$ and $A \in \mathcal{P}(X)$, the distance of x from A is defined by

$$d(x, A) = \inf\{d(x, a) : a \in A\}.$$

Similarly, for $y \in X$ and $B \in \mathcal{P}(X)$

$$d(B, y) = \inf\{d(b, y) : b \in B\}.$$

As usual, $d(x, \emptyset) = d(\emptyset, y) = +\infty$.

Definition 2.8. Let $A, B \in \mathcal{P}(X)$, we define

- $H^*(A, B) = \sup\{d(a, B) : a \in A\}$,
- $H^*(B, A) = \sup\{d(A, b) : b \in B\}$,
- $H(A, B) = \max(H^*(A, B), H^*(B, A))$ (the Hausdorff distance between A and B).

Remark 2.9. Given $\epsilon > 0$, let

$$A_\epsilon = \{x \in X : d(x, A) < \epsilon\}$$

and

$$B_\epsilon = \{x \in X : d(B, x) < \epsilon\}.$$

Then from the above definitions we have

$$H^*(A, B) = \inf\{\epsilon > 0 : A \subset B_\epsilon\},\ H^*(B, A) = \inf\{\epsilon > 0 : B \subset A_\epsilon\}$$

and

$$H(B, A) = \inf\{\epsilon > 0 : B \subset A_\epsilon,\ A \subset B_\epsilon\}.$$

From the definition we can easily prove the following properties:

- $H(A, A) = 0$, for all $A \in \mathcal{P}(X)$,
- $H(A, B) = H(B, A)$, for all $A, B \in \mathcal{P}(X)$,
- $H(A, B) \leq H(A, C) + H(C, B)$, for all $A, B, C \in \mathcal{P}(X)$.

Hence $H(\cdot, \cdot)$ is an extended pseudometric on $P(X)$ (i.e., is a pseudometric which can also take the value $+\infty$). Moreover, we can prove that

$$H(A, B) = 0, \text{ if and only if } \overline{A} = \overline{B}.$$

So $\mathcal{P}_{cl}(X)$ furnished with the Hausdorff distance (H-distance), $H(\cdot, \cdot)$, becomes a metric space.

Lemma 2.10. *If* $\{A_n, A\} \in \mathcal{P}_{cl}(X)$ *and* $A_n \to A$, *then*

$$A = \bigcap_{n\geq 1} \overline{\bigcup_{m\geq n} A_m} = \bigcap_{\epsilon\geq 0} \bigcup_{n\geq 1} \bigcap_{m\geq n} (A_m)_\epsilon.$$

Proof. Let $\epsilon > 0$ be given. Since by hypothesis $A_n \to A$, we can find $n_0(\epsilon) \geq 0$ such that for $m \geq n_0(\epsilon)$, we have $A \subset (A_m)_\epsilon$ and $A_m \subset A_\epsilon$. From this inclusion, we have that

$$A \subset \bigcap_{\epsilon>0} \bigcup_{n\geq 1} \bigcap_{m\geq n} (A_m)_\epsilon$$

and

$$\bigcap_{n\geq 1} \overline{\bigcup_{m\geq n} A_m} \subset A.$$

Hence,

$$\bigcap_{n\geq 1} \overline{\bigcup_{m\geq n} A_m} \subset A \subset \bigcap_{\epsilon>0} \bigcup_{n\geq 1} \bigcap_{m\geq n} (A_m)_\epsilon.$$

Finally, let $x \in \bigcap_{\epsilon\geq 0} \bigcup_{n\geq 1} \bigcap_{m\geq n} (A_m)_\epsilon$. Then for all $\epsilon \geq 0$, there is $n_0(\epsilon) \geq 1$ such that, for $m \geq n_0(\epsilon)$, we have $x \in (A_m)_\epsilon$. Let $n \geq 1$ be given. Then there is $m \geq \max(n, n_0(\epsilon))$ such that $x \in (A_m)_\epsilon \subset (\bigcup_{m\geq n} A_m)_\epsilon$. Since $\epsilon > 0$ was arbitrary, we can deduce that $x \in \overline{\bigcup_{m\geq n} A_m}$, and so $x \in \bigcap_{n\geq 1} \overline{\bigcup_{m\geq n} A_m}$. Thus,

$$A = \bigcap_{n\geq 1} \overline{\bigcup_{m\geq n} A_m} = \bigcap_{\epsilon\geq 0} \bigcup_{n\geq 1} \bigcap_{m\geq n} (A_m)_\epsilon. \qquad \square$$

Now we will check the completeness of the metric space $(\mathcal{P}_{cl}(X), H)$.

Theorem 2.11. *If* (X, d) *is a complete metric space, then so is the space* $(\mathcal{P}_{cl}(X), H)$.

Proof. Let $\{A_n\}_{n\in\mathbb{N}}$ be a Cauchy sequence in $(\mathcal{P}_{cl}(X), H)$. The previous Lemma 2.10 identifies the only possible candidate for a limit of $\{A_n\}_{n\in\mathbb{N}}$. Namely, let $A = \bigcap_{n\geq 1} \overline{\bigcup_{m\geq n} A_m}$. We will now show that

$$A \in \mathcal{P}_{cl}(X) \text{ and } A_n \to A \text{ as } n \to +\infty.$$

First, it is clear that A being the intersection of closed sets, is closed, yet possibly empty. Let $\epsilon > 0$. Then for every $k \geq 0$, we can find $N_k \geq 1$ such that $H(A_n, A_m) \leq \frac{\epsilon}{2^{k+1}}$ for all $n, m \geq N_k$. Pick $n_0 \geq N_0$ and $x_0 \in A_{n_0}$. Then choose $n_1 > \max(n_1, N_1)$ and $x_1 \in A_{n_1}$ with $d(x_0, x_1) < \frac{\epsilon}{2}$ (this is possible, since $d(x_0, A_{n_1}) \leq H(A_{n_0}, A_{n_1}) < \frac{\epsilon}{2}$). Then, if $\{n_k\}_{k\geq 0}$ is a strictly increasing sequence with $n_k \geq N_k$, inductively, we can generate a sequence $\{x_k\}_{k\geq 0} \subset X$ such that $x_k \in A_k$ and $d(x_k, x_{k+1}) < \frac{\epsilon}{2^{k+1}}$. So $\{x_k\}_{k\geq 0}$ is a Cauchy sequence in X and since X is complete, we have that $x_k \to x \in X$. Because $\{n_k\}_{k\geq 0}$ is strictly increasing, given $n \geq 1$, we can find $k_n \geq 1$ such that $n_{k_n} \geq n$. Hence $x_k \in \bigcup_{m\geq n} A_m$ for all $k \geq k_n$ and so $x \in \overline{\bigcup_{m\geq n} A_m}$ for all $n \geq 1$. Thus $x \in A$, which shows that $A \in \mathcal{P}_{cl}(X)$. In addition, we have

$$d(x, x_0) = \lim_{n\to+\infty} d(x_n, x_0) \leq \lim_{n\to+\infty} \sum_{k=1}^{n} d(x_k, x_{k-1}) < \epsilon.$$

So for all $n_0 \geq N_0$ and all $x_0 \in A_{n_0}$, we have obtained an $x \in A$ such that $d(x, x_0) < \epsilon$. Therefore $A_{n_0} \subset A_\epsilon$. We need to show that $A \subset (A_n)_\epsilon$ for all $n \geq N_0$. So let $x \in A$. Then $x \in \overline{\bigcup_{m\geq n_0} A_m}$, and we can find $m \geq N_0$ and $y \in A_m$ such that $d(x, y) < \frac{\epsilon}{2}$. Also, if $n \geq N_0$, we have $d(x, A_n) \leq d(x, A_m) + d(A_m, A_n) < \frac{\epsilon}{2} + \frac{\epsilon}{2} = \epsilon$. So $H^*(A, A_n) < \epsilon$ and this implies that $A \subset (A_n)_\epsilon$ for $n \geq N_0$. Therefore, we conclude that $A_n \to A$. □

Lemma 2.12. *If (X, d) is a complete metric space, then $\mathcal{P}_{cp}(X)$ is a closed subset of $(\mathcal{P}_{cp}(X), H)$; hence, $(\mathcal{P}_{cp}(X), H)$ is a complete metric space.*

Proof. Let $\{A_n\}_{n\geq 1} \subset \mathcal{P}_{cp}(X)$ and assume that $A_n \to A$. Then given $\epsilon > 0$, we can find $n_0(\epsilon) \geq 1$ such that for all $n \geq n_0(\epsilon)$, $H(A_n, A) < \epsilon$ and so $A \subset (A_n)_\epsilon$. But by hypothesis, A_n is compact, and so it is totally bounded. Thus, we can find a finite set $F \subset X$ such that $A_n \subset F_\epsilon$; hence, $(A_n)_\epsilon \subset F_{2\epsilon}$. Therefore, $A \subset F_{2\epsilon}$ which shows that A is totally bounded and closed, and so $A \in \mathcal{P}_{cp}(X)$. From Lemma 2.10, we can easily show that $\mathcal{P}_{cp}(X)$ is complete metric space. □

The next lemma is obvious.

Lemma 2.13. *$\mathcal{P}_{cl,b}(X)$ is a closed subset of $(\mathcal{P}_{cl}(X), H)$. If (X, d) is complete metric space, then so is $\mathcal{P}_{cl,b}(X) = \mathcal{P}_{cl}(X) \cap \mathcal{P}_b(X)$.*

Now we assume that the underlying metric space is a normed space.

Lemma 2.14. *If X is a normed space, then $\mathcal{P}_{cl,cv}(X) = \mathcal{P}_{cl}(X) \cap \mathcal{P}_{cv}(X)$ is a closed subset of $(\mathcal{P}_{cl}(X), H)$.*

Proof. Let $\{A_n, A\}_{n\geq 1} \subset \mathcal{P}_{cl}(X)$, where A_n is convex for every $n \geq 1$ and assume that $A_n \overset{H}{\to} A$. Then from Lemma 2.10, we know that $A = \bigcap_{\epsilon>0} \bigcup_{n\geq 1} \bigcap_{m\geq n} (A_m)_\epsilon$. Observe that for every $m \geq 1$, $(A_m)_\epsilon \subset \mathcal{P}_{cl,cv}$ is convex, hence $C_n^\epsilon = \bigcap_{m\geq n} (A_m)_\epsilon$ is convex. The sequence $\{C_n^\epsilon\}_{n\geq 1} \subset \mathcal{P}_{cl,cv}(X)$ is increasing for $n \geq 1$ for every $\epsilon > 0$. Therefore, $\bigcup_{n\geq 1} C_n^\epsilon = C^\epsilon$ is convex. So finally, $A = \bigcap_{\epsilon>0} C^\epsilon$ is convex; i.e., $A \in \mathcal{P}_{cl,cv}(X)$. □

Combining the previous three Lemmas, we can summarize the situation in a normed space.

Proposition 2.15. *If X is a normed space, then $\mathcal{P}_{cp,cv}(X) \subset \mathcal{P}_{cl,b,cv}(X) \subset \mathcal{P}_{cl,cv}(X)$ and $\mathcal{P}_{cp}(X) \subset \mathcal{P}_{cl,b}(X)$ are closed subspaces of $(\mathcal{P}_{cl}(X), H)$.*

Remark 2.16. If X is a Banach space, then all the above subsets are complete subspaces of the metric space $(\mathcal{P}_{cl}, H)$.

Next, we derive two formulas for the Hausdorff distance. The first formula, known as "Härmondar's formula," concerns sets in $\mathcal{P}_{cl,b,cv}(X)$ and involves the supremum of the support functions of these sets.

Definition 2.17. Let $(X, \|\cdot\|)$ be a normed space, X^* its topological dual, and $A \in \mathcal{P}(X)$. The support function $\sigma(\cdot, A)$ of A is a function from X^* into $\overline{\mathbb{R}} = \mathbb{R} \cup \{+\infty\}$ defined by

$$\sigma(x^*, A) = \sup\{\langle x^*, a\rangle : a \in A\},$$

where the duality bracket $\langle\cdot,\cdot\rangle : X^* \times X \to R$ is defined by $\langle\phi, x\rangle = \phi(x)$.

Lemma 2.18 ([301]). *If X is normed space and $A, B \in \mathcal{P}_{cl,b,cv}(X)$, then*

$$H(A, B) = \sup\{|\langle x^*, A\rangle - \langle x^*, B\rangle : \|x\| \leq 1\}.$$

The second formula for the Hausdorff distance, concerns nonempty subsets of an arbitrary metric space and involves the distance functions from the sets.

Lemma 2.19 ([301]). *If (X, d) is a metric space and $A, B \in P(X)$, then*

$$H(A, B) = \sup\{|d(x, A) - d(x, B)| : x \in X\}.$$

2.3.2 Vietoris Topology

Throughout this section, (X, τ) is a Hausdorff topological space (that is, τ denotes the Haudorff topology on X). Given $A \in \mathcal{P}(X)$, we define

$$A^- = \{B \in \mathcal{P}(X) : A \cap B \neq \emptyset\} \text{ (those sets in } X \text{ that hit } A)$$

and

$$A^+ = \{B \in \mathcal{P}(X) : B \subseteq A\} \text{ (those sets in } X \text{ that "miss" } A^c).$$

Definition 2.20. The "upper Vietoris topology" (denoted by $\widehat{\tau}_{UV}$) is generated by the base

$$\mathcal{L}_{UV} = \{U^+ : U \in \tau\}.$$

- The "lower Vietoris topology" (denoted by $\widehat{\tau}_{LV}$) is generated by the subbase

$$\mathcal{L}_{LV} = \{U^- : U \in \tau\}.$$

- The "Vietoris topology" (denoted by $\widehat{\tau}_V$) is generated by the subbase $\mathcal{L}_{UV} \cup \mathcal{L}_{LV}$.

Remark 2.21. It follows from the above definition, that a basic element for the Vietoris topology $\widehat{\tau}_V$ is given by

$$B(U, V_1, \dots, V_n) = \{A \in \mathcal{P}(X) : A \subseteq U, A \cap V_k \neq \emptyset, k = 1, \dots, n\},$$

where $U, V_1, \dots, V_n \in \tau$.

The Vietoris topology is "natural" in the following sense.

Lemma 2.22. *If $I : X \to \mathcal{P}(X)$ is the injection map defined by $I(x) = \{x\}$, then $I(\cdot)$ is continuous when $\mathcal{P}(X)$ is equipped with the $\widehat{\tau}_V$-topology.*

Proof. Let $U \in \tau$. Then we have

$$I^{-1}(U^+) = \{x \in X : \{x\} \subseteq U\} = U \in \tau.$$

Similarly, if $V_1, \dots, V_n \in \tau$, then

$$I^{-1}(\cap_{k=1}^n V_k^-) = \{x \in X : \{x\} \cap V_k \neq \emptyset, k = 1, \dots, n\} = \cap_{k=1}^n V_k^- \in \tau.$$

Therefore $I(\cdot)$ is continuous into $\mathcal{P}(X)$ with the Vietoris topology $\widehat{\tau}_V$. □

Example 2.23. The Vietoris topology $\widehat{\tau}_V$ in not the finest topology on $\mathcal{P}(X)$ for which $I(\cdot)$ is continuous.

To see this, let X be an infinite set equipped with the cofinite topology τ_c defined by

$$\tau_c = \{U : U \backslash X, \text{is a finite set}\} \cup \{\emptyset, X\}.$$

Then the closed subsets of X are $\emptyset$, X, and finite subsets of X. Let $\mathcal{F}$ denote the family of nonempty, finite subsets of X. Then $I^{-1}(\mathcal{F})$ is an open set in $(\mathcal{P}(X), \widehat{\tau}_V)$ and contains some infinite sets. So $\mathcal{F} \notin \widehat{\tau}_V$ and thus $I(\cdot)$ remains continuous if on $\mathcal{P}(X)$, we consider the stronger topology obtained by $\mathcal{F}$ to the original subset $\mathcal{L}_{UV} \cup \mathcal{L}_{LV}$.

As in the above example, let $\mathcal{F}$ denote the family of nonempty and finite subsets of X.

Proposition 2.24. *The family $\mathcal{F}$ is dense in $(\mathcal{P}(X), \widehat{\tau}_V)$.*

Proof. If $U \in \tau$ is nonempty, then U contains a finite subset and so $U^+ \cap \mathcal{F} \neq \emptyset$. Similarly, if $V_1, \dots, V_n \in \tau$ are nonempty, let $x_k \in V_k$, $k = 1, \dots, n$. Then $\{x_k\}_{k=1}^n \in (\cap_{k=1}^n V_k^-) \cap \mathcal{F}$. Thus, $\mathcal{F}$ intersects every element in the base of $\widehat{\tau}_V$ and so $\mathcal{F}$ is dense as claimed. □

An immediate interesting consequence of the above proposition is the following lemma.

Lemma 2.25 ([301]). *If (X, τ) is a separable Haudorff space, then the space $(\mathcal{P}(X), \widehat{\tau}_V)$ is a separable topological space.*

The next proposition tells us that under some additional, reasonable conditions on X, the topological space $(\mathcal{P}_{cl}(X), \widehat{\tau}_V)$ has nice separation properties (that is, Hausdorff properties).

Lemma 2.26. *If (X, τ) is a regular topological, then $(\mathcal{P}_{cl}(X), \widehat{\tau}_V)$ is a Hausdorff topological space.*

Proof. Let $A, B \in \mathcal{P}_{cl}(X)$ and assume that $A \neq B$. Then $A \cap X \backslash B$ or $B \cap X \backslash A$ is nonempty. Suppose $A \cap X \backslash B \neq \emptyset$ and let $a \in A \cap X \backslash B$. Then since by hypothesis X is regular, we can find U_1, $U_2 \in \tau$ such that $a \in U_1, B \subseteq U_2$ and $U_1 \cap U_2 = \emptyset$. Note the U_1^- and U_2^+ are disjoint elements in $\widehat{\tau}_V$ and $A \in U_1^-$, while $B \in U_2^+$. So indeed $\widehat{\tau}_V$ is a Hausdorff topology. □

Lemma 2.27. *If (X, τ) is a Hausdorff topological space, then (X, τ) is compact if and only if $(\mathcal{P}_{cl}(X), \widehat{\tau}_V)$ is compact.*

In general, there is no simple relationship between the Hausdorff pseudometric (respectively, metric) topology $\widehat{\tau}_H$ and the Vietoris topology $\widehat{\tau}_V$, defined on $\mathcal{P}(X)$ (respectively, on $\mathcal{P}_{cl}(X)$). However, if we restrict ourselves to $\mathcal{P}_{cp}(X)$, then we have the following result.

Lemma 2.28 ([301]). *If (X, d) is a metric space, then on $\mathcal{P}_{cp}(X)$, the Haudorff metric topology $\widehat{\tau}_H$ and the Vietoris topology $\widehat{\tau}_V$ coincide.*

2.3.3 Continuity Concepts and Their Relations

The three Vietoris topologies introduced in Section 2.3.2 lead to corresponding continuity concepts for multifunctions.

Definition 2.29. Let $F : X \to \mathcal{P}(Y)$ be a multifunction (set-valued map).

- If $F : X \to (\mathcal{P}(Y), \tau_{UV})$ is continuous, then $F(\cdot)$ is said to be upper semicontinuous (briefly, u.s.c.)
- If $F : X \to (\mathcal{P}(Y), \tau_{LV})$ is continuous, then $F(\cdot)$ is said to be lower semicontinuous (briefly, l.s.c.)
- If $F : X \to (\mathcal{P}(Y), \tau_V)$ is continuous, then $F(\cdot)$ is said to be continuous (or Vietoris continuous).

We present a local version of the above definition.

Definition 2.30. Let $F : X \to \mathcal{P}(Y)$ be a multifunction (set-valued map).

- F is said to be upper semicontinuous at $x_0 \in X$ if and only if for each open subset U of Y with $F(x_0) \subseteq U$, there exists an open V of x_0 such that for all $x \in V$, we have $F(x) \subseteq U$.
- F is said to be lower semicontinuous at $x_0 \in X$ if the set $\{x \in X : F(x) \cap U \neq \emptyset\}$ is open, for any open set U in Y.

Using the definition of the three Vietoris topologies, we immediately deduce the following results. We recall that a set $\mathcal{M}$ with a preorder $\succeq$ is *directed*, if every finite subset has an upper bound. A generalized sequence is a map

$$\mu \in \mathcal{M} \mapsto x_\mu \in X,$$

where (X, τ) is an topological space. An element $x \in X$ is the *limit* of $(x_\mu)_{\mu \in \mathcal{M}}$ if, for every neighborhood $\mathcal{V}$ of x, there exists $\mu_0 \in \mathcal{M}$ such that x_μ belong to $\mathcal{V}$, for all $\mu \succeq \mu_0$.

Proposition 2.31. *For a multifunction $F : X \to \mathcal{P}(Y)$, the following are equivalent:*

(a) *F u.s.c.*

(b) *$F^+(V)$ is open in X for every $V \subseteq Y$ open.*

(c) *For every closed $C \subseteq Y$, $F^-(C)$ is closed in X.*

(d) $\overline{F^-(D)} \subseteq F^-(\overline{D})$.

(e) *For any $x \in X$, if $\{x_\alpha\}_{\alpha \in J}$ is a generalized sequence, $x_\alpha \to x$, and V is an open subset of Y such that $F(x) \subseteq V$, then there exists $\alpha_0 \in J$ such that, for all $\alpha \in J$ with $\alpha \geq \alpha_0$, we have $F(x_\alpha) \subseteq V$.*

Proof. We proceed by showing $a) \Rightarrow b) \Rightarrow c) \Rightarrow d) \Rightarrow e) \Rightarrow a)$.

(a) $\Rightarrow$ (b). Let W be open in Y, then

$$F^+(W) = \{x \in X : F(x) \subset W\}.$$

We will now show that $F^+(W)$ is an open set in X. Let $x \in F^+(W)$; then $F(x) \subset W$. Since F u.s.c., there exists $V(x) \in \mathcal{N}(x)$ such that

$$F(V(x)) \subset W \Rightarrow V(x) \subset F^-(W).$$

Hence, $F^+(W)$ is open in X.

(b) $\Rightarrow$ (c). Let Q be a closed set in Y, then

$$F^-(Q) = \{x \in X : F(x) \cap Q \neq \emptyset\}$$

and

$$X \backslash F^{-}(Q) = \{x \in X : F(x) \subset X \backslash Q\} = F_{+}^{-1}(X \backslash Q).$$

Since Q is a closed set in Y, $X \backslash Q$ is an open set in Y. From b), we have $F_{+}^{-}(X \backslash Q)$ is open in X. Thus $F^{-}(Q)$ is closed in X.

(c) $\Longrightarrow$ (d). Let D be a subset of Y. Then

$$D \subset \overline{D} \Longrightarrow F^{-}(D) \subset F^{-}(\overline{D}) \Longrightarrow \overline{F^{-}(D)} \subset \overline{F^{-}(\overline{D})}.$$

Since $F^{-}(\overline{D})$ closed, $F^{-}(\overline{D}) = \overline{F^{-}(\overline{D})}$. Thus,

$$\overline{F^{-}(D)} \subset F^{-}(\overline{D}).$$

(d) $\Longrightarrow$ (e). Let $\{x_\alpha\}_{\alpha \in J}$ be a generalized sequence, $x \in X$, $x_\alpha \to x$ and let V be an open set in Y such that $F(x) \subset V$. We will show that there exists $\alpha_0 \in J$ such that for all $\alpha \geq \alpha_0$, we have $F(x_\alpha) \subset V$. Assume that this is not the case. Then for all $\alpha \in J$, there exits $\beta \in J$ such that $\beta \geq \alpha$ and $F(x_\beta) \not\subset V$. This implies that $x_\beta \in F^{-}(Y \backslash V)$, and thus $x_\beta \in \overline{F^{-}(Y \backslash V)}$. Since $x_\alpha \to x$, we can easily show that $x_\beta \to x \in \overline{F^{-}(Y \backslash V)}$. From (d), we have $x \in F^{-}(Y \backslash V)$, which is in contradiction with $F(x) \subset V$.

(e) $\Longrightarrow$ (a). Let $x \in X$ and V be an open set in Y such that $F(x) \subseteq V$. Suppose that for all $V \in \mathcal{N}(x)$, we have $x_v \in V$, such that $F(x_v) \cap Y \backslash V \neq \emptyset$. Let

$$R = \{[x_v, V] \in V \times \mathcal{N}(x) : x_v \in F^{-}(Y \backslash V)\}.$$

We introduce a partial ordering on R by declaring that $[x_v, V] \leq [x_{v'}, V']$ if and only if $V' \subset V$. Our claim is that R with this partial ordering becomes a directed set. Indeed, let $[x_v, V]$, $[x_{v'}, V'] \in R$. Since $V \cap V' \in \mathcal{N}(x)$, there exists $x_{v \cap v'} \in V \cap V'$ such that $x_{v \cap v'} \in F^{-}(Y \backslash V \cap V')$. We consider $[x_{v \cap v'}, V \cap V'] \in R$. It is clear that $[x_v, V] \leq [x_{v \cap v'}, V \cap V']$ and $[x_{v'}, V'] \leq [x_{v \cap v'}, V \cap V']$. Define $\phi : R \to \mathcal{N}(x)$ by $[x_v, V] \to \phi([x_v, V]) = V$. Clearly, $\phi(R)$ is cofinal in $\mathcal{N}(x)$. For any $[x_v, V]$, let $x_{\phi[x_v, V]} = x_v$. We will show that $x_v \to x$. Let $V' \in \mathcal{N}(x)$; then there exist $x_{v'} \in V'$ such that $x_{v'} \in F^{-}(Y \backslash V)$. So for any $[x_v, V] \geq [x_{v'}, V']$, we have $x_v \in V \subset V'$. Hence, $x_v \to x$. Since $F(x) \subset V$, by e), there exists $[x_v, V] \in R$ such that $[x_{v'}, V'] \geq [x_v, V]$ implies $F(x_{v'}) \subset V$. Thus, $x_{v'} \notin F^{-}(Y \backslash V')$ which is a contradiction. □

The corresponding result for lower semicontinuity reads as follows.

Proposition 2.32. *For a multifunction $F : X \to \mathcal{P}(Y)$, the following are equivalent:*

(a) *F l.s.c.*

(b) *For every $V \subseteq Y$ open, $F^{-}(V)$ is open in X.*

(c) *For every closed* $C \subseteq Y$, $F^+(C)$ *is closed in* X.

(d) $\overline{F^+(D)} \subseteq F^+(\overline{D})$.

(e) $F(\overline{A}) \subseteq \overline{F(A)}$, *for every set* $A \subseteq X$.

(g) *For any* $x \in X$, *if* $\{x_\alpha\}_{\alpha \in J}$ *is a generalized sequence,* $x_\alpha \to x$, *then for every* $y \in F(x)$ *there exists a generalized sequence* $\{y_\alpha\}_{\alpha \in J} \subset Y$, $y_\alpha \in F(x_\alpha)$, $y_\alpha \to y$.

Proof. Again, our pattern is $a) \Rightarrow b) \Rightarrow c) \Rightarrow d) \Rightarrow e) \Rightarrow f) \Longrightarrow g) \Longrightarrow a)$.

(a) $\Rightarrow$ (b). Let W be open in Y; then

$$F^-(W) = \{x \in X : F(x) \subseteq W\}.$$

We will now show that $F^-(W)$ is an open set in X. Let $x \in F^-(W)$; then $F(x) \cap W \neq \emptyset$. Since F l.s.c., there exists $V(x) \in \mathcal{N}(x)$ such that

$$F(z) \cap W \neq \emptyset \text{ for all } z \in V(x), \text{ so } V(x) \subseteq F^-(W).$$

Hence, $F^-(W)$ is open in X.

(b) $\Rightarrow$ (c). Let Q be a closed set in Y; then

$$F^+(Q) = \{x \in X : F(x) \subseteq Q\}$$

and

$$X \backslash F^+(Q) = \{x \in X : F(x) \cap X \backslash Q \neq \emptyset\} = F^-(X \backslash Q).$$

Since Q is a closed set in Y, $X \backslash Q$ is an open set in Y. From $b)$, we have $F^-(X \backslash Q)$ is open in X. Thus, $F^+(Q)$ is closed in X.

(c) $\Longrightarrow$ (d). Let D be a set in Y; then we have

$$F^+(D) \subseteq F^+(\overline{D}) \Longrightarrow \overline{F^+(D)} \subseteq \overline{F^+(\overline{D})}.$$

From (c), we obtain

$$\overline{F^+(D)} \subseteq F^+(\overline{D}).$$

(d) $\Longrightarrow$ (e). Let A be a subset in X. We will show that $F(\overline{A}) \subseteq \overline{F(A)}$. Assume that $F(\overline{A}) \not\subseteq \overline{F(A)}$. Then there exists $y \in F(\overline{A})$, such that $y \notin \overline{F(A)}$, and thus there exists $V(y) \in \mathcal{N}(y)$, with $V(y) \cap F(A) = \emptyset$. This implies that

$$A \subseteq F^+(Y \backslash V(y)).$$

From (d), we have

$$A \subseteq \overline{F^+(Y \backslash V(y))} = F^+(Y \backslash V(y)).$$

Since $y \in F(\overline{A})$, there exists $x \in \overline{A}$, such that $y \in F(x)$. Since $x \in \overline{A}$, we have a generalized sequence $\{x_\alpha\}_{\alpha \in J}$ in A, $x_\alpha \to x$. Hence, $x \in F^+(Y \backslash V(y))$, and by definition of F^+ we obtain $F(x) \cap V(y) = \emptyset$, which is a contradiction to $y \in F(x)$.

(e) $\Longrightarrow$ (g). Let $\{x_\alpha\}_{\alpha\in J}$ be a generalized sequence, $x \in X$, $x_\alpha \to x$, and $y \in F(x)$. Set $A = \{x_\alpha : \alpha \in J\}$, where J is a directed set. By e), we have

$$F(\overline{A}) = F(A \cup \{x\}) \subseteq \overline{F(A)}.$$

Let

$$R = \{[x_\alpha, V] \in V \times \mathcal{N}(y) : x_v \in F^-(V)\}.$$

Since $y \in F(\overline{A})$, this implies that $y \in \overline{F(\{x_\alpha : \alpha \in J\})}$. Then $R \neq \emptyset$. We introduce a partial ordering on R, by declaring that $[\alpha, V] \leq [\alpha', V']$ if and only if $V' \subset V$ and $\alpha' \leq \alpha$. Our claim is that R with this partial ordering becomes a directed set. Indeed, let $[\alpha, V]$, $[\alpha', V'] \in R$. Then since J is directed, there exists $\beta \in J$ such that $\alpha \leq \beta$ and $\alpha' \leq \beta$. Also, because $y \in \cap_{\alpha\in J}\overline{\cup_{\beta\geq\alpha} F(x_\beta)}$ and $V \cap V' \in \mathcal{N}(y)$, we can find $\gamma \in J$, $\gamma \geq \alpha$, such that $x_\gamma \in F^-(V \cap V')$. Then $[\alpha, V] \leq [\gamma, V \cap V']$ and $[\beta, V'] \leq [\gamma, V \cap V']$. So R is directed. Define $\phi : R \to \mathcal{N}(y)$ by $[\alpha, V] \to \phi([\alpha, V]) = \alpha$. Clearly, $\phi(R)$ is cofinal in J. For any $[\alpha, V] \in R$, let $y_{\phi([\alpha,V])} \in F(x_\alpha) \cap V$. Also, $x_{\phi([\alpha,V])} = x_\alpha$. Since $\phi(R)$ is cofinal in J, we have $x_{\phi([\alpha,V])} \to x$. We will show that $y_{\phi([\alpha,V])} \to y$. Let $V' \in \mathcal{N}(y)$. Then there exists $\alpha' \in J$ such that $x_{\alpha'} \in F^-(V')$. So for any $[\alpha, V] \geq [\alpha', V']$, we have $y_{\phi([\alpha,V])} \in V \subseteq V'$, which implies that $y_{\phi([\alpha,V])} \to y$.

(g) $\Longrightarrow$ (a). Let $x \in X$ and W be an open set in Y such that $F(x) \cap W \neq \emptyset$. We will show that there exists $V \in \mathcal{N}(x)$, such that $F(z) \cap V \neq \emptyset$, for all $z \in W$. Assume that is not the case. Then for all $V \in \mathcal{N}(x)$, we have $x_v \in V$, such that $F(x_v) \cap V = \emptyset$. Let

$$R = \{[x_v, V] \in V \times \mathcal{N}(x) : x_v \in F^-(V)\},$$

and

$$\phi : R \to \mathcal{N}(x) \text{ by } [x_v, V] \to \phi([x_v, V]) = V.$$

As in Proposition 2.31, we can prove that R is directed and $\phi(R)$ is cofinal in $\mathcal{N}(x)$. For any $[x_v, V]$, let $x_{\phi[x_v,V]} = x_v$. We will show that $x_v \to x$. Let $V' \in \mathcal{N}(x)$; then there exist $x_{v'} \in V'$ such that $x_{v'} \in F^-(V)$. So for any $[x_v, V] \geq [x_{v'}, V']$, we have $x_v \in V \subset V'$. Hence, $x_v \to x$. Since $F(x) \cap W \neq \emptyset$, then there exists $y \in F(x) \cap W$. By g), there exists $y_{\phi([x_v,V])} \in F(x_v)$ such that $y_{\phi([\alpha,V])} \to y$. But $y_{\phi([\alpha,V])} \in F(x_v) \subseteq Y \setminus W$. Thus, $y \in Y \setminus W$, which is contradiction.

Now $[x_{v'}, V'] \geq [x_v, V]$, implies $F(x_{v'}) \subset V$. Thus, $x_{v'} \notin F^-(V')$, which is a contradiction. □

Remark 2.33. In the case where X and Y are topological spaces with countable bases, we may take usual sequences instead of generalized ones in conditions (e) and (g) of Propositions 2.31 and 2.32, respectively.

Example 2.34. The following set-valued mappings are upper semicontinuous:

(1) $F : \mathbb{R} \to \mathcal{P}(\mathbb{R})$ defined by

$$F(x) = \begin{cases} 1, & x > 0, \\ \{-1, 1\} & x = 0, \\ \{-1\} & x < 0. \end{cases}$$

(2) $F : \mathbb{R} \to \mathcal{P}(\mathbb{R})$ defined by

$$F(x) = \begin{cases} x + 1, & x > 0, \\ [-1, 1] & x = 0, \\ x - 1 & x < 0. \end{cases}$$

(3) $F : \mathbb{R} \to \mathcal{P}(\mathbb{R})$ defined by $F(x) = [f(x), g(x)]$, where $f, g : \mathbb{R} \to \mathbb{R}$ are l.s.c and u.s.c. functions, respectively.

Example 2.35. The following set-valued mappings are lower semicontinuous:

(1) $F : \mathbb{R} \to \mathcal{P}(\mathbb{R})$ defined by

$$F(x) = \begin{cases} [a, b], & x \neq 0, \\ \{\alpha\}, & \alpha \in [a, b]. \end{cases}$$

(2) $F : \mathbb{R} \to \mathcal{P}(\mathbb{R})$ defined by

$$F(x) = \begin{cases} [0, |x| + 1], & x \neq 0, \\ \{1\}, & x = 0. \end{cases}$$

(3) $F : \mathbb{R} \to \mathcal{P}(\mathbb{R})$ defined by $F(x) = [f(x), g(x)]$, where $f, g : \mathbb{R} \to \mathbb{R}$ are u.s.c and l.s.c. functions, respectively.

(4) Let $X = Y = [0, 1]$. Define

$$F(x) = \begin{cases} [0, 1], & x \neq \frac{1}{2}, \\ [0, \frac{1}{2}], & x = \frac{1}{2}. \end{cases}$$

In general, the concepts of upper semicontinuity and lower semicontinuity are distinct. The following standard example illustrates this.

Example 2.36. Let $X = Y = \mathbb{R}$. Define

$$F_1(x) = \begin{cases} \{1\}, & x \neq 0, \\ [0, 1], & x = 0, \end{cases} \quad \text{and} \quad F_2(x) = \begin{cases} \{0\}, & x = 0, \\ [0, 1], & x \neq 0. \end{cases}$$

We can easily show that F_1 is u.s.c. but not l.s.c., while F_2 is l.s.c. but not u.s.c.

Another useful continuity notion related to the previous ones, can be defined using the graph of a multifunction.

Definition 2.37. A multifunction is said to be closed if its graph $\text{Gra} F$ is a closed subset of the space $X \times Y$.

Here are some equivalent formulations.

Theorem 2.38. *The following conditions are equivalent:*

(a) *The multifunction F is closed.*

(b) *For every $(x, y) \in X \times Y$ such that $y \notin F(x)$, there exist neighborhoods $V(x)$ of x and $W(y)$ of y such that $F(V(x)) \cap W(y) = \emptyset$.*

(c) *For generalized sequences $\{x_\alpha\}_{\alpha \in J} \subset X$ and $\{y_\alpha\}_{\alpha \in J} \subset Y$, if $x_\alpha \to x$, and $y_\alpha \in F(x_\alpha)$ with $y_\alpha \to y$, then $y \in F(x)$.*

Proof. Our pattern follows $a) \Rightarrow b) \Rightarrow c) \Rightarrow a)$.

(a) $\Rightarrow$ (b). Let $(x, y) \in X \times Y$ be such that $y \notin F(y)$. Then $(x, y) \notin \text{Gra} F$, and this implies that $(x, y) \in X \times Y \text{Gra} F$. Since $\text{Gra} F$ is closed, there exists $(V(x), W(y)) \in \mathcal{N}(x) \times \mathcal{N}(y)$, such that $V(x) \times W(y) \cap \text{Gra} F = \emptyset$. We will show that $F(V(x)) \cap W(y) = \emptyset$. Suppose that there exists $z \in F(V(x)) \cap W(y)$. Then there exists $r \in V(x)$, such that $z \in F(r)$, and this implies that $(r, z) \in \text{Gra} F$, which is a contradiction.

(b) $\Rightarrow$ (c). Let $\{x_\alpha\}_{\alpha \in J}$ be a generalized sequence such that

$$x_\alpha \to x, \ y_\alpha \in F(x_\alpha), \ y_\alpha \to y.$$

Assume that $y \notin F(x)$. Then there exist $(V(x), W(y)) \in \mathcal{N}(x) \times \mathcal{N}(y)$, such that $F(V(x)) \cap W(y) = \emptyset$. Now

$$x_\alpha \to x \Longrightarrow \exists \alpha_0 \in J \text{ such that } \forall \alpha \geq \alpha_0; \text{ we have } x_\alpha \in V(x),$$

and

$$y_\alpha \to y \Longrightarrow \exists \alpha_1 \in J \text{ such that } \forall \alpha \geq \alpha_1; \text{ we have } y_\alpha \in W(y).$$

Since J is is directed, then there exists $\beta \in J$ such that $\alpha_0, \alpha_1 \leq \beta$, and hence for all $\alpha \geq \beta$, we have $x_\alpha \in V(x)$ and $y_\alpha \in W(y)$, with $y_\alpha \in F(x_\alpha)$. Then $F(V(x)) \cap W(y) \neq \emptyset$ which is a contradiction.

(c) $\Rightarrow$ (a). Let $(x_\alpha, y_\alpha) \in \text{Gra} F$, $\alpha \in J$, $x_\alpha \to x, y_\alpha \to y$ and $y_\alpha \in F(x_\alpha)$. From (c), we obtain that $y \in F(x)$. Hence $\text{Gra} F$ is closed. □

Example 2.39. Let $f : Y \to X$ be a continuous surjective map between topological spaces. Then the inverse multifunction $F : X \to \mathcal{P}(Y)$ given by $F(x) = f^{-1}(x)$ is closed.

Next, we give a relationship between u.s.c. and closed multifunctions.

Theorem 2.40. *Let X be a topological space, Y a regular topological space, and $F : X \to \mathcal{P}_{cl}(Y)$ an u.s.c. multifunction. Then F is closed.*

Proof. Let $y \in Y$, $y \notin F(x)$. Since Y is regular, there exist an open neighborhood $W(y)$ of the point y and an open neighborhood W_1 of the set $F(x)$ such that $W(y) \cap F(x) = \emptyset$. Let $V(x)$ be a neighborhood of x such that $F(V(x)) \subset W_1$. Then $F(v(x)) \cap W(y) = \emptyset$ and the statement follows from Theorem 2.38 part b). □

In the next result, we give sufficient conditions for a closed multifunction to be u.s.c. We need the following definition.

Definition 2.41. A multifunction $F : X \to \mathcal{P}(Y)$ is said to be:

(a) compact, if its range $F(X)$ is relatively compact in Y, i.e., $\overline{F(X)}$ is compact in Y;

(b) locally compact, if every point $x \in X$ has a neighborhood $V(x)$ such that the restriction of F to $V(x)$ is compact.

It is clear that (a) $\Longrightarrow$ (b).

Theorem 2.42. *Let $F : X \to \mathcal{P}_{cp}(Y)$ be a closed locally compact multifunction. Then F is u.s.c.*

Proof. Let $x \in X$, let W be an open neighborhood of the set $F(x)$, and $V(x)$ be an open neighborhood of x such that the restriction of F to $V(x)$ is compact. Suppose that the set $Q = \overline{F(v(x))}\backslash W$ is nonempty. Since F is closed, for any $y \in Q$, there exist neighborhoods $\widetilde{W}(y)$ of y and $V_y(x)$ of x such that $F(V_y(x)) \cap \widetilde{W}(y) = \emptyset$. By virtue of the compactness of Q, we can find its finite covering $\widetilde{W}(y_1), \widetilde{W}(y_2), \ldots, \widetilde{W}(y_n)$. But then if we consider the open neighborhood of x defined by $\widetilde{V}(x) = V(x) \cap (\cap_{i=1}^{n} V_{y_1}(x))$, we have $F(\widetilde{V}(x)) \subset W$. □

Example 2.43. The condition of local compactness is essential. The multifunction $F : [-1, 1] \to \mathcal{P}_{cp}(\mathbb{R})$ defined by

$$F(x) = \begin{cases} \{\frac{1}{x}\}, & x \neq 0, \\ \{0\}, & x = 0, \end{cases}$$

is closed but loses its upper semicontinuity at $x = 0$.

Lemma 2.44 ([301]). *If $F : X \to \mathcal{P}(Y)$ has a closed graph and is locally compact (i.e., for every $x \in X$, there exists a $U \in \mathcal{N}(x)$ such that $\overline{F(U)} \in \mathcal{P}_{cp}(Y)$), then $F(\cdot)$ is upper semicontinuous.*

Definition 2.45. A multifunction $F : X \to \mathcal{P}(Y)$ is said be quasicompact if its restriction to any compact subset $A \subset X$ is compact.

Lemma 2.46. *If $G : X \to \mathcal{P}_{cp}(Y)$ is quasicompact and has a closed graph, then G is u.s.c.*

Proof. Assume that G is not u.s.c. at some point x. Then there exists an open neighborhood U of $G(x)$ in Y, a sequence $\{x_n\}$ which converges to x, and for every $l \in \mathbb{N}$ there exists $n_l \in \mathbb{N}$ such that $G(x_{n_l}) \not\subset U$. Then for each $l = 1, 2, \ldots$, there are y_{n_l} such that $y_{n_l} \in G(x_{n_l})$ and $y_{n_l} \notin U$; this implies that $y_{n_l} \in Y \backslash U$. Moreover $\{y_{n_l} : l \in \mathbb{N}\} \subset \overline{G(\{x_n : n \geq 1\})}$. Since G is compact, there exists a subsequence of $\{y_{n_l} : l \in \mathbb{N}\}$ which converges to y. G closed implies that $y \in G(x) \subset U$; but this is a contradiction to the assumption that $y_{n_l} \notin U$ for each n_l. □

2.3.4 Selection Functions and Selection Theorems

The basic connection between "multivalued analysis" and "single-valued analysis" is given by the concept of selection.

Definition 2.47. Let X, Y be nonempty sets and $F : X \to \mathcal{P}(Y)$. The single-valued operator $f : X \to Y$ is called a selection of F if and only if $f(x) \in F(x)$, for each $x \in X$. The set of all selection functions for F is denoted by S_F.

A very famous result is the so-called "Michael selection theorem." We start by proving the following auxiliary results.

Lemma 2.48. *Let (X, d) be a metric space, Y be a Banach space, $F_1 : X \to \mathcal{P}(Y)$ be l.s.c., and $F_2 : X \to \mathcal{P}(Y)$ have an open graph such that $F_1(x) \cap F_2(x) \neq \emptyset$, for each $x \in X$. Then the multivalued operator $F_1 \cap F_2$ is l.s.c.*

Lemma 2.49. *Let (X, d) be a metric space, Y be a Banach space, and $F : X \to \mathcal{P}_{cv}(Y)$ be l.s.c. on X. Then, for each $\epsilon > 0$ there exists a continuous function $f_\epsilon : X \to Y$ such that, for all $x \in X$, we have $f_\epsilon(x) \in V(F(x), \epsilon)$.*

Proof. Since F is l.s.c., we associate to each $x \in X$ and each $y_x \in F(x)$ an open neighborhood U_x of x such that $F(x') \cap B(y_x, \epsilon) \neq \emptyset$, for all $x' \in U_x$. Now X is a paracompact space and so there exists a locally finite refinement $\{U'_x\}_{x \in X}$ of $\{U_x\}_{x \in X}$. Let us recall that $\{\Omega_i\}_{i \in I}$ is a locally finite covering of X if for each $x \in X$ there exists a neighborhood V of x satisfying $\Omega_i \cap V \neq \emptyset$, for all $i = 1, \ldots, m$. Moreover, to each locally finite covering, it is possible to associate a locally Lipschitz partition of unity (see Chapter 4 of Munkres [384]) denoted by $\{\phi_x\}_{x \in X}$. We define $f_\epsilon(t) = \sum_{x \in X} \phi_x(t) y_x$. Then f_ϵ is continuous, being, locally, a finite sum of continuous functions. Moreover, if $\phi_x > 0$ for $x \in U'_c \subset U_x$, then $y_x \in V(F(x), \epsilon)$ implies that $f_\epsilon(t) \in V(F(t), \epsilon)$. □

Theorem 2.50 (Michael's selection theorem). *Let (X, d) be a metric space, Y be a Banach space, and $F : X \to \mathcal{P}_{cl,cv}(Y)$ be l.s.c. on X. Then there exists $f : X \to Y$ which is a continuous selection of F.*

Proof. Let us define inductively a sequence of continuous functions $f_n : X \to Y$, $n = 1, 2 \ldots$, satisfying the following assertions:

(i) for all $x \in X$, $H(f_n(x), F(x)) < \frac{1}{2^n}$, for all $n \in \mathbb{N}$;

(ii) for all $x \in X$, $\|f_n(x) - f(x)\| < \frac{1}{2^{n-2}}$, for each $n = 2, 3, \ldots$.

- **Case $n = 1$.** The conclusion follows from Lemma 2.49 with $\epsilon = \frac{1}{2}$.

- **Case $n + 1$.** Let us suppose that we have defined the mappings $f_1, \ldots, f_n$, and we construct the map f_{n+1} such that (i) and (ii) hold. For this purpose, we consider the multivalued operator F_{n+1} given by, $F_{n+1}(x) = F(x) \cap B(f_n(x), \frac{1}{2^n})$, for each $x \in X$. From (i) we obtain that $F_{n+1}(x) \neq \emptyset$ for all $x \in X$. Using Lemma 2.48, we have that F_{n+1} is l.s.c. From Lemma 2.49, applied to F_{n+1}, we get the existence of a continuous function $f_{n+1} : X \to Y$ such that

$$f_{n+1}(x) \in V^0(F(x), \tfrac{1}{2^{n+1}}).$$

At the same time, we have $H(f_n(x), F(x)) < \frac{1}{2^n}$, so

$$\|f_{n+1}(x) - f_n(x)\| \leq \frac{1}{2^{n-1}}.$$

It follows that $H(f_{n+1}(x), F(x)) < \frac{1}{2^{n+1}}$. At the same time, we have

$$f_{n+1}(x) \in V^0(F(x), \tfrac{1}{2^{n+1}}),$$

which completes the induction. Furthermore, from (ii), we obtain that $(f_n)_{n\in\mathbb{N}}$ is a uniform Cauchy sequence converging to a continuous function $f : X \to Y$. From $i)$ and the fact that $F(x)$ is closed for each $x \in X$, we obtain that $f(x) \in F(x)$, for all $x \in X$. Hence, f is the desired continuous selection and the proof is complete. □

For u.s.c. multifunctions, we have the following approximate selection theorem given by Aubin and Cellina [34].

Theorem 2.51 ([412]). *Let (X, d) be a metric space, Y be a Banach space and $F : X \to \mathcal{P}_{cv}(Y)$ be u.s.c. on X. Then for every $\epsilon > 0$ there exists $f_\epsilon : X \to Y$ that is locally Lipschitzian and satisfies:*

(i) $f_\epsilon(X) \in \mathrm{co}F(X)$,

(ii) $\mathrm{Gra}F_\epsilon \subset V(\mathrm{Gra}F, \epsilon)$,

where co$F$ *is the convex hull of* F.

Let us consider now the selection theorem of Browder [113].

Theorem 2.52. *Let X and Y be Hausdorff topological vector spaces and $K \in \mathcal{P}_{cp}(X)$. Let $F : K \to \mathcal{P}_{cv}(Y)$ be a multivalued operator such that $F^{-1}(y)$ is open, for each $y \in Y$. Then there exists a continuous selection f of F.*

2.3.5 Hausdorff Continuity

When Y is a metric space, by using the Hausdorff pseudometric, we can define three new continuity concepts that, in general, are distinct from ones considered in the previous sections. Throughout this section, X is a Hausdorff topological space and Y is a metric space.

Definition 2.53. A multifunction $F : X \to \mathcal{P}(Y)$ is said to be:

(a) H-upper semicontinuous at $x_0 \in X$, if $H^*(F(x), F(x_0))$ is continuous at x_0; i.e.,

$$\forall \epsilon > 0, \exists U_\epsilon \in \mathcal{N}(x_0) : \forall x \in U_\epsilon \Longrightarrow H^*(F(x), F(x_0)) < \epsilon,$$

where $\mathcal{N}(x)$ is a neighborhood filter of of x.

(b) H-lower semicontinuous at $x_0 \in X$, if $H^*(F(x_0), F(x))$ is continuous at x_0; i.e.,

$$\forall \epsilon > 0, \exists U_\epsilon \in \mathcal{N}(x_0) : \forall x \in U_\epsilon \Longrightarrow H^*(F(x_0), F(x)) < \epsilon.$$

(c) H-continuous at x_0, if it is both H-upper semicontinuous and H-lower semicontinuous at x_0.

We start by comparing these continuity concepts with the Vietoris ones studied earlier.

Proposition 2.54. *If $F : X \to \mathcal{P}(Y)$ is u.s.c., then $F(\cdot)$ is H-u.s.c.*

Proof. Since $F(\cdot)$ is upper semicontinuous, given $\epsilon > 0$ and $x \in X$, we have that $F^+((F(x))_\epsilon) = U \in \mathcal{N}(x)$. So for every $x' \in U$, we have $F(x') \subseteq (F(x))_\epsilon$. Hence, $H^*(F(x'), F(x)) < \epsilon$ for all $x' \in U$, and thus we conclude that $F(\cdot)$ is $H-$upper semicontinuous. □

Example 2.55. A single valued mapping $f : \mathbb{R} \to \mathbb{R}$ is H-u.s.c. (H-l.s.c.) if the set valued mapping F defined by $F(t) = [0, f(t)]$ is upper (lower) semicontinous.

Example 2.56. The converse of Proposition 2.54 is not in general true. We consider the counterexample $F : [0, 1] \to \mathcal{P}(\mathbb{R})$ defined by

$$F(x) = \begin{cases} [0,1], & x \in [0,1), \\ [0,1), & x = 1. \end{cases}$$

It easy to check that $F(\cdot)$ is H-upper semicontinuous but not upper semicontinuous at $x = 1$. Indeed note that $F^+((-1,1)) = \{1\}$ is not an open set.

The second example involves a closed-valued multifunction.

Example 2.57. In the following counterexample, let $F : \mathbb{R} \to \mathcal{P}_{cl}(\mathbb{R}^2)$ be defined by

$$F(x) = \begin{cases} \{[0,z] : z \geq 0\}, & x = 0, \\ \{[x,z] : 0 \leq z \leq \frac{1}{z}\}, & x \neq 0. \end{cases}$$

Then $F(\cdot)$ is H-upper semicontinous but not upper semicontinuous, since for $C = \{[\frac{1}{n}, n] : n \geq 1\} \subset \mathbb{R}^2$ is closed, but $F^-(C)$ is not closed in $\mathbb{R}$.

Proposition 2.58 ([301]). *If $F : X \to \mathcal{P}_{cl}(Y)$ is H-u.s.c., then $F(\cdot)$ is closed.*

For relations between H-u.s.c. multifunctions and single lower semicontinuous functions, we state some interesting results.

Proposition 2.59 ([301]). *If $F : X \to \mathcal{P}_{cl}(Y)$ is H-u.s.c., then for every $v \in Y$, $x \to \phi_v(x) = d(v, F(x))$ is lower semicontinuous.*

Proposition 2.60 ([301]). *If $F : X \to \mathcal{P}_{cl}(Y)$ is H-u.s.c., then $F(\cdot)$ is lower semicontinuous.*

Theorem 2.61 ([301]). *Let $F : X \to \mathcal{P}_{cp}(Y)$. The following conditions are equivalent:*

(a) *F u.s.c. (resp. F l.s.c.).*

(b) *H-upper semicontinuous (resp. H-lower semicontinuous).*

Definition 2.62. A multivalued operator $N : X \to \mathcal{P}_{cl}(X)$ is called

(a) γ-Lipschitz if and only if there exists $\gamma > 0$ such that

$$H(N(x), N(y)) \leq \gamma d(x,y), \quad \text{for each } x,\ y \in X,$$

(b) a contraction if and only if it is γ-Lipschitz with $\gamma < 1$.

Remark 2.63. It clear that, if N is Lipschitz, then N is $H-$continuous.

The following result is are easily deduced from the limit properties.

Lemma 2.64 (see e.g. [35, Lemma 1.1.9]). *Let $(K_n)_{n\in\mathbb{N}} \subset K \subset X$ be a sequence of subsets where K is compact in the separable Banach space X. Then*

$$\overline{co}\,(\limsup_{n\to\infty} K_n) = \bigcap_{N>0} \overline{co}\Big(\bigcup_{n\geq N} K_n\Big),$$

where $\overline{co}\,A$ refers to the closure of the convex hull of A.

Lemma 2.65 ([342]). *Let X be a Banach space. Let $F : [a,b] \times X \longrightarrow P_{cp,c}(X)$ be an L^1-Carathéodory multivalued map with $S_{F,y} \neq \emptyset$ and let Γ be a linear continuous mapping from $L^1([a,b],X)$ into $C([a,b],X)$. Then the operator*

$$\begin{aligned} \Gamma \circ S_F : C([a,b],X) &\longrightarrow \mathcal{P}_{cp,c}(C([a,b],X)), \\ y &\longmapsto (\Gamma \circ S_F)(y) := \Gamma(S_{F,y}) \end{aligned}$$

is a closed graph operator in $C([a,b],X) \times C([a,b],X)$.

Lemma 2.66 (Mazur's Lemma [385, Theorem 21.4]). *Let E be a normed space and $\{x_k\}_{k\in\mathbb{N}} \subset E$ be a sequence weakly converging to a limit $x \in E$. Then there exists a sequence of convex combinations $y_m = \sum_{k=1}^{m} \alpha_{mk} x_k$ with $\alpha_{mk} > 0$ for $k = 1,2,\dots,m$ and $\sum_{k=1}^{m} \alpha_{mk} = 1$, which converges strongly to x.*

2.3.6 Measurable Multifunctions

Throughout this section, (Ω, Σ) is a measurable space and (X,d) a separable metric space. We define several concepts of measurability for a multifunction $F : \Omega \to \mathcal{P}(X)$.

Definition 2.67. A multifunction $F : \Omega \to \mathcal{P}(X)$, is said to be:

(a) Strongly measurable, if for every closed $C \subseteq X$, we have

$$F^{-}(C) = \{\omega \in \Omega : F(\omega) \cap C \neq \emptyset\} \in \Sigma;$$

(b) Measurable, if for every open $U \subseteq X$, we have

$$F^{-}(U) = \{\omega \in \Omega : F(\omega) \cap U \neq \emptyset\} \in \Sigma;$$

(c) $F(\cdot)$ is said to be "K-measurable", if for every compact $K \subseteq X$, we have

$$F^{-}(K) = \{\omega \in \Omega : F(\omega) \cap K \neq \emptyset\} \in \Sigma;$$

(d) Graph measurable, if

$$\text{Gra} F = \{(\omega, x) \in \Omega \times X : x \in F(\omega)\} \in \Sigma \times B(X),$$

where $B(X)$ is the σ-algebra generated by the family of all open sets from X.

Proposition 2.68. *If $F : \Omega \to \mathcal{P}(X)$ is strongly measurable, then $F(\cdot)$ is measurable.*

Proof. Let $U \subseteq X$ be open. Recall that in a metrizable space, every open set in an F_σ−set (i.e., a countable union of closed sets). So $U = \cup_{k\geq 1} C_n$, $C_n \subseteq X$, C_n closed for $n \geq 1$. We have $F^-(U) = F(\cup_{n\geq 1} C_n) = \cup_{n\geq 1} F(C_n) \in \Sigma$. Hence, $F(\cdot)$ is measurable. □

We next state a few popular notions of measurability of multifunctions.

Proposition 2.69. *$F : \Omega \to \mathcal{P}(X)$ is measurable if and only if, for every $x \in X$, $\omega \to d(x, F(\omega)) = \inf\{d(x,x') : x' \in F(\omega)\}$ is a measurable $\overline{\mathbb{R}}_+ = \mathbb{R} \cup \{\infty\}$-valued function.*

Proposition 2.70. *If $F : \Omega \to \mathcal{P}(X)$ is measurable, $F(\cdot)$ is graph measurable.*

Recalling that for $U \subseteq X$ open, we have $A \cap U \neq \emptyset$ if and only if $\overline{A} \cap U \neq \emptyset$, we immediately have the following proposition.

Proposition 2.71. *$F : \Omega \to \mathcal{P}(X)$ is measurable if and only if $\overline{F}(\cdot)$ is measurable.*

As was the case in our topological study of multifunctions, the situation simplifies considerably with compact valued multifunctions.

Proposition 2.72. *If $F : \Omega \to \mathcal{P}_{cp}(X)$, then F is strongly measurable if and only if it is measurable.*

The following two lemmas are needed in this paper. The first one is the celebrated Kuratowski–Ryll–Nardzewski selection theorem.

Lemma 2.73 ([230, Theorem 19.7]). *Let Y be a separable metric space and $F : [a,b] \to \mathcal{P}_{cl}(Y)$ a measurable multivalued. Then F has a measurable selection.*

Definition 2.74. A multivalued map $G : \Omega \to \mathcal{P}_{cl}(X)$ has a Castaing representation if there exists a family measurable single-valued maps $g_n : \Omega \to X$ such that

$$G(\omega) = \overline{\{g_n(\omega) \mid n \in \mathbb{N}\}}.$$

The following result is due to Castaing (see [136]).

Theorem 2.75. *Let X be a separable metric space. Then the multivalued map $G : \Omega \to \mathcal{P}_{cl}(X)$ is measurable if and only if G has a Castaing representation.*

Lemma 2.76 ([498, Lemma 3.2]). *Let $G : [0,b] \to \mathcal{P}(Y)$ be a measurable multivalued map and $u : [a,b] \to Y$ a measurable function. Then for any measurable $v : [a,b] \to (0,+\infty)$, there exists a measurable selection f_v of G such that for a.e. $t \in [a,b]$,*

$$|u(t) - f_v(t)| \leq d(u(t), G(t)) + v(t).$$

Proof. By Theorem 2.75, there is a sequence of measurable selection $\{g_n \mid n \in \mathbb{N}\}$ of G such that

$$G(t) = \overline{\{g_n(t) \mid n \in \mathbb{N}\}}, \text{ for all } t \in [a,b].$$

Set

$$T_n = \big\{t \in [a,b] \,\big|\, |g_n(t) - u(t)| \leq d(u(t), G(t)) + r(t)\big\}.$$

Consider the single-valued map $\Psi_n : [a,b] \to \mathbb{R}^+$ defined by

$$\Psi_n(t) = |g_n(t) - u(t)| - d(u(t), G(t)) + r(t),\ t \in [a,b],$$

It is clear that Ψ_n is measurable map; then

$$\Psi_n^{-1}((-\infty, 0]) = \big\{t \in [a,b] \,\big|\, |g_n(t) - u(t)| \leq d(u(t), G(t)) + r(t)\big\} = T_n.$$

Then the maps T_n, $n \in \mathbb{N}$ are measurable and we can easily show that $[a,b] = \cup_{n=1}^{\infty} T_n$ up to a negligible set. Let $E_1 = T_1$, $E_2 = T_2 \backslash E_1, \ldots, E_n = T_n \backslash \cup_{i=1}^{n-1} E_i, \ldots$. Then $[a,b] = \cup_{i=1}^{\infty} E_i$ up to a negligible set and $\{E_i\}_{i=1}^{\infty}$ is a disjoint sequence of measurable sets. Set

$$g(t) = \sum_{n=1}^{\infty} \chi_{E_n}(t) g_n(t),$$

where χ_{E_n} represents the characteristic function of the set E_n. Then g is a measurable selection of G satisfying the requirement of the lemma. □

Corollary 2.77. *Let $F : [0,b] \to \mathcal{P}_{cp}(Y)$ be a measurable multivalued map and $u : [0,b] \to E$ a measurable function. Then there exists a measurable selection f of F such that for a.e. $t \in [0,b]$,*

$$|u(t) - f(t)| \leq d(u(t), F(t)).$$

Proof. Taking $v(t) = v_n(t) = \frac{1}{n}$ in Lemma 2.76, we get a measurable selection f_n of F such that

$$|u(t) - f_n(t)| \leq d(u(t), F(t)) + 1/n.$$

Using the fact that F has compact values, we may pass to a subsequence if necessary to get that $\{f_n(\cdot)\}$ converges to a measurable function f, yielding our claim. □

By the Mazur Lemma and the above corollary we can easily prove the following corollary.

Corollary 2.78. *Let $G : [0,b] \to \mathcal{P}_{wcp,cv}(E)$ be a measurable multifunction and $g : [0,b] \to E$ a measurable function. Then there exists a measurable selection u of G such that*

$$|u(t) - g(t)| \leq d(g(t), G(t)).$$

Corollary 2.79. *Let E be a reflexive Banach space, $G : [0,b] \to \mathcal{P}_{cl,cv}(E)$ be a measurable multifunction, $g : [0,b] \to E$ be a measurable function, and let there exist $k \in L^1([0,b], E)$ such that*

$$G(t) \subseteq k(t)B(0,1), \ t \in [0,b],$$

where $B(0,1)$ denotes the closed ball in E. Then there exists a measurable selection u of G such that

$$|u(t) - g(t)| \leq d(g(t), G(t)).$$

Definition 2.80. Let $(E, |\cdot|)$ be a Banach space. A multivalued map $F : [a,b] \times E \to \mathcal{P}(E)$ is said to be Carathéodory if

(i) $t \longmapsto F(t,y)$ is measurable for all $y \in E$,

(ii) $y \longmapsto F(t,y)$ is upper semicontinuous for almost each $t \in [a,b]$. If, in addition,

(iii) for each $q > 0$, there exists $\varphi_q \in L^1([a,b], \mathbb{R}_+)$ such that

$$\|F(t,y)\|_{\mathcal{P}} = \sup\{|v| : v \in F(t,y)\} \leq \varphi_q(t), \text{ for all } |y| \leq q \\ and\ a.e.\ t \in [a,b],$$

then F is said to be L^1-Carathéodory.

2.3.7 Decomposable Selection

Consider a measure space $(T, \mathcal{F}, \mu)$, where $\mathcal{F}$ is a $\sigma-$algebra of subsets of T and μ is a nonatomic probability measurable on $\mathcal{F}$. If E is a Banach space, let $L^1(J,E)$ be the Banach space of all functions $u : T \to E$ which are Bochner $\mu-$integrable.

In what follows, we let χ_S denote the characteristic function

$$\chi_S(s) = \begin{cases} 1, & \text{if } s \in S, \\ 0, & \text{if } s \notin S. \end{cases}$$

Definition 2.81. A set $K \subset L^1(J,E)$ is decomposable if, for all, $u, v \in K$, $u\chi_A + v\chi_{T-A} \in K$, whenever $A \in \mathcal{F}$. The collection of all nonempty decomposable subsets of $L^1(T,E)$ is denoted by $D(L^1(T,E))$. For any set $H \subset L^1(T,E)$, the decomposable hull of H is

$$dec[H] = \cap\{K \in D(L^1(T,E)) : H \subset K\}.$$

Definition 2.82. Let Y be a separable metric space and let $N : Y \to \mathcal{P}(L^1([a,b], E))$ be a multivalued operator. We say N has property (BC) if

(1) N is lower semicontinuous (l.s.c.),

(2) N has nonempty closed and decomposable values.

Let $F : [a,b] \times E \to \mathcal{P}(E)$ be a multivalued map with nonempty compact values. Assign to F the multivalued operator

$$\mathcal{F} : C([a,b], E) \to \mathcal{P}(L^1([a,b], E))$$

by letting

$$\mathcal{F}(y) = \{w \in L^1([a,b], E) : w(t) \in F(t, y(t)) \ \ a.e.\ t \in [a,b]\}.$$

The operator $\mathcal{F}$ is called the Niemytzki operator associated with F.

Definition 2.83. Let $F : [a,b] \times E \to \mathcal{P}_{cp}(E)$ be a multivalued function. We say F is of lower semicontinuous type (l.s.c. type) if its associated Niemytzki operator $\mathcal{F}$ is lower semicontinuous and has nonempty closed and decomposable values.

We need the following lemma in the sequel.

Lemma 2.84 ([207]). *Let $F : J \times E \to \mathcal{P}_{cp}(E)$ be a multivalued map and E be a separable Banach space. Assume that*

(i) $F : J \times E \longrightarrow \mathcal{P}(E)$ *is a nonempty compact valued multivalued map such that*
 (a) $(t, y) \mapsto F(t, y)$ *is $\mathcal{L} \otimes \mathcal{B}$ measurable;*
 (b) $y \mapsto F(t, u)$ *is lower semicontinuous for a.e.* $t \in J$;

(ii) *for each $r > 0$, there exists a function $h_r \in L^1(J, \mathbb{R}^+)$ such that*

$$\|F(t,u)\|_{\mathcal{P}} \le h_r(t), \quad \text{for a.e. } t \in J \text{ and for } u \in X \text{ with } |u| \le r.$$

Then F is of l.s.c. type.

Next we state a selection theorem due to Bressan and Colombo.

Theorem 2.85 ([104]). *Let Y be separable metric space and let $N : Y \to \mathcal{P}(L^1(J, E))$ be a multivalued operator that has property (BC). Then N has a continuous selection, i.e., there exists a continuous function (single-valued) $g : Y \to L^1(J, E)$ such that $g(u) \in N(u)$ for every $u \in Y$.*

For additional details on multivalued maps, the books of Aubin and Cellina [34], Aubin and Frankowska [35], Brown et al. [112], Deimling [167], Górniewicz [230, 231], Hu and Papageorgiou [301], Petruşel [412], Smirnov [445], and Tolstonogov [465] are excellent sources.

2.4 Fixed Point Theorems

First, we state a well known result known as the Nonlinear Alternative. By $\overline{U}$ and ∂U we denote the closure of U and the boundary of U, respectively.

Lemma 2.86 (Nonlinear Alternative [184]). *Let X be a Banach space with C a closed and convex subset of X. Assume U is a relatively open subset of C, with $0 \in U$, and $G : \overline{U} \longrightarrow C$ is a compact map. Then either,*

(i) *G has a fixed point in $\overline{U}$, or*

(ii) *there is a point $u \in \partial U$ and $\lambda \in (0,1)$, with $u = \lambda G(u)$.*

There is also a multivalued version of the Nonlinear Alternative.

Lemma 2.87 ([184]). *Let X be a Banach space with $C \subset X$ convex. Assume U is a relatively open subset of C, with $0 \in U$, and let $G : X \to \mathcal{P}_{cp,c}(X)$ be an upper semicontinuous and compact map. Then either,*

(a) *G has a fixed point in $\overline{U}$, or*

(b) *there is a point $u \in \partial U$ and $\lambda \in (0,1)$, with $u \in \lambda G(u)$.*

Lemma 2.88 ([152, 230]). *Let (X,d) be a complete metric space. If $N : X \to \mathcal{P}_{cl}(X)$ is a contraction, then $FixN \neq \emptyset$. Moreover, if N has compact values, then the set $Fix(N)$ is compact.*

Definition 2.89. A multivalued map $F : X \to \mathcal{P}(E)$ is called an admissible contraction with constant $\{k_\alpha\}_{\alpha\in\Lambda}$ if, for each $\alpha \in \Lambda$, there exists $k_\alpha \in (0,1)$ such that

(i) $d_\alpha(F(x), F(y)) \leq k_\alpha d_\alpha(x,y)$ for all $x, y \in X$.

(ii) For every $x \in X$ and every $\varepsilon \in (0,\infty)^\Lambda$, there exists $y \in F(x)$ such that

$$d_\alpha(x,y) \leq d_\alpha(x, F(x)) + \varepsilon_\alpha \quad \text{for every} \quad \alpha \in \Lambda.$$

The following nonlinear alternative is due to Frigon.

Lemma 2.90 (Nonlinear Alternative, [206]). *Let E be a Fréchet space and U an open neighborhood of the origin in E, and let $N : \overline{U} \to \mathcal{P}(E)$ be an admissible multivalued contraction. Assume that N is bounded. Then one of the following statements holds:*

(C1) *N has at least one fixed point;*

(C2) *there exists $\lambda \in [0,1)$ and $x \in \partial U$ such that $x \in \lambda N(x)$.*

2.5 Measures of Noncompactness: MNC

First, we collect some definitions and properties about measures of noncompactness in Banach spaces. More details can be found in [311].

Definition 2.91. Let E be a Banach space and $(\mathcal{A}, \geq)$ a partially ordered set. A map $\beta\colon \mathcal{P}(E) \to \mathcal{A}$ is called a measure of noncompactness (MNC) on E if for every subset $\Omega \in \mathcal{P}(E)$, we have

$$\beta(\overline{co}\,\Omega) = \beta(\Omega).$$

Notice that if D is dense in Ω, then $\overline{co}\,\Omega = \overline{co}\,D$ and hence

$$\beta(\Omega) = \beta(D).$$

Definition 2.92. A measure of noncompactness β is called:

(a) Monotone, if $\Omega_0, \Omega_1 \in \mathcal{P}(E)$, $\Omega_0 \subset \Omega_1$ implies $\beta(\Omega_0) \leq \beta(\Omega_1)$.

(b) Nonsingular, if $\beta(\{a\} \cup \Omega) = \beta(\Omega)$ for every $a \in E$ and $\Omega \in \mathcal{P}(E)$.

(c) Invariant with respect to the union with compact sets, if $\beta(K \cup \Omega) = \beta(\Omega)$ for every relatively compact set $K \subset E$ and $\Omega \in \mathcal{P}(E)$.

(d) Real, if $\mathcal{A} = \overline{\mathbb{R}}_+ = [0, \infty]$ and $\beta(\Omega) < \infty$ for every bounded Ω.

(e) Regular, if the condition $\beta(\Omega) = 0$ is equivalent to the relative compactness of Ω.

(f) Algebraically semiadditive, if

$$\beta(\Omega_0 + \Omega_1) \leq \beta(\Omega_0) + \beta(\Omega_1) \text{ for every } \Omega_0, \Omega_1 \in \mathcal{P}(E).$$

As example of an MNC, one may consider the Hausdorff measure

$$\chi(\Omega) = \inf\{\varepsilon > 0\colon \Omega \text{ has a finite } \varepsilon\text{-net}\}.$$

Recall that a bounded set $A \subset E$ has a finite ε-net if there exits a finite subset $S \subset E$ such that $A \subset S + \varepsilon\overline{B}$ where $\overline{B}$ is a closed ball in E. Other examples can be presented by the following measures of noncompactness defined on the space of continuous functions $C([0,b], E)$ with the value in Banach space E:

(i) The modulus of fiber noncompactness

$$\xi(\Omega) = \sup_{t\in[0,b]} \chi_E(t),$$

where χ_E is the Hausdorff MNC in E and $\Omega(t) = \{y(t) : y \in \Omega\}$;

(ii) The modulus of equicontinuity defined as

$$mod_C(\Omega) = \lim_{\delta\to 0} \sup_{x\in\Omega} \max_{|\tau_2-\tau_1|\leq\delta} |x(\tau_1) - x(\tau_2)|.$$

It should be noted that these MNC's satisfy all above-mentioned properties except regularity.

Definition 2.93. Let $\mathcal{M}$ be a closed subset of a Banach space E and $\beta\colon \mathcal{P}(E) \to (\mathcal{A}, \geq)$ an MNC on E. A multimap $\mathcal{F}\colon \mathcal{M} \to \mathcal{P}_{cp}(E)$ is said to be β–condensing if, for every bounded $\Omega \subset \mathcal{M}$, the inequality

$$\beta(\Omega) \leq \beta(\mathcal{F}(\Omega)),$$

implies the relative compactness of Ω.

Definition 2.94. A sequence $\{v_n\}_{n\in\mathbb{N}} \subset L^1([0,b], E)$ is said to be semicompact, if

(a) it is integrably bounded, i.e., if there exists $\psi \in L^1([0,b], \mathbb{R}^+)$ such that

$$|v_n(t)| \leq \psi(t) \text{ for a.e. } t \in [0,b] \text{ and every } n \in \mathbb{N},$$

(b) the image sequence $\{v_n(t)\}_{n\in\mathbb{N}}$ is relatively compact in E for a.e. $t \in [0,b]$.

The following result follows from the Dunford–Pettis theorem (also see [311, Proposition 4.2.1]).

Lemma 2.95. *Every semicompact sequence is weakly compact in* $L^1([0,b], E)$.

Lemma 2.96 ([311, Theorem 5.1.1]). *Let* $N\colon L^1([a,b], E) \to C([a,b], E)$ *be an abstract operator satisfying the following conditions:*

($\mathcal{S}_1$) *N is ξ–Lipschitz: there exists $\xi > 0$ such that for every $f, g \in L^1([a,b], E)$,*

$$|Nf(t) - Ng(t)| \leq \xi \int_a^b |f(s) - g(s)|ds, \text{ for all } t \in [a,b].$$

($\mathcal{S}_2$) *N is weakly-strongly sequentially continuous on compact subsets: for any compact $K \subset E$ and any sequence $\{f_n\}_{n=1}^{\infty} \subset L^1([a,b], E)$ such that $\{f_n(t)\}_{n=1}^{\infty} \subset K$ for a.e. $t \in [a,b]$, the weak convergence $f_n \rightharpoonup f_0$ implies the strong convergence $N(f_n) \to N(f_0)$ as $n \to +\infty$.*

Then for every semicompact sequence $\{f_n\}_{n=1}^{\infty} \subset L^1([0,b], E)$, the image sequence $N(\{f_n\}_{n=1}^{\infty})$ is relatively compact in $C([a,b], E)$.

Lemma 2.97 ([311, Theorem 5.2.2]). *Let an operator $N\colon L^1([a,b], E) \to C([a,b], E)$ satisfy conditions ($\mathcal{S}_1$) – ($\mathcal{S}_2$) together with*

($\mathcal{S}_3$) *There exits $\eta \in L^1([a,b])$ such that for every integrably bounded sequence $\{f_n\}_{n=1}^{\infty}$, we have*

$$\chi(\{f_n(t)\}_{n=1}^{\infty}) \leq \eta(t) \text{ for a.e. } t \in [a,b],$$

where χ is the Hausdorff MNC.

Then,

$$\chi(\{N(f_n)(t)\}_{n=1}^{\infty}) \leq 2\xi \int_a^b \eta(s)ds, \quad \textit{for all} \quad t \in [a,b],$$

where ξ *is the constant in* $(\mathcal{S}_1)$.

Finally, two useful properties of the fixed point set of β−condensing multimaps are the following (see [311]).

Lemma 2.98. *Let* W *be a convex closed subset of a Banach space* E *and let* $N : W \to \mathcal{P}_{cp,cv}(W)$ *be a closed* β*-condensing multimap where* β *is a nonsingular measure of noncompactness defined on subsets of* W. *Then* $FixN \neq \emptyset$.

Lemma 2.99. *Let* W *be a closed subset of a Banach space* E *and let* $N \colon W \to \mathcal{P}_{cp}(E)$ *be a closed* β*-condensing multimap where* β *is a monotone MNC on* E. *Then* $Fix\ N$ *is compact.*

2.6 Semigroups

2.6.1 C_0-semigroups

Let E be a Banach space and $B(E)$ be the Banach space of linear bounded operators defined on E.

Definition 2.100. A one parameter family $\{T(t) \mid t \geq 0\} \subset B(E)$ is said to be of class C_0 if it satisfies the conditions:

(i) $T(t) \circ T(s) = T(t+s)$ for $t, s \geq 0$,

(ii) $T(0) = I$,

(iii) the map $t \to T(t)(x)$ is strongly continuous, for each $x \in E$, i.e.,

$$\lim_{t \to 0} T(t)x = x, \quad \text{for all } x \in E.$$

A semigroup of bounded linear operators $T(t)$ is uniformly continuous if

$$\lim_{t \to 0} \|T(t) - I\| = 0.$$

Here, I denotes the identity operator in E.

We note that if a semigroup $T(t)$ is of class C_0, then it satisfies the growth condition,

$$\|T(t)\|_{B(E)} \leq M \cdot \exp(\beta t) \quad \text{for } 0 \leq t < \infty, \text{ for some constants } M > 0 \text{ and } \beta.$$

If, in particular $M = 1$ and $\beta = 0$, i.e., $\|T(t)\|_{B(E)} \leq 1$, for $t \geq 0$, then the semigroup $T(t)$ is called a *contraction semigroup*.

Definition 2.101. Let $T(t)$ be a semigroup of class (C_0) defined on E. The *infinitesimal generator* A of $T(t)$ is the linear operator defined by

$$A(x) = \lim_{h\to 0} \frac{T(h)(x) - x}{h}, \quad \text{for } x \in D(A),$$

where $D(A) = \{x \in E \mid \lim_{h\to 0} \frac{T(h)(x)-x}{h} \text{ exists in } E\}$.

Let us recall the following property.

Proposition 2.102. *The infinitesimal generator A is a closed linear and densely defined operator in E. If $x \in D(A)$, then $T(t)(x)$ is a C^1-map and*

$$\frac{d}{dt}T(t)(x) = A(T(t)(x)) = T(t)(A(x)) \quad \text{on } [0,\infty).$$

Theorem 2.103 (Hille and Yosida [410]). *Let A be a densely defined linear operator with domain and range in a Banach space E. Then A is the infinitesimal generator of an uniquely determined semigroup $T(t)$ of class (C_0) satisfying*

$$\|T(t)\|_{B(E)} \le M \exp(\omega t), \quad t \ge 0,$$

where $M > 0$ and $\omega \in \mathbb{R}$, if and only if $(\lambda I - A)^{-1} \in B(E)$ and $\|(\lambda I - A)^{-n}\| \le M/(\lambda-\omega)^n$, $n = 1, 2, \ldots$, for all $\lambda \in \mathbb{R}$.

We say that a family $\{C(t) \mid t \in \mathbb{R}\}$ of operators in $B(E)$ is a strongly continuous cosine family if

(i) $C(0) = I$,

(ii) $C(t+s) + C(t-s) = 2C(t)C(s)$, for all $s, t \in \mathbb{R}$,

(iii) the map $t \mapsto C(t)(x)$ is strongly continuous, for each $x \in E$.

The strongly continuous sine family $\{S(t) \mid t \in \mathbb{R}\}$, associated to the given strongly continuous cosine family $\{C(t) \mid t \in \mathbb{R}\}$, is defined by

$$S(t)(x) = \int_0^t C(s)(x)\, ds, \quad x \in E,\ t \in \mathbb{R}.$$

The infinitesimal generator $A : E \to E$ of a cosine family $\{C(t) \mid t \in \mathbb{R}\}$ is defined by

$$A(x) = \frac{d^2}{dt^2}C(t)(x)|_{t=0}.$$

Here,

$$D(A) = \left\{x \in E : \lim_{h\to 0+} \frac{2}{h^2}[C(h) - h] \text{ exists}\right\},$$

and

$$A(x) = \lim_{h\to 0+} \frac{2}{h^2}[C(h) - I]x \text{ for } x \in D(A)$$

is the infinitesimal generator of a cosine family $\{C(t) \mid t \in \mathbb{R}\}$, and $D(A)$ is the domain of A. Note that

- There exist constants $\omega \geq 0$ and $M \geq 1$ such that

$$\|C(t)\| \leq Me^{\omega|t|} \text{ for } t \in \mathbb{R}.$$

- $D(A)$ is dense in E and A is a closed linear operator. For every $x \in D(A)$ and $t \in \mathbb{R}$, then $C(t)x \in D(A)$ and

$$\frac{d^2}{dt^2}C(t)x = AC(t)x = C(t)Ax.$$

For more details on strongly continuous cosine and sine families, we refer the reader to the books of Goldstein [227], Engel and Nagel [190], Hekkila and Lakshmikantham [271], Fattorini [196], and to the papers of Travis and Webb [468], [469].

2.6.2 Integrated Semigroups

Definition 2.104 ([25]). Let E be a Banach space. An integrated semigroup is a family $(S(t))_{t\geq 0}$ of bounded linear operators on E with the following properties:

(i) $S(0) = 0$;

(ii) $t \to S(t)$ is strongly continuous;

(iii) $S(s)S(t) = \int_0^s (S(t+r) - S(r))dr$, for all $t, s \geq 0$.

Definition 2.105. A family of linear operators $\{S(t)\}_{t\geq 0}$ is called exponentially bounded if there exist constants $M > 0$ and $\omega \geq 0$ such that

$$\|S(t)\| \leq Me^{\omega t},\ t \geq 0.$$

Definition 2.106 ([319]). An operator A is called a generator of an integrated semigroup if there exists $\omega \in \mathbb{R}$ such that $(\omega, \infty) \subset \rho(A)$ ($\rho(A)$ is the resolvent set of A), and there exists a strongly continuous exponentially bounded family $(S(t))_{t\geq 0}$ of bounded operators such that $S(0) = 0$ and $R(\lambda, A) := (\lambda I - A)^{-1} = \lambda \int_0^\infty e^{-\lambda t} S(t)dt$ exists, for all λ with $\lambda > \omega$.

Proposition 2.107 ([25]). *Let A be the generator of an integrated semigroup $(S(t))_{t\geq 0}$. Then*

(i) *for all $x \in E$ and $t \geq 0$,*

$$\int_0^t S(s)x\,ds \in D(A) \quad \text{and} \quad S(t)x = A\int_0^t S(s)x\,ds + tx.$$

(ii) *for all* $x \in D(A)$ *and* $t \geq 0$

$$S(t)x \in D(A),\ AS(t)x = S(t)Ax, \quad \text{and} \quad S(t)x = \int_0^t S(s)Axds + tx;$$

(iii) $R(\lambda, A)S(t) = S(t)R(\lambda, A)$ *for all* $t \geq 0$, $\lambda > \omega$.

Definition 2.108 ([27, 319]).

(i) An integrated semigroup $(S(t))_{t\geq 0}$ is called locally Lipschitz continuous if, for all $\tau > 0$, there exists a constant L such that

$$|S(t) - S(s)| \leq L|t - s|, \quad t, s \in [0, \tau].$$

(ii) An integrated semigroup $(S(t))_{t\geq 0}$ is called nondegenerate if $S(t)x = 0$, for all $t \geq 0$, implies that $x = 0$.

Definition 2.109. A linear (not necessarily densely defined) operator $A : D(A) \subset E \to E$ is said to be the Hille–Yosida operator if there exists $M \geq 0$ and $\omega \geq 0$ such that $(\omega, \infty) \subset \rho(A)$ and

$$\sup\{(\lambda - \omega)^n |(\lambda I - A)^{-n}| : n \in \mathbb{N},\ \lambda > \omega\} \leq M \text{ forall} n \in \mathbb{N} \text{ and } \lambda > \omega.$$

Theorem 2.110 ([319]). *The following assertions are equivalent:*

(i) *A is the generator of a nondegenerate, locally Lipschitz continuous integrated semigroup;*

(ii) *A satisfies the Hille–Yosida condition.*

Proposition 2.111. *Let* $\{S(t)\}_{t\geq 0}$ *be a locally Lipschitz continuous integrated semigroup on E and* $f : [0, b] \to E$ *is Bochner integrable function. Then the function* $F : [0, b] \to E$,

$$F(t) = \int_0^t S(t - s)f(s)ds$$

is continuously differentiable and, moreover,

$$\left\| \frac{dF}{dt}(t) \right\| \leq 2L \int_0^t \|f(s)\|ds \ \text{ for all } \ t \in [0, b],$$

where L is the Lipschitz constant of $S(.)$ *on* $[0, b]$.

- If A is the generator of an integrated semigroup $(S(t))_{t\geq 0}$ which is locally Lipschitz, then from [25, 27], $S(\cdot)x$ is continuously differentiable if and only if $x \in \overline{D(A)}$ and $(S'(t))_{t\geq 0}$ is a C_0 semigroup on $\overline{D(A)}$.

- Let A be a Hille–Yosida operator generating a locally Lipschitz continuous integrated semigroup $\{S(t)\}_{t\geq 0}$, function $F : [0,b] \to E$ be defined as in Proposition 2.111. Then, by applying Proposition 2.107(iii) and using the Lipschitz continuity of $S(.)$,one may verify the following relation (see [27]):

$$R(\lambda, A)\frac{d}{dt}F(t) = \int_0^t S'(t-s)R(\lambda, A)f(s)ds, \qquad t \in [0,b].$$

Moreover,taking into account that $\lim_{\lambda\to\infty} \lambda R(\lambda, A)x = x$ for each $x \in \overline{D(A)}$, we come to the following equality:

$$\frac{d}{dt}F(t) = \lim_{\lambda\to\infty}\int_0^t S'(t-s)\lambda R(\lambda, A)f(s)ds.$$

2.6.3 Examples

Example 2.112. Let $E = C(\overline{\Omega})$, the Banach space of continuous function on $\overline{\Omega}$ with values in $\mathbb{R}$. Define the linear operator A on E by

$$Az = \Delta z, \quad \text{for} \quad D(A) = \{z \in C(\overline{\Omega}) : z = 0 \text{ on } \partial\Omega,\ \Delta z \in C(\overline{\Omega})\},$$

and where $\Delta = \sum_{k=1}^{n} \frac{\partial^2}{\partial x_k^2}$. Now, we have

$$\overline{D(A)} = C_0(\overline{\Omega}) = \{v \in C(\overline{\Omega}) : v = 0 \text{ on } \partial\Omega\} \neq C(\overline{\Omega}).$$

Example 2.113. (An integral semigroup associated with the second order Cauchy problem.) This example generalizes the Cauchy problem for the wave equation. Consider the second order Cauchy problem,

$$u''(t) = Bu(t),\ t \geq 0,\ u(0) = x,\ u'(0) = y, \tag{2.1}$$

in a Banach space X, with linear operator B generating cosine and sine operator-functions $C(\cdot)$ and $S(\cdot)$. The unique solution of (2.1) has the form

$$u(t) = C(t)x + S(t)y.$$

The problem (2.1) can be reduced to a Cauchy problem for the first order system

$$w'(t) = \Phi(t)w(t),\ w(t) = \begin{pmatrix} x \\ y \end{pmatrix},$$

where

$$\Phi = \begin{pmatrix} 0 & I \\ B & 0 \end{pmatrix},\quad w(t) = \begin{pmatrix} u(t) \\ u'(t) \end{pmatrix}.$$

Using cosine and sine operator-functions, we can rewrite w in the form

$$w(t) = \begin{pmatrix} C(t)x + S(t)y \\ C'(t)x + C(t)y \end{pmatrix} \equiv U(t)\begin{pmatrix} x \\ y \end{pmatrix},$$

where the operator $U(t)$ is not defined at some points on $X \times X$ for all $t \geq 0$, because the function $C(\cdot)$ is not necessarily differentiable on X. The operator Φ is the generator of the integrated semigroup

$$w(t) = \begin{pmatrix} S(t) & \int_0^t S(s)ds \\ C(t) - I & S(t) \end{pmatrix}.$$

Conditions (i)–(ii) are satisfied due to the properties of C and S. For more information about cosine and sine families see [227, 368].

2.7 Extrapolation Spaces

Let A_0 be the part of A in $X_0 = \overline{D(A)}$ that is defined by

$$D(A_0) = \{x \in D(A) : Ax \in \overline{D(A)}\}, \text{ and } A_0 x = Ax, \text{ for } x \in D(A_0).$$

Definition 2.114. We say that a linear operator A satisfies the "Hille–Yosida condition" if there exist $M \geq 0$ and $\omega \in \mathbb{R}$ such that $(\omega, \infty) \subset \rho(A)$ and

$$\sup\{(\lambda - \omega)^n |(\lambda I - A)^{-n}| : n \in \mathbb{N}, \lambda > \omega\} \leq M.$$

Lemma 2.115 ([190]). *A_0 generates the strong continuous semigroup $(T_0(t))_{t \geq 0}$ on X_0 and $|T_0(t)| \leq N_0 e^{\omega t}$, for $t \geq 0$. Moreover, $\rho(A) \subset \rho(A_0)$ and $R(\lambda, A_0) = R(\lambda, A)/X_0$, for $\lambda \in \rho(A)$.*

For a fixed $\lambda_0 \in \rho(A)$, we introduce on X_0 a new norm defined by

$$\|x\|_1 = |R(\lambda_0, A_0)x| \text{ for } x \in \overline{D(A_0)}.$$

The completion X_1 of $(X_0, \|\cdot\|_1)$ is called the *extrapolation space of X associated with A*. Note that $\|\cdot\|_1$ and the norm on X_0 given by $|R(\lambda, A_0)x|$, for $\lambda \in \rho(A)$, are extensions $T_1(t)$ to the Banach space X_1, and $(T_1(t))_{t \geq 0}$ is a strongly continuous semigroup on X_1. $(T_1(t))_{t \geq 0}$ is called *the extrapolated semigroup of* $(T_0(t))_{t \geq 0}$, and we denote its generator by $(A_1, D(A_1))$.

Lemma 2.116 ([256]). *The following properties hold:*

(i) $|T(t)|_{B(X_1)} = |T_0(t)|_{L(X_0)}$.

(ii) $D(A_1) = X_0$.

(iii) *$A_1 : X_0 \to X_1$ is the unique continuous extension of $A_0 : D(A_0) \subset (X_0, |\cdot|) \to (X_0, \|\cdot\|_1)$, and $(\lambda - A_1)^{-1}$ is an isometry from $(X_0, |\cdot|) \to (X_0, \|\cdot\|_1)$.*

(iv) *If $\lambda \in \rho(A_0)$, then $(\lambda - A_1)$ is invertible and $(\lambda - A_1)^{-1} \in B(X_1)$. In particular $\lambda \in \rho(A_1)$ and $R(\lambda, A_1)/X_0 = R(\lambda, A_0)$.*

(v) *The space* $X_0 = \overline{D(A)}$ *is dense in* $(X_1, \|\cdot\|_1)$. *Hence, the extrapolation space* X_1 *is also the completion of* $(X, \|.\|_1)$ *and* $X \hookrightarrow X_1$.

(vi) *The operator* A_1 *is an extension of* A. *In particular, if* $\lambda \in \rho(A)$, *then* $R(\lambda, A_1)/X = R(\lambda, A)$ *and* $(\lambda - A_1)X = D(A)$.

Chapter 3

FDEs with Infinite Delay

It is well-known that systems with after effect, with time lag, or with delay, are of great theoretical interest and form an important class with regard to their applications. This class of systems can be described by functional differential equations and inclusions, which are also called differential equations and inclusions with deviating argument. Among functional differential equations and inclusions, one may distinguish some special classes of equations, retarded functional equations, advanced functional equations, and neutral functional equations. In particular, retarded functional differential equations and inclusions describe those systems or processes whose rate of change of state is determined by their past and present states. Such equations are frequently encountered as mathematical models of many dynamical processes in mechanics, control theory, physics, chemistry, biology, medicine, economics, atomic energy, information theory, etc. Especially, since the 1960's, many good books, a number of which are in the Russian literature, have been published on delay differential equations; see, for examples, the books of Burton [117, 118], Èl'sgol'ts [188], Èl'sgol'ts and Norkin [189], Gopalsamy [228], Azbelev et al. [38], Hale [267], Hale and Lunel [268], Kolmanovskii and Myshkis [332], Kolmanovaskii and Nosov [333], Krasovskii [325], Yoshizawa [491], and the references contained therein. The results in this chapter involve existence and uniqueness of solutions for first and second order infinite delay functional differential equations with impulses, in both local and global contexts. Stability issues are also addressed. Examples illustrating the theory are provided.

3.1 First Order FDEs

In this section, we shall establish sufficient conditions for the existence and uniqueness of solutions of impulsive functional differential equations with infinite delay. More precisely, we consider the following impulsive effect problem,

$$y'(t) = f(t, y_t), \qquad \text{a.e. } t \in J = [0,b],\ t \neq t_k,\ k = 1,\dots,m, \tag{3.1}$$

$$y(t_k^+) - y(t_k) = I_k(y(t_k)), \qquad t = t_k,\ k = 1,\dots,m, \tag{3.2}$$

$$y(t) = \phi(t), \qquad t \in (-\infty, 0], \tag{3.3}$$

where $f : J \times B \to \mathbb{R}^n$, $I_k \in C(\mathbb{R}^n, \mathbb{R}^n)$, $k = 1, 2, \dots, m$, are given functions satisfying some assumptions that will be specified later, and $\phi \in B$ where B is called a *phase space* that will also be defined later.

For any function y defined on $(-\infty, b]$ and any $t \in [0, \infty)$, we denote by y_t the element of B defined by $y_t(\theta) = y(t + \theta)$, $\theta \in (-\infty, 0]$. Here, $y_t(\cdot)$ represents the history of the state from time $t - \theta$ up to the present time t.

The notion of the phase space B plays an important role in the study of both qualitative and quantitative theory. A usual choice is a seminormed space satisfying suitable axioms such as those introduced by Hale and Kato [265] (see also Kappel and Schappacher [316] and Schumacher [437]). For a detailed discussion on this topic, we refer the reader to the book by Hino et al. [292]. For the case where the impulses are absent (i.e., $I_k = 0$, $k = 1, \dots, m$), an extensive theory has been developed for the problem (3.1)–(3.3). We refer to Hale and Kato [265], Corduneanu and Lakshmikantham [157], Hino et al. [292], Lakshmikantham et al. [340], and Shin [441].

We will assume that B satisfies the following axioms:

(A) If $y : (-\infty, b] \to \mathbb{R}^n$, $b > 0$, $y_0 \in B$, and $y(t_k^-)$ and $y(t_k^+)$ exist with $y(t_k) = y(t_k^-)$, $k = 1, \dots, m$, then for every t in $[0, b) \backslash \{t_1, \dots, t_m\}$, the following conditions hold:

(i) y_t is in B and y_t is continuous on $[0, b] \backslash \{t_1, \dots, t_m\}$;

(ii) $\|y_t\|_B \leq K(t) \sup\{|y(s)| : 0 \leq s \leq t\} + M(t)\|y_0\|_B$;

(iii) $|y(t)| \leq H \|y_t\|_B$;

where $H \geq 0$ is a constant, $K : [0, \infty) \to [0, \infty)$ is continuous, $M : [0, \infty) \to [0, \infty)$ is locally bounded, and H, K, and M are independent of $y(\cdot)$.

(A-1) For the function $y(\cdot)$ in (A), y_t is a B-valued continuous function on $[0, b) \backslash \{t_1, \dots, t_m\}$.

(A-2) The space B is complete.

Set

$$B_b = \{y : (-\infty, b] \to \mathbb{R}^n \mid y \in PC \cap B\}, \tag{3.4}$$

and let $\|\cdot\|_b$ be the seminorm in B_b defined by

$$\|y\|_b := \|y_0\|_B + \sup\{|y(t)| : 0 \leq s \leq b\}, \; y \in B_b.$$

3.1.1 Examples of Phase Spaces

In this subsection, we present some examples of phase spaces.

Example 3.1. The spaces BC, BUC, C^∞ and C^0.

Let:

BC denote the space of bounded continuous functions defined from $(-\infty, 0]$ to $\mathbb{R}$;

BUC denote the space of bounded uniformly continuous functions defined from $(-\infty, 0]$ to $\mathbb{R}$;

$$C^{\infty} := \left\{\phi \in BC : \lim_{\theta \to -\infty} \phi(\theta) \text{ exist in } \mathbb{R}\right\};$$

$$C^{0} := \left\{\phi \in BC : \lim_{\theta \to -\infty} \phi(\theta) = 0\right\}, \text{ endowed with the uniform norm}$$

$$\|\phi\|_{\infty} = \sup\{|\phi(\theta)| : \theta \in (-\infty, 0]\}.$$

Then, the spaces BUC, C^{∞} and C^{0} satisfy conditions (A)–(A-2), whereas, BC satisfies (A-1), (A-2), but not (A).

Example 3.2. The spaces C_g, UC_g, C_g^{∞} and C_g^0.

Let g be a positive continuous function on $(-\infty, 0]$. We define:

$$C_g := \left\{\phi \in C((-\infty,0],\mathbb{R}) : \frac{\phi(\theta)}{g(\theta)} \text{ is bounded on } (-\infty, 0]\right\};$$

$$C_g^0 := \left\{\phi \in C_g : \lim_{\theta \to -\infty} \frac{\phi(\theta)}{g(\theta)} = 0\right\}, \text{ endowed with the uniform norm}$$

$$\|\phi\|_{\infty} = \sup\left\{\frac{|\phi(\theta)|}{g(\theta)} : \theta \in (-\infty, 0]\right\}.$$

Then, the spaces C_g and C_g^0 satisfy condition (A-2).

If we impose the following condition on the function g:

(g_1) For all $a > 0$, $\displaystyle\sup_{0 \le t \le a} \sup\left\{\frac{g(t+\theta)}{g(\theta)} : -\infty < \theta \le -t\right\} < \infty$,

then, the spaces C_g and C_g^0 satisfy conditions (A) and (A-2).

Example 3.3. The space C_{γ}.

For any real positive constant γ, we define the functional space C_{γ} by

$$C_{\gamma} := \left\{\phi \in C((-\infty,0],\mathbb{R}) : \lim_{\theta \to -\infty} e^{\gamma\theta}\phi(\theta) \text{ exist in } \mathbb{R}\right\}$$

endowed with the following norm

$$\|\phi\|_{\infty} = \sup\{e^{\gamma\theta}|\phi(\theta)| : \theta \le 0\}.$$

Then in the space C_{γ} the axioms (A)–(A-2) are satisfied.

We may consider the following examples of phase spaces satisfying all above properties.

Example 3.4. For $r > 0$, let

$$\begin{aligned}\mathcal{D}([-r,0],E) = \Big\{\psi\colon [-r,0] \to E\colon\ &\psi \text{ is continuous everywhere except}\\ &\text{for a finite number of points } \bar{t} \text{ at which } \psi(\bar{t}^-)\\ &\text{and } \psi(\bar{t}^+) \text{ exist and satisfy } \psi(\bar{t}^-) = \psi(\bar{t})\Big\},\end{aligned}$$

where E is a Banach space. Considering $\mathcal{D}$ as the subspace of the space of measurable functions, we may treat it as the normed space with the norm,

$$\|\psi\|_{\mathcal{D}} = \int_{-r}^{0} \|\psi(\theta)\| d\theta.$$

(1) For $\nu > 0$ let $\mathcal{B} = PC_\nu = \{\psi : (-\infty, 0] \to E$ such that $\psi \in \mathcal{D}([-r, 0], E)$ for each $r > 0$ and $\int_{-\infty}^{0} e^{\nu\theta} \|\psi(\theta)\| d\theta < \infty\}$. Then we set

$$\|\psi\|_{\mathcal{B}} = \int_{-\infty}^{0} e^{\nu\theta} \|\psi(\theta)\| d\theta.$$

(2) (Spaces of "fading memory") Let

$$\begin{aligned} \mathcal{B} = PC_\rho = \Big\{ \psi : (-\infty, 0] \to E \text{ such that } \psi \in \mathcal{D}([-r, 0], E) \\ \text{for some } r > 0 \text{ and } \psi \text{ is Lebesgue measurable on } (-\infty, -r) \\ \text{and there exists a positive Lebesgue integrable function} \\ \rho : (-\infty, -r) \to \mathbb{R}_+ \text{ such that } \rho\psi \in L^1((-\infty, -r), E) \Big\}, \end{aligned}$$

and moreover, there exists a locally bounded function $P : (-\infty, 0] \to \mathbb{R}_+$ such that, for all $\xi \leq 0$ $\rho(\xi + \theta) < P(\xi)\rho(\theta)$ a.e. $\theta \in (-\infty, -r)$. Then

$$\|\psi\|_{\mathcal{B}} = \int_{-\infty}^{-r} \rho(\theta) \|\psi(\theta)\| d\theta + \int_{-r}^{0} \|\psi(\theta)\| d\theta.$$

A simple example of such a space is defined for $\rho(\theta) = e^{\mu\theta}$, $\mu \in \mathbb{R}$.

3.1.2 Existence and Uniqueness on Compact Intervals

In order to define the phase space and the solution of (3.1)–(3.3), we shall consider the space

$$PC = \Big\{ y : [0, b] \to \mathbb{R}^n \mid y(t_k^-) \text{ and } y(t_k^+) \text{ exist with } y(t_k) = y(t_k^-), y_k \in C(J_k, \mathbb{R}^n) \Big\},$$

where y_k is the restriction of y to $J_k = (t_k, t_{k+1}]$, $k = 0, \ldots, m$. Let $\|\cdot\|_{PC}$ be the norm in PC defined by

$$\|y\|_{PC} = \sup\{|y(s)| : 0 \leq s \leq b\}, \ y \in PC.$$

The phase space B for impulsive functional differential equations with infinite delay is a linear space, with seminorm $\|\cdot\|_B$ mapping $(-\infty, 0]$ into a finite-dimensional Banach space $E = \mathbb{R}^n$ or $\mathbb{C}^n$. The first two axioms on B are motivated by the fact that we want solutions of (3.1)–(3.3) to be continuous on $(t_k, t_{k+1}]$ and the left hand limit to exist for every t_k.

Let us start by defining what we mean by a solution of the problem (3.1)–(3.3).

Definition 3.5. A function $y \in B_b$ is said to be a solution of (3.1)–(3.3), if y satisfies (3.1)–(3.3).

In all this section we assume that $t \to f(t, y_t)$ is measurable function.

Theorem 3.6. *Let $f : J \times B \to \mathbb{R}^n$ be a Carathéodory function. Assume that*

(A-3) *There exist a continuous nondecreasing function $\psi : [0, \infty) \longrightarrow (0, \infty)$ and a function $p \in L^1([0, b], \mathbb{R}_+)$ such that*

$$|f(t, x)| \leq p(t)\psi(\|x\|_B) \text{ for a.e. } t \in [0, b] \text{ and each } x \in B,$$

with

$$\int_0^b p(s)ds < \infty \quad \text{and} \quad \int_c^\infty \frac{dx}{\psi(x)} = \infty,$$

where $K_b = \sup\{|K(t)| : t \in [0, b]\}$, $M_b = \sup\{|M(t)| : t \in [0, b]\}$, and $c = M_b\|\phi\|_B + K_b|\phi(0)|$.

Then the initial value problem (3.1)–(3.3) has at least one solution.

Proof. The proof will be given in several steps.

Step 1. Consider the problem,

$$y'(t) = f(t, y_t), \quad \text{a.e.} \quad t \in [0, t_1], \tag{3.5}$$

$$y(t) = \phi(t), \qquad t \in (-\infty, 0]. \tag{3.6}$$

We transform the problem (3.5)–(3.6) into a fixed point problem. Consider the operator $N : B \cap C([0, t_1], \mathbb{R}^n) \longrightarrow B \cap C([0, t_1], \mathbb{R}^n)$ defined by

$$N(y)(t) = \begin{cases} \phi(t), & t \in (-\infty, 0], \\ \phi(0) + \int_0^t f(s, y_s)ds, & t \in [0, t_1]. \end{cases}$$

Let $x(\cdot) : (-\infty, t_1] \to \mathbb{R}^n$ be the function given by

$$x(t) = \begin{cases} \phi(0), & \text{if } t \in [0, t_1], \\ \phi(t), & \text{if } t \in (-\infty, 0]. \end{cases}$$

Set $x_0 = \phi$. For each $z \in C([0, t_1], \mathbb{R}^n) \cap B$, with $z_0 = 0$, we denote by $\bar{z}$ the function given by

$$\bar{z}(t) = \begin{cases} z(t), & \text{if } t \in [0, t_1], \\ 0, & \text{if } t \in (-\infty, 0]. \end{cases}$$

If $y(\cdot)$ satisfies the integral equation,

$$y(t) = \phi(0) + \int_0^t f(s, y_s)ds,$$

we can decompose $y(\cdot)$ as $y(t) = \bar{z}(t) + x(t), 0 \le t \le t_1$. This implies $y_t = \bar{z}_t + x_t$, for every $0 \le t \le t_1$, and the function $z(\cdot)$ satisfies

$$z(t) = \int_0^t f(s, \bar{z}_s + x_s)ds. \tag{3.7}$$

Set

$$C_0 = \{z \in B \cap C([0, t_1], \mathbb{R}^n) : z_0 = 0\}$$

and let $\|\cdot\|_0$ be the norm in C_0 defined by

$$\|z\|_0 = \|z_0\|_B + \sup\{|z(t)| : 0 \le t \le b\} = \sup\{|z(t)| : 0 \le t \le t_1\}, \; z \in C_0.$$

Define the operator $P : C_0 \to C_0$ by

$$(Pz)(t) = \begin{cases} 0, & t \in (-\infty, 0], \\ \displaystyle\int_0^t f(s, \bar{z}_s + x_s)ds, & t \in [0, t_1]. \end{cases}$$

Clearly, that the operator N has a fixed point is equivalent to P having a fixed point, and so we turn our attention to proving that P does in fact have a fixed point. We shall use the Leray–Schauder alternative to prove this.

Claim 1. *P is continuous.*

Let $\{z_n\}$ be a sequence such that $z_n \to z$ in C_0. Then,

$$|(Pz_n)(t) - (Pz)(t)| \le \int_0^{t_1} |f(s, \bar{z}_{n_s} + x_s) - f(s, \bar{z}_s + x_s)|ds.$$

Since f is L^1-Carathéodory, we have, as $n \to \infty$,

$$\|P(z_n) - P(z)\|_0 \le \|f(\cdot, \bar{z}_{n_{(\cdot)}} + x_{(\cdot)}) - f(\cdot, \bar{z}_{(\cdot)} + x_{(\cdot)})\|_{L^1} \to 0.$$

Claim 2. *P maps bounded sets into bounded sets in C_0.*

Indeed, it is enough to show that for any $q > 0$, there exists a positive constant ℓ such that, for each $z \in \mathcal{B}_q = \{z \in C_0 : \|z\|_0 \le q\}$, we have $\|P(z)\|_0 \le \ell$. Let $z \in \mathcal{B}_q$. Since f is an L^1-Carathéodory function, we have for each $t \in [0, t_1]$,

$$\|P(z)\|_\infty \le \int_0^{t_1} h_{q*}(s)ds := \ell,$$

where

$$\|\bar{z}_s + x_s\|_B \le \|\bar{z}_s\|_B + \|x_s\|_B \le K_b q + K_b|\phi(0)| + M_b\|\phi\|_B := q_*.$$

Claim 3. *P maps bounded sets into equicontinuous sets of C_0.*

Let $l_1, l_2 \in [0, t_1]$, $l_1 < l_2$, and let $\mathcal{B}_q$ be a bounded set of C_0 as in Claim 2. Let $z \in \mathcal{B}_q$. Then for each $t \in [0, t_1]$, we have

$$|(Pz)(l_2) - (Pz)(l_1)| \leq \int_{l_1}^{l_2} |f(s, \bar{z}_s + x_s)| ds \leq \int_{l_1}^{l_2} h_{q*}(s) ds.$$

We see that $|(Pz)(l_2) - (Pz)(l_1)|$ tends to zero independently of $z \in \mathcal{B}_q$, as $l_2 - l_1 \to 0$. As a consequence of Claims 1 to 3, together with the Arzelá–Ascoli theorem, we can conclude that $P : C_0 \longrightarrow C_0$ is continuous and completely continuous.

Claim 4. *There exist a priori bounds on solutions.*

Let z be a possible solution of the equation $z = \lambda P(z)$ and $z_0 = \lambda\phi$, for some $\lambda \in (0, 1)$. Then,

$$|z(t)| \leq \int_0^t |f(s, \bar{z}_s + x_s)| ds \leq \int_0^t p(s)\psi(\|\bar{z}_s + x_s\|_B) ds. \tag{3.8}$$

But

$$\begin{aligned} \|\bar{z}_s + x_s\|_B &\leq \|\bar{z}_s\|_B + \|x_s\|_B \\ &\leq K(t)\sup\{|z(s)| : 0 \leq s \leq t\} + M(t)\|z_0\|_B \\ &\quad + K(t)\sup\{|x(s)| : 0 \leq s \leq t\} + M(t)\|x_0\|_B \\ &\leq K_b \sup\{|z(s)| : 0 \leq s \leq t\} + M_b\|\phi\|_B + K_b M|\phi(0)|. \end{aligned} \tag{3.9}$$

If we let $w(t)$ denote the right hand side of (3.9), then we have

$$\|\bar{z}_s + x_s\|_B \leq w(t),$$

and therefore (3.8) becomes

$$|z(t)| \leq \int_0^t p(s)\psi(w(s)) ds, \quad t \in [0, t_1]. \tag{3.10}$$

Using (3.10) in the definition of w, we have that

$$w(t) \leq K_b \int_0^t p(s)\psi(w(s)) ds + M_b\|\phi\|_B + K_b|\phi(0)|, \quad t \in [0, t_1].$$

Denoting by $\beta(t)$ the right hand side of the last inequality, we have

$$\begin{aligned} &w(t) \leq \beta(t), \quad t \in [0, t_1], \\ &\beta(0) = M_b\|\phi\|_B + K_b|\phi(0)|, \end{aligned}$$

and

$$\begin{aligned} \beta'(t) &= K_b p(t)\psi(w(t)) \\ &\leq K_b p(t)\psi(\beta(t)), \quad t \in [0, t_1]. \end{aligned}$$

This implies that for each $t \in [0, t_1]$,

$$\int_{\beta(0)}^{\beta(t)} \frac{ds}{\psi(s)} \leq K_b \int_0^{t_1} p(s)ds < \infty.$$

Thus, by (A-3) there exists a constant K_* such that $\beta(t) \leq K_*$, $t \in [0, t_1]$, and hence $\|\bar{z}_t + x_t\|_B \leq w(t) \leq K_*$, $t \in [0, t_1]$. From (3.10) we have that

$$\|z\|_0 \leq \int_0^{t_1} p(s)\psi(K_*)ds := \widetilde{K}_1.$$

Set

$$U_0 = \{z \in C_0 : \sup\{|z(t)| : 0 \leq t \leq t_1\} < \widetilde{K}_1 + 1\}.$$

Clearly, $P : \overline{U}_0 \to C_0$ is completely continuous. From the choice of U_0, there is no $z \in \partial U_0$ such that $z = \lambda P(z)$, for some $\lambda \in (0, 1)$. As a consequence of the nonlinear alternative of Leray–Schauder type [184], we deduce that P has a fixed point z in U_0. Hence, N has a fixed point y which is a solution to problem (3.5)–(3.6). Denote this solution by y_0.

Step 2. Now consider the problem,

$$y'(t) = f(t, y_t), \quad \text{a.e.} \quad t \in (t_1, t_2], \tag{3.11}$$

$$y(t_1^+) - y(t_1^-) = I_1(y_0(t_1)), \ \ y(t) = y_0(t), \quad t \in (-\infty, t_1]. \tag{3.12}$$

Let

$$C_1 = \{y \in C((t_1, t_2], \mathbb{R}^n) : y(t_1^+) \text{ exists}\}.$$

Set $C_* = B \cap C([0, t_1], \mathbb{R}^n) \cap C_1$. Consider the operator $N_1 : C_* \to C_*$ defined by,

$$N_1(y)(t) = \begin{cases} y_0(t), & (-\infty, t_1], \\ y_0(t_1^-) + I_1(y_0(t_1^-)) + \int_{t_1}^t f(s, y_s)ds, & t \in (t_1, t_2]. \end{cases}$$

Let $x(\cdot) : (-\infty, t_2] \to \mathbb{R}^n$ be the function defined by

$$x(t) = \begin{cases} y_0(t_1^-) + I_1(y_0(t_1)), & \text{if } t \in (t_1, t_2], \\ y_0(t), & \text{if } t \in (-\infty, t_1]. \end{cases}$$

Then $x_{t_1} = y_0$. For each $z \in C_*$ with $z(t_1) = 0$, we denote by $\bar{z}$ the function given by

$$\bar{z}(t) = \begin{cases} z(t), & \text{if } t \in [t_1, t_2], \\ 0, & \text{if } t \in (-\infty, t_1]. \end{cases}$$

If $y(\cdot)$ satisfies the integral equation,

$$y(t) = y_0(t_1^-) + I_1(y_0(t_1^-)) + \int_{t_1}^t f(s, y_s)ds,$$

we can decompose it as $y(t) = \bar{z}(t) + x(t), t_1 \leq t \leq t_2$. This implies $y_t = \bar{z}_t + x_t$, for every $t_1 \leq t \leq t_2$, and the function $z(\cdot)$ satisfies

$$z(t) = \int_{t_1}^{t} f(s, \bar{z}_s + x_s)ds. \tag{3.13}$$

Set

$$C_{t_1} = \{z \in C_* : z_{t_1} = 0\}.$$

Let the operator $P_1 : C_{t_1} \to C_{t_1}$ be defined by,

$$(P_1 z)(t) = \begin{cases} 0, & t \in (-\infty, t_1], \\ \int_{t_1}^{t} f(s, \bar{z}_s + x_s)ds, & t \in [t_1, t_2]. \end{cases}$$

As in Step 1, we can show that P_1 is continuous and completely continuous, and if z is a possible solution of the equation $z = \lambda P_1(z)$ with $z_0 = \lambda y_0$, for some $\lambda \in (0, 1)$, then there exists $K_{*_1} > 0$ such that

$$\|z\|_\infty \leq K_{*_1}.$$

Set

$$U_1 = \{z \in C_{t_1} : \sup\{|z(t)| : t_1 \leq t \leq t_2\} \leq K_{*_1} + 1\}.$$

Again by the nonlinear alternative of Leray–Schauder type [184], we deduce that P_1 has a fixed point z in U_1. Thus, N_1 has a fixed point y which is a solution to problem (3.11)–(3.12). Denote this solution by y_1.

Step 3. We continue this process and taking into account that $y_m := y\big|_{[t_m,b]}$ is a solution to the problem

$$y'(t) = f(t, y_t), \quad \text{a.e.} \quad t \in (t_m, b], \tag{3.14}$$

$$y(t_m^+) = y_{m-1}(t_{m-1}^-) + I_m(y_{m-1}(t_m^-)), \; y(t) = y_{m-1}(t), \; t \in (-\infty, t_{m-1}]. \tag{3.15}$$

The solution y of the problem (3.1)–(3.3) is then defined by

$$y(t) = \begin{cases} y_0(t), & \text{if } t \in (-\infty, t_1], \\ y_1(t), & \text{if } t \in (t_1, t_2], \\ \dots \\ y_m(t), & \text{if } t \in (t_m, b], \end{cases}$$

to complete the proof of the theorem. □

We next introduce some additional conditions that lead to uniqueness of the solution of (3.1)–(3.3).

Theorem 3.7. *Assume the following condition holds:*

(A-4) *There exists* $l \in L^1([0,b],\mathbb{R}_+)$ *and* $f(t,0) \in L^1([0,b],\mathbb{R}^n)$ *such that*

$$|f(t,x) - f(t,\overline{x})| \le l(t)\|x - \overline{x}\|_B \text{ for all } x,\ \overline{x} \in B \text{ and } t \in J.$$

Then the IVP (3.1)–(3.3) *has a unique solution.*

Proof. The proof will given in two steps.

Step 1. We will first prove that the problem (3.5)–(3.6) has a unique solution. To do this, we only need to prove that the operator P defined in Theorem 3.6 has a unique fixed point. We want to show that P is a contraction operator. We will let $\widetilde{l}(t) = K_b l(t)$, $\widehat{l}(t) = \int_0^t \widetilde{l}(s)ds$, and $\|\cdot\|_{B_*}$ denote the Bielecki-type norm on C_0 defined by

$$\|z\|_{B_*} = \sup\{|z(t)|e^{-\tau\widehat{l}(t)} : t \in [0,t_1]\}$$

with $\tau > 1$.

Consider $z, z^* \in C_0$. Then, for each $t \in [0,t_1]$, we have

$$\begin{aligned}
|P(z)(t) - P(z^*)(t)| &\le \int_0^t |f(s,\bar{z}_s + x_s) - f(s,\bar{z}_s^* + x_s)|ds \\
&\le \int_0^t l(s)\|\bar{z}_s - \bar{z}_s^*\|_B ds \\
&\le \int_0^t l(s)K_b \sup_{s\in[0,t]}|z(s) - z^*(s)|ds \\
&\le \int_0^t \widetilde{l}(s)e^{\tau\widehat{l}(s)}e^{-\tau\widehat{l}(s)} \sup_{s\in[0,t]}|z(s) - z^*(s)|ds \\
&\le \int_0^t \widetilde{l}(t)e^{\tau\widehat{l}(t)}ds\|z - z^*\|_{B_*} \\
&\le \frac{1}{\tau}\int_0^t (e^{\tau\widehat{l}(s)})'ds\|z - z^*\|_{B_*} \\
&\le \frac{1}{\tau}e^{\tau\widehat{l}(t)}ds\|z - z^*\|_{B_*}.
\end{aligned}$$

Thus,

$$e^{-\tau\widehat{l}(t)}|P(z)(t) - P(z^*)(t)| \le \frac{1}{\tau}\|z - z^*\|_{B_*}.$$

Therefore,

$$\|P(z) - P(z^*)\|_{B_*} \le \frac{1}{\tau}\|z - z_*\|_{B_*}.$$

As a consequence of the Banach fixed point theorem, we deduce that P has a unique fixed point which is a solution to (3.5)–(3.6). Denote this solution by y_0.

Step 2. Similar to Step 1, we can prove that the problem (3.11)–(3.12) has a unique solution. We denote this solution by y_1. We continue this process and taking into account that y_m is the unique solution of the problem (3.14)–(3.15). A solution y of the problem (3.1)–(3.3) is then defined by

$$y(t) = \begin{cases} y_0(t), & \text{if } t \in (-\infty, t_1], \\ y_1(t), & \text{if } t \in (t_1, t_2], \\ \dots \\ y_m(t), & \text{if } t \in (t_m, b]. \end{cases}$$

Let x and y be two solutions of the problem (3.1)–(3.3). If $t \in (t_k, t_{k+1}]$, $k = 0, \dots, m$, then $x(t) = y(t)$. If $t = t_k^+$, $k = 1, \dots, m$, we have

$$y(t_k^+) - x(t_k^+) = I_k(x(t_k)) - I_k(y(t_k)) = 0.$$

This implies that $x(t_k^+) = y(t_k^+)$. Thus, there is a unique solution of the problem (3.1)–(3.3). □

3.1.3 An Example

In this section, we give an example to illustrate our main results. Let $J := [0,3]\backslash\{1,2\}$, $t_1 = 1$, and $t_2 = 2$. Consider the problem

$$y'(t) = \frac{1}{(t+1)(t+2)(1+y_t^2)} \int_{-\infty}^{-t} e^{\theta+t} e^{\gamma\theta} y(t+\theta)d\theta, \text{ a.e.} \quad t \in J, \quad (3.16)$$

$$y(t_k^+) - y(t_k^-) = b_k y(t_k^-), \qquad k = 1, 2, \quad (3.17)$$

$$y(t) = \phi(t), \qquad t \in (-\infty, 0]. \quad (3.18)$$

Let $\mathcal{D} = \{\psi : (-\infty, 0] \to \mathbb{R}^n \mid \psi$ is continuous everywhere except for a countable number of points $\bar{t}$ at which $\psi(\bar{t}^-)$ and $\psi(\bar{t}^+)$ exist, $\psi(\bar{t}^-) = \psi(\bar{t})\}$, and b_1 and b_2 are real constants. Let γ be a positive real constant and $B_\gamma = PC_\gamma$, where PC_γ is defined in Example 3.4. The norm in B_γ is given by

$$\|\psi\|_\gamma = \int_{-\infty}^{0} e^{\gamma\theta} |\psi(\theta)| d\theta.$$

Here, for $(t, u) \in [0,3] \times B_\gamma$,

$$f(t,u) = \frac{1}{(t+1)(t+2)(u_t^2+1)} \int_{-\infty}^{-t} e^{\theta+t} e^{\gamma\theta} u(t+\theta)d\theta.$$

We have

$$|f(t,u)| \le \frac{e^{-\gamma t}}{(1+t)(2+t)} \int_{-\infty}^{0} e^{\theta} e^{\gamma\theta} |u(\theta)| d\theta \le p(t)[\|u|_\gamma + 1],$$

so (A-3) holds with $\psi(x) = x + 1$ and $p(t) = \frac{e^{-\gamma t}}{(1+t)(2+t)}$. Moreover,

$$\int_0^3 p(t)dt \le \int_0^3 \frac{dt}{1+t} = \ln 4 < \infty \quad \text{and} \quad \int_0^\infty \frac{dx}{\psi(x)} = \infty.$$

Hence, all the conditions of Theorem 3.6 are satisfied and so the problem (3.16)–(3.18) has at least one solution.

3.2 FDEs with Multiple Delays

This section is concerned with the existence and uniqueness of solutions to first order functional differential equations with impulsive effects and multiple delays. In the first subsection, we will consider local existence and uniqueness results for first order impulsive functional differential equations with fixed moments and multiple delays,

$$y'(t) = f(t, y_t) + \sum_{i=1}^{n_*} y(t - T_i), \quad \text{a.e.} \quad t \in J := [0, b]\backslash\{t_1, t_2, \dots, t_m\}, \tag{3.19}$$

$$y(t_k^+) - y(t_k^-) = I_k(y(t_k^-)), \qquad k = 1, \dots, m, \tag{3.20}$$

$$y(t) = \phi(t), \qquad t \in (-\infty, 0], \tag{3.21}$$

where $n_* \in \{1, 2, \dots\}$, and f, I_k, B are as in problem (3.1)–(3.3).

In the second subsection, a recent nonlinear alternative for contraction maps in Fréchet spaces, due to Frigon and Granas [208], is used to investigate the existence and uniqueness of solutions of first order impulsive functional differential equations with multiple delays. More precisely, we consider the problem,

$$y'(t) = f(t, y_t) + \sum_{k=1}^{m} y(t - T_i) \quad \text{a.e.} \quad t \in J_* := [0, \infty)\backslash\{t_1, t_2, \dots\}, \tag{3.22}$$

$$y(t_k^+) - y(t_k^-) = I_k(y(t_k^-)), \qquad k = 1, \dots, \tag{3.23}$$

$$y(t) = \phi(t), \qquad t \in (-\infty, 0], \tag{3.24}$$

where $f : J_* \times B \to \mathbb{R}^n$, $I_k \in C(\mathbb{R}^n, \mathbb{R}^n)$, $k = 1, \dots, 0 = t_0 < t_1 < \dots < t_m < \dots$, $\lim_{n\to\infty} t_n = \infty$, and $y(t_k^+) = \lim_{h\to 0^+} y(t_k + h)$ and $y(t_k^-) = \lim_{h\to 0^-} y(t_k - h)$ represent the right and left hand limits of $y(t)$ at $t = t_k$.

In this section we extend the results in the previous section and some results by Ouahab [397] to infinite delay problems. Our approach here is based on the Leray–Schauder alternative [184], the Banach fixed point theorem, and a nonlinear alternative of Leary–Schauder type in Fréchet spaces due to Frigon and Granas [208].

3.2.1 Existence and Uniqueness Result on a Compact Interval

In this section, we assume that for every $y_t \in B$, the function $t \to f(t, y_t)$ is measurable. In order to prove our main existence results, we will need the following auxiliary result.

Lemma 3.8. *Let $f : B \to \mathbb{R}^n$ be a continuous function and $t \to f(y_t)$ is measurable function. Then y is the unique solution of the initial value problem*

$$y'(t) = f(y_t) + \sum_{i=1}^{n_*} y(t - T_i), \qquad t \in J := [0, b] \backslash \{t_1, t_2, \dots, t_m\}, \tag{3.25}$$

$$y(t_k^+) - y(t_k) = I_k(y(t_k^-)), \qquad k = 1, \dots, m, \tag{3.26}$$

$$y(t) = \phi(t), \qquad t \in (-\infty, 0], \tag{3.27}$$

if and only if y is a solution of the impulsive integral functional differential equation

$$y(t) = \begin{cases} \phi(t), & \text{if } t \in (-\infty, 0], \\ \phi(0) + \sum_{i=1}^{n_*} \int_{-T_i}^{0} \phi(s)ds + \int_0^t f(y_s)d & \\ \quad + \sum_{i=1}^{n_*} \int_0^{t-T_i} y(s)ds + \sum_{0<t_k<t} I_k(y(t_k^-)), & \text{if } t \in [0, b]. \end{cases} \tag{3.28}$$

Proof. Let y be a possible solution of the problem (3.25)–(3.27). Then $y|_{[-r,t_1]}$ is a solution to

$$y'(t) = f(y_t) + \sum_{i=1}^{n_*} y(t - T_i), \quad \text{for } t \in [0, b].$$

Assume that $t_k < t \leq t_{k+1}$, $k = 1, \dots, m$. An integration of the above inequality yields

$$y(t_1^-) - y(0) = \int_0^{t_1} f(y_s)ds + \sum_{i=1}^{n_*} \int_0^{t_1} y(s - T_i)ds,$$

$$y(t_1^-) - y(0) = \int_0^{t_1} f(y_s)ds + \sum_{i=1}^{n_*} \int_{-T_i}^{t_1 - T_i} y(s)ds,$$

$$y(t_2^-) - y(t_1^+) = \int_{t_1}^{t_2} f(y_s)ds + \sum_{i=1}^{n_*} \int_{t_1}^{t_2} y(s - T_i)ds,$$

$$y(t_2^-) - y(t_1^-) = I_1(y(t_1^-)) + \int_{t_1}^{t_2} f(y_s)ds + \sum_{i=1}^{n_*} \int_{t_1 - T_i}^{t_2 - T_i} y(s)ds,$$

$$\dots$$

$$y(t_k^-) - y(t_{k-1}^+) = \int_{t_{k-1}}^{t_k} f(y_s)ds + \sum_{i=1}^{n_*} \int_{t_{k-1}}^{t_k} y(s - T_i)ds,$$

$$y(t_k^-) - y(t_{k-1}^-) = I_k(y(t_k^-)) + \int_{t_{k-1}}^{t_k} f(y_s)ds + \sum_{i=1}^{n_*} \int_{t_{k-1} - T_i}^{t_k - T_i} y(s - T_i)ds,$$

$$y(t) - y(t_k^-) = I_k(y(t_k^-)) + \int_{t_k}^{t} f(y_s)ds + \sum_{i=1}^{n_*} \int_{t_k - T_i}^{t - T_i} y(s)ds.$$

Then,

$$y(t_1) - y(0) = \int_0^{t_1} f(y_s)ds + \sum_{i=1}^{n_*} \int_{-T_i}^{t_1 - T_i} y(s)ds,$$

$$y(t_2) - y(t_1^-) = I_1(y(t_1^-)) + \int_{t_1}^{t_2} f(y_s)ds + \sum_{i=1}^{n_*} \int_{t_1 - T_i}^{t_2 - T_i} y(s)ds,$$

$$\dots$$

$$y(t_k^-) - y(t_{k-1}) = I_k(y(t_k^-)) + \int_{t_{k-1}}^{t_k} f(y_s)ds + \sum_{i=}^{n_*} \int_{t_{k-1-T_i}}^{t_k - T_i} y(s - T_i)ds,$$

$$y(t) - y(t_k^-) = I_k(y(t_k^-)) + \int_{t_k}^{t} f(y_s)ds + \sum_{i=1}^{n_*} \int_{t_k - T_i}^{t - T_i} y(s)ds.$$

Adding these together, we obtain

$$\begin{aligned} y(t) &= y(0) + \sum_{0<t_k<t} I_k(y(t_k^-)) + \int_0^t f(y_s)ds + \sum_{i=1}^{n_*} \int_{-T_i}^{t - T_i} y(s)ds \\ &= \phi(0) + \sum_{0<t_k<t} I_k(y(t_k^-)) + \int_0^t f(y_s)ds + \sum_{i=1}^{n_*} \int_0^{t - T_i} y(s)ds \\ &\quad + \sum_{i=1}^{n_*} \int_{-T_i}^{0} y(s)ds. \end{aligned}$$

Thus,

$$\begin{aligned} y(t) &= \phi(0) + \sum_{0<t_k<t} I_k(y(t_k^-)) + \int_0^t f(y_s)ds + \sum_{i=1}^{n_*} \int_0^{t - T_i} y(s)ds \\ &\quad + \sum_{i=1}^{n_*} \int_{-T_i}^{0} \phi(s)ds. \end{aligned}$$

If y satisfies the integral equation (3.28), then y is solution of the problem (3.25)–(3.27). Let $t \in [0, b]\backslash\{t_1, \dots, t_m\}$ and

$$\begin{aligned} y(t) = \phi(0) + \sum_{0<t_k<t} I_k(y(t_k^-)) + \int_0^t f(y_s)ds \\ + \sum_{i=1}^{n_*} \int_0^{t - T_i} y(s)ds + \sum_{i=1}^{n_*} \int_{-T_i}^{0} \phi(s)ds. \end{aligned}$$

Hence,

$$y'(t) = f(y_t) + \sum_{i=1}^{n_*} y(t - T_i).$$

We can easily prove that $y(t_k^+) - y(t_k^-) = I_k(y(t_k^-)),\ k = 1, \ldots, m.$ □

We are now in a position to state and prove our existence result for the problem (3.19)–(3.21).

Theorem 3.9. *Let $f : J \times B \to \mathbb{R}^n$ be a Carathédory function. Assume that*

(H-1) *There exist a function $p \in L^1(J, \mathbb{R}_+)$ and a continuous nondecreasing function $\psi : [0, \infty) \to [0, \infty)$ such that*

$$|f(t,x)| \leq p(t)\psi(\|x\|_B) \text{ for a.e. } t \in J \text{ and each } x \in B,$$

with

$$\int_c^\infty \frac{du}{u + \psi(u)} = \infty,$$

where $c = K_b[\|\phi\|_B + t_1 n_* |\phi(0)|]$.

Then the IVP (3.19)–(3.21) *has at least one solution on* $(-\infty, b]$.

Proof. We transform the problem (3.19)–(3.21) into a fixed point problem. Consider the operator $N : B_b \to B_b$ defined by

$$N(y)(t) = \begin{cases} \phi(t), & \text{if } t \in (-\infty, 0], \\ \phi(0) + \sum_{i=1}^{n_*} \int_{-T_i}^0 \phi(s)ds + \int_0^t f(s, y_s)ds & \\ \quad + \sum_{i=1}^{n_*} \int_0^{t-T_i} y(s)ds + \sum_{0<t_k<t} I_k(y(t_k^-)), & \text{if } t \in [0, b]. \end{cases}$$

Let $x(\cdot) : (-\infty, b] \to \mathbb{R}^n$ be the function defined by

$$x(t) = \begin{cases} \phi(0), & \text{if } t \in [0, b], \\ \phi(t), & \text{if } t \in (-\infty, 0]. \end{cases}$$

Then $x_0 = \phi$. For each $z \in PC$ with $z(0) = 0$, we denote by $\bar{z}$ the function defined by

$$\bar{z}(t) = \begin{cases} z(t), & \text{if } t \in [0, b], \\ 0, & \text{if } t \in (-\infty, 0]. \end{cases}$$

If $y(\cdot)$ satisfies the integral equation,

$$y(t) = \phi(0) + \sum_{i=1}^{n_*} \int_{-T_i}^0 \phi(s)ds + \int_0^t f(s, y_s)ds + \sum_{i=1}^{n_*} \int_0^{t-T_i} y(s)ds + \sum_{0<t_k<t} I_k(y(t_k^-)),$$

we can decompose $y(\cdot)$ as $y(t) = \bar{z}(t) + x(t), 0 \le t \le b$. This implies $y_t = \bar{z}_t + x_t$, for every $0 \le t \le b$, and the function $z(\cdot)$ satisfies

$$\begin{aligned} z(t) = & \sum_{i=1}^{n_*} \int_{-T_i}^{0} \phi(s)ds + \int_0^t f(s, \bar{z}_s + x_s)ds \\ & + \sum_{i=1}^{n_*} \int_0^{t-T_i} \bar{z}(s) + x(s)ds + \sum_{0<t_k<t} I_k(\bar{z}(t_k^-) + x(t_k^-)). \end{aligned} \tag{3.29}$$

Set

$$C_0 = \{z \in B_b : z_0 = 0\}.$$

For any $z \in C_0$, we have

$$\|z\|_\infty = \|z_0\|_B + \sup\{|z(t)| : t \in [0,b]\} = \sup\{|z(t)| : t \in [0,b]\}.$$

Thus, $(C_0, \|\cdot\|_\infty)$ is a Banach space. Let the operator $P : C_0 \to C_0$ be defined by

$$(Pz)(t) = \begin{cases} 0, & \text{if} \quad t \in (-\infty, 0], \\ \sum_{i=1}^{n_*} \int_{-T_i}^{0} \phi(s)ds + \int_0^t f(s, \bar{z}_s + x_s)ds & \\ \quad + \sum_{i=1}^{n_*} \int_0^{t-T_i} \bar{z}(s) + x(s)ds & \\ \quad + \sum_{0<t_k<t} I_k(\bar{z}(t_k^-) + x(t_k^-)), & \text{if} \quad t \in [0,b]. \end{cases}$$

Clearly, that the operator N has a fixed point is equivalent to P having a fixed point, and so we turn our attention to proving that P has one. We shall use the Leray–Schauder alternative to prove this.

Let z be a possible solution of the equation $z = \lambda P(z)$, for some $\lambda \in (0,1)$. Then, for all $t \in [0, t_1]$,

$$z(t) = \lambda \Big[\sum_{i=1}^{n_*} \int_{-T_i}^{0} \phi(s)ds + \int_0^t f(s, \bar{z}_s + x_s)ds + \sum_{i=1}^{n_*} \int_0^{t-T_i} [\bar{z}(s) + x(s)]ds \Big].$$

Hence,

$$\begin{aligned} |z(t)| & \le \sum_{i=1}^{n_*} \int_{-T_i}^{0} |\phi(s)|ds + \int_0^t |f(s, \bar{z}_s + x_s)|ds + n_* \int_0^t \|\bar{z}(s) + x(s)\|_B ds \\ & \le \sum_{i=1}^{n_*} T_i \|\phi\|_\infty + n_* t_1 |\phi(0)| + n_* \int_0^t |z(s)|ds + \int_0^t p(s)\psi(\|\bar{z}_s + x_s\|_B)ds. \end{aligned} \tag{3.30}$$

But

$$\begin{aligned}\|\bar{z}_s + x_s\|_B &\le \|\bar{z}_s\|_B + \|x_s\|_B \\ &\le K(t)\sup\{|z(s)| : 0 \le s \le t\} + M(t)\|z_0\|_B \\ &\quad + K(t)\sup\{|x(s)| : 0 \le s \le t\} + M(t)\|x_0\|_B \\ &\le K_b \sup\{|z(s)| : 0 \le s \le t\} \\ &\quad + M_b\Big[\|\phi\|_B + \sum_{k=1}^{n_*} T_i\|\phi\|_B\Big] \\ &\quad + K_b\Big[|\phi(0)| + \sum_{k=1}^{n_*} T_i\|\phi\|_\infty\Big].\end{aligned} \tag{3.31}$$

If we call $w(t)$ the right hand side of (3.31), then we have

$$\|\bar{z}_s + x_s\|_B \le w(t),$$

and therefore (3.30) becomes

$$\begin{aligned}|z(t)| \le& \sum_{i=1}^{n_*} T_i\|\phi\|_\infty + [n_* t_1|\phi(0)| + \|\phi\|_\infty] + n_*\int_0^t |z(s)|ds \\ &+ \int_0^t p(s)\psi(w(s))ds, \quad t \in [0, t_1].\end{aligned} \tag{3.32}$$

Using (3.32) in the definition of w, we have that, for $t \in [0, t_1]$,

$$\begin{aligned}w(t) \le& K_b\Big[n_* t_1|\phi(0)| + \|\phi\|_\infty + \sum_{i=1}^{n_*} T_i\|\phi\|_\infty\Big] \\ &+ K_b(n_* + 1)\int_0^t p(s)[\psi(w(s)) + w(s)]ds.\end{aligned}$$

Denoting by $\beta(t)$ the right hand side of the last inequality, we have

$$w(t) \le \beta(t), \quad t \in [0, t_1],$$

$$\beta(0) = K_b n_* t_1\Big[|\phi(0)| + \|\phi\|_\infty + \sum_{i=1}^{n_*} T_i\|\phi\|_\infty\Big],$$

and

$$\beta'(t) = K_b(n_* + 1)p(t)\psi(w(t)) \le K_b(n_* + 1)p(t)[\psi(\beta(t)) + \beta(t)],\ t \in [0, t_1].$$

This implies that, for each $t \in [0, t_1]$,

$$\int_{\beta(0)}^{\beta(t)} \frac{ds}{\psi(s) + s} \le K_b(n_* + 1)\int_0^{t_1} p(s)ds < \infty.$$

Thus, from (H-1), there exists a constant K_* such that $\beta(t) \le K_*$, $t \in [0, t_1]$, and hence $\|\bar{z}_t + x_t\|_B \le w(t) \le K_*$, $t \in [0, t_1]$. From equation (3.31), we have that

$$\|z\|_\infty \le \frac{K_*}{K_b} := \widetilde{K}_1.$$

Now let $t \in (t_1, t_2]$; then

$$z(t) = \lambda\left[\int_{t_1}^{t} f(s, \bar{z}_s + x_s)ds + \sum_{i=1}^{n_*}\int_{t_1-T_i}^{t-T_i} [\bar{z}(s) + x(s)]\, ds\right]$$

and

$$z(t_1^+) = z(t_1) + I_1(\bar{z}(t_1) + x(t_1)).$$

Therefore,

$$\begin{aligned} |z(t_1^+)| &\le |z(t_1)| + |I_1(\bar{z}(t_1) + x(t_1))| \\ &\le \widetilde{K}_1 + \sup\{|I_1(u)| : \|u\| \le \widetilde{K}_1 + [n_* T|\phi(0)| + \|\phi\|_\infty]\}. \end{aligned}$$

In a similar manner, we can show that there exists $\widetilde{K}_2 > 0$ such that

$$\sup\{|z(t)| : t \in [t_1, t_2]\} \le \widetilde{K}_2.$$

Continuing this process and taking into account that

$$z(t) = \lambda\Big[\int_{t_m}^{t} f(s, \bar{z}_s + x_s)ds + \sum_{i=1}^{n_*}\int_{t_m-T_i}^{t-T_i} \bar{z}(s) + x(s)ds\Big],\ t \in (t_m, b],$$

and

$$z(t_m^+) = z(t_m) + I_1(\bar{z}(t_m) + x(t_m)),$$

we obtain that there exists a constant $\widetilde{K}_m$ such that

$$\sup\{|z(t)| : t \in [t_m, b]\} \le \widetilde{K}_m.$$

Consequently, for each possible solution y to $z = \lambda P(z)$ for some $\lambda \in (0, 1)$, we have

$$\|z\|_\infty \le \max\{K_i : i = 1, \dots, m\} := \overline{K}.$$

Set

$$U = \{y \in PC : \|z\|_\infty < \overline{K} + 1\}.$$

Now $P : \overline{U} \to C_0$ is continuous and completely continuous. From this choice of U, there is no $z \in \partial U$ such that $z = \lambda P(z)$, for some $\lambda \in (0, 1)$. As a consequence of the nonlinear alternative of Leray–Schauder type [184], we deduce that P has a fixed point z in U. Hence, N has a fixed point y which is a solution to the problem (3.19)–(3.21). □

Next, we state a uniqueness result for the problem (3.19)–(3.21).

Theorem 3.10. *Assume that (A-4) holds. Then the IVP (3.19)–(3.21) has a unique solution.*

3.2.2 Global Existence and Uniqueness Result

As we know, investigation of many properties of solutions for a given equation, such as stability or oscillation, needs a guarantee of global existence. Thus, it is important and necessary to establish sufficient conditions for global existence of solutions for impulsive differential equations. Global existence results for impulsive differential equations under various conditions conditions were studied by Cheng and Yan [149], Graef and Ouahab [238], Guo [253, 254], Ouahab [397], Marino et al. [366], Stamov and Stamova [450], Weng [480], and Yan [486, 487]. Very recently, Graef and Ouahab [238,240] applied the new nonlinear alternative of Leary–Schauder type in Fréchet spaces mentioned above (see Frigon and Granas [208]) to functional impulsive differential equations with variable times. In [71], Benchohra et al. obtained some results on global existence of first order impulsive functional differential equations with infinite delay and boundary conditions.

This subsection is concerned with an application of a recent nonlinear alternative for contraction maps in Fréchet spaces, due to Frigon and Granas [208], to the existence and uniqueness of solutions of problem (3.22)–(3.24). In order to define the phase space and a solution of (3.22)–(3.24), we shall consider the space

$$PC_* = \Big\{y : [0,\infty) \to \mathbb{R}^n \mid y(t_k^-),\ y(t_k^+) \text{ exist with } y(t_k) = y(t_k^-), \\ y_k \in C(J_k, \mathbb{R}^n),\ k = 1,\dots\Big\},$$

where y_k is the restriction of y to $J_k = (t_k, t_{k+1}]$, $k = 0,\dots$. We will assume that B satisfies the following axioms:

(B) If $y : (-\infty,\infty) \to \mathbb{R}^n$, and $y_0 \in B$, then for every t in $[0,\infty)$, the following conditions hold:

(i) y_t is in B;

(ii) $\|y_t\|_B \le K(t)\sup\{|y(s)| : 0 \le s \le t\} + M(t)\|y_0\|_B$;

(iii) $|y(t)| \le H\|y_t\|_B$;

where $H \ge 0$ is a constant, $K : [0,\infty) \to [0,\infty)$ is continuous, $M : [0,\infty) \to [0,\infty)$ is locally bounded and H, K, and M are independent of $y(\cdot)$.

(B-1) For the function $y(\cdot)$ in (B), y_t is a B-valued continuous function on $[0,\infty)\backslash\{t_1,\dots\}$.

(B-2) The space B is complete.

Set

$$B_* = \{y : (-\infty,\infty) \to \mathbb{R}^n : y \in PC_* \cap B\}, \tag{3.33}$$

and consider

$$B_k = \{y \in B_* : \sup_{t\in J_k^*} |y(t)| < \infty\},\ J_k^* = [0, t_k],$$

where $\|\cdot\|_k$ is the seminorm in B_k defined by

$$\|y\|_k = \|y_0\|_B + \sup\{|y(t)| : 0 \le s \le t_k\},\ y \in B_k.$$

Let us start by defining what we mean by a solution of the problem (3.22)–(3.24).

Definition 3.11. A function $y \in B_*$ is said to be a solution of (3.22)–(3.24) if y satisfies (3.22)–(3.24).

Theorem 3.12 ([398]). *Assume that:*

(B-3) *There exist constants $d_k > 0$, $k = 1, 2, \dots$, such that for all $x, \overline{x} \in \mathbb{R}^n$,*

$$|I_k(x) - I_k(\overline{x})| \le d_k|x - \overline{x}|, \textit{ for each } k = 1, \dots;$$

(B-4) *There exist $l \in L^1([0,\infty), \mathbb{R}^n)$ such that*

$$|f(t,x) - f(t,\overline{x})| \le l(t)\|x - \overline{x}\|_B, \textit{ for } t \in [0,\infty) \textit{and all } x, \overline{x} \in B;$$

(B-5) *There exist a continuous nondecreasing function $\psi : [0,\infty) \longrightarrow (0,\infty)$ and a function $p \in L^1([0,\infty), \mathbb{R}_+)$ such that*

$$|f(t,u)| \le p(t)\psi(\|u\|_B) \textit{ for each } (t,u) \in [0,\infty) \times B,$$

$$\int_c^\infty \frac{du}{\psi(u) + u} = \infty,$$

where

$$K_k = \sup\{|K(t)| : t \in [0, t_k]\},\ k = 1, \dots.$$

If $\sum_{k=1}^{\infty} d_k < 1$, then the initial value problem (3.22)–(3.24) *has a unique solution.*

3.3 Stability

Given a function $f : [0, +\infty) \times B \to \mathbb{R}^n$, we consider the system

$$y'(t) = f(t, y_t), \qquad t \in [t_0, \infty)\backslash\{t_1, t_2, \dots\}, \tag{3.34}$$

$$y(t_k^+) - y(t_k^-) = I_k(y(t_k^-)), \qquad k = 1, \dots. \tag{3.35}$$

The initial condition for system (3.34)–(3.35) is given by

$$y(t_0 + t) = \phi(t), \quad t \in (-\infty, 0], \tag{3.36}$$

where $I_k \in C(\mathbb{R}^n, \mathbb{R}^n)$, $k = 1, 2, \dots, 0 < t_1 < \cdots < t_m < \cdots$, $\lim_{m\to\infty} t_m = \infty$, $\phi \in B$, and $t_0 \in \mathbb{R}_+$. For our stability analysis, we assume that $t \to f(t, y_t)$ is measurable, $f(t,0) = 0$ for all $t \in \mathbb{R}_+$, and $I_k(0) = 0$ for all $k = 1, 2, \dots$. System (3.34)–(3.35) therefore possesses the trivial solution $y(t) = 0$.

Definition 3.13. The trivial solution of system (3.34)–(3.35) is said to be:

(S1) stable if for every $\epsilon > 0$ and $t_0 \in \mathbb{R}_+$, there exists $\delta(t_0, \epsilon)$ such that, if $\phi \in B$ with $\|\phi\|_B \leq \delta$ and $y = y(t_0, \phi)$ is a solution of (3.34)–(3.36), then $\|y_t\|_B \leq \epsilon$ for all $t \geq t_0$;

(S2) uniformly stable if δ is independent of t_0;

(S3) asymptotically stable, if $(S1)$ holds and for every $t_0 \in \mathbb{R}_+$, there exists $\eta = \eta(t_0) > 0$ such that if $\phi \in B$ with $\|\phi\|_B \leq \eta$, then $\lim_{t\to\infty} \|y_t\|_B = 0$;

(S4) uniformly asymptotically stable if $(S2)$ holds and there exists $\eta > 0$ such that, for every $\gamma > 0$, there exists $T = T(\eta, \gamma) > 0$ such that, if $\phi \in B$ with $\|\phi\|_B \leq \eta$, then $\|y_t\|_B \leq \gamma$, for all $t \geq t_0 + T$;

(S5) unstable if $(S1)$ fails to hold.

It is clear that y is the unique solution of the system (3.34)–(3.36) if and only if y is a solution of the impulsive integral functional differential equation

$$y(t) = \begin{cases} \phi(t - t_0), & \text{if} \quad t \in (-\infty, t_0], \\ \phi(0) + \int_{t_0}^{t} f(s, y_s)ds + \sum_{t_0 < t_k < t} I_k(y(t_k^-)), & \text{if} \quad t \in [t_0, \infty). \end{cases} \tag{3.37}$$

3.3.1 Stability Result

In order to give sufficient conditions for the stability of the system (3.34)–(3.36), we will assume that B satisfies the following axioms:

(C) If $y : (-\infty, \infty) \to \mathbb{R}^n$, and for any $t_0 \in \mathbb{R}_+$, $y_{t_0} \in B$, then for every t in $[0, \infty)$, the following conditions hold:

(i) y_t is in B;

(ii) $\|y_t\|_B \leq K(t - t_0) \sup\{|y(s)| : t_0 \leq s \leq t\} + M(t - t_0)\|y_{t_0}\|_B$;

(iii) $|y(t)| \leq H\|y_t\|_B$,

where $H \geq 0$ is a constant, $K : [0, \infty) \to [0, \infty)$ is continuous, $M : [0, \infty) \to [0, \infty)$ is locally bounded and H, K, and M are independent of $y(\cdot)$.

(C-1) For the function $y(\cdot)$ in (C), y_t is a B-valued continuous function on $[0, \infty)\backslash\{t_1, \ldots\}$.

(C-2) The space B is complete.

Theorem 3.14. *Assume that:*

(C-3) *There exist constants* $c_k > 0, k = 1, 2, \ldots$, *such that for all* $x \in \mathbb{R}^n$,

$$|I_k(x)| \leq c_k|x|, \quad \textit{for each} \quad k = 1, \ldots;$$

(C-4) *There exist* $l \in L^1([0,\infty),\mathbb{R}^n)$ *such that*

$$|f(t,x)| \leq l(t)\|x\|_B, \quad \textit{for each} \quad t \in [0,\infty), \quad \textit{and all} \quad x \in B;$$

If

$$\sum_{k=1}^{\infty} c_k < 1,$$

$$K_\infty = \sup\{|K(t)| : t \in [0,\infty)\} < \infty,$$

and

$$M_\infty = \sup\{|M(t)| : t \in [0,\infty)\} < \infty,$$

then the zero solution of the problem (3.34)–(3.36) *is uniformly stable.*

Proof. Let y be a solution of the system (3.34)–(3.36). Then for $t \in [t_0,\infty)$, we have

$$\|y_t\|_B \leq K(t-t_0) \sup_{t_0 \leq s \leq t} |y(s)| + M(t-t_0)\|y_{t_0}\|_B.$$

For $s \in [t_0,t]$, we obtain

$$\begin{aligned}
|y(s)| &\leq |\phi(0)| + \int_{t_0}^{s} |f(u,y_u)|du + \sum_{t_0<t_k<t} |I_k(y(t_k^-))| \\
&\leq |\phi(0)| + \int_{t_0}^{s} l(t)\|y_u\|_B du + \sum_{t_0<t_k<t} c_k \sup_{s\in[t_0,t]} |y(s)| \\
&\leq H\|\phi\|_B + \int_{t_0}^{s} l(u)K(u-t_0) \sup_{u\in[t_0,s]} |y(u)|du \\
&\quad + \|\phi\|_B \int_{t_0}^{s} l(u)M(u-t_0)du + \sum_{t_0<t_k<t} c_k \sup_{u\in[t_0,s]} |y(u)|.
\end{aligned}$$

Hence,

$$\sup_{s\in[t_0,t]} |y(s)| \leq \frac{1}{1-\sum_{k=1}^{\infty} c_k}\left[H\|\phi\|_B + K_\infty \int_{t_0}^{t} l(s) \sup_{s\in[t_0,t]} |y(s)|ds + \|\phi\|_B M_\infty \|l\|_{L^1}\right].$$

By Gronwall's Inequality [267], we obtain

$$\sup_{s\in[t_0,t]} |y(s)| \leq c\|\phi\|_B e^{\|l^*\|_{L^1}},$$

where

$$c = \frac{1}{1-\sum_{k=1}^{\infty} c_k}\left[H + M_\infty \|l\|_{L^1}\right], \text{ and } l^*(t) = K_\infty l(t).$$

Thus,

$$\|y_t\|_B \leq \left(K_\infty c e^{\|l^*\|_{L^1}} + M_\infty\right)\|\phi\|_B.$$

Let $\epsilon > 0$ and $\delta = \frac{\epsilon}{K_\infty c e^{\|l^*\|_{L^1}} + M_\infty}$; then from the above inequality, we obtain that, for every $\phi \in B$ with $\|\phi\|_B \leq \delta$, we have $\|y_t\|_B \leq \epsilon$, i.e., the zero solution is uniformly stable. □

3.4 Second Order Impulsive FDEs

In this section, we investigate the existence and uniqueness of solutions of second order impulsive functional differential equations with infinite delay. More precisely, we consider

$$y''(t) = f(t, y_t), \quad \text{a.e.} \quad t \in J := [0, b],\ t \neq t_k,\ k = 1, \ldots, m, \tag{3.38}$$

$$y(t_k^+) - y(t_k^-) = I_k(y(t_k^-)), \quad t = t_k,\ k = 1, \ldots, m, \tag{3.39}$$

$$y'(t_k^+) - y'(t_k^-) = \overline{I}_k(y(t_k^-)), \quad t = t_k,\ k = 1, \ldots, m, \tag{3.40}$$

$$y(t) = \phi(t), \quad t \in (-\infty, 0],\ y'(0) = \eta, \tag{3.41}$$

where $\eta \in \mathbb{R}^n$, f and B are as in problem (3.1)–(3.3), and $I_k, \overline{I}_k \in C(\mathbb{R}^n, \mathbb{R}^n)$, $k = 1, 2, \ldots, m$, are given functions satisfying some assumptions that will be specified later. Our approach here is based on the Leray–Schauder alternative [184], the Banach fixed point theorem, and the nonlinear alternative of Leary–Schauder type in Fréchet spaces due to Frigon and Granas [208].

Let us start by defining what we mean by a solution of the problem (3.38)–(3.41). Here, the set B_b is given in (3.4) and in all this we assume that $t \to f(t, y_t)$ is measurable function.

Definition 3.15. A function $y \in B_b$ is said to be a solution of (3.38)–(3.41) if y satisfies (3.38)–(3.41).

We will need the following auxiliary result in order to prove our main existence theorems.

Lemma 3.16. *The function y is the unique solution of the problem* (3.38)–(3.41) *if and only if y is a solution of the problem,*

$$y'(t) = \eta + \int_0^t f(s, y_s)ds + \sum_{0<t_k<t} \overline{I}_k(y(t_k^-)), \quad t \in J,\ t \neq t_k,\ k = 1, \ldots, m, \tag{3.42}$$

$$y(t_k^+) - y(t_k^-) = I_k(y(t_k^-)), \quad k = 1, \ldots, m, \tag{3.43}$$

$$y(t) = \phi(t), \quad t \in (-\infty, 0]. \tag{3.44}$$

Proof. Let y be a solution of the problem (3.38)–(3.41). Then

$$y''(t) = f(t, y_t) \text{ for } t \in [0, b]\backslash\{t_1 \ldots, t_m\}.$$

An integration from 0 to t (here $t \in (0, t_1]$) of both sides of the above equality yields

$$y'(t) - y'(0) = \int_0^t f(s, y_s)ds.$$

Thus, for $t \in [0, t_1]$, we have

$$y'(t) = \eta + \int_0^t f(s, y_s)ds.$$

If $t \in (t_1, t_2]$, then we have

$$\begin{aligned} \int_0^{t_1} y''(s)ds + \int_{t_1}^t y''(s)ds &= \int_0^t f(s, y_s)ds \\ y'(t_1^-) - y'(0) + y'(t) - y'(t_1^+) &= \int_0^t f(s, y_s)ds \\ y'(t) - \overline{I}_1(y(t_1^-)) - \eta &= \int_0^t f(s, y_s)ds. \end{aligned}$$

Thus, for $t \in (t_1, t_2]$ we have

$$y'(t) = \eta + \overline{I}_1(y(t_1^-)) + \int_0^t f(s, y_s)ds.$$

Continuing, we obtain

$$y'(t) = \eta + \int_0^t f(s, y_s)ds + \sum_{0<t_k<t} \overline{I}_k(y(t_k^-))$$

for $t \in [0, b]$.

Conversely, we prove that if y satisfies the problem (3.42)–(3.44), then y is a solution of the problem (3.38)–(3.41). It is clear that $y(t) = \phi(t),\ t \in (-\infty, 0]$, and $y(t_k^+) - y(t_k^-) = I_k(y(t_k^-)),\ k = 1, \ldots, m$. Let $t \in [0, b]\backslash\{t_1, \ldots, t_m\}$ and

$$y'(t) = \eta + \int_0^t f(s, y_s)ds + \sum_{0<t_k<t} \overline{I}_k(y(t_k^-)).$$

Then,

$$y''(t) = f(t, y_t), \quad t \in [0, b]\backslash\{t_1, \ldots, t_m\},$$

completing the proof of the theorem. □

3.4.1 Existence and Uniqueness Results

Theorem 3.17. *Let $f : J \times B \to \mathbb{R}^n$ be a Carathéodory function. Assume the condition:*

(H-2) *There exist a continuous nondecreasing function $\psi : [0,\infty) \longrightarrow (0,\infty)$ and a function $p \in L^1([0,b], \mathbb{R}_+)$ such that*

$$|f(t,x)| \le p(t)\psi(\|x\|_B) \text{ for a.e. } t \in [0,b] \text{ and each } x \in B,$$

with

$$\int_c^{\infty} \frac{dx}{\psi(x)} = \infty,$$

where $K_b = \sup\{|K(t)| : t \in [0,b]\}$, $M_b = \sup\{|M(t)| : t \in [0,b]\}$ and $c = M_b\|\phi\|_B + K_b|\phi(0)|$.

Then the initial value problem (3.38)–(3.41) has at least one solution.

Proof. The proof will be given in several steps.

Step 1. Consider the problem,

$$y'(t) = \eta + \int_0^t f(s, y_s)ds, \quad \text{a.e.} \quad t \in [0, t_1], \tag{3.45}$$

$$y(t) = \phi(t), \quad t \in (-\infty, 0]. \tag{3.46}$$

To transform the problem (3.45)–(3.46) into a fixed point problem, consider the operator $N : B \cap C([0,t_1], \mathbb{R}^n) \longrightarrow B \cap C([0,t_1], \mathbb{R}^n)$ defined by,

$$N(y)(t) = \begin{cases} \phi(t), & t \in (-\infty, 0], \\ \phi(0) + t\eta + \int_0^t \int_0^s f(u, y_u)duds, & t \in [0, t_1]. \end{cases}$$

Let $x(\cdot) : (-\infty, t_1] \to \mathbb{R}^n$ be the function defined by

$$x(t) = \begin{cases} \phi(0), & \text{if } t \in [0, t_1], \\ \phi(t), & \text{if } t \in (-\infty, 0]. \end{cases}$$

Then, $x_0 = \phi$. For each $z \in C([0,t_1], \mathbb{R}^n) \cap B$, with $z_0 = 0$, we denote by $\bar{z}$ the function defined by

$$\bar{z}(t) = \begin{cases} z(t), & \text{if } t \in [0, t_1], \\ 0, & \text{if } t \in (-\infty, 0]. \end{cases}$$

If $y(\cdot)$ satisfies the integral equation,

$$y(t) = \phi(0) + t\eta + \int_0^t \int_0^s f(u, y_u)duds,$$

we can decompose $y(\cdot)$ as $y(t) = \bar{z}(t) + x(t), 0 \leq t \leq t_1$. This implies $y_t = \bar{z}_t + x_t$, for every $0 \leq t \leq t_1$, and the function $z(\cdot)$ satisfies

$$z(t) = t\eta + \int_0^t \int_0^s f(u, \bar{z}_u + x_u) du ds. \tag{3.47}$$

Set

$$C_0 = \{z \in C([0, t_1], \mathbb{R}^n) \cap B : z_0 = 0\}.$$

Let the operator $P : C_0 \to C_0$ be defined by

$$(Pz)(t) = t\eta + \int_0^t \int_0^s f(u, \bar{z}_u + x_u) du ds, \quad t \in [0, t_1].$$

Clearly, the operator N having a fixed point is equivalent to P having a fixed point, and so we now show that this is the case. We shall use the Leray–Schauder alternative to prove this. As in Theorem 3.6, we can prove that P is completely continuous.

Let z be a possible solution of $z = \lambda P(z)$ with $z_0 = \lambda\phi$, for some $\lambda \in (0, 1)$. Then

$$|z(t)| \leq t_1|\eta| + t_1 \int_0^t |f(s, \bar{z}_s + x_s)| ds \leq t_1|\eta| + t_1 \int_0^t p(s)\psi(\|\bar{z}_s + x_s\|_B) ds. \tag{3.48}$$

But

$$\begin{aligned} \|\bar{z}_s + x_s\|_B &\leq \|\bar{z}_s\|_B + \|x_s\|_B \\ &\leq K(t) \sup\{|z(s)| : 0 \leq s \leq t\} + M(t)\|z_0\|_B \\ &\quad + K(t) \sup\{|x(s)| : 0 \leq s \leq t\} + M(t)\|x_0\|_B \\ &\leq K_b \sup\{|z(s)| : 0 \leq s \leq t\} + M_b\|\phi\|_B + K_b M|\phi(0)|. \end{aligned} \tag{3.49}$$

If we denote the right hand side of (3.49) by $w(t)$, then we have

$$\|\bar{z}_s + x_s\|_B \leq w(t),$$

and so (3.48) becomes

$$|z(t)| \leq t_1|\eta| + \int_0^t p(s)\psi(w(s)) ds, \quad t \in [0, t_1]. \tag{3.50}$$

Using (3.50) in the definition of w, we have that

$$w(t) \leq t_1 K_b \int_0^t p(s)\psi(w(s)) ds + t_1|\eta| + M_b\|\phi\|_B + K_b|\phi(0)|, \quad t \in [0, t_1].$$

Denoting by $\beta(t)$ the right hand side of the last inequality, we have

$$w(t) \leq \beta(t), \quad t \in [0, t_1],$$

$$\beta(0) = t_1|\eta| + M_b\|\phi\|_B + K_b|\phi(0)|,$$

and

$$\begin{aligned}\beta'(t) &= t_1 K_b p(t)\psi(w(t)) \\ &\leq t_1 K_b p(t)\psi(\beta(t)), \quad t \in [0, t_1].\end{aligned}$$

This implies that, for each $t \in [0, t_1]$,

$$\int_{\beta(0)}^{\beta(t)} \frac{ds}{\psi(s)} \leq t_1 K_b \int_0^{t_1} p(s)ds < \infty.$$

Thus, from (H-2) there exists a constant K_* such that $\beta(t) \leq K_*$ for $t \in [0, t_1]$, and hence $\|\bar{z}_t + x_t\|_B \leq w(t) \leq K_*$ for $t \in [0, t_1]$. From (3.50), we have

$$\|z\|_\infty \leq \int_0^{t_1} p(s)\psi(K_*)ds := \widetilde{K}_1.$$

Set

$$U_0 = \{z \in C_0 |\ \sup\{|z(t)| : 0 \leq t \leq t_1\} < \widetilde{K}_1 + 1\}.$$

We see that $P : \overline{U}_0 \to C_0$ is continuous and completely continuous. From the choice of U_0, there is no $z \in \partial U_0$ such that $z = \lambda P(z)$, for some $\lambda \in (0, 1)$. As a consequence of the nonlinear alternative of Leray–Schauder type [184], we deduce that P has a fixed point z in U_0. Hence, N has a fixed point y which is a solution to problem (3.45)–(3.46). Denote this solution by y_0.

Step 2. Consider next the problem,

$$y'(t) = \int_{t_1}^{t} f(s, y_s)ds + y_0'(t_1^-) + \overline{I}_1(y_0(t_1^-)), \quad \text{a.e.} \quad t \in (t_1, t_2], \tag{3.51}$$

$$y(t_1^+) - y(t_1^-) = I_1(y_0(t_1)), \ \ y(t) = y_0(t), \qquad t \in (-\infty, t_1]. \tag{3.52}$$

Let

$$C_1 = \{y \in C((t_1, b], \mathbb{R}^n) : y(t_1^+) \text{ exists}\}.$$

Set $C_* = B \cap C([0, t_1], \mathbb{R}^n) \cap C_1$. Consider the operator $N_1 : C_* \to C_*$ defined by,

$$N_1(y)(t) = \begin{cases} y_0(t), & (-\infty, t_1], \\ (t - t_1)[y'(t_1^-) + \overline{I}_1(y_0(t_1^-))] & \\ \quad + y_0(t_1^-) + I_1(y_0(t_1^-)) + \int_{t_1}^t \int_{t_1}^s f(u, y_u)duds, & t \in (t_1, t_2]. \end{cases}$$

Let $x(\cdot) : (-\infty, t_2] \to \mathbb{R}^n$ be the function defined by

$$x(t) = \begin{cases} y(t_1^-) + I_1(y_0(t_1)), & \text{if } t \in (t_1, t_2], \\ y_0(t), & \text{if } t \in (-\infty, t_1]. \end{cases}$$

Then $x_{t_1} = y_0$. For each $z \in C_*$ with $z(t_1) = 0$, we denote by $\bar{z}$ the function defined by

$$\bar{z}(t) = \begin{cases} z(t), & \text{if } t \in [t_1, t_2], \\ 0, & \text{if } t \in (-\infty, t_1]. \end{cases}$$

If $y(\cdot)$ satisfies the integral equation,

$$y(t) = y(t_1^-) + I_1(y_0(t_1^-)) + (t - t_1)[y_0'(t_1^-) + \overline{I}_1(y_0(t_1^-))] + \int_{t_1}^{t}\int_{t_1}^{s} f(u, y_u)duds,$$

we can decompose it as $y(t) = \bar{z}(t) + x(t)$, $t_1 \le t \le t_2$. This implies $y_t = \bar{z}_t + x_t$, for every $t_1 \le t \le t_2$, and the function $z(\cdot)$ satisfies

$$z(t) = (t - t_1)[y_0'(t_1^-) + \overline{I}_1(y_0(t_1^-)) + \int_{t_1}^{t}\int_{t_1}^{s} f(u, \bar{z}_u + x_u)dsdu. \tag{3.53}$$

Set

$$C_{t_1} = \{z \in C_* : z_{t_1} = 0\}.$$

Let the operator $P_1 : C_{t_1} \to C_{t_1}$ be defined by,

$$(P_1 z)(t) = \begin{cases} 0, & t \in (-\infty, t_1], \\ (t - t_1)[y_0'(t_1^-) + \overline{I}_1(y_0(t_1^-))] & \\ \quad + \int_{t_1}^{t}\int_{t_1}^{s} f(u, \bar{z}_u + x_u)duds, & t \in (t_1, t_2]. \end{cases}$$

As in Step 1, we can show that P_1 is continuous and completely continuous, and if z is a possible solution of the equations $z = \lambda P_1(z)$ with $z_0 = \lambda y_0$, for some $\lambda \in (0, 1)$, then there exists $K_{*_1} > 0$ such that

$$\|z\|_\infty \le K_{*_1}.$$

Set

$$U_1 = \{z \in C_{t_1} |\ \sup\{|z(t)| : t_1 \le t \le t_2\} \le K_{*_1} + 1\}.$$

Again the nonlinear alternative of Leray–Schauder type [184] implies P_1 has a fixed point z in U_1. Thus, N_1 has a fixed point y which is a solution to problem (3.51)–(3.52). Denote this solution by y_1.

Step 3. We continue this process taking into account that $y_m := y\big|_{[t_m, b]}$ is a solution to the problem,

$$y'(t) = \int_{t_m}^{t} f(s, y_s)ds + y_{m-1}'(t_m^-) + \overline{I}_{m-1}(y(t_m^-)), \quad \text{a.e.} \quad t \in (t_m, b], \tag{3.54}$$

$$y(t_m^+) = y_{m-1}(t_{m-1}^-) + I_m(y_{m-1}(t_m^-)), y(t) = y_{m-1}(t), \qquad t \in (-\infty, t_{m-1}]. \tag{3.55}$$

The solution y of the problem (3.38)–(3.41) is then given by

$$y(t) = \begin{cases} y_0(t), & \text{if } t \in (-\infty, t_1], \\ y_1(t), & \text{if } t \in (t_1, t_2], \\ \dots \\ y_m(t), & \text{if } t \in (t_m, b]. \end{cases}$$

The proof is complete. □

Next we give some additional conditions that lead to uniqueness of solutions of (3.38)–(3.41).

Theorem 3.18. *In addition to* (H-2) *and* (A-4), *assume*

(H-3) *There exist constants* $\overline{d}_k > 0,\ k = 1, \dots, m$, *such that*

$$|\overline{I}_k(y) - \overline{I}_k(x)| \leq \overline{d}_k |x - \overline{x}|, \text{ for each } x, \overline{x} \in \mathbb{R}^n.$$

Then the IVP (3.38)–(3.41) has a unique solution.

Proof. The proof will be given in several steps.

Step 1. We first prove that the problem (3.45)–(3.46) has a unique solution. To this end, we prove that the operator P defined in Theorem 3.17 has a unique fixed point by showing that P is a contraction operator. Indeed, consider $z, z^* \in C_0$. Thus, for each $t \in [0, t_1]$,

$$(Pz)(t) = t\eta + \int_0^t \int_0^s f(u, \bar{z}_u + x_u) du ds.$$

Let $\widetilde{l}(t) = t_1 K_b l(t), \widehat{l}(t) = \int_0^t \widetilde{l}(s) ds$, and $\|\cdot\|_{B_*}$ be the Bielecki-type norm on C_0 defined by

$$\|z\|_{B_*} = \sup\{|z(t)| e^{-\tau \widehat{l}(t)} : t \in [0, t_1]\}, \quad \tau > 1.$$

Then,

$$\begin{aligned} |P(z)(t) - P(z^*)(t)| &\leq \int_0^t \int_0^s |f(u, \bar{z}_u + x_u)) - f(u, \bar{z}_u^* + x_u)| du ds \\ &\leq \int_0^t t_1 l(s) \|\bar{z}_s - \bar{z}_s^*\|_B ds \\ &\leq \int_0^t l(s) K_b \sup_{s \in [0,t]} |z(s) - z^*(s)| ds \\ &\leq \int_0^t \widetilde{l}(s) e^{\tau \widehat{l}(s)} e^{-\tau \widehat{l}(s)} \sup_{s \in [0,t]} |z(s) - z^*(s)| ds \end{aligned}$$

$$\le \int_0^t \widetilde{l}(t)e^{\tau\widehat{l}(t)}ds\|z-z^*\|_{B_*}$$
$$\le \frac{1}{\tau}\int_0^t (e^{\tau\widehat{l}(s)})'ds\|z-z^*\|_{B_*}$$
$$\le \frac{1}{\tau}e^{\tau\widehat{l}(t)}ds\|z-z^*\|_{B_*}.$$

Thus,

$$e^{-\tau\widehat{l}(t)}|P(z)(t)-P(z^*)(t)| \le \frac{1}{\tau}\|z-z^*\|_{B_*},$$

so

$$\|P(z)-P(z^*)\|_{B_*} \le \frac{1}{\tau}\|z-z_*\|_{B_*}.$$

As a consequence of the Banach fixed point theorem, P has a unique fixed point that is a solution to (3.45)–(3.46). Denote this solution by y_0.

Step 2. In the same way, we can prove that the problem (3.51)–(3.52) has a unique solution. We denote this solution by y_1. We continue this process and take into account that y_m is the unique solution of the problem (3.54)–(3.55). A solution y of the problem (3.38)–(3.41) is then defined by

$$y(t) = \begin{cases} y_0(t), & \text{if } t \in (-\infty, t_1], \\ y_1(t), & \text{if } t \in (t_1, t_2], \\ \dots & \\ y_m(t), & \text{if } t \in (t_m, b]. \end{cases}$$

Now, let x, and y be a two solutions of the problem (3.38)–(3.41). If $t \in (t_k, t_{k+1}]$, $k = 0, \dots, m$, then $x(t) = y(t)$. If $t = t_k^+$, $k = 1, \dots, m$, then since I_k, $y(t_k)$, and $x(t_k)$ are continuous functions, we have $y(t_k^+) = x(t_k^+)$, and it follows from (H-3) that

$$|\overline{I}_k(x(t_k)) - \overline{I}_k(y(t_k))| \le \overline{d}_k|x(t_k) - y(t_k)|.$$

Thus,

$$|x'(t_k^+) - y'(t_k^+)| \le 0,\ k = 1, \dots, m,$$

and these inequalities imply that $x'(t_k^+) = y'(t_k^+)$. Therefore, there is a unique solution of the problem (3.38)–(3.41). □

3.5 Global Existence and Uniqueness Result

In this subsection, we are concerned with an application of a recent nonlinear alternative for contraction maps in Fréchet spaces (see Frigon and Granas [208]) to prove the existence and uniqueness of solutions of a problem that has infinitely many impulses and infinite delay. We consider the problem,

$$y''(t) = f(t, y_t) \qquad \text{a.e. } t \in J_* := [0,\infty)\backslash\{t_1, t_2, \ldots\}, \tag{3.56}$$

$$y(t_k^+) - y(t_k^-) = I_k(y(t_k^-)), \qquad k = 1, \ldots, \tag{3.57}$$

$$y'(t_k^+) - y'(t_k^-) = \overline{I}_k(y(t_k^-)), \qquad k = 1, \ldots, \tag{3.58}$$

$$y(t) = \phi(t), \qquad Wt \in (-\infty, 0], \; y'(0) = \eta, \tag{3.59}$$

where $f : J_* \times B \to \mathbb{R}^n$ and $I_k, \overline{I}_k \in C(\mathbb{R}^n, \mathbb{R}^n)$, $k = 1, \ldots, 0 = t_0 < t_1 < \cdots < t_m < \cdots$, $\lim_{n\to\infty} t_n = \infty$, and $y(t_k^+) = \lim_{h\to 0^+} y(t_k + h)$ and $y(t_k^-) = \lim_{h\to 0^-} y(t_k - h)$ represent the right and left hand limits of $y(t)$ at $t = t_k$. For any function y defined on $(-\infty, \infty)$ and any $t \in [0, \infty)$, we denote by y_t the element of B defined by $y_t(\theta) = y(t + \theta)$, $\theta \in (-\infty, 0]$.

In all this section we assume that $t \to f(t, y_t)$ is a measurable function.

3.5.1 Uniqueness Result

In the following definition the set B_* is given in (3.33).

Definition 3.19. A function $y \in B_*$ is said to be a solution of (3.56)–(3.59) if y satisfies (3.56)–(3.59).

Theorem 3.20. *In addition to* (B-3), (B-4), *and* (H-3), *assume that*

(H-4) *There exist a continuous nondecreasing function* $\psi : [0,\infty) \longrightarrow (0,\infty)$ *and a function* $p \in L^1([0,\infty), \mathbb{R}_+)$ *such that*

$$|f(t,u)| \le p(t)\psi(\|u\|_B) \text{ for each } (t,u) \in [0,\infty) \times B,$$

$$\int_c^\infty \frac{du}{\psi(u)} = \infty,$$

where

$$c > 0 \quad \text{and} \quad K_k = \sup\{|K(t)| : t \in [0, t_k]\}, \; k = 1, \ldots .$$

If

$$\sum_{k=1}^\infty d_k < 1 \quad \text{and} \quad \sum_{k=1}^\infty \bar{d}_k < \infty,$$

then the initial value problem (3.56)–(3.59) *has a unique solution.*

Proof. We consider the problem,

$$y'(t) = \eta + \int_0^t f(s, y_s)ds + \sum_{0<t_k<t} \overline{I}_k(y(t_k^-)) \quad \text{a.e.} \quad t \in J_* := [0,\infty)\backslash\{t_1,\ldots\}, \tag{3.60}$$

$$y(t_k^+) - y(t_k^-) = I_k(y(t_k^-)), \qquad k = 1,\ldots, \tag{3.61}$$

$$y(t) = \phi(t), \qquad t \in (-\infty, 0]. \tag{3.62}$$

Similar to what we have in Lemma 3.16, we can easily prove that y is a solution of the problem (3.60)–(3.62) if and only if y is a solution of (3.56)–(3.59). To transform the problem (3.60)–(3.62) into a fixed point problem, consider the operator $\overline{N} : B_* \longrightarrow B_*$ defined by,

$$\overline{N}(y)(t) = \begin{cases} \phi(t), & t \in (-\infty, 0], \\ \phi(0) + t\eta + \int_0^t\int_0^s f(u, y_u)ds & \\ \quad + \sum_{0<t_k<t}[I_k(y(t_k^-)) + (t - t_k)\overline{I}_k(y(t_k^-))], & t \in [0,\infty). \end{cases}$$

Let $x(\cdot) : (-\infty, \infty) \to \mathbb{R}^n$ be the function defined by

$$x(t) = \begin{cases} \phi(0), & \text{if } t \in [0,\infty), \\ \phi(0) + \sum_{0<t_k<t}[I_k(x(t_k^-)) & \\ \quad + (t - t_k)\overline{I}_k(x(t_k^-))], & \text{if } t \in (-\infty, 0]. \end{cases}$$

Then $x_0 = \phi(0)$. For each $z \in PC_* \cap B$, with $z_0 = 0$, we denote by $\bar{z}$ the function defined by

$$\bar{z}(t) = \begin{cases} z(t), & \text{if } t \in [0,\infty), \\ 0, & \text{if } t \in (-\infty, 0]. \end{cases}$$

If $y(\cdot)$ satisfies the integral equation,

$$y(t) = \phi(0) + \sum_{0<t_k<t} [I_k(y(t_k^-)) + (t - t_k)\overline{I}_k(y(t_k^-))] + \int_0^t \int_0^s f(u, y_u)duds,$$

we can decompose $y(\cdot)$ as $y(t) = \bar{z}(t) + x(t), 0 \le t < \infty$. This implies $y_t = \bar{z}_t + x_t$, for every $0 \le t < \infty$, and the function $z(\cdot)$ satisfies

$$\begin{aligned} z(t) &= t\eta + \int_0^t \int_0^s f(u, \bar{z}_u + x_u)duds \\ &\quad + \sum_{0<t_k<t} [I_k(\bar{z}(t_k^-) + x(t_k^-)) + (t - t_k)\overline{I}_k(\bar{z}(t_k^-) + x(t_k^-))]. \end{aligned} \tag{3.63}$$

Let

$$B_*^k = \Big\{z : (-\infty, t_k] \to \mathbb{R}^n \mid z(t_i^-), z(t_i^+) \text{ exist with } z(t_i) = z(t_i^-), \\ z_k \in C(J_i, \mathbb{R}^n),\ i = 1,\ldots,k-1,\ z \in B \text{ and } z_0 = 0\Big\}.$$

For any $z \in B_*^k$, we have

$$\|z\|_k = \|z_0\|_B + \sup\{|z(s)| : 0 \leq s \leq t_k\} = \sup\{|z(s)| : 0 \leq s \leq t_k\}.$$

Thus, $(B_*^k, \|\cdot\|_k)$ is a Banach space. Set

$$\overline{C}_0 = \{z \in B_* : z_0 = 0\}.$$

Now $\overline{C}_0$ is a Fréchet space with a family of seminorms $\|\cdot\|_k$, and we define the operator $\overline{P} : \overline{C}_0 \to \overline{C}_0$ by

$$(\overline{P}z)(t) = \begin{cases} 0, & t \in (-\infty, 0], \\ t\eta + \int_0^t \int_0^s f(u, \bar{z}_u + x_u)duds & \\ \quad + \sum_{0<t_k<t}[I_k(\bar{z}(t_k^-) + x(t_k^-)) & \\ \quad +(t - t_k)\overline{I}_k(\bar{z}(t_k^-) + x(t_k^-))], & t \in [0, \infty). \end{cases}$$

That the operator $\overline{N}$ has a fixed point is equivalent to $\overline{P}$ having a fixed point is clear. We shall use the nonlinear alternative to prove this.

Let z be a possible solution of the problem, $z = \gamma\overline{P}(z)$ for some $0 < \gamma < 1$. For each $t \in [0, t_1]$, we then have

$$z(t) = \gamma\Big[t\eta + \int_0^t \int_0^s f(u, \bar{z}_u + x_u)duds + \sum_{0<t_k<t} [I_k(\bar{z}(t_k^-) + x(t_k^-)) \\ + (t - t_k)\overline{I}_k(\bar{z}(t_k^-) + x(t_k^-))]\Big].$$

As in Theorem 3.17, we can show that there exists $\overline{M}_1 > 0$ such that $\|z\|_1 \leq \overline{M}_1$. Now if $t \in (t_1, t_2]$, then

$$z(t) = \gamma\Big[t\eta + \int_0^t \int_0^s f(u, \bar{z}_u + x_u)duds + I_1(z(t_1^-) + x(t_1^-)) \\ + (t - t_1)\overline{I}_1(z(t_1^-) + x(t_1^-))\Big].$$

Note that

$$|I_1(z(t_1^-) + x(t_1^-)) + (t - t_1)\overline{I}_1(z(t_1^-) + x(t_1^-))| \\ \leq \sup_{x \in B(0,\overline{M}_1)} [|I_1(x)| + (t_2 - t_1)\overline{I}_1(x)] := \overline{M},$$

where

$$B(0, \overline{M}_1) = \{x \in \mathbb{R}^n : |x| \leq \overline{M}_1\}.$$

Hence,

$$|z(t)| \leq M_1 + \overline{M} + t_2 \int_{t_1}^t p(s)\psi(\|\bar{z}_s + x_s\|_B)ds,$$

where

$$\|\bar{z}_s + x_s\|_B \leq K_2 \sup_{s\in[0,t]} |z(s)| + K_2 \sup_{s\in[0,t]} |x(s)| + M_2\|x_0\|_B := h(t),$$

$h(t_*) = K_2 \sup_{s\in[0,t_*]} |z(s)| + K_2 \sup_{s\in[0,t_*]} |x(s)| + M_2\|x_0\|_B$. By the previous inequality, for $t \in [0, t_2]$ we have

$$\begin{aligned} h(t) \leq K_2 t_2 \int_0^t p(s)\psi(h(s))ds &+ K_2[\overline{M}_1 + (t_2 - t_1)\overline{M}] \\ &+ K_2 \sup_{s\in[0,t_*]} |x(s)| + M_k\|x_0\|_B. \end{aligned}$$

Let us denote the right-hand side of the above inequality by $v(t)$. Then we have

$$v(0) = K_2[\overline{M}_1 + (t_2 - t_1)\overline{M}] + K_2 \sup_{s\in[0,t_*]} |x(s)| + M_2\|x_0\|_B$$

and

$$v'(t) = K_2 t_2 p(t)\psi(h(t)), \quad \text{a.e.} \quad t \in [0, t_2].$$

Using the increasing character of ψ, we get

$$v'(t) \leq K_2 t_2 p(t)\psi(h(t)) \quad \text{a.e.} \quad t \in [0, t_2].$$

Then, for each $t \in [0, t_2]$, we have

$$\int_{v(0)}^{h(t)} \frac{du}{\psi(u)} \leq K_2 t_2 \int_0^\infty p(s)ds < \infty.$$

Consequently, by (H-4), there exists a constant M_* such that $v(t) \leq M_*$, $t \in [0, t_2]$, and hence

$$\|z\|_2 \leq M_*/K_2 := \overline{M}_2.$$

We continue this process and take into account that $t \in [0, t_{k+1}]$, $k > 2$. Then,

$$\begin{aligned} z(t) = \gamma\Bigg[t\eta + \int_{t_k}^t \int_{t_k}^s f(u, \bar{z}_u + x_u)duds \\ + \sum_{0<t_k<t} [I_k(z(t_k^-) + x(t_k^-) + (t - t_k)\overline{I}_k(z(t_k^-) + x(t_k^-))\Bigg]. \end{aligned}$$

We obtain that there exists a constant $\overline{M}_{k+1}$ such that

$$\sup\{|z(t)| : t \in [0, t_{k+1}]\} \leq \overline{M}_{k+1}.$$

Let

$$Y = \{z \in \overline{C}_0 : \sup\{|z(t)| : 0 \leq t \leq t_k\} \leq \overline{M}_k + 1 \text{ for all } k = 1, 2, \ldots\}.$$

Clearly, Y is a closed subset of C_0. We shall show that $\overline{P} : Y \to B_*^k$ is a contraction map. Indeed, consider $z, z^* \in Y$. Then we have, for each $t \in [0, t_k]$ and $k \in \{1, 2, \ldots\}$,

$$\begin{aligned}
|\overline{P}(z)(t) - \overline{P}(z^*)(t)| &\leq \int_0^t |f(s, \bar{z}_s + x_s)) - f(s, \bar{z}_s^* + x_s)| ds \\
&\quad + \sum_{i=1}^{i=k} |I_i(\bar{z}(t_i^-) + x(t_i)) - I_i(\bar{z}^*(t_i^-) + x(t_i^-))| \\
&\quad + \sum_{0<t_k<t}^{i=k} (t - t_k)|I_i(\bar{z}(t_i^-) + x(t_i)) \\
&\quad - I_i(\bar{z}^*(t_i^-) + x(t_i^-))| \\
&\leq \int_0^t l(s)\|\bar{z}_s - \bar{z}_s^*\|_B ds + \sum_{i=1}^{i=k} d_i |z(t_i^-) - z^*(t_i^-)| \\
&\quad + \sum_{0<t_k<t} (t - t_k)\overline{d}_k |z(t_i^-) - z^*(t_i^-)| \\
&\leq \int_0^t l(s) K_k \sup_{s\in[0,t]} |z(s) - z^*(s)| ds \\
&\quad + \sum_{i=1}^{i=k} d_i |z(t_i^-) - z^*(t_i^-)| \\
&\quad + \sum_{0<t_k<t} (t - t_k)\overline{d}_k |z(t_i^-) - z^*(t_i^-)| \\
&\leq \int_0^t \frac{1}{\tau}\widehat{L}(s) e^{\tau\widehat{L}(s)} e^{-\tau\widehat{L}(s)} \sup_{s\in[0,t]} |z(s) - z^*(s)| ds \\
&\quad + \sum_{i=1}^{i=k} d_i e^{\tau\widehat{L}(t)} e^{-\tau\widehat{L}(t)} \sup_{s\in[0,t_k]} |z(s) - z^*(s)| \\
&\quad + \sum_{0<t_k<t} (t - t_k)\overline{d}_k e^{\tau\widehat{L}(t_k)} e^{-\tau\widehat{L}(t_k)} \times \\
&\quad \sup_{s\in[0,t_k]} |z(s) - z^*(s)|.
\end{aligned}$$

Since $(t-s) \leq e^{(t-s)}$ for $0 \leq s \leq t$, we see that $(t-s)e^s \leq e^t$. Let $\widehat{L}(t) = \int_0^t l_*(s) ds$, and $l_*(t) = \max\{1, K_k l(t)\}$. Assume that τ is sufficiently large such that,

$$t - s = \int_s^t d\xi \leq \int_s^t l_*(\xi) d\xi = \widehat{L}(t) - \widehat{L}(s).$$

Then,

$$\tau(t - s) \leq \tau\widehat{L}(t) - \tau\widehat{L}(s).$$

This implies that

$$\tau(t-s)e^{\tau\widehat{L}(s)} \leq e^{\tau\widehat{L}(t)}$$

so

$$(t-s)e^{\tau\widehat{L}(s)} \leq \frac{e^{\tau\widehat{L}(t)}}{\tau},$$

Thus,

$$\begin{aligned} e^{-\tau\widehat{L}(t)}|\overline{P}(z)(t)-\overline{P}(z^*)(t)| &\leq \frac{1}{\tau}\|z-z^*\|_{B^k_*} + \sum_{i=1}^{i=k} d_i\|z-z^*\|_{B^k_*} \\ &\quad + \frac{1}{\tau}\sum_{i=1}^{i=k}\overline{d}_i\|z-z^*\|_{B^k_*}, \\ &\leq \left(\frac{1}{\tau}+\sum_{i=1}^{i=k} d_i + \frac{1}{\tau}\sum_{i=1}^{i=k}\overline{d}_i\right)\|z-z^*\|_{B^k_*}. \end{aligned}$$

Therefore,

$$\|\overline{P}(z)-\overline{P}(z^*)\|_{B^k_*} \leq \left(\frac{1}{\tau}+\sum_{i=1}^{i=k} d_i + \frac{1}{\tau}\sum_{i=1}^{i=k}\overline{d}_i\right)\|z-z_*\|_{B^k_*},$$

where $\|\cdot\|_{B^k_*}$ is the Bielecki-type norm on B^k_* defined by

$$\|z\|_{B^k_*} = \max\{|z(t)|e^{-\tau\widehat{L}(t)} : t \in [0,t_k]\}$$

with $\tau > 1$. From the choice of Y, there is no $z \in \partial Y^n$ such that $z = \lambda\overline{P}(z)$, for some $\lambda \in (0,1)$. As a consequence of the nonlinear alternative type [184], we deduce that $\overline{P}$ has a unique fixed point which is a solution to (3.56)–(3.59). □

3.5.2 Example

In this section, we give an example to illustrate our main results. Consider the problem

$$\begin{aligned} y''(t) &= \frac{e^{-\gamma t}y_t}{(t+1)(t+2)}\int_{-\infty}^{-t} e^{\gamma\theta}y(\theta+t)d\theta, && \text{a.e. } t \in \mathbb{R}_+, && (3.64) \\ y(t_k^+)-y(t_k^-) &= b_k y(t_k^-), && k=1,\ldots, && (3.65) \\ y'(t_k^+)-y'(t_k^-) &= \bar{b}_k y(t_k^-), && k=1,\ldots, && (3.66) \\ y(t) &= \phi(t), && t \in (-\infty,0]. && (3.67) \end{aligned}$$

Let $\mathcal{D} = \{\psi : (-\infty,0] \to \mathbb{R}^n \mid \psi$ is continuous everywhere, except for a countable number of points $\bar{t}$ at which $\psi(\bar{t}^-)$ and $\psi(\bar{t}^+)$ exist, $\psi(\bar{t}^-) = \psi(\bar{t})\}$, and b_k and

$\bar{b}_k, k = 1, \ldots,$ are real numbers with

$$\sum_{k=1}^{\infty} b_k < 1 \quad \text{and} \quad \sum_{k=1}^{\infty} \bar{b}_k < \infty.$$

Let γ be a positive real constant and B_γ is the phase space defined in Example 3.4. The norm in B_γ is given by

$$\|y\|_\gamma = \int_{-\infty}^{0} e^{\gamma\theta}\|y(\theta)\|d\theta.$$

Then, $(B_\gamma, \|\cdot\|_\gamma)$ is a Banach space that will be the phase space for our example.

We want to show that $f(t,u) = \frac{1}{(t+1)(t+2)}\int_{-\infty}^{-t} e^{\gamma\theta}u(t+\theta)d\theta$ satisfies

$$|f(t,u)| \le p(t)\psi(\|u\|_{B_\gamma}) \quad \text{for} \quad (t,u) \in [0,\infty) \times B_\gamma$$

with $\psi(x) = 1 + x$ and $p(t) = \frac{1}{(1+t)(2+t)}$. We have

$$|f(t,u)| \le p(t)[\|u\|_{B_\gamma} + 1],$$

and

$$\int_0^\infty p(t)dt \le \int_0^\infty \frac{dt}{(1+t)(2+t)} = \log 2 < \infty \quad \text{and} \quad \int_0^\infty \frac{dx}{\psi(x)} = \infty.$$

Let x, $y \in B_\gamma$; then we have

$$|f(t,x) - f(t,y)| \le \frac{1}{(t+1)}\|x - y\|_{B_\gamma}.$$

Hence, the hypotheses of Theorem 3.20 are satisfied, and our problem has a unique solution.

3.5.3 Stability

We consider the system

$$y''(t) = f(t, y_t), \qquad t \in [t_0, \infty)\backslash\{t_1, t_2, \ldots\}, \tag{3.68}$$

$$y(t_k^+) - y(t_k^-) = I_k(y(t_k^-)), \qquad k = 1, \ldots, \tag{3.69}$$

$$y'(t_k^+) - y'(t_k^-) = \bar{I}_k(y(t_k^-)), \qquad k = 1, \ldots, \tag{3.70}$$

$$y(t - t_0) = \phi(t), \qquad t \in (-\infty, 0], \; y'(t_0) = 0, \tag{3.71}$$

where f, I_k, ϕ, I_k are as in problem (3.34)–(3.36), and $\bar{I}_k \in C(\mathbb{R}^n, \mathbb{R}^n)$. Assume that $f(t,0) = 0$, $I_k(0) = 0$ and $\bar{I}_k(0) = 0$, for all $k = 1, \ldots$.

It is clear that y is the unique solution of the system (3.68)–(3.71) if and only if y is a solution of impulsive integral functional differential equation

$$y(t) = \begin{cases} \phi(t - t_0), & \text{if } t \in (-\infty, t_0] \\ \phi(0) + \int_{t_0}^{t} (t - s) f(s, y_s) ds & \\ \quad + \sum_{t_0 < t_k < t} [I_k(y(t_k^-)) + (t - t_k)\bar{I}_k(y(t_k^-))], & \text{if } t \in [t_0, \infty). \end{cases} \tag{3.72}$$

Theorem 3.21. *In addition to* (C-3) *and* (C-4) *assume that*

(C-5) *There exist constants* $\overline{d}_k > 0, k = 1, 2, \ldots,$ *such that for all* $x \in \mathbb{R}^n$,

$$|\bar{I}_k(x)| \le \overline{d}_k |x|, \; x \in \mathbb{R}^n, \quad \sum_{k=1}^{\infty} \overline{d}_k < \infty.$$

If

$$\sum_{k=1}^{\infty} c_k < 1, \quad |K(t)| < Re^{-\tau t} \text{ for all } t \ge t_0, \quad \tau \gg \sum_{k=1}^{\infty} \overline{d}_k,$$

and

$$M_\infty = \sup\{|M(t)| : t \in [0, \infty)\} < \infty,$$

then the zero solution of the problem (3.68)–(3.71) is stable.

Proof. Let y be a solution of the system (3.68)–(3.71). Then for $t \in [t_0, \infty)$, we have

$$\|y_t\|_B \le K(t - t_0) \sup_{t_0 \le s \le t} |y(s)| + M(t - t_0)\|y_{t_0}\|_B.$$

For $s \in [t_0, t]$, we have

$$\begin{aligned} |y(s)| &\le |\phi(0)| + \int_{t_0}^{s} (s - u)|f(u, y_u)| du \\ &\quad + \sum_{t_0 < t_k < s} |I_k(y(t_k^-))| + \sum_{t_0 < t_k < s} (s - t_k)|\bar{I}_k(y(t_k^-))| \\ &\le |\phi(0)| + \int_{t_0}^{s} (s - u) l(u) \|y_u\|_B du + \sum_{t_0 < t_k < t} c_k \sup_{s \in [t_0, t]} |y(s)| \\ &\quad + \sum_{t_0 < t_k < t} \frac{\tau(s - t_k)}{\tau} \overline{d}_k |y(t_k^-)| \end{aligned}$$

$$\le H\|\phi\|_B + \int_{t_0}^{s} \frac{1}{\tau} e^{\tau(s-u)} l(u) K(u-t_0) \sup_{u\in[t_0,s]} |y(u)| du$$
$$+ \|\phi\|_B \int_{t_0}^{s} l(u) M(u-t_0) du$$
$$+ \sum_{t_0<t_k<s} c_k \sup_{s\in[t_0,s]} |y(s)| + \sum_{t_0<t_k<s} \frac{1}{\tau} e^{\tau(s-t_k)} \overline{d}_k \sup_{u\in[0,s]} |y(s)|.$$

Hence,

$$\sup_{s\in[t_0,t]} e^{-\tau s}|y(s)| \le \frac{1}{1-\sum_{k=1}^{\infty}[c_k - \frac{1}{\tau}\overline{d}_k]}$$
$$\times \left[H\|\phi\|_B + R\int_{t_0}^{t} \frac{1}{\tau} l(s) \sup_{u\in[t_0,s]} e^{-u\tau}|y(u)| du + \|\phi\|_B M_\infty \|l\|_{L^1} \right].$$

By Gronwall's inequality [267], we obtain

$$\sup_{s\in[t_0,t]} e^{-\tau s}|y(s)| \le c\|\phi\|_B e^{\|l^*\|_{L^1}},$$

where

$$c = \frac{1}{1-\sum_{k=1}^{\infty}[c_k - \frac{1}{\tau}d_k]} [H + M_\infty \|l\|_{L^1}] \quad \text{and} \quad l^*(t) = \frac{1}{\tau} R l(t).$$

Thus,

$$\|y_t\|_B \le \left(Re^{\tau t_0} c e^{\|l^*\|_{L^1}} + M_\infty\right) \|\phi\|_B.$$

Let $\epsilon > 0$ be given and choose $\delta = \frac{\epsilon}{Re^{\tau t_0} c e^{\|l^*\|_{L^1}} + M_\infty}$; then from the above inequality, we obtain that for every $\phi \in B$,

$$\|\phi\|_B \le \delta \quad \text{implies} \quad \|y_t\|_B \le \epsilon.$$

Thus, the system is stable. □

Chapter 4

Boundary Value Problems on Infinite Intervals

4.1 Introduction

The history of boundary value problems on infinite intervals starts at the end of the 1800's with the pioneering work of A. Kneser [324] about monotone solutions and their derivatives on $[0, \infty)$ for second-order ordinary differential equations. The Kneser-type results were then followed by A. Mambriani [364] in 1929 and others from the beginning of the 1950's; see [249, 270, 425–427, 438, 448, 481], and the references therein. Many papers have been devoted to the study of ordinary impulsive boundary value problems. Several existence results have been presented for these equations based on different techniques: upper and lower solutions, monotone iterations, degree theory, fixed points theorems, topological transversality, etc.; see for instance [126, 127, 200, 210, 212, 221, 241, 242, 252, 258, 304, 388, 428, 495], and the references therein.

Recently, boundary value problems on infinite intervals appear in many problems of practical interest; for example, in linear elasticity, nonlinear fluid flow, and foundation engineering; see [7, 251, 252, 350, 351, 370, 371] and the references therein.

In this chapter, we study the existence and uniqueness of solutions to first order boundary value problems for impulsive functional differential equations with infinite delay,

$$y'(t) = f(t, y_t), \quad \text{a.e.} \quad t \in J := [0, \infty),\ t \neq t_k,\ k = 1, \ldots, \tag{4.1}$$

$$y(t_k^+) - y(t_k^-) = I_k(y(t_k^-)), \quad t = t_k,\ k = 1, \ldots, \tag{4.2}$$

$$Ay(t) - y_\infty = \phi(t), \quad t \in (-\infty, 0], \tag{4.3}$$

where f, B are as in problem (3.1)–(3.3), and $I_k \in C(\mathbb{R}^n, \mathbb{R}^n)$, $k \in \mathbb{N}$, $\lim_{t\to\infty} y(t) = y_\infty$, $A > 1$, and $\phi \in B$.

Fixed point arguments such as the Banach contraction principle, fixed point index theory and monotone iterative techniques have been applied to first and second order impulsive differential equations. Our goal here is to give existence results for the above problem by using the nonlinear alternative of Leray–Schauder type, followed by a uniqueness result. The chapter concludes with an example. In all this chapter we assume that and $t \to f(t, y_t)$ is a measurable function for every $y_t \in B$.

4.1.1 Existence Result

Let us start by defining what we mean by a solution of the problem (4.1)–(4.3). Set

$$B_* = \{y : (-\infty, \infty) \to \mathbb{R}^n : y \in PC \cap B\}.$$

Definition 4.1. A function $y \in B_*$ is said to be a solution of (4.1)–(4.3) if y satisfies (4.1)–(4.3).

In the proof of the existence result for the problem (4.1)–(4.3) we need the following auxiliary lemma.

Lemma 4.2. *Let* $f : C((-\infty, 0], \mathbb{R}^n) \to \mathbb{R}^n$ *be measurable,*

$$\int_0^\infty |f(y_s)|ds < \infty, \quad \text{and} \quad \sum_{k=1}^\infty I_k(x) < \infty, \quad x \in \mathbb{R}^n.$$

Then y *is a solution of the impulsive integral equation*

$$y(t) = \begin{cases} \frac{\phi(0)}{A(A-1)} + \frac{\phi(t)}{A} \\ \quad + \frac{1}{A-1}\left[\int_0^\infty f(y_s)ds + \sum_{k=1}^\infty I_k(y(t_k))\right], & t \in (-\infty, 0], \\ \frac{\phi(0)}{A-1} + \frac{1}{A-1}\left[\int_0^\infty f(y_s)ds + \sum_{k=1}^\infty I_k(y(t_k))\right] \\ \quad + \int_0^t f(y_s)ds + \sum_{0<t_k<t} I_k(y(t_k)), & t \in [0, \infty), \end{cases} \tag{4.4}$$

where $\lim_{t\to\infty} y(t) = y_\infty$, *if and only if* y *is a solution of the impulsive boundary value problem*

$$y'(t) = f(y_t); \qquad t \in [0, \infty),\ t \neq t_k,\ k = 1, \dots, \tag{4.5}$$

$$y(t_k^+) - y(t_k^-) = I_k(y(t_k^-)), \quad k = 1, \dots, \tag{4.6}$$

$$Ay(t) - y_\infty = \phi(t), \qquad t \in (-\infty, 0]. \tag{4.7}$$

Proof. Let y be a solution of the impulsive integral equation (4.4); then for $t \in [0, \infty)$ and $t \neq t_k$, $k = 1, \dots,$ we have

$$y(t) = \frac{\phi(0)}{A-1} + \frac{1}{A-1}\left[\int_0^\infty f(y_s)ds + \sum_{k=1}^\infty I_k(y(t_k))\right] + \int_0^t f(y_s)ds + \sum_{0<t_k<t} I_k(y(t_k)).$$

Thus $y'(t) = f(y_t)$ for $t \in [0, \infty)$ and $t \neq t_k$, $k = 1, \dots$. From the definition of y we can prove that

$$y(t_k^+) - y(t_k^-) = I_k(y(t_k)), \quad \text{for } k = 1, \dots .$$

Finally, we show that $Ay(t) - y_\infty = \phi(t)$ for $t \in (-\infty, 0]$. We have

$$\lim_{t\to\infty} y(t) = \frac{\phi(0)}{A-1} + \frac{A}{A-1}\left[\int_0^\infty f(y_s)ds + \sum_{k=1}^{\infty} I_k(y(t_k))\right],$$

and for $t \in (-\infty, 0]$,

$$y(t) = \frac{\phi(0)}{A(A-1)} + \frac{1}{A-1}\left[\int_0^\infty f(y_s)ds + \sum_{k=1}^{\infty} I_k(y(t_k))\right] + \frac{\phi(t)}{A}.$$

Hence,

$$\begin{aligned} Ay(t) - \lim_{t\to\infty} y(t) &= \frac{\phi(0)}{A-1} + \frac{A}{A-1}\left[\int_0^\infty f(y_s)ds + \sum_{k=1}^{\infty} I_k(y(t_k))\right] \\ &\quad + \phi(t) - \frac{\phi(0)}{A-1} - \frac{A}{A-1}\left[\int_0^\infty f(y_s)ds + \sum_{k=1}^{\infty} I_k(y(t_k))\right] \\ &= \phi(t), \quad t \in (-\infty, 0]. \end{aligned}$$

Now let y be a solution of the problem (4.5)–(4.7). Then,

$$y'(t) = f(y_t) \quad \text{for } t \in [0, t_1].$$

An integration from 0 to $t \le t_1$ of the above equality yields

$$y(t) - y(0) = \int_0^t f(y_s)ds.$$

Thus, for $t \in [0, t_1]$ we have

$$y(t) = y(0) + \int_0^t f(y_s)ds.$$

If $t \in (t_1, t_2]$, then we have

$$\begin{aligned} \int_0^{t_1} y'(s)ds + \int_{t_1}^t y'(s)ds, &= \int_0^t f(y_s)ds \\ y(t_1) - y(0) + y(t) - y(t_1^+) &= \int_0^t f(s)ds, \\ y(t) - I_1(y(t_1)) &= y(0) + \int_0^t f(y_s)ds. \end{aligned}$$

Hence, for $t \in (t_1, t_2]$, we have

$$y(t) = y(0) + \int_0^t f(y_s)ds + I_1(y(t_1)).$$

Continuing, for $t \in [0, \infty)$, we obtain

$$y(t) = y(0) + \int_0^t f(y_s)ds + \sum_{0<t_k<t} I_k(y(t_k)). \tag{4.8}$$

Since $\lim_{t\to\infty} y(t) = y_\infty$, we get

$$y_\infty = y(0) + \int_0^\infty f(y_s)ds + \sum_{k=1}^\infty I_k(y(t_k)).$$

Thus,

$$y(0) = y_\infty - \int_0^\infty f(y_s)ds - \sum_{k=1}^\infty I_k(y(t_k)).$$

By (4.7), we have $y_\infty = Ay(0) - \phi(0)$, and hence

$$y(0) = \frac{\phi(0)}{A-1} + \frac{1}{A-1}\left[\int_0^\infty f(y_s)ds + \sum_{k=1}^\infty I_k(y(t_k))\right]. \tag{4.9}$$

Substituting (4.9) into (4.8) yields

$$\begin{aligned} y(t) = \frac{\phi(0)}{A-1} &+ \frac{1}{A-1}\left[\int_0^\infty f(s)ds + \sum_{k=1}^\infty I_k(y(t_k))\right] \\ &+ \int_0^t f(y_s)ds + \sum_{0<t_k<t} I_k(y(t_k)) \end{aligned}$$

for $t \in [0, \infty)$. Now from (4.7) and (4.9), we have, for $t \in (-\infty, 0]$,

$$\begin{aligned} y(t) &= \frac{\phi(t)}{A} + \frac{1}{A}\left[y(0) + \int_0^\infty f(y_s)ds + \sum_{k=1}^\infty I_k(y(t_k))\right] \\ &= \frac{\phi(0)}{A(A-1)} + \frac{1}{A-1}\left[\int_0^\infty f(y_s)ds + \sum_{k=1}^\infty I_k(y(t_k))\right] + \frac{\phi(t)}{A}. \end{aligned}$$

□

Theorem 4.3. *Let $f : J \times B \to \mathbb{R}^n$ be an Carathéodory function. Assume that:*

(D-1) *There exist positive constants c_k, $k = 1, \ldots,$ such that*

$$|I_k(y)| \le c_k, \text{ for all } y \in \mathbb{R}^n, \text{ and } \sum_{k=1}^\infty c_k < \infty;$$

(D-2) *There exists $p \in L^1([0, +\infty), \mathbb{R}_+)$ such that*

$$|f(t, y)| \le p(t) \text{ for a.e. } t \in [0, \infty) \text{ and each } y \in B.$$

Then the boundary value problem (4.1)–(4.3) *has at least one solution.*

Proof. We transform the problem (4.1)–(4.3) into a fixed point problem. Consider the operator $N : B \longrightarrow B$ defined by,

$$N(y)(t) = \begin{cases} \frac{\phi(0)}{A(A-1)} + \frac{1}{A-1}\left[\int_0^\infty f(s, y_s)ds + \sum_{k=1}^\infty I_k(y(t_k))\right] + \frac{\phi(t)}{A}, \\ \qquad\qquad t \in (-\infty, 0], \\ \frac{\phi(0)}{A-1} + \frac{1}{A-1}\left[\int_0^\infty f(s, y_s)ds + \sum_{k=1}^\infty I_k(y(t_k))\right] \\ \quad + \int_0^t f(s, y_s)ds + \sum_{0<t_k<t} I_k(y(t_k)), \qquad t \in [0, \infty). \end{cases}$$

Let $x(\cdot) : (-\infty, +\infty) \to \mathbb{R}^n$ be the function defined by,

$$x(t) = \begin{cases} \frac{\phi(0)}{A-1} + \frac{1}{A-1}\left[\int_0^\infty f(s, x_s)ds + \sum_{k=1}^\infty I_k(x(t_k))\right], & \text{if } t \in [0, \infty), \\ \frac{\phi(0)}{A(A-1)} + \frac{1}{A-1}\left[\int_0^\infty f(s, x_s)ds + \sum_{k=1}^\infty I_k(x(t_k))\right]\frac{\phi(t)}{A}, \\ & \text{if } t \in (-\infty, 0]. \end{cases}$$

Then,

$$x_0 = \frac{\phi(0)}{A-1} + \frac{1}{A-1}\left[\int_0^\infty f(s, x_s)ds + \sum_{k=1}^\infty I_k(x(t_k))\right].$$

For each $z \in C([0, \infty), \mathbb{R}^n)$ with $z(0) = 0$, we denote by $\bar{z}$ the function defined by

$$\bar{z}(t) = \begin{cases} z(t), & \text{if } t \in [0, \infty), \\ 0, & \text{if } t \in (-\infty, 0]. \end{cases}$$

If $y(\cdot)$ satisfies the integral equation,

$$y(t) = \frac{\phi(0)}{A-1} + \frac{1}{A-1}\left[\int_0^\infty f(s, y_s)ds + \sum_{k=1}^\infty I_k(y(t_k))\right] + \int_0^t f(s, y_s)ds + \sum_{0<t_k<t} I_k(y(t_k)),$$

we can decompose $y(\cdot)$ as $y(t) = \bar{z}(t) + x(t), 0 \le t < \infty$. This implies $y_t = \bar{z}_t + x_t$, for every $0 \le t < \infty$, and the function $z(\cdot)$ satisfies

$$z(t) = \int_0^t f(s, \bar{z}_s + x_s)ds + \sum_{0<t_k<t} I_k(\bar{z}(t_k) + x(t_k)). \tag{4.10}$$

Set

$$C_0 = \{z \in B : z_0 = 0\}.$$

Let the operator $P : C_0 \to C_0$ be defined by

$$(Pz)(t) = \begin{cases} 0, & t \le 0, \\ \int_0^t f(s, \bar{z}_s + x_s)ds + \sum_{0<t_k<t} I_k(\bar{z}(t_k) + x(t_k)), & t \in [0, \infty). \end{cases}$$

Note that the operator N having a fixed point is equivalent to P having a fixed point. We shall use the Leray–Schauder alternative to prove this about P.

Step 1. *P is continuous.*

Let $\{z_n\}$ be a sequence such that $z_n \to z$ in C_0. Then

$$\begin{aligned}|(Pz_n)(t) - (Pz)(t)| \leq & \int_0^\infty |f(s, \bar{z}_{n_s} + x_s) - f(s, \bar{z}_s + x_s)| ds \\ & + \sum_{k=1}^{\infty} |I_k(z_{n_s} + x_s) - I_k(z_s + x_s)|.\end{aligned}$$

On the other hand, for any $\varepsilon > 0$, we can choose a positive integer m such that $\sum_{k=m+1}^{\infty} c_k < \varepsilon$. Then, choose an integer n_0 such that, for $k = 1, \ldots, m$,

$$\sum_{k=1}^{m} |I_k(\bar{z}_n(t_k) + x(t_k)) - I_k(\bar{z}(t_k) + x(t_k))| < \varepsilon \quad \text{for all} \quad n > n_0. \tag{4.11}$$

From (4.11), we have, for all $n > n_0$,

$$\sum_{k=1}^{\infty} |I_k(\bar{z}_n(t_k) + x(t_k)) - I_k(\bar{z}(t_k) + x(t_k))| < \varepsilon + \sum_{k=m+1}^{\infty} c_k < 2\varepsilon.$$

Thus,

$$\lim_{n \to \infty} \sum_{k=1}^{\infty} |I_k(\bar{z}_n(t_k) + x(t_k)) - I_k(\bar{z}(t_k) + x(t_k))| = 0. \tag{4.12}$$

Since f is L^1-Carathéodory, we have

$$\begin{aligned}\|P(z_n) - P(z)\|_\infty \leq & |f(\cdot, \bar{z}_{n_{(\cdot)}} + x_{(\cdot)}) - f(\cdot, \bar{z}_{(\cdot)} + x_{(\cdot)})|_{L^1} \\ & + \sum_{k=1}^{\infty} |I_k(\bar{z}_n(t_k) + x(t_k)) - I_k(\bar{z}(t_k) + x(t_k))| \\ & \to 0 \text{ as } n \to \infty.\end{aligned}$$

Step 2. *P maps bounded sets into bounded sets in C_0.*

Indeed, it is enough to show that for any $q > 0$, there exists a positive constant ℓ such that, for each $z \in \mathcal{B}_q = \{z \in C_0 : \|z\|_\infty \leq q\}$, we have $\|P(z)\|_\infty \leq \ell$. Let $z \in \mathcal{B}_q$. By (D-2), we have for each $t \in [0, \infty)$,

$$|(Pz)(t)| \leq \int_0^\infty p(s) ds + \sum_{k=1}^{\infty} c_k := \ell.$$

Step 3. *P maps bounded sets into equicontinuous sets of C_0.*

Let $l_1, l_2 \in [0,\infty)$, $l_1 < l_2$ and $\mathcal{B}_q$ be a bounded set in C_0 as in Claim 2. Let $z \in \mathcal{B}_q$. Then for each $t \in [0,\infty)$, we have

$$\begin{aligned}|(Pz)(l_2) - (Pz)(l_1)| &\le \int_{l_1}^{l_2} |f(s, \bar{z}_s + x_s)|ds \\ &\quad + \sum_{0<t<l_2-l_1} |I_k(\bar{z}(t_k) + x(t_k))| \\ &\le \int_{l_1}^{l_2} p(s)ds + \sum_{0<t<l_2-l_1} c_k.\end{aligned}$$

We see that $|(Pz)(l_2) - (Pz)(l_1)|$ tends to zero independently of $z \in \mathcal{B}_q$, as $l_2 - l_1 \to 0$. As a consequence of Steps 1–3, together with the fact that $P(\mathcal{B}_q)$ is equiconvergent at ∞, and the Arzelá–Ascoli theorem, we can conclude that $P : C_0 \to C_0$ is continuous and completely continuous.

Step 4. *A priori bounds on solutions.*

Let z be a possible solution of the equation $z = \lambda P(z)$ with $z_0 = \lambda\phi$ for some $\lambda \in (0,1)$. Then,

$$\begin{aligned}|z(t)| &\le \int_0^t |f(s, \bar{z}_s + x_s)|ds + \sum_{0<t_k<t} |I_k(\bar{z}(t_k)) + x(t_k))| \\ &\le \int_0^t p(s)ds + \sum_{k=1}^{\infty} c_k.\end{aligned}$$

This implies that, for each $t \in [0,\infty)$,

$$\|z\|_\infty \le \|p\|_{L^1} + \sum_{k=1}^{\infty} c_k := K_1.$$

Set

$$U = \{z \in C_0 : \sup\{|z(t)| : 0 \le t \le b\} < K_1 + 1\}.$$

We see that $P : \overline{U} \to C_0$ is continuous and completely continuous. From this choice of U, there is no $z \in \partial U$ such that $z = \lambda P(z)$, for some $\lambda \in (0,1)$. As a consequence of the nonlinear alternative of Leray–Schauder type [184], we deduce that P has a fixed point z in U. Hence, N has a fixed point y that is a solution to problem (4.1)–(4.3). □

4.1.2 Uniqueness Result

Next, we give sufficient conditions for existence and uniqueness of the solutions of the problem the (4.1)–(4.3). For more details about this problem, see Graef and Ouahab [242].

We consider

$$B_k = \{y \in B_* : \sup_{t\in J_k^*} |y(t)| < \infty\},\ J_k^* = (-\infty, t_k],$$

and let $\|\cdot\|_k$ be the seminorm in B_k defined by

$$\|y\|_k = \|y_0\|_B + \sup\{|y(t)| : 0 \le s \le t_k\},\ y \in B_k.$$

Now, we are in position to give our result.

Theorem 4.4. *Let $f : J \times B \to \mathbb{R}^n$ be an Carathéodory function. Assume that:*

(D-3) *There exist constants $d_k > 0$ such that*

$$|I_k(x) - I_k(\overline{x})| \le d_k|x - \overline{x}|, \text{ for } k = 1, \dots, \text{ for all } x, \overline{x} \in \mathbb{R}^n;$$

(D-4) *There exist $l \in L^1([0,\infty), \mathbb{R}^n)$ such that*

$$|f(t,x) - f(t,\overline{x})| \le l(t)\|x - \overline{x}\|_B, \text{ for } t \in [0,\infty), \text{ and } x, \overline{x} \in B;$$

(D-5) *There exist a function $p \in L^1([0,\infty), \mathbb{R}_+)$ and positive constants c_k, $k = 1, \dots,$ such that*

$$|f(t,u)| \le p(t) \text{ for each } (t,u) \in [0,\infty) \times B,$$

and

$$|I_k(z)| \le c_k, \text{ for all } z \in \mathbb{R}^n, \text{ and } \sum_{k=1}^{\infty} c_k < \infty.$$

If $\sum_{k=1}^{\infty} d_k < 1$, then the initial value problem (4.1)–(4.3) *has a unique solution.*

Proof. To transform the problem (4.1)–(4.3) into a fixed point problem, consider the operator $N : B_* \longrightarrow B_*$ defined by,

$$N(y)(t) = \begin{cases} \frac{\phi(0)}{A(A-1)} + \frac{1}{A-1}\left[\int_0^\infty f(s,y_s)ds + \sum_{k=1}^{\infty} I_k(y(t_k^-))\right] + \frac{\phi(t)}{A}, & t \in (-\infty, 0], \\ \frac{\phi(0)}{A-1} + \frac{1}{A-1}\left[\int_0^\infty f(s,y_s)ds + \sum_{k=1}^{\infty} I_k(y(t_k^-))\right] \\ \quad + \int_0^t f(s,y_s)ds + \sum_{0<t_k<t} I_k(y(t_k^-)), & t \in [0,\infty). \end{cases}$$

Let $x(\cdot) : (-\infty,\infty) \to \mathbb{R}^n$ be the function defined by

$$x(t) = \begin{cases} \frac{\phi(0)}{A-1} + \frac{1}{A-1}\left[\int_0^\infty f(s,x_s)ds + \sum_{k=1}^{\infty} I_k(x(t_k^-))\right], & t \in [0,\infty), \\ \frac{\phi(0)}{A(A-1)} + \frac{1}{A-1}\left[\int_0^\infty f(s,x_s)ds + \sum_{k=1}^{\infty} I_k(x(t_k^-))\right] + \frac{\phi(t)}{A}, & t \in (-\infty, 0]. \end{cases}$$

Then

$$x_0 = \frac{\phi(0)}{A-1} + \frac{1}{A-1}\left[\int_0^\infty f(s, x_s)ds + \sum_{k=1}^{\infty} I_k(x(t_k^-))\right].$$

For each $z \in C([0,\infty), \mathbb{R}^n)$ with $z(0) = 0$, we denote by $\bar{z}$ the function defined by

$$\bar{z}(t) = \begin{cases} z(t), & \text{if } t \in [0,\infty), \\ 0, & \text{if } t \in (-\infty, 0]. \end{cases}$$

If $y(\cdot)$ satisfies the integral equation,

$$y(t) = \frac{\phi(0)}{A-1} + \frac{1}{A-1}\left[\int_0^\infty f(s, y_s)ds + \sum_{k=1}^{\infty} I_k(y(t_k^-))\right] + \int_0^t f(s, y_s)ds + \sum_{0<t_k<t} I_k(y(t_k^-)),$$

we can decompose $y(\cdot)$ as $y(t) = \bar{z}(t) + x(t)$, $0 \le t < \infty$. Thus, $y_t = \bar{z}_t + x_t$, for every $0 \le t < \infty$, and the function $z(\cdot)$ satisfies

$$z(t) = \int_0^t f(s, \bar{z}_s + x_s)ds + \sum_{0<t_k<t} I_k(\bar{z}(t_k^-) + x(t_k^-)). \tag{4.13}$$

Let

$$B_*^k = \{z \in B^k : z_0 = 0\}.$$

For any $z \in B_*^k$, we have

$$\|z\|_k = \|z_0\|_B + \sup\{|z(s)| : 0 \le s \le t_k\} = \sup\{|z(s)| : 0 \le s \le t_k\}.$$

Thus, $(B_*^k, \|\cdot\|_k)$ is a Banach space. Set

$$C_0 = \{z \in B_* : z_0 = 0\}.$$

Observe that C_0 is a Fréchet space with a family of seminorms $\|\cdot\|_k$. Let the operator $P : C_0 \to C_0$ be defined by

$$(Pz)(t) = \begin{cases} 0, & t \le 0, \\ \int_0^t f(s, \bar{z}_s + x_s)ds + \sum_{0<t_k<t} I_k(\bar{z}(t_k^-) + x(t_k^-)), & t \in [0,\infty). \end{cases}$$

The operator N having a fixed point is equivalent to P having a fixed point, and so we use the nonlinear alternative to prove this.

Let z be a possible solution of the problem $z = \gamma P(z)$ for some $0 < \gamma < 1$. For $t \in [0,\infty)$, we have

$$z(t) = \int_0^t f(s, \bar{z}_s + x_s)ds + \sum_{0<t_k<t} I_k(\bar{z}(t_k^-) + x(t_k^-)).$$

This implies, by (D-5), that for each $t \in [0, t_k]$, we have

$$\begin{aligned}|z(t)| &\le \int_0^t p(s)ds + \sum_{i=1}^{i=k} I_i(\bar{z}(t_i) + x(t_i))| \\ &\le \int_0^t p(s)ds + \sum_{i=1}^{i=k} c_i := M_k^*.\end{aligned}$$

Hence,

$$\sup\{|z(t)| : t \in [0, t_k]\} \le M_k^*.$$

Set

$$Y = \{z \in C_0 : \|z\|_{B_*^k} : 0 \le t \le t_k\} \le M_k^* \quad \text{for all} \quad k = 1, 2 \ldots\},$$

where $\widehat{l}(t) = \int_0^t \widetilde{l}(s)ds$, $\widetilde{l}(t) = K_k l(t)$, and $\|\cdot\|_{B_*^k}$ is the Bielecki-type norm on B_*^k defined by

$$\|z\|_{B_*^k} = \sup\{e^{-\tau \widehat{l}(t)}|z(t)| : t \in [0, t_k]\}, \quad \tau > 1.$$

Clearly, Y is a closed subset of C_0. We shall show that $P : Y \to B_*^k$ is a contraction map. Indeed, consider $z, z^* \in Y$. Then we have for each $t \in [0, t_k]$ and $k = 1, \ldots,$

$$\begin{aligned}|P(z)(t) - P(z^*)(t)| &\le \int_0^t |f(s, \bar{z}_s + x_s)) - f(s, \bar{z}_s^* + x_s)|ds \\ &\quad + \sum_{i=1}^{i=k} |I_i(\bar{z}(t_i^-) + x(t_i)) - I_i(\bar{z}^*(t_i^-) + x(t_i^-))| \\ &\le \int_0^t l(s)\|\bar{z}_s - \bar{z}_s^*\|_B ds + \sum_{i=1}^{i=k} d_i |z(t_i^-) - z^*(t_i^-)| \\ &\le \int_0^t l(s) K_k \sup_{s \in [0,t]} |z(s) - z^*(s)| ds + \sum_{i=1}^{i=k} d_i |z(t_i^-) - z^*(t_i^-)| \\ &\le \int_0^t \widetilde{l}(s) e^{\tau \widehat{l}(s)} e^{-\tau \widehat{l}(s)} \sup_{s \in [0,t]} |z(s) - z^*(s)| ds \\ &\quad + \sum_{i=1}^{i=k} d_i e^{\tau \widehat{l}(t)} e^{-\tau \widehat{l}(t)} \sup_{s \in [0,t_k]} |z(s) - z^*(s)| \\ &\le \int_0^t \widetilde{l}(t) e^{\tau \widehat{l}(t)} ds \|z - z^*\|_{B_*^k} + \sum_{i=1}^{i=k} d_i e^{\tau \widehat{l}(t)} \|z - z^*\|_{B_*^k} \\ &\le \frac{1}{\tau} \int_0^t (e^{\tau \widehat{l}(s)})' ds \|z - z^*\|_{B_*^k} + \sum_{i=1}^{i=k} d_i e^{\tau \widehat{l}(t)} \|z - z^*\|_{B_*^k}\end{aligned}$$

$$\leq \frac{1}{\tau} e^{\tau \widehat{l}(t)} ds \|z - z^*\|_{B_*^k} + \sum_{i=1}^{i=k} d_i e^{\tau \widehat{l}(t)} \|z - z^*\|_{B_*^k}.$$

Thus,

$$\begin{aligned} e^{-\tau \widehat{l}(t)} \|P_1(z)(t) - P(z^*)(t)\| &\leq \frac{1}{\tau} \|z - z^*\|_{B_*^k} + \sum_{i=1}^{i=k} d_i \|z - z^*\|_{B_*^k} \\ &\leq \left(\frac{1}{\tau} + \sum_{i=1}^{i=k} d_i\right) \|z - z^*\|_{B_*^k}. \end{aligned}$$

Therefore,

$$\|P(z) - P(z^*)\|_{B_*^k} \leq \left(\frac{1}{\tau} + \sum_{i=1}^{i=k} d_i\right) \|z - z_*\|_{B_*^k}.$$

From the choice of Y, there is no $z \in \partial Y^n$ such that $z = \lambda P(z)$, for some $\lambda \in (0, 1)$. As a consequence of the Nonlinear Alternative of Section 2.4, we deduce that P has a unique fixed point that is a solution to (4.1)–(4.3). □

4.1.3 Example

In this subsection, we give an example to illustrate the usefulness of our main results. Consider the boundary value problem:

$$y'(t) = \frac{\|y_t\|_\gamma}{(t+1)(t+2)(1+\|y_t\|_\gamma)}, \quad \text{a.e.} \quad t \in J := [0, \infty)]/\{t_1, t_2, \ldots\}, \tag{4.14}$$

$$y(t_k^+) - y(t_k^-) = b_k y(t_k^-), \qquad k = 1, \ldots, m, \tag{4.15}$$

$$y(t) - Ay_\infty = \phi(t), \qquad t \in (-\infty, 0], \tag{4.16}$$

where b_k, $k = 1, 2, \ldots, m$, and $A >$ are real numbers. Let $\mathcal{D} = \{\psi : (-\infty, 0] \to \mathbb{R}$, ψ is continuous everywhere except for a countable number of points $\bar{t}$ at which $\psi(\bar{t}^-)$ and $\psi(\bar{t}^+)$ exist, and $\psi(\bar{t}^-) = \psi(\bar{t})\}$. Let $(B_\gamma, \|\cdot\|)$ be the Banach (phase) space defined in Example 3.4. With

$$f(t, x) = \frac{\|x\|_\gamma}{(t+1)(t+2)(1+\|x\|_\gamma)}, \quad (t, x) \in [0, \infty) \times B_\gamma,$$

we see that

$$|f(t, u)| \leq p(t),$$

where $p(t) = \frac{1}{(1+t)(2+t)}$, and

$$\int_0^\infty p(t)dt = \int_0^\infty \frac{dt}{(1+t)(2+t)} = \ln 2 < \infty.$$

Let $x, y \in B_\gamma$; then we have

$$\begin{aligned}|f(t,x) - f(t,y)| &= \frac{1}{(t+1)(t+2)}\left|\frac{\|x\|_\gamma}{1+\|x\|_\gamma} - \frac{\|y\|_\gamma}{1+\|y\|_\gamma}\right| \\ &= \frac{\big|\|x\|_\gamma - \|y\|_\gamma\big|}{(t+1)(t+2)(1+\|x\|_\gamma)(1+\|y\|_\gamma)} \\ &\le \frac{\|x-y\|_\gamma}{(t+1)(t+2)(1+|x|_\gamma)(1+|y|_\gamma)} \\ &\le \frac{\|x-y\|_\gamma}{(t+1)(t+2)}.\end{aligned}$$

Hence, the conditions (D-4) and (D-5) of Theorem 4.4 are satisfied.

Assume that $\sum_{k=1}^{\infty} b_k < 1$. Then, by Theorem 4.4, the problem (4.14)–(4.16) has a unique solution.

Chapter 5

Differential Inclusions

5.1 Introduction

It is well known that in $C([0,\infty),\mathbb{R}^n)$, the distance between two trajectories of an ordinary differential equation $y' = f(x)$ (with a Lipschitz continuous vector field) is majorized by the distance between two initial points multiplied by an exponential function of time. The object of this chapter is to prove a Filippov type theorem and a Filippov–Wazewski type theorem for impulsive differential inclusions. More precisely, we consider the problem

$$y' \in F(t,y(t)), \qquad \text{a.e.} \quad t \in J := [0,b],\ t \neq t_k,\ \ k = 1,\dots,m, \tag{5.1}$$

$$y(t_k^+) - y(t_k) = I_k(y(t_k^-)), \qquad k = 1,\dots,m, \tag{5.2}$$

$$y(0) = y_0, \tag{5.3}$$

where $F : J \times \mathbb{R}^n \to \mathcal{P}(\mathbb{R}^n)$ is a multivalued map, $y_0 \in \mathbb{R}^n$, and $I_k \in C(\mathbb{R}^n,\mathbb{R}^n)$, $k = 1,2,\dots,m$. In the case where the impulses are absent (i.e., $I_k \equiv 0$, $k = 1,2,\dots,m$) and the problem (5.1)–(5.3) reduces to an autonomous control system, that is, $F(t,y) := f(t,U)$, where f is a single valued map and U is a control set, a property of Filippov's theorem first studied by H. M. Soner. Generalizations of Filippov's theorem were considered by Frankowska [202] and Zhu [498].

5.1.1 Filippov's Theorem

In order to define a solution of (5.1)–(5.3), we consider the space

$$PC(J,\mathbb{R}^n) = \big\{y : J \to \mathbb{R}^n \mid y(t) \text{ is continuous everywhere except for some } t_k \text{ at which } y(t_k^-) \text{ and } y(t_k^+) \text{ exist and } y(t_k^-) = y(t_k),\ k = 1,2,\dots,m\,\big\}.$$

Clearly, $PC(J,\mathbb{R}^n)$ is a Banach space with the norm

$$\|y\|_{PC} = \sup\{|y(t)| : t \in J\}.$$

We begin by defining what is meant by a solution of the problem (5.1)–(5.3).

Definition 5.1. A function $y \in PC \cap AC(t_k,t_{k+1})$, $k = 0,\dots,m$, is said to be a solution of the impulsive differential inclusion (5.1)–(5.3) if y satisfies the differential inclusion $y'(t) \in F(t, y(t))$ a.e. on $J \setminus \{t_1, \dots, t_m\}$, and the conditions $y(t_k^+) - y(t_k) = I_k(y(t_k^-))$, $k = 1, 2, \dots, m$, and $y(0) = y_0$.

We will consider a map generated by (5.1)–(5.3) by associating with each initial point $y_0 \in \mathbb{R}^n$ the set

$$S_{[0,b]}(y_0) = \{y \mid y \text{ is a solution to (5.1)–(5.3) on } [0,b] \text{ and } y(0) = y_0\}.$$

We now present a Filippov type result for the problem (5.1)–(5.3).

Theorem 5.2. *Consider a function $x \in PC \cap AC(t_k, t_{k+1})$, $k = 1, \dots, m$, and assume that the following conditions hold.*

(E-1) *The function $F : J \times \mathbb{R}^n \to \mathcal{P}_{cl}(\mathbb{R}^n)$ satisfies:*
   (a) *for all $u \in \mathbb{R}^n$, $t \mapsto F(t, u)$ is measurable,*
   (b) *the map $\gamma : t \to d(x'(t), F(t, x(t))$ is integrable.*

(E-2) *There exists a function $p \in L^1(J, \mathbb{R}_+)$ such that*

$$H(F(t,u), F(t,z)) \le p(t)|u - z| \text{ for all } x, \overline{x} \in \mathbb{R}^n.$$

(E-3) *There exist constants $c_k \ge 0$, $k = 1, 2, \dots, m$, such that*

$$|I_k(u) - I_k(z)| \le c_k|u - z| \text{ for each } u, z \in \mathbb{R}^n.$$

If $\sum_{k=1}^m c_k < 1$, then for every $y_0 \in \mathbb{R}^n$, there exists $y \in S(y_0)$ with $|y(0) - x(0)| < \delta$ such that

$$|x(t) - y(t)| \le \eta(t) + \frac{\sum_{k=1}^m c_k \delta}{1 - \sum_{k=1}^m c_k} + \frac{\sum_{k=1}^m c_k |x(t_k)|}{1 - \sum_{k=1}^m c_k},$$

and

$$|x'(t) - y'(t)| \le p(t)\eta(t) + \gamma(t),$$

for all $t \in [0, b]$, where

$$\eta(t) = \frac{\delta}{1 - \sum_{k=1}^m c_k} \exp\left(\frac{m(t)}{1 - \sum_{k=1}^m c_k}\right) + \int_0^t \gamma(s) \exp\left(\frac{m(t) - m(s)}{1 - \sum_{k=1}^m c_k}\right) ds$$

and

$$m(t) = \exp \int_0^t p(s) ds.$$

Proof. Let $f_0 = x'$, $y_0(t) = x(0) + \int_0^t f_0(s)ds + \sum_{0<t_k<t} I_k(y_0(t_k))$, and $y_0(t_k) = x(t_k)$, $k = 1, 2, \dots, m$. Let $U_1 : [0, b] \to \mathcal{P}(\mathbb{R}^n)$ be given by $U_0(t) = F(t, y_0(t)) \cap (f_0(t) + (\gamma(t)))B(0, 1)$. Since g and γ are measurable, Theorem III.4.1 in [136] tells us that the ball $(f_0(t) + \gamma(t))B(0, 1))$ is measurable. Moreover, $F(t, y_0(t))$ is measurable and U_1 is nonempty. From Lemma 2.76 above, there exists a function u which is a measurable selection for $F(t, y_0(t))$ such that

$$|u(t) - \bar{g}(t)| \leq d(\bar{g}(t), F(t, y_0(t)) = \gamma(t).$$

Then $u \in U_1(t)$, proving our claim. We deduce that the intersection multivalued operator $U_1(t)$ is measurable (see [35, 136, 230]). Since $U_1(t)$ is measurable (see Proposition III.4 in [136]), there exists a function $t \to f_1(t)$, which is a measurable selection for U_1, and so

$$y_1(t) = y_0 + \int_0^t f_1(s)ds + \sum_{0<t_k<t} I_k(y_1(t_k)),\ y_1(t_k) = y_0(t_k),\ k = 1, \dots, m.$$

Then, we have

$$\begin{aligned} |y_1(t) - y_0(t)| &\leq |y(0) - x(0)| + \int_0^t |f_0(s) - f_1(s)|\,ds \\ &\quad + \sum_{0<t_k<t} |I_k(y_1(t_k^-)) - I_k(y_0(t_k^-))| \\ &\leq |y(0) - x(0)| + \int_0^t \gamma(s)ds + \sum_{0<t_k<t} c_k |y_1(t_k) - y_0(t_k)|, \end{aligned}$$

and so

$$\begin{aligned} |y_1(t) - y_0(t)| &\leq \frac{1}{1 - \sum_{k=1}^{m} c_k} \left[|x(0) - y(0)| + \int_0^t \gamma(s)ds \right] \\ &\leq \frac{1}{1 - \sum_{k=1}^{m} c_k} \left[\delta + \int_0^t \gamma(s)ds \right]. \end{aligned}$$

Define the set-valued map $U_2(t) = F(t, y_1(t)) \cap B(f_1(t), p(t)|y_1(t) - y_0(t)|)$. It follows that there exists a measurable selection $f_2(t) \in U_2(t)$ so that

$$y_2(t) = y_0 + \int_0^t f_2(s)ds + \sum_{0<t_k<t} I_k(y_2(t_k)), \quad y_2(t_k) = y_1(t_k), \quad k = 1, \dots, m.$$

Hence,

$$|y_2(t) - y_1(t)| \leq \frac{1}{1 - \sum_{k=1}^{m} c_k} \int_0^t |f_2(s) - f_1(s)|\,ds$$

$$
\begin{aligned}
&\le \frac{1}{1-\sum\limits_{k=1}^{m} c_k} \int_0^t p(s)|y_1(s)-y_0(s)|\,ds \\
&\le \frac{1}{\left(1-\sum\limits_{k=1}^{m} c_k\right)^2} \int_0^t p(s)\left[\delta + \int_0^s \gamma(u)du\right] ds \\
&\le \frac{\delta}{\left(1-\sum\limits_{k=1}^{m} c_k\right)^2} \int_0^t p(s)ds + \frac{1}{\left(1-\sum\limits_{k=1}^{m} c_k\right)^2} \int_0^t p(s)\int_0^s \gamma(u)duds \\
&\le \frac{\delta}{\left(1-\sum\limits_{k=1}^{m} c_k\right)^2} \int_0^t p(s)e^{\int_s^t p(\tau)d\tau} ds \\
&\quad + \frac{1}{\left(1-\sum\limits_{k=1}^{m} c_k\right)^2} \int_0^t p(s)e^{\int_s^t p(\tau)d\tau} \int_0^s \gamma(u)duds \\
&\le \frac{-\delta}{\left(1-\sum\limits_{k=1}^{m} c_k\right)^2} \int_0^t (e^{\int_s^t p(\tau)d\tau})' ds \\
&\quad + \frac{-1}{\left(1-\sum\limits_{k=1}^{m} c_k\right)^2} \int_0^t (e^{\int_s^t p(\tau)d\tau})' \int_0^s \gamma(u)duds \\
&\le \frac{\delta}{\left(1-\sum\limits_{k=1}^{m} c_k\right)^2} e^{\int_0^t p(s)ds} \\
&\quad + \frac{1}{\left(1-\sum\limits_{k=1}^{m} c_k\right)^2} \int_0^t \gamma(s)[e^{\int_s^t p(\tau)d\tau} - 1]ds \\
&\le \frac{\delta}{\left(1-\sum\limits_{k=1}^{m} c_k\right)^2} m(t) + \frac{1}{\left(1-\sum\limits_{k=1}^{m} c_k\right)^2} \int_0^t \gamma(s)[m(t)-m(s)]ds.
\end{aligned}
$$

We consider $U_3(t) = F(t, y_2(t)) \cap B(f_2(t), p(t)|y_2(t)-y_1(t)|)$. By Proposition III.4 in [136], there exists a measurable selection $f_3(t) \in U_3(t)$ so that

$$
y_3(t) = y_0 + \int_0^t f_3(s)ds + \sum_{0<t_k<t} I_k(y_3(t_k)),\ y_3(t_k) = y_2(t_k),\ k = 1, \dots, m.
$$

Then,

$$
\begin{aligned}
|y_3(t) - y_2(t)| &\le \frac{1}{1 - \sum\limits_{k=1}^{m} c_k} \int_0^t |f_2(s) - f_3(s)|\, ds \\
&\le \frac{1}{1 - \sum\limits_{k=1}^{m} c_k} \int_0^t p(s)|y_2(s) - y_1(s)|\, ds \\
&\le \frac{1}{\Big(1 - \sum\limits_{k=1}^{m} c_k\Big)^3} \int_0^t p(s)[\delta m(s) \\
&\quad + \int_0^s \gamma(u)[m(s) - m(u)]du]ds \\
&\le \frac{\delta}{\Big(1 - \sum\limits_{k=1}^{m} c_k\Big)^3} \int_0^t p(s) e^{2\int_0^s p(\tau)d\tau} ds \\
&\quad + \frac{1}{\Big(1 - \sum\limits_{k=1}^{m} c_k\Big)^3} \int_0^t p(s) e^{2\int_0^s p(\tau)d\tau} \int_0^s \gamma(u)[m(s) - m(u)]duds \\
&\le \frac{\delta}{2\Big(1 - \sum\limits_{k=1}^{m} c_k\Big)^3} \int_0^t (e^{2\int_s^t p(\tau)d\tau})' ds \\
&\quad + \frac{1}{\Big(1 - \sum\limits_{k=1}^{m} c_k\Big)^3} \int_0^t (e^{\int_0^s p(\tau)d\tau})' \int_0^s \gamma(u)[m(s) - m(u)]duds \\
&\le \frac{\delta}{2\Big(1 - \sum\limits_{k=1}^{m} c_k\Big)^3} m^2(t) \\
&\quad + \frac{1}{\Big(1 - \sum\limits_{k=1}^{m} c_k\Big)^3} m(t) \int_0^t \gamma(s)[m(t) - m(s)]ds \\
&\quad - \frac{1}{\Big(1 - \sum\limits_{k=1}^{m} c_k\Big)^3} \int_0^t p(s) m^2(s) \int_0^s \gamma(u)duds \\
&= \frac{\delta}{2\Big(1 - \sum\limits_{k=1}^{m} c_k\Big)^3} m^2(t)
\end{aligned}
$$

$$+\frac{1}{\left(1-\sum_{k=1}^{m}c_k\right)^3}\int_0^t \gamma(s)[m^2(t)-m(t)m(s)]ds$$
$$-\frac{1}{2\left(1-\sum_{k=1}^{m}c_k\right)^3}\int_0^t (m^2(s))'\int_0^s \gamma(u)duds$$
$$=\frac{\delta}{2\left(1-\sum_{k=1}^{m}c_k\right)^3}m^2(t)$$
$$+\frac{1}{\left(1-\sum_s k=1^m c_k\right)^3}\int_0^t \gamma(s)[m^2(t)-m(t)m(s)]ds$$
$$-\frac{1}{2\left(1-\sum_{k=1}^{m}c_k\right)^3}\left[m^2(t)\int_0^t \gamma(s)ds-\int_0^t m^2(s)\gamma(s)ds\right]$$
$$=\frac{\delta}{2\left(1-\sum_{k=1}^{m}c_k\right)^3}m^2(t)$$
$$+\frac{1}{2\left(1-\sum_{k=1}^{m}c_k\right)^3}\int_0^t \gamma(s)[m(t)-m(s)]^2ds.$$

Thus, we have

$$|y_3(t)-y_2(t)|\leq\frac{\delta}{2\left(1-\sum_{k=1}^{m}c_k\right)^3}m^2(t)+\frac{1}{2\left(1-\sum_{k=1}^{m}c_k\right)^3}\int_0^t \gamma(s)[m(t)-m(s)]^2ds.$$

Proceeding by induction, we have

$$|y_{n+1}(t)-y_n(t)|\leq\frac{\delta m^n(b)}{(n+1)!\left(1-\sum_{k=1}^{m}c_k\right)^{n+1}} \tag{5.4}$$

$$+\frac{\int_0^b \gamma(s)[m(b)-m(s)]^n ds}{(n+1)!\left(1-\sum_{k=1}^{m}c_k\right)^{n+1}}. \tag{5.5}$$

We deduce that $\{y_n\}$ is a Cauchy sequence in PC, converging uniformly to a function $y\in PC$. From the definition of U_n, $n\in\mathbb{N}$,

$$|f_{n+1}(t)-f_n(t)|\leq p(t)|y_n(t)-y_{n-1}(t)|,\ \text{for}\ n\in\mathbb{N}, \text{a.e.}\ t\in[0,b].$$

Hence, for almost every $t \in [0,b]$, the sequence $\{f_n(t)\}$ is Cauchy in $\mathbb{R}^n$, and so $\{f_n(t)\}$ converges almost everywhere to a measurable function $\{f(\cdot)\}$ in $\mathbb{R}^n$. Moreover, since $f_0 = x'$, and by the last inequality, we obtain

$$\begin{aligned}|f_n(t) - f_0(t) + f_0(t)| &\le |f_n(t) - f_0(t)| + |f_0(t)| \\ &\le \sum_{i=1}^{i=n} |f_i(t) - f_{i-1}(t)| + |f_0(t)| \\ &\le \sum_{k=1}^{n} p(t)|y_k(t) - y_{k-1}(t)| + |f_0(t)| \\ &\le p(t)\sum_{k=1}^{\infty} |y_k(t) - y_{k-1}(t)| + |f_0(t)| \\ &\le p(t)A + \gamma(t),\end{aligned}$$

where

$$A = \delta \frac{1}{1 - \sum_{k=1}^{m} c_k} \exp\left(\frac{m(b)}{1 - \sum_{k=1}^{m} c_k}\right) + \int_0^b \gamma(s) \exp\left(\frac{m(b) - m(s)}{1 - \sum_{k=1}^{m} c_k}\right) ds.$$

Then,

$$\text{for all } n \in \mathbb{N}, \ \ |y_n(t)| \le Ap(t) + \gamma(t). \tag{5.6}$$

From (5.6), we conclude that f_n converges to f in $L^1([0,b], \mathbb{R}^n)$. Consequently,

$$y(t) = y_0 + \int_0^t f(s)ds + \sum_{0<t_k<t} I_k(y(t_k)), \ t \in [0,b],$$

is a solution of the problem (5.1)–(5.2) with the condition $y(0) = y_0$. Thus, $y \in S_{[0,b]}(y_0)$.

Next, we prove that

$$|x(t) - y(t)| \le \eta(t) + \frac{\sum_{k=1}^{m} c_k \delta}{1 - \sum_{k=1}^{m} c_k} + \frac{\sum_{k=1}^{m} c_k |x(t_k)|}{1 - \sum_{k=1}^{m} c_k}, \text{ for } t \in [0,b].$$

We have,

$$
\begin{aligned}
|x(t)-y(t)| &= \left| x(0)+\int_0^t x'(s)ds - y(0) - \int_0^t f(s)ds - \sum_{0<t_k<t} I_k(y(t_k)) \right| \\
&\le |x(0)-y(0)| + \int_0^t |f(s)-f_0(s)|ds + \sum_{0<t_k<t} |I_k(y(t_k))| \\
&\le |x(0)-y(0)| + \int_0^t |f(s)-f_0(s)|ds \\
&\quad + \sum_{0<t_k<t} c_k|y(t_k)-x(t_k)| + \sum_{0<t_k<t} c_k|x(t_k)| \\
&\le |x(0)-y(0)| + \int_0^t |f(s)-f_n(s)|ds + \int_0^t |f_n(s)-f_0(s)|ds \\
&\quad + \sum_{0<t_k<t} c_k|y(t_k)-x(t_k)| + \sum_{0<t_k<t} c_k|x(t_k)| \\
&\le |x(0)-y(0)| + \int_0^t |f(s)-f_n(s)|ds \\
&\quad + \int_0^t p(s)\delta \exp\left(\frac{m(s)}{1-\sum\limits_{k=1}^m c_k} \right) ds \\
&\quad + \int_0^t p(s) \int_0^s \gamma(\tau) \exp\left(\frac{m(t)-m(\tau)}{1-\sum\limits_{k=1}^m c_k} \right) d\tau ds \\
&\quad + \sum_{0<t_k<t} c_k|y(t_k)-x(t_k)| + \sum_{0<t_k<t} c_k|x(t_k)|.
\end{aligned}
$$

Hence,

$$
\begin{aligned}
|x(t)-y(t)| \le \frac{1}{1-\sum\limits_{k=1}^m c_k} \Bigg[& \delta + \int_0^t |f(s)-f_n(s)|ds \\
& + \int_0^t p(s)\delta \exp\left(\frac{m(s)}{1-\sum\limits_{k=1}^m c_k} \right) ds \\
& + \int_0^t p(s) \int_0^s \gamma(\tau)d\tau \exp\left(\frac{m(t)-m(s)}{1-\sum\limits_{k=1}^m c_k} \right) ds + \sum_{0<t_k<t} c_k|x(t_k)| \Bigg]
\end{aligned}
$$

$$
\begin{aligned}
&\leq \frac{1}{1-\sum\limits_{k=1}^{m} c_k}\delta + \frac{1}{1-\sum\limits_{k=1}^{m} c_k}\Bigg[\int_0^t |f(s)-f_n(s)|ds \\
&\quad + \exp\left(\frac{m(t)}{1-\sum\limits_{k=1}^{m} c_k}\right)\delta \\
&\quad - \delta - \int_0^t [\exp\left(\frac{m(t)-m(s)}{1-\sum\limits_{k=1}^{m} c_k}\right)]' \int_0^s \gamma(\tau)d\tau ds + \sum_{0<t_k<t} c_k|x(t_k)|\Bigg] \\
&\leq \frac{\delta}{1-\sum\limits_{k=1}^{m} c_k} + \frac{1}{1-\sum\limits_{k=1}^{m} c_k}\int_0^b |f(s)-f_n(s)|ds \\
&\quad + \exp\left(\frac{m(t)}{1-\sum\limits_{k=1}^{m} c_k}\right)\delta - \delta + \int_0^t \exp\left(\frac{m(t)-m(s)}{1-\sum\limits_{k=1}^{m} c_k}\right)\gamma(s)ds \\
&\quad + \frac{1}{1-\sum\limits_{k=1}^{m} c_k}\sum_{k=1}^{m} c_k|x(t_k)| \\
&\leq \frac{\sum\limits_{k=1}^{m} c_k}{1-\sum\limits_{k=1}^{m} c_k}\delta + \frac{1}{1-\sum\limits_{k=1}^{m} c_k}\int_0^b |f(s)-f_n(s)|ds \\
&\quad + \eta(t) + \frac{\sum\limits_{k=1}^{m} c_k}{1-\sum\limits_{k=1}^{m} c_k}|x(t_k)|.
\end{aligned}
$$

Letting $n \to \infty$ yields

$$
|x(t)-y(t)| \leq \eta(t) + \frac{\sum\limits_{k=1}^{m} c_k\delta}{1-\sum\limits_{k=1}^{m} c_k} + \frac{\sum\limits_{k=1}^{m} c_k|x(t_k)|}{1-\sum\limits_{k=1}^{m} c_k}.
$$

Finally, we prove that

$$|x'(t) - y'(t)| \leq p(t)\eta(t) + \gamma(t), \quad \text{a.e. } t \in [0,b].$$

$$\begin{aligned}
|x'(t) - y'(t)| &= |f_0(t) - f(t)| \\
&\leq |f_0(t) - f_n(t) + f_n(t) - f(t)| \\
&\leq |f_0(t) - f_n(t)| + |f_n(t) - f(t)| \\
&\leq \sum_{k=1}^{n} p(t)|y_k(t) - y_{k-1}(t)| + |f_n(t) - f(t)| \\
&\leq p(t)\sum_{k=1}^{\infty} |y_k(t) - y_{k-1}(t)| + |f_n(t) - f(t)| \\
&\leq |f_n(t) - f(t)| + p(t)\sum_{n=0}^{\infty} \frac{\delta m^n(t)}{n!\Big(1 - \sum_{k=1}^{m} c_k\Big)^{n+1}} \\
&\quad + p(t)\sum_{n=0}^{\infty} \frac{\int_0^t \gamma(s)[m(s) - m(u)]^n ds}{n!\Big(1 - \sum_{k=1}^{m} c_k\Big)^{n+1}} \\
&\quad + |f_1(t) - f_0(t)| \\
&\leq p(t)\eta(t) + \gamma(t) + |f_n(t) - f(t)|.
\end{aligned}$$

Thus,

$$|x'(t) - y'(t)| \leq p(t)\eta(t) + \gamma(t).$$

□

For our next result, set

$$APC([0,b], \mathbb{R}^n) = \{y : [0,b] \to \mathbb{R}^n : y \in AC(J_k, \mathbb{R}^n)\},$$

where $J_k = [t_k, t_{k+1}]$, $k = 0, \ldots, m$, which is a Banach space with the norm

$$\|y\|_{APC} = \sup_{t\in[0,b]} |y(t)| + \int_0^b |y'(t)|dt.$$

Corollary 5.3. *Assume that the conditions* (E-2), (E-3) *and the first part of* (E-1) *hold. In addition, assume that there exists* $f \in L^1(J, \mathbb{R}_+)$ *such that*

$$H(0, F(t,0)) \leq f(t) \text{ for a.e. } t \in J.$$

Then the set-valued map $S_{[0,b]} : \mathbb{R}^n \to \mathcal{P}(APC)$ *is Lipschitz with Lipschitz constant*

$$L = \frac{1 - \Big(1 - \sum_{k=1}^{m} c_k\Big)^2}{1 - \sum_{k=1}^{m} c_k} + \Big(2 - \sum_{k=1}^{m} c_k\Big)\exp\left(\frac{m(b)}{1 - \sum_{k=1}^{m} c_k}\right).$$

Moreover, if F has compact and convex values, then $S_{[0,b]}(\cdot) \in \mathcal{P}_{cp}(PC)$, that is, $S_{[0,b]}$ is compact in PC.

Proof. We show that

$$H(S_{[0,b]}(x_0), S_{[0,b]}(y_0)) \leq L|x_0 - y_0|, \text{ for all } x_0, y_0 \in \mathbb{R}^n.$$

Let $x_0, y_0 \in \mathbb{R}^n$ and $y \in S_{[0,b]}(y_0)$. Then there exists $f \in L^1([0,b], \mathbb{R}^n)$ such that $f(t) \in F(t, y(t))$, and

$$y(t) = y_0 + \int_0^t f(s)ds + \sum_{0<t_k<t} I_k(y(t_k)).$$

Then, $\gamma(t) = d(y'(t), F(t, y(t))) = 0$. From Theorem 5.2, there exist $x \in S_{[0,b]}(x_0)$ and $f_1(t) \in F(t, x(t))$, such that

$$x(t) = x_0 + \int_0^t f_1(s) + \sum_{0<t_k<t} I_k(x(t_k))$$

and

$$\begin{aligned}
|x(t) - y(t)| &\leq \frac{|x_0 - y_0|}{1 - \sum_{k=1}^m c_k} + \eta(t) \\
&\leq \frac{|x_0 - y_0|}{1 - \sum_{k=1}^m c_k} + |x_0 - y_0| \exp\left(\frac{m(t)}{1 - \sum_{k=1}^m c_k}\right) \\
&\leq \left(\frac{1}{1 - \sum_{k=1}^m c_k} + \exp\left(\frac{m(t)}{1 - \sum_{k=1}^m c_k}\right)\right) |x_0 - y_0|.
\end{aligned}$$

Also, from Theorem 5.2, we have

$$\begin{aligned}
\int_0^t |x'(s) - y'(s)|ds &\leq \int_0^t p(s) \exp\left(\frac{m(s)}{1 - \sum_{k=1}^m c_k}\right) ds \\
&= \left(1 - \sum_{k=1}^m c_k\right) \int_0^t \frac{1}{m(s)} \left[\exp\left(\frac{m(s)}{1 - \sum_{k=1}^m c_k}\right)\right]' |x_0 - y_0| ds \\
&\leq (1 - \sum_{k=1}^m c_k) \left(\exp\left(\frac{m(t)}{1 - \sum_{k=1}^m c_k}\right) - 1\right) |x_0 - y_0|.
\end{aligned}$$

By an analogous relation, obtained by interchanging the roles of x_0 and y_0, it follows that

$$H(S_{[0,b]}(x_0), S_{[0,b]}(y_0)) \leq \left[\frac{1-\left(1-\sum\limits_{k=1}^{m} c_k\right)^2}{1-\sum\limits_{k=1}^{m} c_k} + \left(2-\sum_{k=1}^{m} c_k\right)\exp\left(\frac{m(b)}{1-\sum\limits_{k=1}^{m} c_k}\right)\right]|x_0 - y_0|.$$

Now, we shall prove that for every $x_0 \in \mathbb{R}^n$, $S_{[0,b]}(x_0)$ is compact. As in the proof of Theorem 5.2 we can see that $S_{[0,b]}(x_0) \neq \emptyset$. Let $y_n \in S_{[0,b]}(x_0)$; then there exist $g_n \in S_{F,y_n}$ such that, for each $t \in J$,

$$y_n(t) = x_0 + \int_0^t g_n(s)ds + \sum_{0<t_k<t} I_k(y_n(t_k^-)), \quad t \in [0,b].$$

Since there exists $f \in L^1([0,b], \mathbb{R}^n)$ such that $f(t) \in F(t,0)$ a.e. $t \in [0,b]$, by (E-2), we have

$$\begin{aligned} |g_n(t)| &\leq H_d(F(t,0), F(t, y_n(t))) + |f(t)| \\ &\leq p(t)|y_n(t)| + |f(t)| \\ &\leq Mp(t) + |f(t)|, \text{ a.e. } t \in [0,b], \end{aligned} \tag{5.7}$$

where $|y_n(t)| \leq M$ for all $n \in \mathbb{N}$ for some $M > 0$. For $t \in [0, t_1]$, we have

$$y_n(t) = x_0 + \int_0^t g_n(s)ds,$$

and hence

$$|y_n(t)| \leq |x_0| + \int_0^t |g_n(s)|ds. \tag{5.8}$$

By the above inequality and (5.7), we obtain

$$|y_n(t)| \leq |x_0| + M\int_0^t p(s)ds + \int_0^t |f(s)|ds, \quad t \in [0, t_1]. \tag{5.9}$$

Hence,

$$\sup\{|y_n(t)| : t \in [0,t_1]\} \leq |x_0| + M\|p\|_{L^1} + \|f\|_{L^1} := K_0, \text{ for all } n \geq 1.$$

Let $t \in (t_1, t_2]$; then

$$y_n(t) = \int_{t_1}^t g_n(s)ds + y_n(t_1) + I_1(y_n(t_1)).$$

It is clear that

$$y_n(t_1^+) = y_n(t_1) + I_1(y_n(t_1)).$$

Thus,

$$\begin{aligned} |y_n(t_1^+)| &\le |y_n(t_1)| + |I_1(y_n(t_1))| \\ &\le K_0 + \sup\{|x| : x \in B(0, K_0)\}, \end{aligned}$$

where

$$B(0, K_0) = \{x \in \mathbb{R}^n : |x| \le K_0\}.$$

Analogous to what we did above, we can show that there exists $K_1 > 0$ such that

$$\sup\{|y(t)| : t \in [t_1, t_2]\} \le K_1.$$

We continue this process and also take into account that

$$y_n(t) = \int_{t_m}^{t} g_n(s)ds + y_n(t_m) + I_m(y(t_m)),\ t \in (t_m, b].$$

We obtain that there exists a constant K_m such that

$$\sup\{|y_n(t)| : t \in [t_m, b]\} \le K_m.$$

Consequently, for each $n \ge 1$ we have

$$\|y_n\|_{PC} \le \max\{K_i : i = 1, \dots, m\} := \overline{K}.$$

As in Theorem 4.3, we can prove that $\{y_n : n \ge 1\}$, are equicontinuous sets in PC. By the Arzela–Ascoli theorem, we can conclude that $\{y_n : n \ge 1\}$ is compact in PC, so there exists y such that $y_n \to y$. We shall prove that $y \in S_{[0,b]}(x_0)$. We need to show that there exists $g \in S_{F,y}$ such that for each $t \in J$,

$$y(t) = x_0 + \int_0^t g(s)ds + \sum_{0<t_k<t} I_k(y(t_k^-)).$$

Since $g_n(t) \in (Mp(t) + f(t))B(0, 1)$ then there exists a subsequence of g_n converging to g.

It remains to prove that $g(t) \in F(t, y(t))$, for a.e. $t \in J$. Lemma 2.66 yields the existence of $\alpha_i^n \ge 0$, $i = n, \dots, k(n)$ such that $\sum_{i=1}^{k(n)} \alpha_i^n = 1$ and the sequence of convex combinations $v_n(\cdot) = \sum_{i=1}^{k(n)} \alpha_i^n g_i(\cdot)$ converges strongly to g in L^1. Since F

takes convex values, using Lemma 2.64, we obtain that

$$
\begin{aligned}
g(t) &\in \bigcap_{n\geq 1} \overline{\{v_n(t)\}}, \quad \text{a.e. } t \in J,\\
&\subset \bigcap_{n\geq 1} \overline{co}\{g_k(t),\ k \geq n\}\\
&\subset \bigcap_{n\geq 1} \overline{co}\{\bigcup_{k\geq n} F(t, y_k(t))\}\\
&= \overline{co}(\limsup_{k\to\infty} F(t, y_k(t))).
\end{aligned} \tag{5.10}
$$

Since $F(t,.)$ is u.s.c. with compact values, then for every $\epsilon > 0$ there exists $n \in \mathbb{N}$ such that for each $k \geq n$ we have

$$F(t, y_k(t)) \subset F(t, y(t)) + \epsilon B(0, 1)$$

This, together with (5.10), implies that

$$v(t) \in \overline{co}\, F(t, y(t)) + \epsilon B(0, 1).$$

Since ϵ is arbitrary and $F(.,.)$ has convexe values, we deduce that

$$g(t) \in F(t, y(t)) \text{ for a.e. } t \in J.$$

By the Lebesgue dominated convergence theorem and the continuity of I_k, $k = 1, 2, \ldots, m$, we have

$$
\begin{aligned}
&\Big\| \Big(y_n(t) - x_0 - \sum_{0<t_k<t} I_k(y_n(t_k^-)) - \int_0^t g_n(s)ds\Big)\\
&\quad - \Big(y(t) - x_0 - \sum_{0<t_k<t} I_k(y(t_k^-)) - \int_0^t g(s)ds\Big)\Big\|_{PC} \to 0, \text{ as } n \to \infty.
\end{aligned}
$$

Therefore,

$$y(t) = x_0 + \sum_{0<t_k<t} I_k(y(t_k^-)) + \int_0^t g(s)ds, \quad t \in [0, b]. \qquad \square$$

5.1.2 Relaxation Theorem

In this subsection, we examine to what extent the convexification of the right-hand side of the inclusion introduces new solutions. More precisely, we want to find out if the solutions of the nonconvex problem are dense in those of the convex one. Such a result is known in the literature as a relaxation theorem and has important implications in optimal control theory. It is well-known that in order to have optimal state-control pairs, the system has to satisfy certain convexity requirements. If these conditions are not present, then in order to guarantee existence of optimal solutions we need to pass to an augmented system with convex structure by introducing the so-called relaxed (generalized, chattering) controls. The resulting relaxed problem has a solution. The relaxation theorems tell us that the relaxed optimal state can be approximated by the original states, which are generated by a more economical set of controls that are much simpler to build. In particular, "strong relaxation" theorems imply that this approximation can be achieved using states generated by bang-bang controls. More precisely, we compare trajectories of (5.1)–(5.3) to those of the relaxation impulsive differential inclusion,

$$y'(t) \in \overline{co}F(t, y(t)), \quad \text{a.e. } t \in J =: [0,b]\backslash\{t_1,\dots,t_m\}, \tag{5.11}$$

$$y(t_k^+) - y(t_k) = I_k(y(t_k^-)), \quad k = 1,\dots,m, \tag{5.12}$$

$$y(0) = y_0. \tag{5.13}$$

The following result is known.

Theorem 5.4. ([288]) *Let* $U : [0,b] \to \mathcal{P}_{cl}(\mathbb{R}^n)$ *be a measurable, integrable bounded set-valued map and let* $t \to d(0, U(t))$ *be an integrable map. Then the integral* $\int_0^b U(t)dt$ *is convex, the mapping* $t \to \overline{co}U(t)$ *is measurable, and for every* $\epsilon > 0$, *and every measurable selection* u *of* $\overline{co}U(t)$, *there exists a measurable selection* $\overline{u}$ *of* U *such that*

$$\sup_{t\in[0,b]}\left|\int_0^t u(s)ds - \int_0^t \overline{u}(s)ds\right| \le \epsilon$$

and

$$\overline{\int_0^t \overline{co}U(t)dt} = \overline{\int_0^t U(t)dt} = \int_0^t \overline{co}U(t)dt.$$

Our relaxation result in this subsection is the following.

Theorem 5.5. *Assume that the conditions* (E-1)–(E-3) *are satisfied. If* $\sum_{k=1}^m c_k < 1$, *then for every* $\delta > 0$, *there exists a solution* y *to the problem* (5.1)–(5.3) *on* $[0,b]$ *satisfying*

$$x(0) = y(0) \text{ and } \|x - y\|_{PC} \le \delta.$$

This implies that $S^{co}_{[0,b]}(y_0) = \overline{S_{[0,b]}(y_0)}$, *where*

$$S^{co}_{[0,b]} = \{y \mid y \text{ is a solution to (5.11)–(5.13) on } [0,b],\ y(0) = y_0\}.$$

Proof. Let y be a solution of the problem (5.11)–(5.13). Let $\epsilon > 0$ be given and let y' be an integral solution of $t \to \overline{\text{conv}} F(t, y(t))$. Let

$$\delta = \frac{\epsilon}{\exp\left(\dfrac{m(b)}{1 - \sum\limits_{k=1}^{m} c_k}\right)}.$$

From Theorem 5.4, there exists a measurable selection f of $t \to F(t, y(t))$ such that

$$\sup_{t\in[0,b]} \left| \int_0^t f(s)ds - \int_0^t y'(s)ds \right| \le \delta.$$

Set

$$x(t) = x(0) + \int_0^t f(s)ds + \sum_{0<t_k<t} I_k(x(t_k)),\ x(t_k) = y(t_k),\ k = 1, 2, \dots, m.$$

Then,

$$d(x'(t), F(t, x(t)) \le p(t)|x(t) - y(t)| \le p(t)\delta,\ a.e\ t \in [0, b].$$

From Theorem 5.2, there exists a solution y_* of the problem (5.1)–(5.3) such that

$$\begin{aligned} |y_*(t) - y(t)| &\le \frac{1}{1 - \sum_{k=1}^{m} c_k} \int_0^t \delta p(s) \exp\left(\frac{m(t) - m(s)}{1 - \sum\limits_{k=1}^{m} c_k}\right) ds \\ &\le \delta \exp\left(\frac{m(b)}{1 - \sum\limits_{k=1}^{m} c_k}\right) \\ &\le \epsilon \end{aligned}$$

for all $t \in [0, b]$. Thus, $S^{co}_{[0,b]}(y_0) = \overline{S_{[0,b]}(y_0)}$. □

5.2 Functional Differential Inclusions

In this section, we give a Filippov type theorem for first order impulsive functional differential inclusions of the form

$$y'(t) \in F(t, y_t), \quad \text{a.e. } t \in [0, b] \setminus \{t_1, t_2, \dots, t_m\}, \tag{5.14}$$

$$y(t_k^+) - y(t_k^-) = I_k(y(t_k^-)), \quad k = 1, 2, \dots, m, \tag{5.15}$$

$$y(t) = \phi(t), \quad t \in [-r, 0], \tag{5.16}$$

where $r > 0$, $F : J \times \mathcal{D} \to \mathcal{P}(\mathbb{R}^n)$ is a multivalued map,

$$\mathcal{D} = \big\{\psi : [-r,0] \to \mathbb{R}^n \mid \psi \text{ is continuous everywhere except for a finite number of points } \bar{t} \text{ at which } \psi(\bar{t}^-) \text{ and } \psi(\bar{t}^+) \text{ exist, and } \psi(\bar{t}^-) = \psi(\bar{t})\big\},$$

$\phi \in \mathcal{D}, 0 < r < \infty$, and $I_k \in C(\mathbb{R}^n, \mathbb{R}^n)$.

For every $t \in [0,b]$, the history function $y_t : [-r,0] \to \mathbb{R}^n$ is defined as usual by

$$y_t(\theta) = y(t+\theta), \text{ for } \theta \in [-r,0].$$

We assume that the histories y_t belong to $\mathcal{D}$. In the space $\mathcal{D}$, we consider the L^1-norm:

$$\|\phi\|_D := \int_{-r}^{0} \|\phi(\theta)| d\theta.$$

In order to define a solution of (5.14)–(5.16), we consider the space

$$\Omega = \{y : J_1 \to \mathbb{R}^n : y \in \mathcal{D} \cap PC(J,\mathbb{R}^n)\},\ J_1 = [-r,0] \cup J,$$

which is a Banach space with the norm

$$\|y\|_\Omega = \sup\{|y(t)| : t \in [0,b]\}.$$

Definition 5.6. A function $y : [-r,b] \to \mathbb{R}^n$ is said to be a solution of (5.14)–(5.16) if there exists $v \in L^1([0,b],\mathbb{R}^n)$ such that $v(t) \in F(t,y_t)$ a.e. $t \in [0,b]$ satisfying the equation

$$y(t) = \phi(0) + \int_0^t v(s)ds + \sum_{0<t_k<t} I_k(y(t_k)), \quad t \in [0,b],$$

with $y(t) = \phi(t)$ for $t \in [-r,0]$.

Consider the map associating with each initial point $\phi \in \mathcal{D}$ the set

$$S_{[0,b]}(\phi) = \big\{y \mid y \text{ is a solution to (5.14)–(5.16 on } [-r,b],\ y_0 = \phi\big\}.$$

5.2.1 Filippov's Theorem for FDIs

Here, we present a Filippov's result for the problem (5.14)–(5.16).

Theorem 5.7. *Let $x \in \Omega \cap AC(t_k,t_{k+1})$, $k = 1,2,\dots,m$, and assume the following conditions hold.*

(E-4) *The function $F : J \times \mathcal{D} \to \mathcal{P}_{cl}(\mathbb{R}^n)$ satisfies:*
(a) *for all $u \in \mathcal{D}$, $t \mapsto F(t,u)$ is measurable;*
(b) *the map $t \to \gamma(t) = d(x'(t), F(t,x_t))$ is integrable.*

(E-5) *There exists a function* $p \in L^1(J, \mathbb{R}_+)$ *such that*

$$H(F(t,u), F(t,z)) \le p(t)\|u - z\|_\infty \text{ for all } u,\ z \in \mathcal{D}.$$

(E-6) *There exist constants* $c_k \ge 0$, $k = 1, \dots, m$, *such that*

$$|I_k(u) - I_k(z)| \le c_k |u - z| \text{ for each } \ u, z \in \mathbb{R}^n.$$

If $\sum_{k=1}^m c_k < 1$, *then there exists* $y \in S(\psi)$ *such that*

$$\|x_t - y_t\|_\infty \le \eta(t) + \frac{\sum\limits_{k=1}^m c_k \delta}{1 - \sum\limits_{k=1}^m c_k} + \frac{\sum\limits_{k=1}^m c_k |x(t_k)|}{1 - \sum\limits_{k=1}^m c_k}, \quad \text{for all } t \in [0, b]$$

and

$$|x'(t) - y'(t)| \le p(t)\eta(t) + \gamma(t),$$

where

$$\eta(t) = \delta \frac{1}{1 - \sum\limits_{k=1}^m c_k} \exp\left(\frac{m(t)}{1 - \sum\limits_{k=1}^m c_k} \right) + \int_0^t \gamma(s) \exp\left(\frac{m(t) - m(s)}{1 - \sum\limits_{k=1}^m c_k} \right) ds,$$

and

$$m(t) = \exp \int_0^t p(s)ds, \text{ with } \|\phi - \psi\|_\infty \le \delta.$$

Proof. Let $f_0 = x'$ and

$$y^0(t) = \begin{cases} \phi(t), & t \in [-r, 0], \\ \phi(0) + \int_0^t f_0(s)ds + \sum_{0<t_k<t} I_k(y^0(t_k)), & t \in [0, b], \\ \ y^0(t_k) = x(t_k), & k = 1, \dots, m. \end{cases}$$

Let $U_1 : [0, b] \to \mathcal{P}(\mathbb{R}^n)$ given by $U_1(t) = F(t, y_t^0) \cap B(x'(t), \gamma(t))$. The multivalued operator $U_1(t)$ is measurable (see Proposition III.4 in [136]), and so there exists a function $t \to f_1(t)$, which is a measurable selection for U_1. Consider

$$y^1(t) = \begin{cases} \psi(t), & t \in [-r, 0], \\ \psi(0) + \int_0^t f_1(s)ds + \sum_{0<t_k<t} I_k(y^1(t_k)), & t \in [0, b], \\ \ y^1(t_k) = y^0(t_k), & k = 1, \dots, m. \end{cases}$$

Then, for every $t \in [0, b]$, we have

$$\begin{aligned}|y^1(t) - y^0(t)| &\le |\phi(0) - \psi(0)| + \int_0^t |f_0(s) - f_1(s)|\, ds \\ &\quad + \sum_{0<t_k<t} |I_k(y^1(t_k^-)) - I_k(y^0(t_k^-))| \\ &\le \|\psi - \phi\|_\infty + \int_0^t \gamma(s)ds + \sum_{0<t_k<t} c_k |y^1(t_k) - y^0(t_k)|.\end{aligned}$$

Let $\mu(t) = \sup\{|y^1(s) - y^0(s)| : s \in [-r, t]\}$, $t \in [0, b]$. Thus,

$$\begin{aligned}|y^1(t) - y^0(t)| &\le \frac{1}{1 - \sum\limits_{k=1}^{m} c_k}\left[\|\psi - \phi\|_\infty + \int_0^t \gamma(s)ds\right] \\ &\le \frac{1}{1 - \sum\limits_{k=1}^{m} c_k}\left[\delta + \int_0^t \gamma(s)ds\right].\end{aligned}$$

Define the set-valued map $U_2(t) = F(t, y_t^1) \cap B(f_1(t), p(t)\|y_t^1 - y_t^0\|_\infty)$. Again, we deduce that there exists a measurable selection $f_2(t) \in U_2(t)$. Next, consider

$$y^2(t) = \begin{cases} \psi(t), & t \in [-r, 0], \\ \psi(0) + \int_0^t f_2(s)ds + \sum_{0<t_k<t} I_k(y^2(t_k)), & t \in [0, b], \\ y^2(t_k) = y^1(t_k), & k = 1, \dots, m. \end{cases}$$

Hence,

$$\begin{aligned}|y^2(t) - y^1(t)| &\le \frac{1}{1 - \sum\limits_{k=1}^{m} c_k}\int_0^t |f_2(s) - f_1(s)|\, ds \\ &\le \frac{1}{1 - \sum\limits_{k=1}^{m} c_k}\int_0^t p(s)\|y_s^1 - y_s^0\|_\infty\, ds \\ &\le \frac{1}{\left(1 - \sum\limits_{k=1}^{m} c_k\right)^2}\int_0^t p(s)\left[\delta + \int_0^s \gamma(u)du\right] ds \\ &\le \frac{\delta}{\left(1 - \sum\limits_{k=1}^{m} c_k\right)^2}\int_0^t p(s)ds + \frac{1}{\left(1 - \sum\limits_{k=1}^{m} c_k\right)^2}\int_0^t p(s)\int_0^s \gamma(u)duds\end{aligned}$$

$$\le \frac{\delta}{\left(1-\sum_{k=1}^{m} c_k\right)^2} \int_0^t p(s) e^{\int_s^t p(\tau)d\tau} ds$$

$$+ \frac{1}{\left(1-\sum_{k=1}^{m} c_k\right)^2} \int_0^t p(s) e^{\int_s^t p(\tau)d\tau} \int_0^s \gamma(u) du ds$$

$$\le \frac{-\delta}{\left(1-\sum_{k=1}^{m} c_k\right)^2} \int_0^t (e^{\int_s^t p(\tau)d\tau})' ds$$

$$+ \frac{-1}{\left(1-\sum_{k=1}^{m} c_k\right)^2} \int_0^t (e^{\int_s^t p(\tau)d\tau})' \int_0^s \gamma(u) du ds$$

$$= \frac{\delta}{\left(1-\sum_{k=1}^{m} c_k\right)^2} e^{\int_0^t p(s)ds}$$

$$+ \frac{1}{\left(1-\sum_{k=1}^{m} c_k\right)^2} \int_0^t \gamma(s)[e^{\int_s^t p(\tau)d\tau} - 1] ds$$

$$\le \frac{\delta}{\left(1-\sum_{k=1}^{m} c_k\right)^2} m(t)$$

$$+ \frac{1}{\left(1-\sum_{k=1}^{m} c_k\right)^2} \int_0^t \gamma(s)[m(t) - m(s)] ds.$$

We consider $U_3(t) = F(t, y_t^2) \cap B(f_2(t), p(t)\|y_t^2 - y_t^1\|_\infty)$. Once again, there exists a measurable selection $f_3(t) \in U_3(t)$. Consider

$$y^3(t) = \begin{cases} \psi(t), & t \in [-r, 0], \\ \psi(0) + \int_0^t f_3(s)ds + \sum_{0<t_k<t} I_k(y^3(t_k)), & t \in [0, b], \\ y^3(t_k) = y^2(t_k), & k = 1, \ldots, m. \end{cases}$$

We have

$$|y^3(t) - y^2(t)| \le \frac{1}{1-\sum_{k=1}^{m} c_k} \int_0^t |f_2(s) - f_3(s)| \, ds$$

$$
\begin{aligned}
&\leq \frac{1}{1-\sum\limits_{k=1}^{m} c_k} \int_0^t p(s)\|y_s^2 - y_s^1\|_\infty \, ds \\
&\leq \frac{1}{\left(1-\sum\limits_{k=1}^{m} c_k\right)^3} \int_0^t p(s)[\delta m(s) + \int_0^s \gamma(u)[m(s)-m(u)]du]ds \\
&\leq \frac{\delta}{\left(1-\sum\limits_{k=1}^{m} c_k\right)^3} \int_0^t p(s) e^{2\int_0^s p(\tau)d\tau} ds \\
&\quad + \frac{\delta}{\left(1-\sum\limits_{k=1}^{m} c_k\right)^3} \int_0^t p(s) e^{2\int_0^s p(\tau)d\tau} \int_0^s \gamma(u)[m(s)-m(u)]duds \\
&\leq \frac{\delta}{2\left(1-\sum\limits_{k=1}^{m} c_k\right)^3} \int_0^t (e^{2\int_s^t p(\tau)d\tau})' ds \\
&\quad + \frac{1}{\left(1-\sum\limits_{k=1}^{m} c_k\right)^3} \int_0^t (e^{\int_0^s p(\tau)d\tau})' \int_0^s \gamma(u)[m(s)-m(u)]duds \\
&\leq \frac{\delta}{2\left(1-\sum\limits_{k=1}^{m} c_k\right)^3} m^2(t) + \frac{1}{\left(1-\sum\limits_{k=1}^{m} c_k\right)^3} m(t) \\
&\quad \times \int_0^t \gamma(s)[m(t)-m(s)]ds - \frac{1}{\left(1-\sum\limits_{k=1}^{m} c_k\right)^3} \int_0^t p(s)m^2(s) \\
&\quad \times \int_0^s \gamma(u) ds du \\
&= \frac{\delta}{2(1-\sum\limits_{k=1}^{m} c_k)^3} m^2(t) \\
&\quad + \frac{1}{(1-\sum\limits_{k=1}^{m} c_k)^3} \int_0^t \gamma(s)[m^2(t)-m(t)m(s)]ds \\
&\quad - \frac{1}{2\left(1-\sum\limits_{k=1}^{m} c_k\right)^3} \int_0^t (m^2(s))' \int_0^s \gamma(u) ds du
\end{aligned}
$$

$$
\begin{aligned}
&= \frac{\delta}{2\left(1-\sum_{k=1}^{m} c_k\right)^3} m^2(t) \\
&\quad + \frac{1}{\left(1-\sum_{k=1}^{m} c_k\right)^3} \int_0^t \gamma(s)[m^2(t) - m(t)m(s)]ds \\
&\quad - \frac{1}{2\left(1-\sum_{k=1}^{m} c_k\right)^3} \left[m^2(t) \int_0^t \gamma(s)ds - \int_0^t m^2(s)\gamma(s)ds\right] \\
&= \frac{\delta}{2\left(1-\sum_{k=1}^{m} c_k\right)^3} m^2(t) \\
&\quad + \frac{1}{2\left(1-\sum_{k=1}^{m} c_k\right)^3} \int_0^t \gamma(s)[m(t) - m(s)]^2 ds.
\end{aligned}
$$

Hence, we obtain

$$
\begin{aligned}
|y^3(t) - y^2(t)| \le & \frac{\delta}{2\left(1-\sum_{k=1}^{m} c_k\right)^3} m^2(t) \\
& + \frac{1}{2\left(1-\sum_{k=1}^{m} c_k\right)^3} \int_0^t \gamma(s)[m(t) - m(s)]^2 ds.
\end{aligned}
$$

An induction argument then shows that

$$
|y^{n+1}(t) - y^n(t)| \le \frac{\delta m^n(t)}{\left(1-\sum_{k=1}^{m} c_k\right)^{n+1}} + \frac{\int_0^t \gamma(s)[m(t) - m(s)]^n ds}{\left(1-\sum_{k=1}^{m} c_k\right)^{n+1}}.
$$

It follows that $\{y^n\}$ is a Cauchy sequence in Ω that converges uniformly to a function $y \in PC$. From the definition of U_n, $n \in \mathbb{N}$,

$$
|f_{n+1}(t) - f_n(t)| \le p(t)\|y_t^n - y_t^{n-1}\|_\infty, \text{ for } n \in \mathbb{N}, \text{ a.e. } t \in [0, b].
$$

Hence, for almost every $t \in [0, b]$, the sequence $\{f_n(t)\}$ is Cauchy in $\mathbb{R}^n$, and so $\{f_n(t)\}$ converges almost everywhere to a measurable function $\{f(\cdot)\}$ in $\mathbb{R}^n$. Moreover, since $f_0 = x'$, the last inequality yields

$$\begin{aligned}
|f_n(t) - f_0(t) + f_0(t)| &\leq |f_n(t) - f_0(t)| + |f_0(t)| \\
&\leq \sum_{i=1}^{i=n} |f_i(t) - f_{i-1}(t)| + |f_0(t)| \\
&\leq \sum_{k=1}^{n} p(t)|y^k(t) - y^{k-1}(t)| + |f_0(t)| \\
&\leq p(t) \sum_{k=1}^{\infty} |y^k(t) - y^{k-1}(t)| + |f_0(t)| \\
&\leq p(t)A + \gamma(t),
\end{aligned}$$

where

$$A = \delta \exp\left(\frac{m(b)}{1 - \sum\limits_{k=1}^{m} c_k}\right) + \int_0^b \gamma(s) \exp\left(\frac{m(b) - m(s)}{1 - \sum\limits_{k=1}^{m} c_k}\right) ds.$$

Then,

$$\text{for all } n \in \mathbb{N}, \ |y_n(t)| \leq Ap(t) + \gamma(t). \tag{5.17}$$

By (5.17), we see that f_n converges to f in $L^1([0,b], \mathbb{R}^n)$. Consequently,

$$y(t) = y_0 + \int_0^t f(s)ds + \sum_{0<t_k<t} I_k(y(t_k)), \ t \in [0,b],$$

is a solution of the problem (5.14)–(5.15) with the condition that $y(t) = \psi(t)$, $t \in [-r, 0]$. Hence, $y \in S_{[0,b]}(\psi)$.

Now, we prove that

$$|x(t) - y(t)| \leq \eta(t) + \sum_{k=1}^{m} c_k |I_k(y(t_k))|, \text{ for } t \in [0,b].$$

We have

$$\begin{aligned}
|x(t) - y(t)| &= \left|\phi(0) + \int_0^t x'(s)ds - \psi(0) - \int_0^t f(s)ds - \sum_{0<t_k<t} I_k(y(t_k))\right| \\
&\leq |\phi(0) - \psi(0)| + \int_0^t |f(s) - f_0(s)|ds + \sum_{0<t_k<t} |I_k(y(t_k))| \\
&\leq |\phi(0) - \psi(0)| + \int_0^t |f(s) - f_0(s)|ds \\
&\quad + \sum_{0<t_k<t} c_k|y(t_k) - x(t_k)| + \sum_{0<t_k<t} c_k|x(t_k)|
\end{aligned}$$

$$
\begin{aligned}
&\le \|\phi-\psi\|_\infty + \int_0^t |f(s)-f_n(s)|ds \\
&\quad + \int_0^t |f_n(s)-f_0(s)|ds + \sum_{0<t_k<t} c_k|y(t_k)-x(t_k)| \\
&\quad + \sum_{0<t_k<t} c_k|x(t_k)| \\
&\le \|\phi-\psi\|_\infty + \int_0^t |f(s)-f_n(s)|ds \\
&\quad + \int_0^t p(s)\delta \exp\left(\frac{m(s)}{1-\sum\limits_{k=1}^m c_k}\right) ds \\
&\quad + \int_0^t p(s)\int_0^s \gamma(\tau)d\tau \exp\left(\frac{m(t)-m(s)}{1-\sum\limits_{k=1}^m c_k}\right) ds \\
&\quad + \sum_{0<t_k<t} c_k|y(t_k)-x(t_k)| + \sum_{0<t_k<t} c_k|x(t_k)|.
\end{aligned}
$$

It follows that

$$
\begin{aligned}
|x(t)-y(t)| &\le \frac{1}{1-\sum\limits_{k=1}^m c_k}\Bigg[\delta + \int_0^t |f(s)-f_n(s)|ds \\
&\quad + \int_0^t p(s)\delta \exp\left(\frac{m(s)}{1-\sum\limits_{k=1}^m c_k}\right) ds \\
&\quad + \int_0^t p(s)\int_0^s \gamma(\tau)d\tau \\
&\quad \times \exp\left(\frac{m(t)-m(s)}{1-\sum\limits_{k=1}^m c_k}\right) ds + \sum_{0<t_k<t} c_k|x(t_k)|\Bigg] \\
&\le \frac{1}{1-\sum\limits_{k=1}^m c_k}\delta + \frac{1}{1-\sum\limits_{k=1}^m c_k}\Bigg[\int_0^t |f(s)-f_n(s)|ds \\
&\quad + \exp\left(\frac{m(t)}{1-\sum\limits_{k=1}^m c_k}\right)\delta - \delta\Bigg]
\end{aligned}
$$

$$
\begin{aligned}
&-\int_0^t \left[\exp\left(\frac{m(t)-m(s)}{1-\sum\limits_{k=1}^{m} c_k}\right)\right]' \int_0^s \gamma(\tau)d\tau ds \\
&\left.+\frac{\sum\limits_{0<t_k<t} c_k|x(t_k)|}{1-\sum\limits_{k=1}^{m} c_k}\right] \\
\leq\ & \frac{\delta}{1-\sum\limits_{k=1}^{m} c_k}+\frac{1}{1-\sum\limits_{k=1}^{m} c_k}\left[\int_0^b |f(s)-f_n(s)|ds\right. \\
&+\exp\left(\frac{m(t)}{1-\sum\limits_{k=1}^{m} c_k}\right)\delta \\
&-\delta+\int_0^t \exp\left(\frac{m(t)-m(s)}{1-\sum\limits_{k=1}^{m} c_k}\right)\gamma(s)ds \\
&+\frac{1}{1-\sum\limits_{k=1}^{m} c_k}\sum_{k=1}^{m} c_k|x(t_k)| \\
\leq\ & \frac{\sum\limits_{k=1}^{m} c_k}{1-\sum\limits_{k=1}^{m} c_k}\delta+\frac{1}{1-\sum\limits_{k=1}^{m} c_k}\int_0^b |f(s)-f_n(s)|ds \\
&+\eta(t)+\frac{\sum\limits_{k=1}^{m} c_k}{1-\sum\limits_{k=1}^{m} c_k}|x(t_k)|.
\end{aligned}
$$

As $n \to \infty$, we conclude that

$$
|x(t)-y(t)| \leq \eta(t)+\frac{\sum\limits_{k=1}^{m} c_k\delta}{1-\sum\limits_{k=1}^{m} c_k}+\frac{\sum\limits_{k=1}^{m} c_k|x(t_k)|}{1-\sum\limits_{k=1}^{m} c_k}.
$$

From Theorem 5.2, we easily have that

$$
|x'(t)-y'(t)| \leq p(t)\eta(t)+\gamma(t), \quad \text{a.e. } t \in [0,b]. \qquad \square
$$

5.2.2 Some Properties of Solution Sets

We shall examine some properties of the map $S_{[0,b]}(\phi)$, where ϕ is the initial condition of the problem (5.14)–(5.16). Set

$$\Omega = AC([-r,b],\mathbb{R}^n) = \{y : [0,b] \to \mathbb{R}^n : y \in \mathcal{D} \cap AC(J_k,\mathbb{R}^n)\},$$

where $J_k = [t_k, t_{k+1}]$, $k = 0,\dots,m$, which is a Banach space with the norm

$$\|y\|_{APC} = \sup_{t\in[-r,b]} |y(t)| + \int_0^b |y'(t)|dt.$$

By the same approach used for Corollary 5.3, we can easily prove the following lemma.

Lemma 5.8. *Assume that the conditions* (E-5) *and* (E-6) *hold and there exists* $f \in L^1(J,\mathbb{R}_+)$ *such that*

$$H(0,F(t,0)) \le f(t) \text{ for a.e. } t \in J.$$

Then the set-valued map $S_{[0,b]} : \mathcal{D} \to \mathcal{P}(\Omega AC)$ *is Lipschitz with Lipschitz constant*

$$L = \frac{1-\left(1-\sum\limits_{k=1}^{m} c_k\right)^2}{1-\sum\limits_{k=1}^{m} c_k} + 2\exp\left(\frac{m(b)}{1-\sum\limits_{k=1}^{m} c_k}\right).$$

Moreover, if F *has compact and convex values, then* $S_{[0,b]}(\phi)$ *is compact.*

The next lemma is for the case where F is an L^1-Carathéodory multivalued map.

Lemma 5.9. *Let* $F : [0,b]\times\mathcal{D} \to \mathcal{P}_{cp}(\mathbb{R}^n)$ *be an* L^1*-Carathéodory multivalued map. Assume that*

(i) *for every* $\phi \in \mathcal{D}$, *the solution set is not empty (i.e.,* $S_{[0,b]}(\phi) \ne \emptyset$).

Then $S_{[0,b]}(\phi)$ *is a closed set in* Ω. *Moreover, if there exist a continuous nondecreasing function* $\psi : [0,\infty) \longrightarrow (0,\infty)$ *and* $p \in L^1(J,\mathbb{R}_+)$ *such that*

$$\|F(t,u)\|_{\mathcal{P}} \le p(t)\psi(\|u\|_\infty) \text{ for a.e. } t \in J \text{ and each } u \in \mathcal{D},$$

with

$$\int_0^b p(s)ds < \int_1^\infty \frac{du}{\psi(u)},$$

then $S_{[0,b]}(\phi)$ *is a compact set in* Ω, *and the multivalued map* $\phi \to S_{[0,b]}(\phi)$ *is upper semicontinuous.*

Proof. Consider the operator $N : \Omega \to \mathcal{P}(\Omega)$ defined by

$$N(y) = \left\{ h \in \Omega : h(t) = \begin{cases} \phi(t), & \text{if } t \in [-r,0], \\ \phi(0) + \int_0^t f(s)ds & \\ \quad + \sum_{0<t_k<t} I_k(y(t_k^-)), & \text{if } t \in [0,b], \end{cases} \right\}$$

where $f \in S_{F,y} = \{f \in L^1([0,b],\mathbb{R}^n) : f(t) \in F(t,y_t)\ a.e.\ t \in [0,b]\}$. Using the fact that F has compact values and is L^1-Carathéodory, we can easily show that N is upper semicontinuous, completely continuous and there exists $y \in \Omega$ such that $y \in N(y)$. Also $FixN = S_{[0,b]}(\phi)$; for more details, see Theorem 4.31 [178].

Next, we prove that $S_{[0,b]}(\phi)$ is a compact set in Ω. We can easily prove that there exists $L > 0$ such that

$$\|y\|_\Omega \leq L, \text{ for all } y \in S_{[0,b]}(\phi).$$

Hence, we will only prove that $S_{[0,b]}(\phi)$ is an equicontinuous set in Ω. Let $l_1, l_2 \in J$, $l_1 < l_2$, and let $y \in S_{[0,b]}(\phi)$. Then there exists a selection function $g \in S_{F,y}$ such that, for each $t \in [0,b]$, we have

$$y(t) = \phi(0) + \int_0^t g(s)ds + \sum_{0<t_k<t} I_k(y(t_k)).$$

Then,

$$|y(l_2) - y(l_1)| \leq \psi(L) \int_{l_1}^{l_2} p(s)ds + \sum_{l_1 \leq t_k < l_2} \sup\{|I_k(z)| : z \in B(0,L)\}.$$

As $l_2 \longrightarrow l_1$, the right-hand side of the above inequality tends to zero. The equicontinuity for the cases $l_1 < l_2 \leq 0$ and $l_1 < 0 \leq l_2$ is obvious. As a consequence of the Arzelà–Ascoli theorem, we conclude that $S_{[0,b]}(\phi)$ is a compact set in Ω.

Finally, we prove that the multivalued map $\phi \to S_{[0,b]}(\phi)$ is upper semicontinous. Since $S_{[0,b]}(\cdot)$ is quasicompact, from Lemma 2.46, we prove only that $S_{[0,b]}$ is closed (i.e., has a closed graph). Let $\phi_n \longrightarrow \phi$, $y^n \in S_{[0,b]}(\phi_n)$, and $y^n \longrightarrow y_*$. We shall show that $y_* \in S_{[0,b]}(\phi)$. Since $y_n \in S_{[0,b]}(\phi_n)$, there exists a selection $g_n \in S_{F,y^n}$ such that

$$y^n(t) = \begin{cases} \phi_n(t), & t \in [-r,0], \\ \phi_n(0) + \int_0^t g_n(s)ds + \sum_{0<t_k<t} I_k(y^n(t_k)), & t \in [0,b]. \end{cases}$$

Set

$$z^n(t) = \begin{cases} \phi(t), & t \in [-r,0], \\ \phi(0) + \int_0^t g_n(s)ds + \sum_{0<t_k<t} I_k(z^n(t_k)), & t \in [0,b]. \end{cases}$$

Note that $z^n(t) = y^n(t)$ for $t \in [0, b]$. By same idea as in Corollary 5.3, we can prove that there exists $g \in S_{F,y}$, such that

$$z_*(t) = \begin{cases} \phi(t), & t \in [-r, 0], \\ \phi(0) + \int_0^t g(s)ds + \sum_{0<t_k<t} I_k(y_*(t_k)), & t \in [0, b], \end{cases}$$

where we have $z_*(t) = y_*(t)$ for $t = t_k$, and where

$$\|z^n - z_*\|_\Omega \to 0 \; as \; n \to \infty.$$

Since

$$\|z^n - y^n\|_\Omega \le \|\phi_n - \phi\|_\infty \to 0 \; as \; n \to \infty,$$

we have

$$y_*(t) = \begin{cases} \phi(t), & t \in [-r, 0], \\ \phi(0) + \int_0^t g(s)ds + \sum_{0<t_k<t} I_k(y_*(t_k)), & t \in [0, b]. \end{cases}$$

This implies that $y_* \in S_{[0,b]}(\phi)$. □

5.3 Upper Semicontinuity without Convexity

It is well known that the initial value problem for the impulsive differential inclusions

$$y'(t) \in F(y), \quad \text{a.e.} \quad t \in J := [0, b] \backslash \{t_1, t_2, \dots, t_m\}, \tag{5.18}$$

$$y(t_k^+) - y(t_k) = I_k(y(t^-)), \quad k = 1, \dots, m, \tag{5.19}$$

$$y(0) = y_0, \tag{5.20}$$

where $F : J \times \mathbb{R}^n \to \mathcal{P}(\mathbb{R}^n)$ is a multivalued map with nonempty compact values, may not have solutions in case F is upper semicontinuous, but not necessarily convex-valued. Let

$$F(y) = \begin{cases} 1, & y < 0, \\ \{+1, -1\} & y = 0, \\ -1, & y > 0, \end{cases}$$

$I_k \in C(\mathbb{R}, \mathbb{R})$, $y_0 \in \mathbb{R} \backslash \{-1, 1\}$, and $t_k \in \mathbb{R}, k = 1, \dots, m$. It is clear that F is u.s.c. and the Cauchy problem has no solutions. So the upper semicontinuity of F is not sufficient to obtain the existence of solutions for (5.18)–(5.20) with non convex values of F. On other hand, we can present a classical problem from existence theory for upper semicontinuous differential inclusions without convexity. More precisely, we consider the impulsive problem,

$$y'(t) \in \left\{ \begin{array}{ll} -1 - y, & y < 0 \\ {[-1, +1]}, & y = 0, \\ +1 - y, & y > 0, \end{array} \right\}, \quad t \in [0, \infty) \setminus \{t_1, \dots\},$$

$$y(t_k^+) = y(t_k) + b_k,\ k = 1, \dots,$$

and the Cauchy condition

$$y(0) = 0,$$

where $t_k \in \mathbb{R}_+$, such that $t_1 < t_2 < \cdots$, $\lim\limits_{k\to\infty} t_k = \infty$, and $b_k \in \mathbb{R}_+$. It is clear that

$$F(y(t)) = \begin{cases} -1 - y, & y < 0, \\ [-1, +1], & y = 0, \\ +1 - y, & y > 0, \end{cases}$$

is upper semicontinuous and not convex. A solution of the above problem is defined by

$$y(t) = \begin{cases} 0, & \text{if } t \in [0, t_1], \\ 1 + b_1 e^{t_1 - t} - e^{t_1 - t}, & \text{if } t \in (t_1, t_2], \\ 1 + e^{t_2 - t}[1 + b_1 e^{t_1 - t_2} - e^{t_1 - t_2} + b_2] - e^{t_2 - t}, & \text{if } t \in (t_2, t_3], \\ \dots \\ \vdots \end{cases}$$

In this current section, we present a class of existence results for impulsive differential inclusions with upper semicontinuous, not necessarily convex-valued right hand side.

5.3.1 Nonconvex Theorem and Upper Semicontinuity

We first recall that the subdifferential of a convex function $V : \mathbb{R}^n \to \mathbb{R}$ at x is the set

$$\partial V(x) = \{v \in \mathbb{R} : V(y) - V(x) \geq \langle v, y - x\rangle\ \forall y \in \mathbb{R}^n\}.$$

In order to define the phase space and a solution of (5.18)–(5.20), we consider the space

$$H^1([0, b], \mathbb{R}^n) = \{y : [0, b] \to \mathbb{R}^n : y \in AC(J, ; \mathbb{R}^n),\ y' \in L^2(J, \mathbb{R}^n)\},$$

which is a Banach space equipped with the norm

$$\|y\|_{H^1} = \|y\|_{L^2} + \|y'\|_{L^2}, \text{for each } y \in H^1(J, \mathbb{R}^n).$$

We define

$$PC(J, \mathbb{R}^n) = \big\{y\ :\ y \text{ is a map from } J \text{ into } \mathbb{R}^n \text{ such that } y(t) \text{ is continuous at } t \neq t_k, \text{ left continuous at } t = t_k, \text{ and } y(t_k^+) \text{ exist, } k = 1, 2, \dots, m\big\}$$

which is a Banach space with the norm

$$\|y\|_{PC} = \sup\{|y(t)| : t \in [0,b]\}, \text{ for each } y \in PC,$$

and define

$$\Omega^1(J,\mathbb{R}^n) = PC(J,\mathbb{R}^n) \cap \bigcup_{k=0}^{m} H^1(J_k,\mathbb{R}^n),$$

which is a Banach space with the norm

$$\|y\|_{\Omega^1} = \|y\|_{PC} + \|y'\|_{H^1}.$$

Definition 5.10. A function $y \in \Omega^1$ is said be a solution of (5.18)–(5.20) if there exists a function $v \in L^2(J,\mathbb{R}^n)$ such that $v(t) \in F(y(t))$ a.e. on J such that $y'(t) = v(t)$ $a.e.\ t \in [0,b]$, and conditions (5.19) and (5.20) are satisfied.

Lemma 5.11. *Let $x \in PC[0,b],\mathbb{R}^n) \cap A(J_k,\mathbb{R}^n)$, $J_k = [t_{k-1},t_{k+1}]$, $k = 1,\dots,m$, $t_0 = 0, t_{m+1} = b$, with $x'(t) \in \partial V(x(t))$ a.e. on $[0,b]$. Then*

$$(V \circ x)'(t) = \|x'(t)\|^2, \quad \textit{for a.e. } t \in [0,b].$$

Now we are in a position to present an existence result for the above problem.

Theorem 5.12. *Let $F : \mathbb{R}^n \to \mathbb{R}^n$ be upper semicontinous, compact valued and such that there exists $V : \mathbb{R}^n \to \mathbb{R}$ that is convex and continuous with*

$$F(x) \subseteq \partial V, \textit{ for each } x \in \mathbb{R}^n.$$

Suppose there exist a continuous nondecreasing function $\psi : [0,\infty) \longrightarrow (0,\infty)$ and $p \in L^2(J,\mathbb{R}_+)$ such that

$$\|F(x)\| \le p(t)\psi(|x|) \textit{ for a.e. } t \in J \textit{ and each } x \in \mathbb{R}^n,$$

with

$$\int_1^\infty \frac{du}{\psi(u)} = \infty.$$

Then the impulsive Cauchy problem (5.18)–(5.20) *has at least one solution.*

Proof. It is clear that there exist $M', M > 0$ such that

$$\sup\{|y(t)|, t \in [0,b]\} \le M \quad \text{and} \quad \|y\|_{\Omega^1} < M',$$

where y is solution of the problem (5.18)–(5.20). On $[0,b]$ we define a sequence of approximate solutions y_n by

$$\begin{aligned} y_0(t) &= y_0, && t \in [0,b],\\ y_n(t) &= y_0 + \int_0^t f_n(s)ds + \sum_{0<t_k<t} I_k(y_{n-1}(t_k)), && t \in [0,b],\ n \in \mathbb{N}, \end{aligned}$$

where $f_n(t) \in F(y_{n-1}) \cap B(0, M)$. For all $r_1, r_2 \in [0, b]$, we have

$$|y_n(r_2) - y_n(r_1)| \leq \psi(M) \int_{r_1}^{r_2} p(s)ds + \sum_{0<t_k<r_2-r_1} \sup\{|I_k(u)| : u \in B(0, M)\}.$$

The right-hand side tends to zero as $r_2 - r_1 \to 0$. So the sequence $\{y_n\}$ is equicontinuous in PC. Notice also that $\int_0^b |y_n(t)|^2 dt = \int_0^b |f_n(t)|^2 dt$, and therefore, the sequence $\{y_n\}$ is bounded in $L^2([0, b], \mathbb{R}^n)$. Hence, there exists a subsequence, still denoted by $\{y_n\}_{n\in\mathbb{N}}$, and $y \in PC$ such that y_n converges to y in the supremum norm topology, and y_n' converges in the weak topology of L^2. Since $y_n'(t) = f_n(t) \in F(y_{n-1}(t))$, then $(y_{n-1}, y_n') \in \text{Gra}(F)$. Thus,

$$\lim_{n\to\infty} d((y_n(t), y_n'(t)), \text{Gra}(F)) \leq \lim_{n\to\infty} |y_n(t) - y_{n-1}(t)| = 0. \tag{5.21}$$

The weak convergence of y_n' to y' and Mazur's Lemma 2.66 imply

$$y'(t) \in \cap_n \overline{co}\{y_k(t) : k \geq n\}, \text{ a.e. on } [0, b].$$

Fix any $t \in [0, b] \setminus \{t_1, \ldots, t_m\}$ and consider $\xi \in \mathbb{R}^n$. Then, the last relation above yields

$$\begin{aligned}\langle y'(t), \xi\rangle &\leq \inf_n \sup_{k\geq n} \langle y_k'(t), \xi\rangle \\ &\leq \lim_{n\to\infty} \sup \sigma(coF(y_n(t)), \xi) \leq \sup \sigma(coF(y(t)), \xi),\end{aligned}$$

where the last inequality follows from the u.s.c. of coF. Thus, by the convexity and the closedness of the values of coF, we obtain $y'(t) \in coF(y(t))$. Hence, we can show that y is solution of the convex problem

$$y'(t) \in coF(y), \quad \text{a.e. } t \in [0, b], \tag{5.22}$$

$$y(t_k^+) - y(t_k) = I_k(y(t^-)), \quad k = 1 \ldots, m, \tag{5.23}$$

$$y(0) = y_0. \tag{5.24}$$

Since $F(x) \subseteq \partial V(x)$, we have

$$y'(t) \in \partial V(y), \quad \text{a.e. } t \in [0, b], \tag{5.25}$$

$$y(t_k^+) - y(t_k) = I_k(y(t^-)), \quad k = 1 \ldots, m, \tag{5.26}$$

$$y(0) = y_0. \tag{5.27}$$

From Lemma 5.11 and (5.25), we have that

$$\frac{d}{dt}(V(y(t)) = \langle y'(t), y'(t)\rangle \quad \text{a.e. on } [0, b],$$

therefore,

$$V(y(b)) - V(y_0) + \sum_{k=1}^{m} V(y(t_k)) - V(y(t_k) + I_k(y(t_k)) = \int_0^b |y'(s)|^2 ds. \tag{5.28}$$

From the definition of sub-differential, we obtain

$$V(y) \geq V(y_n(t)) + \langle y_n'(t), z - y_n(t)\rangle, \text{ for each } z \in \mathbb{R}^n,$$

and so

$$\begin{aligned} V(y_0)) &\geq V(y_n(t)) + \langle y_n'(t), y_0) - y_n(t)\rangle \\ &= V(y_n(t)) - \langle y_n'(t), y_n(t) - y_0 - \sum_{0<t_k<t} I_k(y_n(t_k))\rangle \\ &= V(y_n(t)) - \int_0^t \langle y_n'(s), y_n'(s)\rangle ds, \text{ for } t \in [0, t_1]. \end{aligned}$$

Thus,

$$\int_0^{t_1} |y_n'(s)|^2 ds \leq V(y_n(t_1)) - V(y_0). \tag{5.29}$$

For $t \in (t_1, t_2]$,

$$\begin{aligned} V(y_n(t_1) + I_1(y_n(t_1)) &\geq V(y_n(t)) + \langle y_n'(t), y_n(t_1) + I_1(y_n(t_1)) - y_n(t)\rangle \\ &= V(y_n(t)) - \langle y_n'(t), y_n(t) - y_n(t_1) - I_1(y_n(t_1))\rangle \\ &= V(y_n(t)) - \int_{t_1}^t \langle y_n'(s), y_n'(s)\rangle ds, \end{aligned}$$

so

$$\int_{t_1}^{t_2} |y_n'(s)|^2 ds \leq V(y_n(t_1) + I_1(y_n(t_1)) - V(y_n(t_2)). \tag{5.30}$$

We continue this process and so for $t \in (t_m, b]$,

$$\int_{t_m}^{b} |y_n'(s)|^2 ds \leq V(y_n(t_m) + I_1(y_n(t_m)) - V(y_n(b)). \tag{5.31}$$

Adding these together, we obtain

$$\int_0^b |y_n'(s)|^2 ds \leq \sum_{k=1}^m V(y_n(t_m) + I_1(y_n(t_m)) - V(y_n(t_k)) + V(y_n(b)) - V(y_0). \tag{5.32}$$

From (5.28) and the continuity of V, we have

$$\overline{\lim_{n\to\infty}} \int_0^b |y_n'(s)|^2 ds = \lim_{n\to\infty} \sup \|y_n'\|_{L^2}^2 \leq \int_0^b |y'(s)|^2 ds = \|y'\|_{L^2}^2. \tag{5.33}$$

By the weak lower semicontinuity of the norm,

$$\|y'\|_{L^2}^2 \leq \lim_{n\to\infty} \inf \|y_n'\|_{L^2}^2. \tag{5.34}$$

Hence,

$$\lim_{n\to\infty} \|y_n'\|_{L^2}^2 = \lim_{n\to\infty} \inf \|y_n'\|_{L^2}^2 = \lim_{n\to\infty} \sup \|y_n'\|_{L^2}^2 := \|y'\|_{L^2}^2. \tag{5.35}$$

By a result of Yosida [492], y_n' converges to y' strongly in $L^2([0,b],\mathbb{R}^n)$. Thus there exists a subsequence y_n' that converges pointwise a.e. to y'. Recalling (5.21),

$$d((y(t), y'(t)), \mathrm{Gra}(F)) = 0 \text{ for a.e. } t \in [0,b]. \tag{5.36}$$

Since the graph of F is closed,

$$y'(t) \in F(y(t)) \quad \text{for a.e.} \quad t \in [0,b].$$

Therefore, the problem (5.18)–(5.20) has at least one solution. □

5.3.2 An Application

Let $K \subset \mathbb{R}^n$ be closed and $x \in \mathbb{R}^n$. We define the best approximation map as

$$\pi_K(x) = \{y \in K : |y - x| = d(y, K)\}.$$

Lemma 5.13. *Let $\pi_K : \mathbb{R}^n \to \mathcal{P}(\mathbb{R}^n)$, $x \to \pi_K(x)$, be a best approximation map. Then π_K is upper semicontinuous and there exists $a, b > 0$ such that*

$$|y| \le a|x| + b, \textit{ for each } y \in \pi_K(x).$$

Proof. Let $y_m, x_m \in \mathbb{R}^n$ such that $x_m \to x$, $y_m \to y$ and $y_m \in \pi_K(x_m)$. Now

$$|y - x| \le |y_m - y| + |y_m - x_m| + |x_m - x|,$$

so

$$|y - x| - d(x_m, K) \le |y_m - y| + |x_m - x|. \tag{5.37}$$

$$d(x_m, K) - |x - y| \le |y_m - y| + |x_m - x|. \tag{5.38}$$

From (5.37) and (5.38), we obtain

$$|d(x_m, K) - |x - y|| \le |y_m - y| + |x_m - x| \to 0 \text{ as } m \to \infty.$$

Since $x \to d(x, K)$ is a continuous function, $|y - x| = d(x, K)$, and this implies that $y \in \pi_K(x)$. Hence, π_K has closed graph. Finally, we prove that π_K is locally compact. Set

$$B(0, q) = \{x \in \mathbb{R}^n : |x| \le q\}.$$

We can easily show that $\pi_K(B(0,q))$ is compact. Thus, π_K is upper semicontinuous. Let $y \in \pi_K(x)$; then

$$|y| \le |y - x| + |x| = d(x, K) + |x| \le 2|x| + d(0, K).$$

We consider the impulsive problem,

$$x'(t) \in \pi_K(x), \quad \text{a.e. } t \in J := [0,b]\backslash\{t_1, t_2, \ldots, t_m\}, \tag{5.39}$$

$$x(t_k^+) - x(t_k) = I_k(x(t_k^-)), \quad k = 1, \ldots, m, \tag{5.40}$$

$$x(0) = x_0, \tag{5.41}$$

where $I_k \in C(\mathbb{R}^n, \mathbb{R}^n)$, $k = 1, \ldots, m$. In [107], it is proved that the map $\pi_K(.)$ is contained in the subdifferential of a convex and continuous function $V : \mathbb{R}^n \to \mathbb{R}$. Then, from Lemma 5.13 and Theorem 5.12, we can prove that the problem (5.39)–(5.41) has at least one solution. □

5.4 Inclusions with Dissipative Right Hand Side

Let $(E, \|\cdot\|)$ be a real Banach space. It is known that $(\mathcal{P}_{cp}(E), H))$ is a complete metric space if H is a Hausdorff metric.

Definition 5.14. A multivalued map $F : \mathbb{R} \times E \to \mathcal{P}_{cp}(E)$ is said to be

(i) Upper semicontinuous at $x_0 \in \mathbb{R} \times E$ if for every $\epsilon > 0$ there exists $\delta > 0$ such that for all $x \in B(x_0, \delta)$ we have $F(x) \subset F(x_0) + B(0, \epsilon)$.

(ii) Strongly dissipative if there exist $c > 0$ and $r \geq 2$ such that for all $(t, x, y) \in \mathbb{R} \times E \times E$,

$$\langle F(t,x) - F(t,y), x - y\rangle_- \leq -c\|x - y\|^r,$$

where

$$\langle x, y\rangle_- = \|y\| \lim_{h \to 0^-} \frac{\|y + hx\| - \|y\|}{h},$$

is the lower semi-inner product on E.

5.4.1 Existence and Uniqueness Result

Let $(E, \|\cdot\|)$ be a real and reflexive Banach space, and $F : \mathbb{R} \times E \to \mathcal{P}_{cpc}(E)$ be u.s.c. We consider the impulsive differential inclusion,

$$y'(t) \in F(t, y(t)) \quad \text{a.e. } t \in \mathbb{R}_+, \tag{5.42}$$

$$y(t_k^+) - y(t_k^-) = I_k(y(t_k)), \quad k = 1, \ldots, \tag{5.43}$$

$$y(0) = y_0, \tag{5.44}$$

where $0 < t_1 < \cdots,\ \lim_{k \to \infty} t_k = \infty$, and $I_k \in C(E, E)$.

Theorem 5.15. *Suppose the following hypotheses hold:*

(E-7) *There exists $c > 0$ such that*

$$\langle F(t,x) - F(t,y), x - y\rangle_- \leq -c\|x - y\|^2.$$

(E-8) *There exist a function $p \in L^1([0,\infty), \mathbb{R}_+)$ and a continuous nondecreasing function $\psi : [0,\infty) \to [0,\infty)$ such that*

$$\|F(t,x)\| \leq p(t)\psi(\|x\|) \text{ for a.e. } t \in J \text{ and each } x \in E,$$

with

$$\int_1^\infty \frac{ds}{\psi(s)} = \infty.$$

(E-9) *There exist $c_k > 0$ such that*

$$\|I_k(x) - I_k(y)\| \leq c_k\|x - y\|, \quad \text{for all } x, y \in E,$$

$$\prod_{k=1}^{\infty} \sqrt{(1 + c_k)} < \infty, \quad \text{and} \quad \sum_{k=1}^{\infty} |I_k(0)| < \infty.$$

Then the initial value problem (5.42)–(5.44) *has a unique solution on* $[0,\infty)$. *Moreover, the solution of the problem* (5.42)–(5.44) *is stable.*

We proceed via an auxiliary lemma for the proof of the above theorem. Let $x(t,0,x_0)$ and $y(t,0,y_0)$ $(x_0 \neq y_0)$ be two solutions of the the inclusion (5.42)–(5.43). Then, we have the following lemma using hypotheses (E-7)–(E-9).

Lemma 5.16. *Assume that conditions* (E-7)–(E-9) *hold. Then*

(i) *For every two solutions of the problem* (5.42)–(5.43), *the following estimation is satisfied:*

$$\|x(t,0,x_0) - y(t,0,y_0)\| \leq e^{-ct}\|x_0 - y_0\| \prod_{0<t_k<t} \sqrt{(1 + c(t_k))},$$

$$c(t_k) = c_k.$$

(ii) *There exists $M > 0$ such that, for every solution y of the problem (5.42)–(5.44), we have*

$$\sup\{\|y(t)\| : t \in [0,\infty)\} \leq M.$$

Proof. Let $x(t,0,x_0)$ and $y(t,0,y_0)$ be two solutions of the problem (5.42)–(5.44). Then $x|_{[0,t_1]}$ and $y|_{[0,t_1]}$ are solutions to

$$x'(t) \in F(t,x(t)),\ x(0) = x_0,\ y'(t) \in F(t,y(t)),\ y(0) = y_0 \quad \text{a.e} \quad t \in [0,t_1].$$

$$\langle x'(t) - y'(t), x(t) - y(t)\rangle_- \leq -c\|x(t) - y(t)\|^2.$$

Notice that

$$\frac{1}{2}\frac{d}{dt}\|x(t)-y(t)\|^2 \leq -c\|x(t)-y(t)\|^2,$$

and integrating, we obtain

$$\|x(t)-y(t)\| \leq \|x_0-y_0\|e^{-ct}, \text{ for each } t \in [0,t_1].$$

Now $x|_{(t_1,t_2]}$ and $y|_{(t_1,t_2]}$, respectively, are solutions of the problems

$$\begin{aligned} x'(t) &\in F(t,x(t)),\ x(t_1^+) = x(t_1) + I_k(y(t_1)),\ t \in (t_1,t_2] \\ y'(t) &\in F(t,y(t)),\ y(t_1^+) = y(t_1) + I_k(y(t_1)),\ t \in (t_1,t_2]. \end{aligned}$$

Let $t \in (t_1,t_2]$; then

$$\langle x'(t)-y'(t), x(t)-y(t)\rangle_- \leq -c\|x(t)-y(t)\|^2.$$

Thus,

$$\frac{1}{2}\frac{d}{dt}\|x(t)-y(t)\|^2 \leq -c\|x(t)-y(t)\|^2.$$

Integrating the above equations gives

$$\begin{aligned} \|x(t)-y(t)\|^2 &\leq \|x(t_1^+)-y(t_1^+)\|^2 - 2c\int_0^t \|x(s)-y(s)\|^2 ds \\ &\leq (1+c_1)\|x(t_1)-y(t_1)\|^2 - 2c\int_{t_1}^t \|x(s)-y(s)\|^2 ds \\ &\leq (1+c_1)e^{-2ct_1}\|x_0-y_0\|^2 - 2c\int_{t_1}^t \|x(s)-y(s)\|^2 ds. \end{aligned}$$

By Gronwall's inequality, we have

$$\|x(t)-y(t)\|^2 \leq (1+c_1)e^{-2ct}\|x_0-y_0\|^2.$$

Hence,

$$\|x(t)-y(t)\| \leq \sqrt{(1+c_1)}e^{-ct}\|x_0-y_0\|.$$

We continue this process and also take into account that $x|_{(t_m,t_{m+1}]}$ and $y|_{(t_m,t_{m+1}]}$ are solutions of the problems

$$\begin{aligned} x'(t) &\in F(t,x(t)),\ x(t_1^+) = x(t_1) + I_k(y(t_1)),\ t \in (t_m,t_{m+1}] \\ y'(t) &\in F(t,y(t)),\ y(t_1^+) = y(t_1) + I_k(y(t_1)),\ t \in (t_1,t_{m+1}]. \end{aligned}$$

We obtain that for $t \in (t_m,t_{m+1}]$,

$$\|x(t)-y(t)\| \leq \prod_{k=1}^{m}\sqrt{(1+c_k)}e^{-ct}\|x_0-y_0\|.$$

Finally by (E-8) and (E-9), we can show that there exists $M > 0$ such that

$$\sup\{\|y(t)\| : t \in [0,\infty)\} \leq M.$$

This completes the proof of the lemma. □

Proof of Theorem 5.15. We consider the impulsive problem

$$y'(t) \in F_1(t, y(t)) \quad \text{a.e. } t \in I_a := [0,a], \tag{5.45}$$

$$y(t_k^+) - y(t_k^-) = I_k(y(t_k)), \quad k = 1,\dots,m, \tag{5.46}$$

$$y_n(0) = \frac{ny_0}{n+1} := a_n, \tag{5.47}$$

where $F_1(t, y(t)) = F(t, y(t)) \cap \overline{B}(0, M)$, $t_m = \sup\{t_k |\ k = 1,\dots,m\} < a$ and $n \in \mathbb{N}$.

For every initial condition, we define on $[0,a]$ the solution function $y_n(\cdot, 0, a_n)$ given by

$$ly_n(t) = a_n + \int_0^t f_n(s)ds + \sum_{0<t_k<t} I_k(y_n(t_k)), t \in [0,a],$$

$$y_n(t_k) = y_{n-1}(t_k),$$

where $f_n(t) \in F(t, y_n(0)) \cap B(0, M)$. Let $n, m \in \mathbb{N}$, $n > m$. Then from Lemma 5.16, we have

$$\begin{aligned}
\|y_n(t,0,a_n) - y(t,0,a_m)\| &\leq e^{-ct}\|x_0\|\left|\frac{n}{n+1} - \frac{m}{m+1}\right| \prod_{k=1}^{\infty}\sqrt{(1+c_k)} \\
&\leq \|x_0\| \prod_{k=1}^{\infty}\sqrt{(1+c_k)}\left|\frac{n}{n+1} - \frac{m}{m+1}\right| \\
&\leq 2\|x_0\| \prod_{k=1}^{\infty}\sqrt{(1+c_k)}\frac{n}{n+1}.
\end{aligned}$$

Hence,

$$\sup\{\|y_n(t,0,a_n) - y(t,0,a_m)\| : t \in [0,a]\} \leq 2\|x_0\| \prod_{k=1}^{\infty}\sqrt{(1+c_k)}\frac{n}{n+1}.$$

Therefore, the sequence $y_n(t) = y(t, 0, a_n)$ converges uniformly in every compact subset of $\mathbb{R}_+$ to a function in $PC([0,\infty, E)$ that we denote by $x(t, x_0)$. From Lemma 5.16, we have

$$\sup\{\|y(t)\| : t \in [0,\infty)\} \leq M.$$

Now we have to show that the function $y(t,0,y_0)$ is a solution of the problem (5.42)–(5.44). Put $I_\alpha = [0,\alpha]$, with $n > \alpha$, and denote $y(t,0,y_0)$ by $y(t)$. By the definition of $y_n(t)$, we have $y_n'(t) \in F(t, y_n(t))$, a.e. $t \in I_\alpha$.

Notice also that $\int_0^\alpha |y_n(t)|^2 dt = \int_0^\alpha |f_n(t)|^2 dt$ and therefore the sequence $\{y_n\}$ is bounded in $L^2(I_a, E)$.

It is clear that $y_n' \in L^\infty(I_\alpha, E) \subset L^2(I_\alpha, E)$, and since the Banach space $L^2(I_\alpha, E)$ is reflexive, $\{y_n'\}$ converges in the weak topology of L^2. Hence, for every $z \in L^2(I_\alpha, E^*)$, we have

$$\langle z, y_n'\rangle \to \langle z, u\rangle \text{ as } n \to \infty.$$

Consequently,

$$\int_{I_\alpha} \langle z(s), y_n'(s)\rangle ds \to \int_{I_\alpha} \langle z(s), u(s)\rangle ds.$$

So

$$\begin{aligned} y_n(t) &= y_n(0) + \int_0^t y_n'(s)ds + \sum_{0<t_k<t} I_k((y_n(t_k)) \\ &\to y(0) + \int_0^t y'(s)ds + \sum_{0<t_k<t} I_k((y(t_k)). \end{aligned}$$

Then y_n converges uniformly to

$$u_1(t) = y(0) + \int_0^t u(s)ds + \sum_{0<t_k<t} I_k((y(t_k))$$

in $PC(I_\alpha, E)$. But $y_n(t) \to y(t)$, and hence

$$y(t) = y(0) + \int_0^t u(s)ds + \sum_{0<t_k<t} I_k(y(t_k)).$$

We show that $y'(t) = u(t)$ a.e. $t \in I_\alpha$. Since $L^2 \hookrightarrow L^1$, we conclude that $y_n' \to y'$ in L^2 for the weak topology $\sigma(L^2(I_\alpha, E), L^2(I_\alpha, E^*))$. Let $\epsilon > 0$ and set

$$S_\epsilon = \{z \in L^2 : z(t) \in F(t, x(t)) + \overline{B}(0,\epsilon), \text{ a.e. } t \in I_\alpha\}.$$

Since $y_n \to y$ and F is u.s.c., $y_n' \in S_\epsilon$ for large n. Hence, $S_\epsilon \neq \emptyset$, closed and convex, so S_ϵ is weakly closed. Since $y_n' \in S_\epsilon$ and $y_n' \to y'$, we see that $y'(t) \in \overline{F(t, y(t))} = F(t, y(t))$ a.e. $t \in I_\alpha$. Since α is arbitrary, the problem (5.42)–(5.44) has at least one solution.

Finally we show that the solutions is unique. Let x and y be two solutions of the problem (5.42)–(5.44). By using Lemma 5.16 we have

$$\|x(t) - y(t)\| \le \prod_{k=1}^{\infty}(1 + c_k)\|x(0) - y(0)\|, \text{ for every } t \in \mathbb{R}_+.$$

Thus, $x(t) = y(t)$ for all $t \in \mathbb{R}_+\backslash\{t_1, t_2, \ldots\}$, and since the I_k are continuous functions, we have

$$x(t_k^+) = x(t_k) + I(x(t_k)) = y(t_k) + I_k(y(t_k)) = y(t_k^+),\ k = 1, 2\ldots .$$

For stability of solutions, let $x(t, 0, x_0)$ and $y(t, 0, y_0)$ be two solutions of the problem (5.42)–(5.44) with initial conditions x_0, y_0, respectively. Then, from Lemma 5.16, we have

$$\sup\{\|x(t) - y(t)\| : t \in [0, \infty)\} \le \prod_{k=1}^{\infty}(1 + c_k)\|x_0 - y_0\| < \epsilon,$$

where ϵ is an positive number. Hence, for $\eta := \epsilon/[\prod_{k=1}^{\infty}(1 + c_k)]$, for every x_0, y_0 with $\|x_0 - y_0\| \le \eta$, we have $\sup\{\|x(t) - y(t)\| : t \in [0, \infty)\} \le \epsilon$. i.e., solutions are stable. □

5.5 Directionally Continuous Selection and IDIs

Consider an impulsive differential inclusion of the form,

$$y'(t) \in F(t, y(t)), \quad \text{a.e.} \quad t \in J\backslash\{t_1, \ldots, t_m\}, \tag{5.48}$$

$$y(t_k^+) - y(t_k) = I_k(y(t_k^-)), \quad k = 1, \ldots, m, \tag{5.49}$$

$$y(0) = y_0, \tag{5.50}$$

where $J = [0, b]$, $F : J \times \mathbb{R}^n \to \mathcal{P}(\mathbb{R}^n)$ is a multivalued map, $I_k \in \mathcal{F}C(\mathbb{R}^n, \mathbb{R}^n)$, $\mathcal{F}C(\mathbb{R}^n, \mathbb{R}^n) = \{g : \mathbb{R}^n \to \mathbb{R}^n | \mathrm{Dom}(g) = \mathbb{R}^n\}$, and $\mathrm{Dom}(g) = \{x \in \mathbb{R}^n | g(x) < \infty\}$. A Carathéodory solution of (5.48)–(5.50) can be obtained by constructing a selection $f(t, y) \in F(t, y)$ and by solving the impulsive differential equation,

$$y'(t) = f(t, y(t)), \quad \text{a.e.} \quad t \in J\backslash\{t_1, \ldots, t_m\} \tag{5.51}$$

$$y(t_k^+) - y(t_k) = I_k(y(t_k^-)), \quad k = 1, \ldots, m, \tag{5.52}$$

$$y(0) = y_0. \tag{5.53}$$

The aim of the present section is to show that the naive approach of selecting $f(t, x) \in F(t, x)$ and then solving (5.51)–(5.53) can be pursued in nonconvex cases as well.

5.5.1 Directional Continuity

In the following, a set $\Gamma \subset \mathbb{R}^n$ will be called a *cone* if Γ is a nonempty closed convex subset of $\mathbb{R}^n$ such that

(i) $x \in \Gamma, \lambda \ge 0 \Rightarrow \lambda x \in \Gamma$

(ii) $\Gamma \cap (-\Gamma) = \{0\}$.

We now introduce the basic concept of directional topology.

Definition 5.17. Let Γ be a cone in $\mathbb{R}^n$ and let Y be a metric space. A map $f : \mathbb{R}^n \to Y$ is Γ-continuous at a point $x_0 \in \mathbb{R}^n$ if and only if for every $\epsilon > 0$ there exists $\delta > 0$ such that $d(f(x), f(x_0)) < \epsilon$ for all $x \in B(x_0, \delta) \cap (x_0 + \Gamma)$. We say that f is Γ-continuous on a set A if f is Γ-continuous at every point $x_0 \in A$.

Now, we set some lemmas concerning directional continuity.

Lemma 5.18 ([102]). *Let $f : \Gamma \to \mathbb{R}^n$ be a function. Then:*

(a) *f is Γ-continuous at x_0 if and only if $\lim_{n\to\infty} f(x_n) = f(x_0)$ for every sequence x_n tending to x_0 such that $x_n - x_0 \in \Gamma$ for all $n \geq 1$.*

(b) *If f is Γ-continuous at x_0, then f is also Γ'-continuous at x_0 for every cone $\Gamma' \subset \Gamma$.*

(c) *If $(f_n)_{n\geq 1}$ is a Γ-continuous sequence which converges uniformly to f, then f is Γ-continuous.*

Next we state a selection theorem due to Bressan [102]

Theorem 5.19. *Let F be a lower semicontinuous map with nonempty closed values from $\mathbb{R}^n$ into a complete metric space Y. Then, for every cone $\Gamma \subset \mathbb{R}^n$, F admits a Γ-continuous selection.*

To prove our existence result, we first give an auxiliary theorem due to Bressan [102].

Theorem 5.20. *Let $f : [0, b] \times \mathbb{R}^n \to \mathbb{R}^n$ be Γ^M–continuous and locally bounded, where*

$$\Gamma^M = \{(t, x) \in \mathbb{R} \times \mathbb{R}^n : |x| \leq tM\}.$$

Consider the substitution operator $g : C([0, b], \mathbb{R}^n) \to L^1([0, b], \mathbb{R}^n)$ defined by

$$g(u)(t) = f(t, u(t)). \tag{5.54}$$

Then the restriction of g to K_L is continuous for every $L > M$, where

$$K_L = \Big\{u \in C([0, b], \mathbb{R}^n) : \|u(t) - u(t')\| \leq (M + 1)|t - t'|, \text{ for all } t, t' \in [0, b]\Big\}.$$

Here is our existence result.

Theorem 5.21. *Let $F : [0, b] \times \mathbb{R}^n \to \mathcal{P}_{cl}(\mathbb{R}^n)$ be such that,*

(E-10) *$t \to F(t, x)$ is measurable for each $x \in \mathbb{R}^n$;*

(E-11) *for every $t \in [0, b]$, $x \to F(t, x)$ is l.s.c.;*

(E-12) *there exits* $M > 0$ *such that*

$$F(t,x) \cap \overline{B}(0,M) \neq \emptyset \textit{ for all } (t,x) \in \mathbb{R} \times \mathbb{R}^n.$$

Then the problem (5.48)–(5.50) *has at least one Carathéodory solution on* $[0,b]$.

Proof. We consider the problem

$$y'(t) \in F_1(t, y(t)), \qquad \text{a.e. } t \in J = [0,b]\backslash\{t_1,\dots,t_m\}, \tag{5.55}$$

$$\Delta y|_{t=t_k} = y(t_k^+) - y(t_k) = I_k(y(t_k^-)), \quad k = 1,\dots,m, \tag{5.56}$$

$$y(0) = y_0, \tag{5.57}$$

where $F_1 : \mathbb{R} \times \mathbb{R}^n \to \mathcal{P}(\mathbb{R}^n)$, is a multivalued map defined by

$$F_1(t,x) = F(t,x) \cap \overline{B}(0,M) \text{ for all } (t,y) \in \mathbb{R} \times \mathbb{R}^n.$$

We can easily show that F_1 is lower semicontinuous, and by definition of F_1, we have

$$F_1(t,y) \subset B(0,M), \text{ for all } t \in [0,b] \text{and } \|y\| \leq Mb.$$

Let $(t_m, x_m)_{m\geq 1} \in \mathbb{R} \times \mathbb{R}^n$ be a sequence converging to (t,x). We consider the multivalued map $F_2 : \mathbb{R} \times \mathbb{R}^n \to \mathcal{P}(\mathbb{R}^n)$ defined by

$$(t,x) \to F_2(t,x) = \overline{B}(0,M).$$

For every $m \geq 1$, we have $F(t_m, x_m) = \overline{B}(0,M)$. Let $y \in F_2(t,x)$; then $y \in F(t_m, x_m)$. Set $y_m = y$, $m \geq 1$. This implies that $y_m \to y \in F_2(t,x)$ as $m \to \infty$. Then F_1 l.s.c. From Theorem 5.19, F_1 has a Γ^{M+1}–continuous selection $f(t,x) \in F(t,x)$ for all $(t,x) \in \mathbb{R} \times \mathbb{R}^n$. Consider the problem

$$y'(t) = f(t, y(t)), \qquad t \in [0,b], \tag{5.58}$$

$$y(t_k^+) - y(t_k) = I_k(y(t_k^-)), \quad k = 1,\dots,m, \tag{5.59}$$

$$y(0) = y_0. \tag{5.60}$$

Remark 5.22. Notice that if $y \in PC$ is a solution of the problem (5.58)–(5.60), then y is a solution to the problem (5.48)–(5.50).

The existence of a solution for the above problem will be given in several steps.

Step 1. Consider the problem,

$$y'(t) = f(t, y(t)) \quad \text{a.e.} \quad t \in [0, t_1], \tag{5.61}$$

$$y(0) = y_0. \tag{5.62}$$

Transform the problem (5.61)–(5.62) into a fixed point problem. Consider the operator $N : K_M \longrightarrow K_M$ defined by,

$$N(y)(t) = y_0 + \int_0^t f(s, y(s))ds, \quad t \in [0, t_1],$$

where $K_M = \{y \in C([0, t_1], \mathbb{R}^n) : y(0) = y_0, \ \|y(t) - y(t')\| \leq M|t - t'| \text{ for all } t, t' \in [0, t_1]\}$. From Theorem 5.20 $f(\cdot, \cdot)$ is a continuous function. By the definition of N, we can easily show that $N(K_M) \subset K_M$, N is completely continuous, and K_M is a compact, convex subset of $C([0, t_1], \mathbb{R}^n)$. Therefore, by Schauder's theorem, N has at least one fixed point that is a solution of the problem (5.61)–(5.62). We denote this solution by y_1.

Step 2. Now consider the problem,

$$y'(t) = f(t, y(t)), \quad \text{a.e.} \quad t \in (t_1, t_2], \tag{5.63}$$

$$y(t_1^+) = y_1(t_1) + I_1(y_1(t_1)). \tag{5.64}$$

To transform this problem into a fixed point problem, consider the operator $N_1 : K_M^1 \to K_M^1$ defined by,

$$N_1(y)(t) = y_1(t_1) + I_1(y_1(t_1)) + \int_{t_1}^t f(s, y(s))ds,$$

where $K_M^1 = \{y \in C^1 : y(t_1^+)) = y_1(t_1) + I_1(y_1(t_1)) \text{ and } \|y(t) - y(t')\| \leq M|t - t'| \text{ for all } t, t' \in (t_1, t_2]\}$. It is clear that K_M^1 is a compact convex subset in C^1 and $N_1(K_M^1) \subset K_M^1$, where

$$C^1 = \{y \in C((t_1, t_2]) : y(t_1^+) \text{ exists}\}.$$

Since f is a continuous function on K_M^1, we can show that N_1 is completely continuous. Hence, by Schauder's theorem, N_1 has at least one fixed point, which is a solution of the problem (5.63)–(5.64). We denote this solution by y_2.

Step 3. We continue this process and taking into account that $y_m := y\big|_{[t_m, b]}$ is a solution to the problem

$$y(t) = f(t, y(t)), \quad \text{a.e.} \quad t \in (t_m, b], \tag{5.65}$$

$$y(t_m^+) = y_{m-1}(t_m) - I_m(y_{m-1}(t_m)). \tag{5.66}$$

A solution y of the problem (5.51)–(5.53) is then defined by

$$y(t) = \begin{cases} y_1(t), & \text{if } t \in [0, t_1], \\ y_2(t), & \text{if } t \in (t_1, t_2], \\ \dots \\ y_{m+1}(t), & \text{if } t \in (t_m, b]. \end{cases}$$

□

Chapter 6

Differential Inclusions with Infinite Delay

In this chapter, we establish sufficient conditions for membership in some classes of impulsive functional differential inclusions with unbounded delay. More precisely, in the first section, we will consider, for example, first order functional differential inclusions of the form,

$$y'(t) \in F(t, y_t), \qquad \text{a.e. } t \in J =: [0, b] \backslash \{t_1, \dots, t_m\}, \tag{6.1}$$

$$y(t_k^+) - y(t_k) = I_k(y(t_k^-)), \qquad k = 1, \dots, m, \tag{6.2}$$

$$y(t) = \phi(t), \qquad t \in (-\infty, 0], \tag{6.3}$$

where I_k, ϕ, and B are as in problem (3.1)–(3.3) and $F : J \times B \to \mathcal{P}(\mathbb{R}^n)$ is a multivalued map. We will give three theorems: one for the convex case and two for the nonconvex. In the second section, we consider first order impulsive functional differential inclusions with infinite delay,

$$y'(t) \in F(t, y_t), \qquad \text{a.e. } t \in J := [0, \infty),\ t \neq t_k,\ k = 1, \dots, \tag{6.4}$$

$$y(t_k^+) - y(t_k^-) = I_k(y(t_k^-)), \qquad t = t_k,\ k = 1, \dots, \tag{6.5}$$

$$Ay(t) - y_\infty = \phi(t), \qquad t \in (-\infty, 0], \tag{6.6}$$

where I_k, B, ϕ, A, y_∞,and F are as in (3.1)–(3.3), (4.1)–(4.3) and (6.1)–(6.3). In all this chapter we assume that the multifunction $t \to F(t, y_t)$ is measurable.

6.1 Existence Results

In this section, we give sufficient conditions for existence of solutions for the first order functional differential inclusions (6.1)–(6.3). First, we assume that F has convex values. Before stating and proving our result for this case, we give the definition of its solution.

Definition 6.1. A function $y \in B_b$ is said to be solution of (6.1)–(6.3) if there exists a function $v \in L^1([0, b], \mathbb{R}^n)$ such that $v(t) \in F(t, y_t)$ a.e. $t \in [0, b]$, and v satisfies $y'(t) = v(t)$ a.e. on J, and the conditions (6.2)–(6.3) are satisfied.

Theorem 6.2 (Convex case). *Suppose the following hypotheses are satisfied:*

(F-1) *The function $F : [0, b] \times B \longrightarrow \mathcal{P}(\mathbb{R}^n)$ is a nonempty, compact, convex, multivalued map such that:*

(a) $(t,\cdot) \mapsto F(t,\cdot)$ *is measurable;*
(b) $x \mapsto F(t,x)$ *is upper semicontinuous for a.e.* $t \in [0,b]$.

(F-2) *There exist a continuous nondecreasing function* $\psi : [0,\infty) \longrightarrow (0,\infty)$ *and* $p \in L^1([0,b],\mathbb{R}_+)$ *such that*

$$\|F(t,x)\|_{\mathcal{P}} \leq p(t)\psi(\|x\|_B) \text{ for a.e. } t \in [0,b] \text{ and each } x \in B,$$

with

$$K_b \int_0^b p(s)ds < \int_0^\infty \frac{dx}{\psi(x)}.$$

Then the impulsive initial value problem (6.1)–(6.3) has at least one solution on $(-\infty,b]$.

Proof. Consider the operator $N : B_b \to \mathcal{P}(B_b)$ defined by

$$N(y) = \left\{ h \in B_b : h(t) = \begin{cases} \phi(t), & t \in (-\infty,0], \\ \phi(0) + \int_0^t v(s)ds \\ \quad + \sum_{0<t_k<t} I_k(y(t_k^-)), & t \in [0,b], \end{cases} \right\}$$

where

$$v \in S_{F,y} = \{v \in L^1([0,b],\mathbb{R}^n) : v(t) \in F(t,y_t) \text{ for a.e. } t \in [0,b]\}.$$

First, we will prove that N has a fixed point.

Let $x(\cdot) : (-\infty,b] \to \mathbb{R}^n$ be the function defined by

$$x(t) = \begin{cases} \phi(0), & \text{if } t \in [0,b], \\ \phi(t), & \text{if } t \in (-\infty,0]. \end{cases}$$

Then $x_0 = \phi$. For each $z \in B \cap PC$ with $z_0 = 0$, we denote by $\bar{z}$ the function defined by

$$\bar{z}(t) = \begin{cases} z(t), & \text{if } t \in [0,b], \\ 0, & \text{if } t \in (-\infty,0]. \end{cases}$$

If $y(\cdot)$ satisfies the integral equation

$$y(t) = \phi(0) + \int_0^t v(s)ds + \sum_{0<t_k<t} I_k(y(t_k^-)),$$

we can decompose it as $y(t) = z(t) + x(t)$, $0 \leq t \leq b$. This implies $y_t = \bar{z}_t + x_t$, for every $0 \leq t \leq b$, and the function $z(\cdot)$ satisfies

$$z(t) = \int_0^t v(s)ds + \sum_{0<t_k<t} I_k(\bar{z}(t_k^-) + x(t_k^-)), \tag{6.7}$$

where $v(t) \in F(t, \bar{z}_t + x_t)$ a.e $t \in [0, b]$. Let the operator $P : C_0 \to \mathcal{P}(C_0)$ be defined by

$$P(z) = \left\{ h \in C_0 \mid h(t) = \begin{cases} 0, & t \in (-\infty, 0], \\ \int_0^t v(s)ds \\ \quad + \sum_{0<t_k<t} I_k(\bar{z}(t_k^-) + x(t_k^-)), & t \in [0, b], \end{cases} \right\}$$

where

$$v \in S_{F,z} = \{v \in L^1([0,b], \mathbb{R}^n) : v(t) \in F(t, \bar{z}_t + x_t) \text{ for a.e. } t \in [0,b]\}.$$

Clearly, the operator N having a fixed point is equivalent to P having one, so it suffices to prove that P has a fixed point.

Let $z \in \lambda P(z)$ be a possible solution to our problem for some $\lambda \in (0,1)$. As in the proof of Theorem 3.6, we can show that there exists $M > 0$ such that

$$\|z\|_0 := \sup\{|z(t)| : t \in [0,b]\} \le M.$$

Set

$$U_0 = \{z \in C_0 | \, \|z\|_0 < M + 1\}.$$

We now show that P satisfies the assumptions of the Leray–Schauder alternative for multivalued maps. The proof will be given in several steps.

Step 1. *$P(z)$ is convex for each $z \in C_0$.*

Indeed, if h_1, h_2 belong to $P(z)$, then there exist $v_1, v_2 \in S_{F,\bar{z}+x}$ such that, for each $t \in J$, we have

$$h_i(t) = \int_0^t v_i(s)ds + \sum_{0<t_k<t} I_k(\bar{z}(t_k^-) + x(t_k^-)), \quad i = 1, 2.$$

Let $0 \le d \le 1$. Then for each $t \in J$, we have

$$(dh_1 + (1-d)h_2)(t) = \int_0^t [dv_1(s) + (1-d)v_2(s)]ds + \sum_{0<t_k<t} I_k(\bar{z}(t_k^-) + x(t_k)).$$

Since $S_{F,\bar{z}+x}$ is convex (because F has convex values),

$$dh_1 + (1-d)h_2 \in P(z).$$

Step 2. *P maps bounded sets into bounded sets in C_0.*

It suffices to show that there exists a positive constant ℓ such that, for each $w \in \mathcal{B}_q = \{z \in C_0 : \|z\|_0 \le q\}$, we have $\|P(z)\|_0 \le \ell$.
Let $h \in N(u)$. Then there exists $v \in S_{F,\bar{z}+x}$ such that

$$h(t) = \int_0^t v(s)ds + \sum_{0<t_k<t} I_k(\bar{z}(t_k^-) + x(t_k^-)).$$

From (F-1), we obtain for $t \in J$,

$$\begin{aligned}|h(t)| &\leq \psi(q_*)\int_0^b p(s)ds + \sum_{0<t_k<t} |I_k(\bar{z}(t_k^-) + x(t_k^-))| \\ &\leq \psi(q_*)\|p\|_{L^1} + \sum_{k=1}^{m} \sup\{|I_k(u)| : |u| \leq \bar{q}\} := \ell,\end{aligned}$$

where

$$\|\bar{z}_s + x_s\|_B \;\leq\; K_b q + M_b\|\phi\|_B + K_b|\phi(0)| := q_*$$

and

$$|\bar{z}(t_k) + x(t_k)| \leq |\bar{z}(t_k)| + |x(t_k)\| \leq q + |\phi(0)| := \bar{q}, \text{ for each } t_k \in J.$$

Step 3. *P maps bounded sets into equicontinuous sets of C_0.*

Let $\bar{t}_1, \bar{t}_2 \in J, \bar{t}_1 < \bar{t}_2$, and $\mathcal{B}_q$ be a bounded set of C_0 as in Step 2. Then,

$$|h(\bar{t}_2) - h(\bar{t}_1)| \leq \psi(q_*)\int_{\bar{t}_1}^{\bar{t}_2} p(s)ds + \sum_{0<t_k<\bar{t}_2-\bar{t}_1} |I_k(\bar{z}(t_k^-, x) + x(t_k^-))|.$$

The right-hand side tends to zero as $\bar{t}_2 - \bar{t}_1 \to 0$.

As a consequence of Steps 2 and 3, together with the Arzelà–Ascoli theorem, we conclude that $P : C_0 \longrightarrow C_0$ is completely continuous.

Step 4. *P has a closed graph.*

Let $z_n \longrightarrow z_*, h_n \in P(z_n)$, and $h_n \longrightarrow h_*$. We shall prove that $h_* \in P(z_*)$. Now $h_n \in P(z_n)$ means that there exists $v_n \in S_{F,\bar{z}_n+x}$ such that, for each $t \in J$,

$$h_n(t) = \int_0^t v_n(s)ds + \sum_{0<t_k<t} I_k(\bar{z}(t_k) + x(t_k)).$$

We must show that there exists $v_* \in S_{F,\bar{z}_*+x}$ such that for each $t \in J$

$$h_*(t) = \int_0^t v_*(s)ds + \sum_{0<t_k<t} I_k(\bar{z}(t_k) + x(t_k)).$$

Clearly, since I_k are continuous functions we have that, as $n \to \infty$,

$$\left\|\left(h_n - \sum_{0<t_k<t} I_k(\bar{z}_n(t_k) + x(t_k))\right) - \left(h_* - \sum_{0<t_k<t} I_k(\bar{z}_*(t_k) + x(t_k))\right)\right\|_\infty \to 0.$$

Consider the linear continuous operator

$$\Gamma : L^1(J, \mathbb{R}^n) \longrightarrow C_0$$

defined by

$$v \longmapsto \Gamma(v)(t) = \int_0^t v(s)dsd\tau.$$

From Lemma 2.65, it follows that $\Gamma \circ S_F$ is a closed graph operator. Moreover, we have that

$$\left(h_n(t) - \sum_{0<t_k<t} I_k(\overline{z}_n(t_k) + x(t_k))\right) \in \Gamma(S_{F,\overline{z}_n+x}).$$

Since $z_n \longrightarrow z_*$, it follows from Lemma 2.65 that

$$h_*(t) = \int_0^t v_*(s)ds + \sum_{0<t_k<t} I_k(\overline{z}(t_k) + x(t_k)),$$

for some $v_* \in S_{F,z_*+x}$.

Clearly, $P : \overline{U}_0 \to \mathcal{P}(C_0)$ is completely continuous. From the choice of U_0, there is no $u \in \partial U_0$ such that $z \in \lambda P(z)$ for some $\lambda \in (0,1)$. As a consequence of the nonlinear alternative of Leray Schauder type (Lemma 2.87), we deduce that P has a fixed point z in $\overline{U}_0$. Hence, N has a fixed point y that is a solution to the problem (6.1)–(6.3). □

Now we give sufficient conditions for the existence of solutions to the problem (6.1)–(6.3) in case F has nonconvex values.

Theorem 6.3. (Nonconvex case) *Suppose that condition (F-2) is satisfied, and assume that*

(F-3) $F : J \times B \longrightarrow \mathcal{P}(\mathbb{R}^n)$ *is a nonempty compact valued multivalued map such that:*
  (a) $t \mapsto F(t,x)$ *is* $\mathcal{L} \otimes \mathcal{B}$ *measurable;*
  (b) $x \mapsto F(t,x)$ *is lower semicontinuous for a.e.* $t \in J$.

Then the impulsive initial value problem (6.1)–(6.3) *has at least one solution.*

Proof. Let $P : C_0 \to \mathcal{P}(C_0)$ where P is defined in Theorem 6.2. We shall show that P has at least one fixed point. We define the Niemytzki operator associated to F,

$$\mathcal{F} : C_0 \to \mathcal{P}(L^1(J,\mathbb{R}^n)),$$

by letting

$$\mathcal{F}(z) = \left\{v \in L^1(J,\mathbb{R}^n) : v(t) \in F(t,\bar{z}_t + x_t) \text{ for a.e. } t \in J\right\}$$

be a selection set of P. From (F-2) and (F-3) and by Lemma 2.2 in Frigon [205], it follows that F is of lower semicontinuous type. Then, from Theorem 2.85, there exists a continuous function $f : C_0 \to L^1(J, \mathbb{R}^n)$ such that $f(z) \in \mathcal{F}(\bar{z}+x)$ for all $y \in C_0$. Consider the problem,

$$
\begin{aligned}
z'(t) &= f(\bar{z}_t + x_t), && \text{a.e. } t \in J,\ t \neq t_k\ k = 1, \dots, m, && (6.8)\\
\Delta z(t_k) &= I_k(\bar{z}(t_k) + x(t_k)), && k = 1, \dots, m, && (6.9)\\
z(t) &= 0, && t \in (-\infty, 0], && (6.10)
\end{aligned}
$$

and consider the operator $P_1 : C_0 \to C_0$ defined by

$$
(P_1 z)(t) = \begin{cases} 0, & t \in (-\infty, 0],\\ \int_0^t f(\bar{z}_s + x_s)ds + \sum_{0<t_k<t} I_k(\bar{z}(t_k) + x(t_k)), & t \in J. \end{cases}
$$

Clearly, if $z \in C_0$ is fixed point of P_1, then z is fixed fixed point of P. Thus, there exists $y \in B_b$ such that y is a fixed point of the operator N, where N is defined in Theorem 6.2.

From (F-2) and (F-3), similar to the proof of Theorem 3.6, we can show that P_1 is completely continuous and there exists $M > 0$ such that for every solution of $z = \lambda P_1(z)$, for some $\lambda \in (0, 1)$, we have $\|z\|_0 \leq M$. Set

$$U = \{z \in C_0 : \|z\|_0 < M + 1\}.$$

We see that U is an open set in C_0. From the choice of U there is no $z \in \partial U$ such that $z = \lambda P_1(z)$ for some $\lambda \in (0, 1)$. As a consequence of the nonlinear alternative of Leray Schauder type (Lemma 2.86), we deduce that P_1 has a fixed point z in $\overline{U}$. Hence, N has a fixed point y that is a solution to the problem (6.1)–(6.3). □

In this next part we present a second result for the problem (6.1)–(6.3) with a non-convex valued right-hand side. Our considerations are based on a fixed point theorem for contraction multivalued operators given by Covitz and Nadler.

Theorem 6.4. *Suppose that*

(F-4) $F : J \times B \longrightarrow \mathcal{P}_{cp}(\mathbb{R}^n)$; $(t, u) \longmapsto F(t, u)$ *is measurable for each* $u \in B$.

(F-5) *There exists a function* $l \in L^1(J, \mathbb{R}^+)$ *such that, for a.e.* $t \in J$ *and all* u, $\overline{u} \in B$,

$$H(F(t,u), F(t,\overline{u})) \leq l(t)\|u - \overline{u}\|_B,$$

and

$$H(0, F(t,0)) \leq l(t) \ \text{for a.e. } t \in J.$$

Then the IVP (6.1)–(6.3) *has at least one solution.*

Proof. The proof will be given in several steps.

Step 1. Consider the problem,

$$y'(t) \in F(t, y_t), \quad a.e. \quad t \in [0, t_1], \tag{6.11}$$

$$y(t) = \phi(t), \qquad t \in (-\infty, 0]. \tag{6.12}$$

Transform the problem (6.11)–(6.12) into a fixed point problem. Consider the operator $N : B \cap C([0, t_1], \mathbb{R}^n) \to \mathcal{P}(B \cap C([0, t_1], \mathbb{R}^n))$ defined by,

$$N(y) = \left\{ h \in B \cap C([0, t_1], \mathbb{R}^n) | h(t) = \begin{cases} \phi(t), & t \in (-\infty, 0], \\ \phi(0) + \int_0^t v(s)ds, & t \in [0, b], \end{cases} \right.$$

where

$$v \in S_{F,y} = \{v \in L^1([0, t_1], \mathbb{R}^n) : v(t) \in F(t, y_t) \text{ for a.e. } t \in [0, t_1]\}.$$

We will prove that N has a fixed point.

Let $x(\cdot) : (-\infty, t_1] \to \mathbb{R}^n$ be the function defined by

$$x(t) = \begin{cases} \phi(0), & \text{if } t \in [0, t_1], \\ \phi(t), & \text{if } t \in (-\infty, 0]. \end{cases}$$

Then $x_0 = \phi$. For each $z \in B \cap C([0, t_1], \mathbb{R}^n)$ with $z_0 = 0$, we denote by $\bar{z}$ the function defined by

$$\bar{z}(t) = \begin{cases} z(t), & \text{if } t \in [0, t_1], \\ 0, & \text{if } t \in (-\infty, 0]. \end{cases}$$

If $y(\cdot)$ satisfies the integral equation,

$$y(t) = \phi(0) + \int_0^t v(s)ds,$$

we can decompose it as $y(t) = z(t) + x(t), 0 \le t \le b$, which implies $y_t = \bar{z}_t + x_t$, for every $0 \le t \le t_1$. The function $z(\cdot)$ satisfies

$$z(t) = \int_0^t v(s)ds, \tag{6.13}$$

where $v(t) \in F(t, \bar{z}_t + x_t)$ a.e. $t \in [0, t_1]$. Set

$$C_0 = \{z \in B \cap C([0, t_1], \mathbb{R}^n) : z_0 = 0\}.$$

Let $\| \cdot \|_0$ be the norm in C_0 defined by

$$\begin{aligned} \|z\|_0 &= \|z_0\|_B + \sup\{|z(t)| : 0 \le t \le b\} \\ &= \sup\{|z(t)| : 0 \le t \le t_1\}, \ z \in C_0. \end{aligned}$$

Let the operator $P : C_0 \to \mathcal{P}(C_0)$ be defined by

$$P(z) = \left\{ h \in C_0 | h(t) = \begin{cases} 0, & \text{if } t \in (-\infty, 0], \\ \int_0^t v(s)ds, & \text{if } t \in [0, t_1], \end{cases} \right\}$$

where

$$v \in S_{F,z} = \{v \in L^1([0, t_1], \mathbb{R}^n) : v(t) \in F(t, \overline{z}_t + x_t) \text{ for a.e. } t \in [0, t_1]\}.$$

Let $\widehat{l}(t) = \int_0^t \widetilde{l}(s)ds$, $\widetilde{l}(t) = K_b l(t)$, and $\|\cdot\|_{B_*}$ be the Bielecki-type norm on C_0 defined by

$$\|z\|_{B_*} = \sup\{e^{-\tau \widehat{l}(t)}|z(t)| : t \in [0, t_1]\}, \quad \tau > 1.$$

We shall show that P satisfies the assumptions of Lemma 2.88. The proof will be given in two claims.

Claim 1. $P(z) \in \mathcal{P}_{cl}(C_0)$ *for each* $z \in C_0$.

Indeed, let $(z_n)_{n \geq 0} \in P(z)$ such that $z_n \longrightarrow \tilde{z}$ in C_0. Then there exists $v_n \in S_{F,\overline{w}+z}$ such that, for each $t \in [0, t_1]$,

$$z_n(t) = \int_0^t v_n(s)ds.$$

Using the fact that F has compact values and from (F-4) and (F-5), we may pass to a subsequence if necessary to get that v_n converges to v in $L^1([0, t_1], \mathbb{R}^n)$ and hence $v \in S_{F,\overline{z}+x}$. Then, for each $t \in [0, t_1]$,

$$z_n(t) \longrightarrow \tilde{z}(t) = \int_0^t v(s))ds.$$

So $\tilde{z} \in P(z)$.

Claim 2. *There exists* $\gamma < 1$, *such that*

$$H(P(z), P(z_*)) \leq \gamma \|z - z_*\|_0 \text{ for each } z, z_* \in C_0.$$

Let $z, z_* \in C_0$ and $h \in P(z)$. Then there exists $v(t) \in F(t, \overline{z}_t + x_t)$ such that, for each $t \in [0, t_1]$,

$$h(t) = \int_0^t v(s)ds.$$

From (F-5) it follows that

$$H(F(t, \overline{z}_t + x_t), F(t, \overline{z}_{*t} + x_t) \leq l(t)\|\overline{z}_t - \overline{z_{*t}}\|_B.$$

Hence, there is $u \in F(t, \overline{z}_t + x_t)$ such that

$$\|v(t) - u| \leq l(t)\|\overline{z}_t - \overline{z}_{*t}\|_B, \quad t \in [0, t_1].$$

Consider $U : [0, t_1] \to \mathcal{P}(\mathbb{R}^n)$ given by

$$U(t) = \{u \in C_0 : |v(t) - u| \le l(t)\|z_t - z_{*t}\|_B\}.$$

Since the multivalued operator $V(t) = U(t) \cap F(t, \overline{z}_{*t} + x_t)$ is measurable (see Proposition III.4 in [136]), there exists a function $\overline{v}(t)$, which is a measurable selection for V. So, $\overline{v}(t) \in F(t, z_{*t})$ and

$$|v(t) - \overline{v}(t)| \le l(t)\|z_t - z_{*t}\|_B, \text{ for each } t \in [0, t_1].$$

Let us define for each $t \in [0, t_1]$,

$$\overline{h}(t) = \int_0^t \overline{v}(s)ds.$$

Then, we have

$$\begin{aligned}
|h(t) - \overline{h}(t)| &\le \int_0^t l(s)\|z_s - z_{*s}\|_B\, ds \\
&\le \int_0^t \widetilde{l}(s)e^{\tau\widehat{l}(s)}e^{-\tau\widehat{l}(s)} \sup_{s\in[0,t]} |z(s) - z^*(s)|ds \\
&\le \int_0^t \widetilde{l}(t)e^{\tau\widehat{l}(t)}ds\|z - z^*\|_{B_*} \\
&\le \frac{1}{\tau}\int_0^t (e^{\tau\widehat{l}(s)})'ds\|z - z^*\|_{B_*} \\
&\le \frac{1}{\tau}e^{\tau\widehat{l}(t)}ds\|z - z^*\|_{B_*}.
\end{aligned}$$

Thus,

$$e^{-\tau\widehat{l}(t)}|h_1(t)(t) - h_2(t)| \le \frac{1}{\tau}\|z - z^*\|_{B_*}.$$

Therefore,

$$\|h - \overline{h}\|_{B_*} \le \frac{1}{\tau}\|z - z_*\|_{B_*}.$$

By an analogous relation, obtained by interchanging the roles of h and h_*, it follows that

$$H(P(z), P(z_*)) \le \frac{1}{\tau}\|z - z_*\|_{B_*}.$$

Hence, P is a contraction and thus, by Lemma 2.88, P has a fixed point z, so Then the problem (6.11)–(6.12) has at least one solution. Denote this solution by y_0.

Step 2. Consider now the problem,

$$y'(t) \in F(t, y_t), \quad \text{a.e. } t \in (t_1, t_2], \tag{6.14}$$

$$y(t_1^+) = y_0(t_1^-) + I_1(y_0(t_1)), \; y(t) = y_0(t), \quad t \in (-\infty, t_1]. \tag{6.15}$$

Let

$$C_1 = \big\{y \in C((t_1, t_2], \mathbb{R}^n) : y(t_1^+) \text{ exists}\big\}.$$

Set $C_* = B \cap C([0,t_1],\mathbb{R}^n) \cap C_1$ and consider the operator $N_1 : C_* \to \mathcal{P}(C_*)$ defined by,

$$N_1(y) = \left\{ h \in C_* | h(t) = \begin{cases} y_0(t), & t \in (-\infty, t_1], \\ y_0(t_1^-) + I_1(y_0(t_1^-)) + \int_{t_1}^t v(s)ds, & t \in (t_1, t_2], \end{cases} \right\}$$

where

$$v \in S_{F,y} = \{v \in L^1([0,t_1],\mathbb{R}^n) | v \in F(t, y_t), \text{ a.e}t \in [t_1, t_2]\}.$$

Let $x(\cdot) : (-\infty, t_2] \to \mathbb{R}^n$ be the function defined by

$$x(t) = \begin{cases} y(t_1^-) + I_1(y_0(t_1)), & \text{if } t \in (t_1, t_2], \\ y_0(t), & \text{if } t \in (-\infty, t_1]. \end{cases}$$

Then $x_{t_1} = y_0$. For each $z \in C_*$ with $z(t_1) = 0$, we denote by $\bar{z}$ the function defined by

$$\bar{z}(t) = \begin{cases} z(t), & \text{if } t \in [t_1, t_2], \\ 0, & \text{if } t \in (-\infty, t_1]. \end{cases}$$

If $y(\cdot)$ satisfies the integral equation,

$$y(t) = y(t_1^-) + I_1(y_0(t_1^-)) + \int_{t_1}^t v(s)ds,$$

we can decompose it as $y(t) = \bar{z}(t) + x(t), t_1 \le t \le t_2$, which implies $y_t = \bar{z}_t + x_t$, for every $t_1 \le t \le t_2$. Also, the function $z(\cdot)$ satisfies

$$z(t) = \int_{t_1}^t v(s)ds. \tag{6.16}$$

Set

$$C_{t_1} = \{z \in C_* : z_{t_1} = 0\}$$

and let the operator $P_1 : C_{t_1} \to \mathcal{P}(C_{t_1})$ be defined by

$$P_1(z) = \left\{ h \in C_{t_1} | h(t) = \begin{cases} 0, & t \in (-\infty, t_1] \\ \int_{t_1}^t v(s)ds, & t \in [t_1, t_2], \end{cases} \right\}$$

where

$$v \in S_{F,\bar{z}+x} := \{v \in L^1([t_1,t_2],\mathbb{R}^n) | v \in F(t, \bar{z}_t + x_t), \text{ a.e } t \in [t_1, t_2]\}.$$

Similar to Step 1, we can prove that the problem (6.14)–(6.15) has at least one solution, which we denote by y_1.

Step 3. We continue this process and take into account that $y_m := y\big|_{[t_m,b]}$ is a solution to the problem

$$y'(t) = f(t, y_t), \quad \text{a.e.} \quad t \in (t_m, b], \tag{6.17}$$

$$y(t_m^+) = y_{m-1}(t_{m-1}^-) + I_m(y_{m-1}(t_m^-)), \; y(t) = y_{m-1}(t), \quad t \in (-\infty, t_{m-1}]. \tag{6.18}$$

The solution y of the problem (6.1)–(6.3) is then defined by

$$y(t) = \begin{cases} y_0(t), & \text{if } t \in (-\infty, t_1], \\ y_1(t), & \text{if } t \in (t_1, t_2], \\ \dots \\ y_m(t), & \text{if } t \in (t_m, b]. \end{cases}$$

□

6.2 Boundary Differential Inclusions

In this section, we investigate the existence of solutions of the problem (6.4)–(6.6), where the right hand side is convex or nonconvex. First, we give the definition of a solution of the problem (6.4)–(6.6)

Definition 6.5. A function $y \in B_b$ is said to be solution of (6.4)–(6.6) if there exists a function $v \in L^1([0,\infty), \mathbb{R}^n)$ such that $v(t) \in F(t, y_t)$ a.e. $t \in [0,\infty)$, v satisfies $y'(t) = v(t)$ a.e. on $[0,\infty)$, and conditions (6.5)–(6.6) are satisfied.

Now, we are able to state and prove our main theorem.

Theorem 6.6. (Convex case) *In addition to conditions (D-1) and (F-1), assume that*

(F-6) *There exists a function* $p \in L^1([0,\infty), \mathbb{R}_+)$ *such that*

$$\|F(t,x)\|_{\mathcal{P}} \le p(t) \text{ for a.e. } t \in [0,\infty) \text{ and each } x \in B.$$

Then the boundary value problem (6.4)–(6.6) *has at least one solution on* $(-\infty,\infty)$.

Proof. Consider the operator $N : B_0 \longrightarrow \mathcal{P}(B_0)$ defined by

$$N(y) = \left\{ h \in B_0 \mid h(t) = \begin{cases} \frac{\phi(0)}{A(A-1)} + \frac{\phi(t)}{A} + \frac{1}{A-1}\big[\int_0^\infty v(s)ds \\ + \sum_{k=1}^\infty I_k(y(t_k))\big], & t \in (-\infty, 0], \\ \frac{\phi(0)}{A-1} + \frac{1}{A-1}\big[\int_0^\infty v(s)ds + \sum_{k=1}^\infty I_k(y(t_k))\big] \\ + \int_0^t v(s)ds + \sum_{0<t_k<t} I_k(y(t_k)), & t \in [0,\infty), \end{cases} \right\}$$

where B_0 is defined in Subsection 3.1.1 and

$$v \in S_{F,y} := \{v \in L^1([0,\infty), \mathbb{R}^n) | v(t) \in F(t, y_t), \text{ a.e} t \in [0,\infty)\}.$$

Let $x(\cdot) : (-\infty, +\infty) \to \mathbb{R}^n$ be the function defined by

$$x(t) = \begin{cases} \frac{\phi(0)}{A-1} + \frac{1}{A-1}\big[\int_0^\infty v(s)ds + \sum_{k=1}^\infty I_k(x(t_k))\big], & t \in [0,\infty), \\ \frac{\phi(0)}{A(A-1)} + \frac{1}{A-1}\big[\int_0^\infty v(s)ds + \sum_{k=1}^\infty I_k(x(t_k))\big] + \frac{\phi(t)}{A}, & t \in (-\infty, 0]. \end{cases}$$

Then,

$$x_0 = \frac{\phi(0)}{A-1} + \frac{1}{A-1}\left[\int_0^\infty v(s)ds + \sum_{k=1}^{\infty} I_k(x(t_k))\right].$$

For each $z \in B \cap C([0,\infty), \mathbb{R}^n)$, with $z(0) = 0$, we denote by $\bar{z}$ the function defined by

$$\bar{z}(t) = \begin{cases} z(t), & \text{if } t \in [0,\infty), \\ 0, & \text{if } t \in (-\infty, 0]. \end{cases}$$

If $y(\cdot)$ satisfies the integral equation,

$$y(t) = \frac{\phi(0)}{A-1} + \frac{1}{A-1}\left[\int_0^\infty v(s)ds + \sum_{k=1}^{\infty} I_k(y(t_k))\right] + \int_0^t v(s)ds + \sum_{0<t_k<t} I_k(y(t_k)),$$

we can decompose $y(\cdot)$ as $y(t) = \bar{z}(t) + x(t), 0 \le t < \infty$. This implies $y_t = \bar{z}_t + x_t$ for every $0 \le t < \infty$, and the function $z(\cdot)$ satisfies

$$z(t) = \int_0^t v(s)ds + \sum_{0<t_k<t} I_k(\bar{z}(t_k) + x(t_k)), \tag{6.19}$$

where

$$S_{F,\bar{z}+x} := \{v \in L^1([0,\infty), \mathbb{R}^n) |\ v(t) \in F(t, \bar{z}_t + x_t), \text{ a.e} t \in [0,\infty)\}.$$

Let the operator $P : C_0 \to \mathcal{P}(C_0)$ be defined by

$$P(z) = \left\{h \in C_0 \mid h(t) = \begin{cases} 0, & t \in (-\infty, 0], \\ \int_0^t v(s)ds + \sum_{0<t_k<t} I_k(\bar{z}(t_k) + x(t_k)), & t \in [0,\infty), \end{cases}\right\}$$

where $v \in S_{F,\bar{z}+x}$ and C_0 are as in Subsection 3.1.1. Obviously, that the operator N has a fixed point is equivalent to P having a fixed point, and so we turn to proving that P has a fixed point. As in Theorem 6.2 we can prove that the operator P is completely continuous, and from Theorem 4.3, each possible solution of the problem $z \in \lambda P(z)$, $\lambda \in (0,1)$, is a priori bounded by a constant M. Set

$$U := \{z \in B_0 |\ \|z\|_0 < M + 1\}.$$

From the choice of U_0 there is no $u \in \partial U_0$ such that $u \in \lambda N(u)$, for some $\lambda \in (0,1)$. As a consequence of the nonlinear alternative of Leray Schauder type [184], we deduce that N has a fixed point z in U which is a solution of the problem (6.4)–(6.6). □

For the next part, we study the case where F is not necessarily convex valued. Our approach here is again based on a nonlinear alternative of Leray–Schauder type [184], a fixed point theorem, combined with a selection theorem due to Bressan and Colombo [104] for lower semicontinuous multivalued operators, and a fixed point theorem for contractive multivalued operators given by Covitz and Nadler.

Theorem 6.7 (Nonconvex). *Suppose that* (D-1), (F-6), *and*

(F-7) $F : [0,\infty) \times B \longrightarrow \mathcal{P}(\mathbb{R}^n)$ *is a nonempty compact valued multivalued map such that:*
   (a) $t \mapsto F(t,x)$ *is* $\mathcal{L} \otimes \mathcal{B}$ *measurable,*
   (b) $x \mapsto F(t,x)$ *is lower semicontinuous for a.e.* $t \in [0,\infty)$,

hold. Then problem (6.4)–(6.6) *has at least one solution.*

Proof. Analogous to the proof of Theorem 6.3, we can prove that there exists a continuous function $f : C_0 \to L^1([0,\infty),\mathbb{R}^n)$ where C_0 is defined in Theorem 4.3. Now we consider the impulsive boundary value problem,

$$z'(t) = f(\bar{z}_t + x_t), \quad \text{a.e. } t \in [0,\infty)\backslash\{t_1,\ldots,\}, \tag{6.20}$$

$$\Delta z(t_k) = I_k(\bar{z}(t_k) + x(t_k)), \quad k \in \mathbb{N}, \tag{6.21}$$

$$z(t) = 0, \quad t \in (-\infty,0]. \tag{6.22}$$

Consider the operator $P : C_0 \to C_0$, defined by

$$(Pz)(t) = \begin{cases} 0, & t \in (-\infty,0], \\ \int_0^t f(\overline{z}_s + x_s)ds + \sum_{0<t_k<t} I_k(\overline{z}(t_k,x) + x(t_k)), & t \in [0,\infty). \end{cases}$$

As in Theorem 4.3, we can prove that P is completely continuous and there exists $M > 0$ such that for every solution of $z = \lambda P(z)$, for some $\lambda \in (0,1)$, we have $\|z\|_0 \le M$. Set

$$U = \{z \in C_0 : \|z\|_0 < M + 1\}.$$

Clearly, U is an open set in C_0. From the choice of U, there is no $z \in \partial U$ such that $z = \lambda P(z)$, for some $\lambda \in (0,1)$. As a consequence of the nonlinear alternative of Leray Schauder type [184] we deduce that P has a fixed point z in $\overline{U}$. Hence, N has a fixed point y that is a solution to (6.4)–(6.6). □

Here we shall prove a second result for the boundary value problem (6.4)–(6.6) in the case where the right hand side satisfies a Lipschitz condition.

Theorem 6.8. *In addition to* (D-3), *assume that*

(F-8) $F : [0,\infty) \times B \longrightarrow \mathcal{P}_{cp}(\mathbb{R}^n)$; $(t,u) \longmapsto F(t,u)$ *is measurable for each* $u \in B$.

(F-9) *There exists a function* $l \in L^1([0,\infty),\mathbb{R}^+)$ *such that, for a.e.* $t \in [0,\infty)$ *and all* x, $y \in B$,

$$H(F(t,x),F(t,y)) \le l(t)\|x-y\|_B,$$

and

$$H(0, F(t,0)) \le l(t) \;\; \textit{for a.e.} \;\; t \in [0,\infty).$$

Then (6.4)–(6.6) *has at least one solution.*

Proof. For the proof, see Theorem 6.3. □

Chapter 7

Impulsive FDEs with Variable Times

7.1 Introduction

The theory of impulsive differential equations with variable times is relatively less developed due to the difficulties created by the state-dependent impulses. Recently, some interesting results have been done by Bajo and Liz [41], Frigon and O'Regan [211–213], Kaul et al. [334], Kaul and Liu [317,318], Lakshmikantham et al. [336,339] and Liu and Ballinger [43]. Very recently, impulsive functional differential equations with variable times were studied by Benchohra et al. [72] and Benchohra and Ouahab [81]. The main theorems of the present chapter extend the problem (3.38)–(3.41) considered by Benchohra et al. [67, 72], when the delay is finite and the impulse times are constant or variable. In particular, the following problem is studied,

$$y'(t) = f(t, y_t), \quad \text{a.e.} \quad t \in J = [0, b],\ t \neq \tau_k(y(t)),\ k = 1, \dots, m, \tag{7.1}$$

$$y(t^+) = I_k(y(t)), \quad t = \tau_k(y(t)),\ k = 1, \dots, m, \tag{7.2}$$

$$y(t) = \phi(t), \quad t \in (-\infty, 0], \tag{7.3}$$

where $f : J \times B \to \mathbb{R}^n$, $\tau_k : \mathbb{R}^n \to \mathbb{R}$, $I_k : \mathbb{R}^n \to \mathbb{R}^n$, $k = 1, 2, \dots, m$, are given functions satisfying some assumptions that will be specified later, $\phi \in B$, and where B, called a *phase space*, was defined in the Chapter 1. In the second subsection, we consider the first order impulsive neutral functional equation with infinite delay. More precisely, we consider the initial value problem with impulse effects,

$$\frac{d}{dt}[y(t) - g(t, y_t)] = f(t, y_t) \quad t \in J,\ t \neq \tau_k(y(t)),\ k = 1, \dots, m, \tag{7.4}$$

$$y(t^+) = I_k(y(t)), \quad t = \tau_k(y(t)),\ k = 1, \dots, m, \tag{7.5}$$

$$y_0 = \phi \in B, \tag{7.6}$$

where f, I_k, B are as in problem (7.1)–(7.3), and $g : J \times B \to \mathbb{R}^n$ is a given function. In all this chapter we assume that $t \to f(t, y_t)$ is a measurable function.

7.1.1 Existence Results

Let us start by defining what we mean by a solution of problem (7.1)–(7.3).

Definition 7.1. A function $y \in B_b$, is said to be a solution of (7.1)–(7.3) if y satisfies (7.1)–(7.3).

In all this chapter the functions $t \to f(t, y_t)$ and $t \to g(t, y_t)$ are measurable functions.

Theorem 7.2. *Let $f : J \times B \to \mathbb{R}^n$ be an L^1-Carathéodory function. Assume that:*

(G-1) *The functions $\tau_k \in C^1(\mathbb{R}^n, \mathbb{R})$ for $k = 1, \ldots, m$. Moreover,*

$$0 < \tau_1(x) < \ldots < \tau_m(x) < b \text{ for all } x \in \mathbb{R}^n;$$

(G-2) *There exist a continuous nondecreasing function $\psi : [0, \infty) \to (0, \infty)$ and $p \in L^1([0, b], \mathbb{R}_+)$ such that*

$$|f(t, x)| \leq p(t)\psi(\|x\|_B) \text{ for a.e. } t \in [0, b] \text{ and each } x \in B,$$

with

$$K_b \int_0^b p(s)ds < \int_c^\infty \frac{dx}{\psi(x)},$$

where $K_b = \sup\{|K(t)| : t \in [0, b]\}$, $M_b = \sup\{|M(t)| : t \in [0, b]\}$, and $c = M_b\|\phi\|_B + K_b|\phi(0)|$.

(G-3) *For all $(t, x) \in [0, b] \times \mathbb{R}^n$ and for all $y_t \in B$, we have*

$$\langle \tau_k'(x), f(t, y_t) \rangle \neq 1 \text{ for } k = 1, \ldots, m,$$

where $\langle \cdot, \cdot \rangle$ denotes the scalar product in $\mathbb{R}^n$;

(G-4) *For all $x \in \mathbb{R}^n$,*

$$\tau_k(I_k(x)) \leq \tau_k(x) < \tau_{k+1}(I_k(x)) \text{ for } k = 1, \ldots, m.$$

Then the initial value problem (7.1)–(7.3) has at least one solution.

Proof. The proof employs exactly the same ideas as in the paper of Benchohra et al. [71]. □

7.1.2 Neutral Functional Differential Equations

This subsection is concerned with the existence of solutions for initial value problems for first order neutral functional differential equations with impulsive effects at variable times given in (7.4)–(7.6).

Neutral differential equations arise in many areas of applied mathematics, and such equations have received much attention in recent years. A good guide to the literature for neutral functional differential equations is the book by Hale and Lunel [268].

In the last few years, impulsive neutral functional differential equations with bounded delay, where the impulses are constant or variable, were studied by Benchohra et al. [70], Benchohra and Ouahab [81]. The goal of this section is to extend those results.

Theorem 7.3. *Let $f : J \times B \to \mathbb{R}^n$ be an L^1-Carathéodory function. In addition to* (G-1) *and* (G-4), *assume that the following conditions hold.*

(G-5) *The function g is continuous and completely continuous, and for any bounded set $Q \subseteq C((-\infty, b], \mathbb{R}^n)$, the set $\{t \to g(t, x_t) : x \in Q\}$ is equicontinuous in $C([0, b], \mathbb{R}^n)$, and there exist constants c_1, $c_2 \geq 0$ such that*

$$|g(t,u)| \leq c_1 \|u\|_B + c_2, \ \ t \in [0,b], \ u \in B;$$

(G-6) *There exist a continuous nondecreasing function $\psi : [0, \infty) \longrightarrow (0, \infty)$ and $p \in L^1(J, \mathbb{R}_+)$ such that*

$$|f(t,x)| \leq p(t)\psi(\|x\|_B), \ \ \text{for a.e. } t \in [0,b] \text{ and each } y \in \mathbb{R}^n$$

with

$$\frac{K_b}{1 - K_b c_1} \int_0^b p(s)ds < \int_{\bar{c}}^{\infty} \frac{dx}{\psi(x)},$$

where $c_1 K_b < 1$,

$$\bar{c} = \frac{1}{1 - c_1 K_b} [K_b |g(0, \phi)| + c_2 K_b + \alpha]$$

and

$$\alpha = M_b \|\phi\|_B + K_b |\phi(0)| + M_b \|\phi\|_B;$$

(G-7) *For all $t \in [0, b]$ and for all $y_t \in B$ we have*

$$\langle \tau_k'(y(t) - g(t, y_t)), f(t, y_t) \rangle \neq 1 \text{ for } k = 1, \ldots, m,$$

where $\langle \cdot, \cdot \rangle$ denotes the scalar product in $\mathbb{R}^n$;

(G-8) *g is a nonnegative function.*

Then the IVP (7.4)–(7.6) has at least one solution.

Proof. The proof employs exactly the same ideas as in the paper of Benchohra et al. [71]. □

7.2 Impulsive Hyperbolic Differential Inclusions with Infinite Delay

In this section, we shall be concerned with the existence of solutions of the second order impulsive hyperbolic differential inclusion with infinite delay,

$$\frac{\partial^2 u(t,x)}{\partial t \partial x} \in F(t, x, u_{(t,x)}), \quad \text{a.e.} \ (t,x) \in J_a \times J_b, t \neq t_k, \quad t \neq \tau_k(u(t,x)), k = 1, \ldots, m, \tag{7.7}$$

$$u(t_k^+, x) = I_k(u(t_k, x)), \ \ t \neq \tau_k(u(t,x)), \quad k = 1, \ldots, m, \tag{7.8}$$

$$u(t,x) = \phi(t,x), \ \ (t,x) \in (-\infty, a] \times [-b, b] \backslash (0, a] \times (0, b], \tag{7.9}$$

where $F : J_a \times J_b \times B \to \mathcal{P}(\mathbb{R}^n)$ is a multivalued map with compact values, $J_a \times J_b := [0,a] \times [0,b]$, B is a phase space, $I_k \in C(\mathbb{R}^n, \mathbb{R}^n)$, $\phi \in B$, $\phi(t_k^+, 0) = \lim_{h\to 0^+} \phi(t_k + h, 0) = \lim_{h\to 0^-} \phi(t_k - h, 0) = \phi(t_k, 0)$, $u(t_k^+, y) = \lim_{(h,x)\to(0^+,y)} u(t_k + h, x)$ and $u(t_k^-, y) = \lim_{(h,x)\to(0^-,y)} u(t_k - h, x)$, $k = 1, \ldots, m$, and $\mathbb{R}^n$ is Euclidean space with norm $|\cdot|$. For each $u \in B$ and each $(t,x) \in J_a \times J_b$ the function $u_{(t,x)} : (-\infty, 0] \times [-b, 0] \to \mathbb{R}^n$ is defined by

$$u_{(t,x)}(s,y) = u(s+t, y+x), \quad \text{for each } (s,t) \in (-\infty, 0] \times [-b, 0].$$

In the last three decades, several papers have been devoted to the study of hyperbolic partial differential equations with local and nonlocal initial conditions; see for instance [120–123,338], and the reference therein. For similar results with set-valued right hand side, we refer the reader to the papers by Benchohra and Ntouyas [75,76], Byszewski and Papageorgiou [124], Liu, and Zhang [349], and Papageorgiou [404].

Impulsive partial differential equations and inclusions were studied by Benchohra et al. [58,59,192] and Rogovchenko et al. [432]. Recently, in the case where the impulses are absent, Kamont and Koziel [313] studied mixed problems for hyperbolic functional differential equations with unbounded delay. Very recently, Henderson and Ouahab [281,282] studied impulsive hyperbolic functional differential inclusions with infinite delay.

Our goal here is to give existence results in both the convex and nonconvex cases for the above problem by using the nonlinear alternative of Leray–Schauder type [184], and Covitz and Nadler's [152] theorem for contraction multivalued maps.

7.3 Existence Results

We begin this section with a description of the phase space needed for our problem. Then in separate subsections we consider the convex and nonconvex cases.

7.3.1 Phase Spaces

In order to define the phase space and a solution of (7.7)–(7.9), we shall consider the space

$$PC = \{u : (-\infty, a] \times [-b, b] \to \mathbb{R}^n \mid u(t_k^-, \cdot),\ u(t_k^+, \cdot) \text{ exist with } u(t_k, \cdot) = u(t_k^-, \cdot), \quad u_k \in C(\Gamma_k, \mathbb{R}^n)\}$$

where u_k is the restriction of u to $\Gamma_k = (t_k, t_{k+1}) \times J_b$, $k = 0, \ldots, m$.

We will assume that B satisfies the following axioms:

(A) If $u : (-\infty, a] \times [-b, b] \to \mathbb{R}^n$, $a, b > 0$, $u_{(0,x)} \in B$ for $x \in [0, b]$, and $u(t_k^-, \cdot)$, $u(t_k^+, \cdot)$, exist with $u(t_k, \cdot) = u(t_k^-, \cdot)$, $k = 1, \ldots, m$, then for every t in $[0, a) \backslash \{t_1 \ldots, t_m\}$, the following conditions hold:

(i) $u_{(t,x)}$ is in B, and $u_{(t,x)}$ is continuous on $[0,a]\backslash\{t_1,\ldots,t_m\}\times[0,b]$,
(ii) $\|u_{(t,x)}\|_B \leq K(t)\sup\{|u(s,x)| : (s,x)\in[0,t]\times[0,b]\} + M(t)\|u_{(0,x)}\|_B$,
(iii) $|u(t,x)| \leq H\|u_{(t,x)}\|_B$,

where $H \geq 0$ is a constant, $K:[0,\infty)\to[0,\infty)$ is continuous, $M:[0,\infty)\to[0,\infty)$ is locally bounded and H,K,M are independent of $u(\cdot,\cdot)$.

(A-1) For the function $u(\cdot,\cdot)$ in (A), $u_{(t,x)}$ is a B-valued continuous function on $[0,a)\backslash\{t_1,\ldots,t_m\}\times[0,b]$.

(A-2) The space B is complete.

Set

$$B_{a,b} = \{y : (-\infty,a]\times[-b,b]\to\mathbb{R}^n : u\in PC\cap B\},$$

and let $\|\cdot\|_{a,b}$ be the seminorm in $B_{a,b}$ defined by

$$\|u\|_{a,b} := \|u_{(0,x)}\|_B + \sup\{|u(t,x)| : (t,x)\in[0,a]\times[0,b]\},\ u\in B_{a,b}.$$

Let us start by defining what we mean by a solution of problem (7.7)–(7.9). In all this part we assume that $(t,x)\to F(t,x,u_{t,x})$ is a measurable multifunction.

Definition 7.4. A function $u\in B_{a,b}$, is said to be a solution of (7.7)–(7.9) if there exists $v\in L^1(J_a\times J_b)$ such that $v(t,x)\in F(t,x,u_{(t,x)})$ is satisfied a.e on $J_a\times J_b$, $\frac{\partial^2 u(t,x)}{\partial t\partial x} = v(t,x)$ a.e on $J_a\times J_b$, and v satisfies the conditions (7.8)–(7.9).

We need the following auxiliary result.

Lemma 7.5 ([402]). *Let $J_1=[x_0,N]$ and $J_2=[y_0,R]$, and let $\psi\in C(\mathbb{R}_+,\mathbb{R}_+)$ be a nondecreasing function with $\psi(u)>0$ for $u>0$, and $u,a\in C(J_1\times J_2,\mathbb{R}_+)$, $b\in C(J_1\times J_2\times J_1\times J_2,\mathbb{R}_+)$ for $x_0\leq s\leq x\leq X$, $y_0\leq t\leq y\leq Y$, $\alpha\in C^1(J_1,J_1)$, $\beta\in C^1(J_2,J_2)$ be nondecreasing with $\alpha(x)\leq x$ on J_1, $\beta(y)\leq y$ on J_2, and $k\geq 0$ be a constant. If*

$$\begin{aligned} u(x,y) \leq k &+ \int_{\alpha(x_0)}^{\alpha(x)}\int_{\beta(y_0)}^{\beta(y)} \big[a(s,t)\psi(u(s,t)) \\ &+ \int_{\alpha(x_0)}^{s}\int_{\beta(y_0)}^{t} b(s,t,\sigma,\eta)\psi(u(\sigma,\eta))d\eta d\sigma\big]dtds, \end{aligned}$$

for $(x,y)\in J_1\times J_2$, then for $x_0\leq x\leq x_1\leq N$, $y_0\leq y\leq y_1\leq R$,

$$u(x,y)\leq G^{-1}[G(k)+A(x,y)],$$

where G^{-1} is the inverse function of

$$G(r)=\int_{r_0}^{r}\frac{ds}{\psi(s)},\qquad r>0,\ r_0>0,$$

and

$$A(x,y) = \int_{\alpha(x_0)}^{\alpha(x)} \int_{\beta(y_0)}^{\beta(y)} \Big[a(s,t) + \int_{\alpha(x_0)}^{s} \int_{\beta(y_0)}^{t} b(s,t,\sigma,\eta)d\eta d\sigma\Big] dt ds,$$

for $(x,y) \in J_1 \times J_2$.

Theorem 7.6. *Assume that* F *is* L^1*–Carathéodory and the following conditions hold:*

(G-9) *There exist a function* $p \in C(J_a \times J_b, \mathbb{R}_+)$ *and a continuous nondecreasing function* $\psi : [0,\infty) \longrightarrow (0,\infty)$, *such that*

$$\|F(t,x,u)\|_{\mathcal{P}} \leq p(t,x)\psi(\|u\|_B) \ \text{ a.e. } (t,x) \in J_a \times J_b \text{ and } u \in B;$$

(G-10) *The functions* $\tau_k \in C^1(\mathbb{R}^n, \mathbb{R})$ *for* $k = 1,\dots,m$. *Moreover,*

$$0 < \tau_1(x) < \cdots < \tau_m(x) < a \ \text{ for all } x \in \mathbb{R}^n;$$

(G-11) *For all* $u \in B$ *and all* $v \in S_{F,u}$, *we have*

$$\left\langle \tau_k'(x), \int_{\bar{t}}^{t} v(s,x)ds \right\rangle \neq 1 \text{ for } \text{ for all } (t,\bar{t},x) \in J_a \times J_a \times \mathbb{R}^n,$$

and $k = 0,\dots,m$, *where* $\langle\cdot,\cdot\rangle$ *denotes the scalar product in* $\mathbb{R}^n$;

(G-12) *For all* $x \in \mathbb{R}^n$

$$\tau_k(I_k(x)) \leq \tau_k(x) < \tau_{k+1}(I_k(x)) \text{ for } k = 1,\dots,m.$$

Then the IVP (7.7)–(7.9) *has at least one solution.*

Proof. The proof will be given in several steps.

Step 1. Consider the following problem

$$\frac{\partial^2 u(t,x)}{\partial t \partial x} \in F(t,x,u_{t,x}), \quad \text{a.e. } (t,x) \in J_a \times J_b, \tag{7.10}$$

$$u(t,x) = \phi(t,x), \quad (t,x) \in (-\infty,a] \times [-b,b] \backslash (0,a] \times (0,b]. \tag{7.11}$$

A solution to the problem (7.10)–(7.11) is a fixed point of the operator $N : C(J_a \times J_b, \mathbb{R}^n) \to \mathcal{P}(C(J_a \times J_b, \mathbb{R}^n))$ defined by

$$N(u) = \left\{ h \in B_{a,b} : h(t,x) = \left\{ \begin{array}{l} \phi(t,x), \\ \qquad \text{if } (t,x) \in \mathcal{D}, \\ z_0(t,x) + \int_0^t \int_0^x v(s,y)dsdy, \\ \qquad \text{if } (t,x) \in J_a \times J_b, \ v \in S_{F,u}, \end{array} \right. \right\}$$

where $\mathcal{D} = (-\infty, a]\times[-b, b]\backslash(0, a]\times(0, b]$ and $z_0(t, x) := \phi(t, 0)+\phi(0, y)-\phi(0, 0)$. Let $z(\cdot,\cdot) : (-\infty, a] \times (-\infty, b] \to \mathbb{R}^n$ be the function defined by

$$z(t,x) = \begin{cases} \phi(t,0) + \phi(0,y) - \phi(0,0), & \text{if } (t,x) \in [0,a] \times [0,b], \\ \phi(t,x), & \text{if } (t,x) \in \mathcal{D}. \end{cases}$$

Then $z_{(0,0)} = \phi$. For each $w \in B_{a,b}$ with $w(0, x) = 0$, we denote by $\overline{w}$ the function defined by

$$\overline{w}(t,x) = \begin{cases} w(t,x), & \text{if } (t,x) \in [0,a] \times [0,b], \\ 0, & \text{if } (t,x) \in \mathcal{D}. \end{cases}$$

If $u(\cdot,\cdot)$ satisfies the integral equation,

$$u(t,x) = z_0(t,x) + \int_0^t \int_0^x v(s,y)dsdy,$$

we can decompose $u(\cdot,\cdot)$ as $u(t,x) = z(t,x) + w(t,x)$, $(t,x) \in [0,a] \times [0,b]$. This implies $u_{(t,x)} = w_{(t,x)} + z_{(t,x)}$ for every $(t,x) \in [0,a]\times[0,b]$, and the function $w(\cdot,\cdot)$ satisfies

$$w(t,x) = \int_0^t \int_0^x v(s,y)dsdy, \tag{7.12}$$

where $v \in S_{F,\overline{w}+z}$. Set

$$C_{(0,0)} = \{w \in B_{a,b} : \overline{w}_{(0,x)} = 0, \text{ for each } x \in [-b,b]\}.$$

For any $w \in C_{(0,0)}$, we have

$$\begin{aligned} \|\overline{w}\|_\infty &= \|\overline{w}_{(0,0)}\|_B + \sup\{|\overline{w}(t,x))| : (t,x) \in J_a \times J_b\} \\ &= \sup\{|w(t,x)| : (t,x) \in J_a \times J_b\}. \end{aligned}$$

Thus, $(C_{(0,0)}, \|\cdot\|_\infty)$ is a Banach space.

Let the $P : C_{(0,0)} \to \mathcal{P}(C_{(0,0)})$ be defined by

$$P(w) = \left\{ h \in C_{(0,0)} : h(t,x) = \begin{cases} 0, & (t,x) \in (-\infty,0] \times [-b,b], \\ \int_0^t \int_0^x v(s,y)dsdy, \\ \quad (t,x) \in J_a \times J_b, v \in S_{F,\overline{w}+z}. \end{cases} \right.$$

That the operator N has a fixed point is equivalent to P having a fixed point is clear, so we turn our attention to proving that P has a fixed point. We shall use the Leray–Schauder alternative to do this. The proof will be given in several steps.

Claim 1. *$P(w)$ is convex for each $w \in C_{(0,0)}$.*

Indeed, if h_1, h_2 belong to $P(w)$, then there exist v_1, $v_2 \in S_{F,\overline{w}+z}$ such that for each $(t,x) \in J_a \times J_b$, we have

$$h_i(t,x) = \int_0^t \int_0^x v_i(s,y)dsdy, \quad i = 1,2.$$

Let $0 \leq d \leq 1$. Then for each $(t,x) \in J_a \times J_b$, we have

$$(dh_1 + (1-d)h_2)(t,x) = \int_0^t \int_0^x [dv_1(s,y) + (1-d)v_2(s,y)]dsdy.$$

Since $S_{F,\overline{w}+z}$ is convex (because F has convex values) it follows that

$$dh_1 + (1-d)h_2 \in P(w).$$

Claim 2. *P maps bounded sets into bounded sets in $C_{(0,0)}$.*

Indeed, it is enough to show that there exists a positive constant ℓ such that for each $w \in \mathcal{B}_q = \{w \in C_{(0,0)} : \|w\|_\infty \leq q\}$ we have $\|P(w)\|_\infty \leq \ell$. Let $h \in N(u)$. Then there exists $v \in S_{F,\overline{w}+z}$ such that

$$h(t,x) = \int_0^t \int_0^x v(s,y)dsdy.$$

Since F is L^1–Carathéodory we have, for each $(t,x) \in J_a \times J_b$,

$$|h(t,x)| \leq \int_0^a \int_0^b |\varphi_{q*}(t,x)|ds \leq \|\varphi_{q*}\|_{L^1} := \ell,$$

where

$$\begin{aligned}\|\overline{w}_{(s,x)} + z_{(s,x)}\|_B &\leq \|\overline{w}_{(s,x)}\|_B + \|z_{(s,x)}\|_B \\ &\leq K_a q + M_a\|\phi\|_B + K_a[\|\phi(.,0)\|_\infty \\ &\quad + \|\phi(0,.)\|_\infty + |\phi(0,0)|] := q_*\end{aligned}$$

with

$$K_a = \sup\{|K(t)| : t \in [0,a]\}, \text{ and } M_a = \sup\{|M(t)| : t \in [0,a]\}.$$

Claim 3. *P* maps bounded sets into equicontinuous sets of $C_{(0,0)}$.

Let $(\bar{t}_1, x_1), (\bar{t}_2, x_2) \in J_a \times J_b$, $\bar{t}_1 < \bar{t}_2$, $x_1 < x_2$, and let $\mathcal{B}_q$ be a bounded set of $C_{(0,0)}$ as in Claim 2. Then,

$$|h(\bar{t}_2, x_2) - h(\bar{t}_1, x_1)| \leq \int_0^{\bar{t}_2} \int_{x_1}^{x_2} \varphi_{q*}(t,s)dtds + \int_{\bar{t}_1}^{\bar{t}_2} \int_0^{x_1} \varphi_{q*}(t,s)dtds.$$

The right-hand side tends to zero as $\bar{t}_2 - \bar{t}_1 \to 0$ and $x_2 - x_1 \to 0$.

As a consequence of Claims 2 and 3 along with the Arzelà–Ascoli theorem, we conclude that $P : C_{(0,0)} \longrightarrow C_{(0,0)}$ is completely continuous.

Claim 4. *P has a closed graph.*

Let $w_n \longrightarrow u_*$, $h_n \in P(w_n)$, and $h_n \longrightarrow h_*$. We shall prove that $h_* \in \mathcal{P}(w_*)$. Now $h_n \in \mathcal{P}(w_n)$ means that there exists $v_n \in S_{F,\overline{w}_n+z}$ such that for each $(t,x) \in J_a \times J_b$,

$$h_n(t,x) = \int_0^t \int_0^x v_n(s,y)dsdy.$$

We must prove that there exists $v_* \in S_{F,\overline{w}_*+z}$ such that for each $(t,x) \in J_a \times J_b$,

$$h_*(t,x) = \int_0^t \int_0^x v_*(s,y)dsdy.$$

Consider the continuous linear operator

$$\Psi : L^1(J_a \times J_b, \mathbb{R}^n) \longrightarrow C_{(0,0)}$$

defined by

$$v \longmapsto \Psi(v)(t,x) = \int_0^t \int_0^x v(s,\tau)dsd\tau.$$

From Lemma 2.65, it follows that $\Psi \circ S_F$ is a closed graph operator. Moreover, we have that

$$h_n(t,x) \in \Psi(S_{F,\overline{w}_n+z}).$$

Since $w_n \longrightarrow w_*$, it follows from Lemma 2.65 that

$$h_*(t,x) = \int_0^t \int_0^x v_*(s,y)dsdy.$$

for some $v_* \in S_{F,u_*+z}$.

Claim 5. *A priori bounds on solutions.*

Let $w \in C_{(0,0)}$ be a possible solution to $w \in \lambda P(w)$ for some $\lambda \in (0,1)$. Then there exists $v \in S_{F,\overline{w}+z}$ such that for each $(t,x) \in J$,

$$w(t,x) = \int_0^t \int_0^x v(s,y)dsdy.$$

By (G-9)–(G-10), for each $(t,x) \in J_a \times J_b$, we have

$$\begin{aligned} |w(t,x)| &\le \int_0^t \int_0^x |p(s,\tau)|\psi(\|\overline{w}_{(s,\tau)} + z_{(s,\tau)}\|_B]dsd\tau \\ &\le \int_0^t \int_0^x |p(s,\tau)|\psi(\|\overline{w}_{(s,\tau)} + z_{(s,\tau)}\|_B]dsd\tau \qquad (7.13) \\ &\le \int_0^t \int_0^x |p(s,\tau)|\psi(\|\overline{w}_{(s,\tau)} + z_{(s,\tau)}\|_B]dsd\tau. \end{aligned}$$

But

$$\begin{aligned}\|\overline{w}_{(s,x)} + z_{(s,x)}\|_B &\leq \|\overline{w}_{(s,x)}\|_B + \|z_{(s,x)}\|_B\\ &\leq K(t)\sup\{|w(s,x)| : (s,x) \in [0,t] \times [0,b]\}\\ &\quad + K(t)\sup\{|z(s,x)| : (s,x) \in [0,t] \times [0,b]\}\\ &\quad + M(t)\|\overline{w}_{(0,0)}\|_B + M(t)\|z_{(0,0)}\|_B\\ &\leq K_a \sup\{|w(s,x)| : (s,x) \in [0,t] \times [0,b]\}\\ &\quad + M_a\|\phi\|_B + K_b[\|\phi(\cdot,0)\|_\infty + \|\phi(0,\cdot)\|_\infty + |\phi(0,0)|].\end{aligned}$$

Hence,

$$\begin{aligned}\|\overline{w}_{(s,x)} + z_{(s,x)}\|_B \leq\; & K_a\|w\|_\infty + M_a\|\phi\|_B + K_b[\|\phi(\cdot,0)\|_\infty\\ &+ \|\phi(0,\cdot)\|_\infty + |\phi(0,0)|],\end{aligned} \tag{7.14}$$

where

$$\|w\|_\infty := \sup\{|w(s,x)| : (s,x) \in [0,t] \times [0,b]\}.$$

If we name $\mu(t,x)$ the right hand side of (7.14), then we have

$$\|\overline{w}_{(s,x)} + z_{(s,x)}\|_B \leq \mu(t,x),$$

and therefore (7.13) becomes

$$\begin{aligned}|w(t,x)| \leq & \int_0^t \int_0^x |p(s,\tau)|\psi(\mu(t,x))ds d\tau\\ &+ M_a\|\phi\|_B + K_b[\|\phi(\cdot,0)\|_\infty\\ &+ \|\phi(0,\cdot)\|_\infty + |\phi(0,0)|].\end{aligned} \tag{7.15}$$

Using (7.15) in the definition of μ, we have that

$$\begin{aligned}\mu(t,x) \leq_a & \int_0^t \int_0^x |p(s,\tau)|\psi(\mu(t,x))ds d\tau\\ &+ M_a\|\phi\|_B + K_b[\|\phi(.,0)\|_\infty + \|\phi(0,.)\|_\infty + |\phi(0,0)|].\end{aligned}$$

From Lemma 7.5, we get, for each $(t,x) \in [0,a] \times [0,b]$,

$$\mu(t,x) \leq G^{-1}[G(k) + \int_0^t \int_0^x p_*(s,x)ds dx],$$

where

$$p_*(s,x) = K_a p(s,x),$$

and

$$k = M_a\|\phi\|_B + K_b[\|\phi(\cdot,0)\|_\infty + \|\phi(0,\cdot)\|_\infty + |\phi(0,0)|].$$

Thus,

$$\|w\|_\infty \leq G^{-1}[G(k) + \int_0^a \int_0^b p_*(s,x)dsdx] := M.$$

Set

$$U = \{w \in C_{(0,0)} : \|w\|_\infty < M/K_a + 1\}.$$

It is clear that $P : \overline{U} \to P(C_{(0,0)})$ is completely continuous. From the choice of U, there is no $u \in \partial U$ such that $w \in \lambda P(w)$ for some $\lambda \in (0,1)$. As a consequence of the nonlinear alternative of Leray Schauder type (Lemma 2.87), we conclude that P has a fixed point w in $\overline{U}$. Hence, N has a fixed point u that is a solution to problem (7.10)–(7.11).

Denote this solution by u_1. Define the function

$$r_{k,1}(t,x) = \tau_k(u_1(t,x)) - t \quad \text{for } t \geq 0.$$

Condition (G-10) implies

$$r_{k,1}(0,0) \neq 0 \text{ for } k = 1,\ldots,m.$$

If

$$r_{k,1}(t,x) \neq 0 \text{ on } J_a \times J_b \text{ for } k = 1,\ldots,m,$$

i.e.,

$$t \neq \tau_k(u_1(t,x)) \text{ on } J_a \times J_b \text{ for } k = 1,\ldots,m,$$

then u_1 is a solution of the problem (7.7)–(7.9).

It remains to consider the case when

$$r_{1,1}(t,x) = 0 \text{ for some } (t,x) \in J_a \times J_b.$$

Now, since

$$r_{1,1}(0,0) \neq 0$$

and $r_{1,1}$ is continuous, there exist $t_1 > 0$ and $x_1 > 0$ such that

$$r_{1,1}(t_1,x_1) = 0, \text{ and } r_{1,1}(t,x) \neq 0 \text{ for all } (t,x) \in [0,t_1) \times [0,x_1].$$

Thus, by (G-12) we have

$$r_{1,1}(t_1,x_1) = 0 \text{ and } r_{1,1}(t,x) \neq 0 \text{ for all } (t,x) \in [0,t_1) \times [0,x_1] \cup (x_1,b].$$

Suppose that there exists $(\bar{t},\bar{x}) \in [0,t_1) \times [0,x_1) \cup (x_1,b]$ such that $r_{1,1}(\bar{t},\bar{x}) = 0$. The function $r_{1,1}$ attains a maximum at some point $(s,\bar{s}) \in [0,t_1] \times J_b$. Since

$$\frac{\partial^2 u(t,x)}{\partial t \partial x} \in F(t,x,u_{1(t,x)}), \ a.e. \ (t,x) \in J_a \times J_b,$$

there exists $v(\cdot,\cdot) \in L^1(J_a \times J_b)$ with $v(t,x) \in F(t,x,u_{1_{(t,x)}})$, a.e. $(t,x) \in J_a \times J_b$ such that

$$\frac{\partial^2 u(t,x)}{\partial t \partial x} = v(t,x) \quad \text{a.e.} \quad t \in J_a \times J_b.$$

Now,

$$\frac{\partial u_1(t,x)}{\partial t} \text{ and } \frac{\partial u_1(t,x)}{\partial x} \text{ exist.}$$

Then,

$$\frac{\partial r_{1,1}(s,\bar{s})}{\partial t} = \tau_1'(u_1(s,\bar{s})\frac{\partial u_1(s,\bar{s})}{\partial t} - 1 = 0.$$

Since

$$\frac{\partial u_1(t,x)}{\partial t} = \int_0^t v(s,x,u_1(s,x))ds,$$

we have

$$\tau_1'(u_1(s,\bar{s}))\int_0^s v(\tau,\bar{s})d\tau - 1 = 0.$$

Therefore,

$$\left\langle \tau_1'(u_1(s,\bar{s})), \int_0^s v(\tau,\bar{s})d\tau \right\rangle = 1,$$

which contradicts (G-11). From (G-11) we have

$$r_{k,1}(t,x) \neq 0 \text{ for all } t \in [0,t_1) \times J_b, \text{ and } k = 1,\dots,m.$$

Step 2. Consider the problem

$$\frac{\partial^2 u(t,x)}{\partial t \partial x} \in F(t,x,u_{(t,x)}), \qquad \text{a.e.} \quad t \in (t_1,a] \times J_b, \tag{7.16}$$

$$u(t_1^+,x) - u(t_1^-,x) = I_1(u_1(t_1,x)),$$

$$u(t,x) = u_1(t,x), \qquad (t,x) \in (-\infty,t_1] \times J_b. \tag{7.17}$$

We transform the problem (7.16)–(7.17) into a fixed point problem. Consider the operator $N_1 : C((-\infty,a] \times J_b, \mathbb{R}^n) \to C((-\infty,a] \times J_b, \mathbb{R}^n)$ defined by

$$N_1(u) = \Bigg\{ h \in C((-\infty,a] \times J_b, \mathbb{R}^n) : h(t,x)$$

$$= \begin{cases} u_1(t,x), & (t,x) \in \mathcal{D}_1, \\ u_1(t_1,x) + I_1(u_1(t_1,x)) + \displaystyle\int_{t_1}^t \int_0^x v(s,y)dsdy, & (t,x) \in (t_1,a] \times (0,b], \end{cases}$$

where $\mathcal{D}_1 = (-\infty,a] \times [-b,b] \backslash (t_1,a] \times (0,b]$ and $v \in S_{F,u}$. $z_0(t,x) := \phi(t,0) + \phi(0,y) - \phi(0,0)$. Let $z(\cdot,\cdot) : (-\infty,a] \times (-\infty,b] \to \mathbb{R}^n$ be the function defined by

$$z(t,x) = \begin{cases} u_1(t,0) + u_1(t_1,y) - u(t_1,0), & \text{if } (t,x) \in [t_1,a] \times [0,b], \\ u_1(t,x), & \text{if } (t,x) \in \mathcal{D}_1. \end{cases}$$

Then $z_{(t_1,0)} = u_1$. For each $w \in B_{a,b}$ with $w(t_1,x) = 0$, we denote by $\overline{w}$ the function defined by

$$\overline{w}(t,x) = \begin{cases} w(t,x), & \text{if } (t,x) \in [t_1,a] \times [0,b], \\ 0, & \text{if } (t,x) \in \mathcal{D}_1. \end{cases}$$

If $u(\cdot,\cdot)$ satisfies the integral equation,

$$u(t,x) = \int_{t_1}^{t}\int_0^x v(s,y)dsdy + u_1(t_1^-,x) + I_1(u_1(t_1,x)),$$

we can decompose $u(\cdot,\cdot)$ as $u(t,x) = z(t,x) + w(t,x)$, $(t,x) \in [t_1,a] \times [t_1,b]$, and this implies $u_{(t,x)} = w_{(t,x)} + z_{(t,x)}$, for every $(t,x) \in [t_1,a] \times [0,b]$. Also, the function $w(\cdot,\cdot)$ satisfies

$$w(t,x) = \int_{t_1}^{t}\int_{t_1}^{x} v(s,y)dsdy + I_1(\overline{w}(t_1,x) + z(t_1,x)), \tag{7.18}$$

where $v \in S_{F,\overline{w}+z}$. Set

$$C_{(t_1,t_1)} = \{w \in B_{a,b} : \overline{w}_{(t_1,x)} = 0, \text{ for each } x \in [-b,b]\}.$$

For any $w \in C_{(t_1,t_1)}$ we have

$$\begin{aligned} \|\overline{w}\|_\infty &= \|\overline{w}_{(t_1,t_1)}\|_B + \sup\left\{|\overline{w}(t,x))| : (t,x) \in J_a \times J_b\right\} \\ &= \sup\left\{|w(t,x)| : (t,x) \in J_a \times J_b\right\}. \end{aligned}$$

Thus $(C_{(t_1,t_1)}, \|\cdot\|_\infty)$ is a Banach space.

Let the $P_1 : C_{(t_1,0)} \to \mathcal{P}(C_{(t_1,0)})$ be defined by

$$P_1(w) = \left\{ h \in C_{(t_1,t_1)} : h(t,x) = \begin{cases} 0, \ (t,x) \in (-\infty,t_1] \times [-b,b], \\ \int_{t_1}^{t}\int_0^x v(s,y)dsdy + I_1(\overline{w}(t_1,x) + z(t_1,x)), \\ \qquad (t,x) \in J_a \times J_b, \\ \qquad\qquad v \in S_{F,\overline{w}+z}. \end{cases} \right.$$

Now N_1 having a fixed point is equivalent to P_1 having one, so we next prove this is so.

As in Step 1, we can show that N_1 is completely continuous, and each possible solution of (7.16)–(7.17) is a priori bounded by a constant M_2. Set

$$U_2 := \{u \in C_{(t_1,0)} : \|u\|_\infty < M_2 + 1\}.$$

From the choice of U_2 there is no $u \in \partial U_2$ such that $u = \lambda N_1(u)$ for some $\lambda \in (0,1)$. As a consequence of the nonlinear alternative of Leray Schauder type (Lemma 2.87) we deduce that N_1 has a fixed point u in U_2 that is a solution of (7.16)–(7.17). Denote this solution by u_2. Define

$$r_{k,2}(t,x) = \tau_k(u_2(t,x)) - t \quad \text{for } (t,x) \in [t_1,a] \times J_b.$$

If

$$r_{k,2}(t,x) \neq 0 \text{ on } (t_1,a] \times J_b \text{ and for all } k = 1,\ldots,m,$$

then

$$u(t,x) = \begin{cases} u_1(t,x), & \text{if } (t,x) \in [0,t_1) \times J_b, \\ u_2(t,x), & \text{if } (t,x) \in [t_1,a] \times J_b, \end{cases}$$

is a solution of the problem (7.16)–(7.17).

It remains to consider the case where

$$r_{2,2}(t,x) = 0, \text{ for some } (t,x) \in (t_1,a] \times J_b.$$

By (G-12), we have

$$\begin{aligned} r_{2,2}(t_1^+,x_1) &= \tau_2(u_2(t_1^+,x_1)) - t_1 \\ &= \tau_2(I_1(u_1(t_1,x_1)) - t_1 \\ &> \tau_1(u_2(t_1,x_1)) - t_1 \\ &= r_{1,1}(t_1,x_1) = 0. \end{aligned}$$

Since $r_{2,2}$ is continuous and (G-10) holds, there exists $t_2 > t_1$ and $x_2 > x_1$ such that

$$r_{2,2}(u_2(t_2,x_2)) = 0,$$

and

$$r_{2,2}(t,x) \neq 0 \text{ for all } (t,x) \in (t_1,t_2) \times J_b.$$

It is clear from (G-10) that

$$r_{k,2}(t,x) \neq 0 \text{ for all } (t,x) \in (t_1,t_2) \times J_b, \quad k = 2,\ldots,m.$$

Suppose now that there is $(s,\bar{s}) \in (t_1,t_2] \times [0,x_2) \cup (x_2,b]$ such that

$$r_{1,2}(s,\bar{s}) = 0.$$

From (G-12) it follows that

$$\begin{aligned} r_{1,2}(t_1^+,x_1)) &= \tau_1(u_2(t_1^+,x_1)) - t_1 \\ &= \tau_1(I_1(u_2(t_1,x_1)) - t_1 \\ &\leq \tau_1(u_1(t_1,x_1) - t_1 \\ &= r_{1,1}(t_1,x_1) = 0. \end{aligned}$$

Thus, the function $r_{1,2}$ attains a nonnegative maximum at some point $(s_1,\bar{s}_1) \in (t_1,a] \times [0,x_2) \cup (x_2,b]$. Since

$$\frac{\partial^2 u_1(t,x)}{\partial t \partial x} \in F(t,x,u_1(t,x)),$$

there exists $v(t,x) \in F(t,x,u_1(t,x))$ a.e. $(t,x) \in [t_1,a] \times J_b$ such that

$$\frac{\partial^2 u_2(t,x)}{\partial t \partial x} = v(t,x), \quad (t,x) \in [t_1,a] \times J_b.$$

Then, we have

$$r'_{1,2}(t,x) = \tau'_1(u_2(t,x))\frac{\partial u_2(t,x)}{\partial t} - 1 = 0.$$

Therefore,

$$\left\langle \tau'_1(u_2(s_1,\bar{s}_1)), \int_{t_1}^{s_1} v(s,\bar{s}_1)ds \right\rangle = 1,$$

which contradicts (G-12).

Step 3. We continue this process and we take into account that $u_m := u\big|_{[t_m,a]\times J_b}$ is a solution to the problem

$$\frac{\partial^2 u(t,x)}{\partial t \partial x} \in F(t,x,u_{(t,x)}), \quad \text{a.e.} \quad t \in (t_m,a] \times (0,b],$$
$$u(t_m^+,x) = I_m(u_{m-1}(t_m^-,x)).$$

The solution u of the problem (7.7)–(7.9) is then defined by

$$u(t,x) = \begin{cases} u_0(t,x), & \text{if } t \in [0,t_1] \times J_b, \\ u_1(t,x), & \text{if } t \in (t_1,t_2) \times J_b, \\ \dots \\ u_m(t,x), & \text{if } t \in (t_m,a] \times J_b. \end{cases}$$

□

7.3.2 The Nonconvex Case

Let $\mathcal{A}$ be a subset of $J_a \times J_b \times \mathbb{R}^n$. Then $\mathcal{A}$ is $\mathcal{L} \otimes \mathcal{B}$ measurable if $\mathcal{A}$ belongs to the σ-algebra generated by all sets of the form $\mathcal{N} \times \beta$ where $\mathcal{N}$ is Lebesgue measurable in $J_a \times J_b$ and β is Borel measurable in $\mathbb{R}^n$. A subset $\mathcal{I}$ of $L^1(J_a \times J_b, \mathbb{R}^n)$ is *decomposable* if, for all $u, v \in \mathcal{I}$ and measurable set $\mathcal{N} \subset J_a \times J_b$, the function $u\chi_{\mathcal{N}} + v\chi_{J_a \times J_b - \mathcal{N}} \in \mathcal{I}$, where $\chi_{J_a \times J_b}$ stands for the characteristic function of $J_a \times J_b$.

Theorem 7.7. *In addition to conditions* (G-9)–(G-10), *assume that*

(G-13) $F : J_a \times J_b \times B \longrightarrow \mathcal{P}(\mathbb{R}^n)$ *is a nonempty compact valued multivalued map such that:*
(a) $(t,x,u) \mapsto F(t,x,u)$ *is* $\mathcal{L} \otimes \mathcal{B}$ *measurable;*
(b) $u \mapsto F(t,x,u)$ *is lower semicontinuous for a.e.* $(t,x) \in J_a \times J_b$.

Then, the impulsive hyperbolic initial value problem (7.7)–(7.9) *has at least one solution.*

Proof. Let $\mathcal{F} : C_{(0,0)} \to \mathcal{P}(C_{(0,0)})$ be defined by

$$\mathcal{F}(w) = \{v \in L^1([0,a] \times [0,b]) : v(t,x) \in F(t,x,\bar{w}_{(t,x)} + z_{(t,x)}\},$$

where $C_{(0,0)}$ and $\bar{w}, z$ are as in Theorem 7.6. Next we establish the properties of $\mathcal{F}(\cdot)$. Analogous results can be found in Frigon [205], Kandilakis–Papageorgiou [314], and Halidias and Papageorgiou [269]. Now we prove that $\mathcal{F}(\cdot)$ has nonempty, closed, decomposable values and is l.s.c. For the nonemptiness, note that if we let $w \in C_{(0,0)}$, then from condition (G-13), we have $F(\cdot,\cdot,\overline{w}_{(\cdot,\cdot)} + z_{(\cdot,\cdot)})$ is a measurable multifunction. Then, there exists a sequence of measurable functions $\{f_n(t,x)\}_{n\in\mathbb{N}}$ such that $F(t,x,\overline{w}_{(t,x)} + z_{(t,x)}) = \overline{\{f_n(t,x)\}}$. Using the fact that F has compact values and from condition (G-9), we pass to a subsequence, if necessary, to get f_{n_k} converges to f in $L^1(J_a \times J_b, \mathbb{R}^n)$. Therefore, for every $w \in C_{(0,0)}$, $\mathcal{F}(w) \neq \emptyset$. The closedness and decomposability of the values of $\mathcal{F}(\cdot)$ are easy to check. To prove that $\mathcal{F}(\cdot)$ is l.s.c., we need to show that, for every $v \in L^1(J_a \times J_b, \mathbb{R}^n)$, $d(v,\mathcal{F}(\cdot))$ is an upper semicontinuous $\mathbb{R}_+$-valued function defined on $C_{(0,0)}$. To this end, we have

$$\begin{aligned} d(v,\mathcal{F}(w)) &= \inf\{\|v-u\|_{L^1} : u \in \mathcal{F}(w)\} \\ &= \inf\left\{ \int_0^a \int_0^b |v(t,x) - u(t,x)|dtdx : \right. \\ &\qquad\qquad \left. u(t,x) \in F(t,x,\overline{w}_{(t,x)} + z_{(t,x)}) \right\} \\ &= \int_0^a \int_0^b d(v(t,x), F(t,x,\overline{w}_{(t,x)} + z_{(t,x)})dtdx \end{aligned}$$

(see Hiai–Umegak [288] or Hu and Papageorgiou [301]). We shall show that for every $\alpha \geq 0$, the superlevel set

$$U_\alpha = \{w \in C_{0,0} : d(u,\mathcal{F}(w)) \geq \alpha\}$$

is closed. For this purpose let $\{w_n\}_{n\geq 1} \subset U_\alpha$ and assume that $w_n \to w$ as $n \to \infty$. By virtue of (G-13)(b), $(t,x) \to d(u(t,x), F(t,x,\overline{w}_{(t,x)} + z_{(t,x)})$, is an upper semicontinuous $\mathbb{R}_+$-valued function. So by Fatou's Lemma, we have

$$\begin{aligned} \alpha \leq \overline{\lim}\, d(u,\mathcal{F}(w_n)) &= \overline{\lim} \int_0^a \int_0^b d(u(t,x), F(t,x,\overline{w}_{n_{(t,x)}})dtdx \\ &\leq \int_0^a \int_0^b \overline{\lim} d(u(t,x), \overline{w}_{n_{(t,x)}} + z_{(t,x)})dtdx \\ &\leq \int_0^a \int_0^b d(u(t,x), F(t,x,\overline{w}_{(t,x)} + z_{(t,x)})dtdx = d(u,\mathcal{F}(w)). \end{aligned}$$

Therefore, $w \in U_\alpha$ and this proves the lower semicontinuity of $\mathcal{F}(\cdot)$. This allows us to apply Theorem 2.85 of Bressan and Colombo [104] and obtain

$$f : C_{(0,0)} \to L^1(J_a \times J_b, \mathbb{R}^n)$$

is a continuous map such that $f(w) \in \mathcal{F}(w)$, for every $w \in C_{(0,0)}$.

We consider the problem,

$$\frac{\partial^2 u(t,x)}{\partial t \partial x} = f(\overline{u}(t,x) + z_{(t,x)}), \quad \text{a.e.} \quad (t,x) \in J_a \times J_b, \tag{7.19}$$

$$u(0,x) = \phi(x), \qquad x \in J_b. \tag{7.20}$$

Clearly, if $u \in C_{(0,0)}$ is a solution of the problem (7.19)–(7.20), then u is a solution to the problem (7.7)–(7.9). To transform the problem (7.19)–(7.20) into a fixed point problem, consider the operator $P_1 : C_{(0,0)} \to C_{(0,0)}$ defined by,

$$P_1(w) = \begin{cases} 0, & (t,x) \in (-\infty, 0] \times [-b,b], \\ \displaystyle\int_0^t \int_0^x f(\overline{w}_{(s,x)} + z_{(s,x)}) ds dy, & (t,x) \in J_a \times J_b. \end{cases}$$

We shall show that P_1 is completely continuous. First, we have that P_1 is continuous. Next, let $\{u_n\}$ be a sequence such that $u_n \longrightarrow u$ in $C_{(0,0)}$. Then

$$\begin{aligned} |P_1(u_n(t,x)) - P_1(u(t,x))| &\le \int_0^t \int_0^x |g(u_n(s,\tau)) - g(u(s,\tau))| ds d\tau \\ &\le \int_0^a \int_0^b |g(u_n(s,\tau)) - g(u(s,\tau))| ds d\tau. \end{aligned}$$

Since the functions g and I_k, $k = 1, \dots, m$, are continuous,

$$\begin{aligned} \|P_1(u_n) - P_1(u)\|_\infty &\le \|f(u_{n(\cdot,\cdot)} + z_{(\cdot,\cdot)}) - f(\overline{u}(\cdot,\cdot) + z_{(\cdot,\cdot)})\|_{L^1} \\ &\to 0 \text{ as } n \to \infty. \end{aligned}$$

As a consequence of Steps 2, 3 and 4 together with Theorem 7.6, P_1 is completely continuous, and there exists $b^* > 0$ such that, for every solution of the problem (7.19)–(7.20), we have

$$\|u\|_\infty \le b^*.$$

Set

$$U = \{u \in C_{(0,0)} : \|u\|_\infty < b^* + 1\}.$$

From the choice of U there is no $u \in \partial U$ such that $u = \lambda P_1(u)$, for some $\lambda \in (0,1)$. As a consequence of the nonlinear alternative of Leray Schauder type (Lemma 2.86), we deduce that P_1 has a fixed point u in U that in turn is a solution of the problem (7.19)–(7.20). By the same argument used in the proof of Theorem 7.6, we can complete the proof of this theorem. □

Chapter 8

Neutral Differential Inclusions

The results of this chapter extend the first two sections of Chapter 5 in that they are devoted to a Filippov's Theorem and to a relaxed problem in the context of neutral differential inclusions.

Given a real separable Banach space E with norm $|\cdot|$, we will consider in this chapter the impulsive problem for first-order impulsive neutral differential inclusions,

$$\frac{d}{dt}[y(t) - g(t, y_t)] \in F(t, y_t), \quad \text{a.e.} \quad t \in J\backslash\{t_1, \ldots, t_m\}, \tag{8.1}$$

$$\Delta y_{t=t_k} = I_k(y(t_k^-)), \qquad k = 1, \ldots, m, \tag{8.2}$$

$$y(t) = \phi(t), \qquad t \in [-r, 0], \tag{8.3}$$

where $0 < r < \infty$, $0 = t_0 < t_1 < \cdots < t_m < t_{m+1} = b$, $J = [0, b]$,

$$\mathcal{D} = \{\psi\colon [-r, 0] \to E\colon \psi \text{ is continuous everywhere except for a finite number of points } \bar{t} \text{ at which } \psi(\bar{t}^-) \text{ and } \psi(\bar{t}^+) \text{ exist and satisfy } \psi(\bar{t}^-) = \psi(\bar{t})\},$$

$F\colon J \times \mathcal{D} \to \mathcal{P}(E)$ is a multifunction, $g : J \times \mathcal{D} \to E$, $\phi \in \mathcal{D}$, $\lim_{t\to t_k^+} g(t, \psi) = \lim_{t\to t_k^-} g(t, \psi) = g(t_k, \psi)$, for $\psi \in \mathcal{D}$, $I_k \in C(E, E)$, $k = 1, \ldots, m$, and $\Delta y|_{t=t_k} = y(t_k^+) - y(t_k^-)$. The notations $y(t_k^+) = \lim_{h\to 0^+} y(t_k + h)$ and $y(t_k^-) = \lim_{h\to 0^+} y(t_k - h)$ stand for the usual right and the left hand limits of the function y at $t = t_k$, respectively. For any function y defined on $[-r, b]$ and any $t \in J$, y_t refers to the element of $\mathcal{D}$ such that

$$y_t(\theta) = y(t + \theta), \quad \theta \in [-r, 0];$$

thus the function y_t represents the history of the state from time $t - r$ up to the present time t.

8.1 Filippov's Theorem

Let $J_k = (t_k, t_{k+1}]$, $k = 0, \ldots, m$, and let y_k be the restriction of a function y to J_k. In order to define solutions for the problem (8.1)–(8.3), consider the space

$$PC = \{y\colon [0, b] \to E \mid y_k \in C(J_k, E),\ k = 0, \ldots, m, \text{ such that } y(t_k^-) \text{ and } y(t_k^+) \text{ exist and satisfy } y(t_k^-) = y(t_k) \text{ for } k = 1, \ldots, m\}.$$

Endowed with the norm

$$\|y\|_{PC} = \max\{\|y_k\|_\infty, \quad k = 0, \dots, m\},$$

PC is a Banach space. Moreover, if

$$\Omega = \{y: [-r, b] \to E, \ y \in PC \cap \mathcal{D}\},$$

then Ω is a Banach space with the norm

$$\|y\|_\Omega = \sup\{\|y(t)\| : t \in [-r, b]\},$$

In the space $\mathcal{D}$, we consider the L^1–norm

$$\|y\|_{\mathcal{D}} = \int_{-r}^{0} |y(\theta)| d\theta.$$

Definition 8.1. A function $y \in \Omega$ is said to be a solution of problem (8.1)–(8.3) if there exists $v \in L^1(J, E)$ such that $v(t) \in F(t, y_t)$ a.e. on J, $y(t) = \phi(t)$, $t \in [-r, 0]$, and

$$\begin{aligned} &\frac{d}{dt}[y(t) - g(t, y_t)] = v(t), \quad \text{a.e.} \quad t \in [0, b], \\ &y(t_k^+) - y(t_k) = I_k(y(t_k)), \qquad k = 1, \dots, m. \end{aligned}$$

Let $\psi \in \mathcal{D}, \overline{g} \in L^1(J, E)$, and let $x \in \Omega$ be a solution of the impulsive differential problem with a semilinear equation,

$$\begin{cases} \frac{d}{dt}[x(t) - g(t, x_t)] = \overline{g}(t), & \text{a.e. } t \in J \backslash \{t_1, \dots, t_m\}, \\ \Delta x_{t=t_k} = I_k(x(t_k^-)), & k = 1, \dots, m, \\ x(t) = \psi(t), & t \in [-r, 0]. \end{cases} \tag{8.4}$$

We will make the following three covering assumptions for this section.

(H-1) The function $F: J \times \mathcal{D} \to \mathcal{P}_{cl}(E)$ satisfies
(a) for all $y \in \mathcal{D}$, the map $t \mapsto F(t, y)$ is measurable;
(b) the map $\gamma : t \mapsto d(g(t), F(t, x_t))$ is integrable.

(H-2) There exists a function $p \in L^1(J, \mathbb{R}^+)$ such that

$$H_d(F(t, z_1), F(t, z_2)) \leq p(t)\|z_1 - z_2\|_\infty, \text{ for all } z_1, z_2 \in \mathcal{D}.$$

(H-3) There exists $c \in (0, 1)$ such that

$$\|g(t, \phi) - g(t, \psi)\| \leq c\|\phi - \psi\|_\infty, \text{ for all } \phi, \psi \in \mathcal{D}.$$

Remark 8.2. From Assumptions (H-1)(a) and (H-3), it follows that the multifunction $t \mapsto F(t, x_t)$ is measurable, and by Lemmas 1.4 and 1.5 in [201], it follows that $\gamma(t) = d(g(t), F(t, x_t))$ is measurable (also see the Remark on p. 400 in [35]).

Let $P(t) = \int_0^t p(s)ds$, and δ and ϵ be positive constants. Define the family of functions $(\eta_k(t))_{t\geq 0}$ ($k = 0, \ldots, m$) by

$$\eta_0(t) = \frac{1}{1-c}\left(2\delta \exp\left(\frac{e^{P(t)}}{1-c}\right)\right.$$
$$\left. + \int_0^t [\gamma(s)+\epsilon]\left(\exp\left(\frac{e^{P(s)}-e^{P(u)}}{1-c}\right)+1\right)ds\right), \quad t \in [0, t_1],$$

$$H_0(t) = \frac{1}{1-c}(2\delta \exp\left(\frac{e^{P(t)}}{1-c}\right) + \int_0^t [\gamma(s)+\epsilon]\exp\left(\frac{e^{P(t)}-e^{P(s)}}{1-c}\right)ds),$$

and for $k = 1, \ldots, m$,

$$\bar{R}_k(t_k) = |x(t_k) - y_1(t_k)| + c\|x_{t_k} - (y_k)_{t_k}\|_\infty + |I_k(x(t_k)) - I_k(y_k(t_k))|,$$

$$\eta_k(t) = \frac{1}{1-c}\left(\bar{R}_k(t_k)\exp\left(\frac{e^{P(t)}}{1-c}\right)\right.$$
$$\left. + \int_{t_k}^t [\gamma(s)+\epsilon]\left(\exp\left(\frac{e^{P(t)}-e^{P(s)}}{1-c}\right)+1\right)ds\right),$$

and

$$H_k(t) := \frac{1}{1-c}\left(\bar{R}_k(t_k)\exp\left(\frac{e^{P(t)}}{1-c}\right)\right.$$
$$\left. + \int_{t_k}^t [\gamma(s)+\epsilon]\exp\left(\frac{e^{P(t)}-e^{P(s)}}{1-c}\right)ds\right).$$

Theorem 8.3. *Let* (H-1)–(H-3) *hold and set* $\gamma_k := \gamma_{|J_k}$. *Then, for every* $\phi \in \mathcal{D}$ *with* $\|\phi - \psi\|_\infty \leq \delta$, *problem* (8.1)–(8.3) *has at least one solution* y *satisfying the estimates*

$$\|y_t - x_t\|_\infty \leq \frac{1}{1-c}\sum_{0<t_k<t}[|x(t_k) - y_k(t_k)| + \|x_{t_k} - (y_k)_{t_k}\|_\infty]$$
$$+ \frac{1}{1-c}\sum_{0<t_k<t}|I_k(x(t_k)) - I_k(y_k(t_k))| + \frac{1}{1-c}\sum_{k=0}^m \eta_k(t),$$

and

$$\left|\frac{d}{dt}[y(t) - g(t, y_t)] - \bar{g}(t)\right| \leq p(t)\sum_{0<t_k<t} H_k(t) + \sum_{0<t_k<t}[\gamma_k(t) + \epsilon],$$

for a.e. $t \in [0, b]$.

Proof. We are going to study problem (8.1)–(8.3) in the respective intervals $[-r, t_1]$, $(t_1, t_2]$, ..., $(t_m, b]$. The proof will be given in three steps and then continued by induction in a fourth step. Let $\phi \in \mathcal{D}$ be such that $\|\phi - \psi\|_\infty \leq \delta$.

Step 1. In this first step, we construct a sequence of functions $(y_n)_{n\in\mathbb{N}}$ which will be shown to converge to some solution of problem (8.1)–(8.3) on the interval $[-r, t_1]$, namely, to

$$\begin{cases} \frac{d}{dt}[y(t) - g(t, y_t)] \in F(t, y_t), & t \in J_0 = [0, t_1], \\ y(t) = \phi(t), & t \in [-r, 0]. \end{cases} \tag{8.5}$$

Let $f_0 = \overline{g}$ on $[-r, t_1]$ and $y^0(t) = x(t)$, $t \in [0, t_1]$, i.e.,

$$y^0(t) = \begin{cases} \psi(t), & t \in [-r, 0], \\ \psi(0) - g(0, \psi) + g(t, y_t) + \displaystyle\int_0^t f_0(s)ds, & t \in [0, t_1]. \end{cases}$$

Then define the multivalued map $U_1\colon [0, t_1] \to \mathcal{P}(E)$ by $U_1(t) = F(t, y_t^0) \cap (g(t) + (\gamma(t) + \epsilon))B(0, 1)$. Since g and γ are measurable, Theorem III.4.1 in [136] tells us that the ball $(g(t) + (\gamma(t) + \epsilon)B(0, 1))$ is measurable. Moreover, $F(t, y_t^0)$ is measurable (see Remark 8.2) and U_1 is nonempty. Indeed, since $v = \epsilon$ is a measurable function, from Lemma 2.76, there exists a function u which is a measurable selection of $F(t, y_t^0)$ such that

$$|u(t) - \bar{g}(t)| \leq d(\bar{g}(t), F(t, y_t^0)) + \epsilon = \gamma(t) + \epsilon.$$

Then $u \in U_1(t)$, proving our claim. We deduce that the intersection multivalued operator $U_1(t)$ is measurable (see [35, 136, 230]). By Lemma 2.77 (Kuratowski–Ryll-Nardzewski selection theorem), there exists a function $t \to f_1(t)$ which is a measurable selection for U_1. Consider

$$y^1(t) = \begin{cases} \phi(t), & t \in [-r, 0], \\ \phi(0) - g(0, \phi) + g(t, y_t^1) + \int_0^t f_1(s)ds, & t \in [0, t_1], \end{cases}$$

where y^1 is a solution of the problem

$$\begin{cases} \frac{d}{dt}[y(t) - g(t, y_t)] = f_1(t), & \text{a.e.} \quad t \in [0, t_1], \\ y(t) = \phi(t), & t \in [-r, 0]. \end{cases} \tag{8.6}$$

Then, for each $t \in [0, t_1]$, we have

$$\begin{aligned} |y^1(t) - y^0(t)| &\leq |\phi(0) - \psi(0)| + |g(0, \phi) - g(0, \psi)| \\ &\quad + |g(0, y_t^0) - g(0, y_t^1)| + \int_0^t |f_0(s) - f_1(s)|\, ds \\ &\leq 2\|\phi - \psi\|_\infty + c\|y_t^0 - y_t^1\|_\infty + \int_0^t |f_0(s) - f_1(s)| ds. \end{aligned} \tag{8.7}$$

Hence,

$$\sup_{s\in[-r,t]}\{|y^1(s)-y^0(s)|\}\leq \frac{1}{1-c}\left[2\delta+\int_0^t \gamma(s)ds+\epsilon t\right].$$

Then Lemma 1.4 in [201] tells us that $F(t,y_t^1)$ is measurable. The ball $f_1(t)+p(t)\|y_t^1-y_t^0\|_\infty B(0,1)$ is also measurable by Theorem III.4.1 in [136]. The set $U_2(t)=F(t,y_t^1)\cap f_1(t),(f_1(t)+p(t)\|y_t^1-y_t^0\|_\infty)B(0,1)$ is nonempty. Indeed, From (H-2), it follows that

$$H(F(t,y_t^1),F(t,y_t^0))\leq p(t)\|y_t^1-y_t^0\|_\infty.$$

Thus, there exists a measurable selection u of $F(t,y_t^1)$ such that

$$|u(t)-f_1(t)|\leq p(t)\|y_t^0-y_t^1\|_\infty,$$

i.e., $u\in U_2(t)$, proving our claim.

Now, since the intersection multivalued operator U_2 defined above is measurable (see [35, 136, 230]), there exists a measurable selection $f_2(t)\in U_2(t)$. Hence,

$$|f_1(t)-f_2(t)|\leq p(t)\|y_t^1-y_t^0\|_\infty. \tag{8.8}$$

Define

$$y^2(t)=\begin{cases}\phi(t), & t\in[-r,0],\\ \phi(0)-g(0,\phi)+g(t,y_t^2)+\int_0^t f_2(s)ds, & t\in[0,t_1],\end{cases}$$

where y^2 is a solution of the problem,

$$\begin{cases}\frac{d}{dt}[y(t)-g(t,y_t)]=f_2(t), & \text{a.e.} \quad t\in J=[0,t_1],\\ y(t)=\phi(t), & t\in[-r,0].\end{cases} \tag{8.9}$$

Using (8.7) and (8.8), a simple integration by parts yields the estimate

$$\begin{aligned}|y^2(t)-y^1(t)|&\leq |g(t,y_t^1)-g(t,y_t^2)|+\int_0^t|f_2(s)-f_1(s)|ds\\
&\leq c\|y_t^1-y_t^2\|+\frac{1}{1-c}\int_0^t p(s)\left(2\delta+\int_0^s\gamma(u)du+\epsilon s\right)ds\\
&=c\|y_t^1-y_t^2\|+\frac{1}{1-c}\\
&\quad\times\left(2\delta\int_0^t p(s)ds+\int_0^t p(s)ds\int_0^s[\gamma(u)+\epsilon]du\right)\\
&\leq c\|y_t^1-y_t^2\|+\frac{1}{1-c}\\
&\quad\times\left(2\delta\int_0^t p(s)e^{P(s)}ds+\int_0^t p(s)e^{P(s)}ds\int_0^s e^{-P(u)}\gamma(u)du\right)\\
&\leq c\|y_t^1-y_t^2\|+\frac{1}{1-c}\left(2\delta e^{P(t)}+\int_0^t[\gamma(s)+\epsilon][e^{P(t)}-e^{P(s)}]ds\right)\end{aligned}$$

for every $t \in [-r, t_1]$. Thus,

$$|y^2(t) - y^1(t)| \leq \frac{1}{(1-c)^2}\left(2\delta e^{P(t)} + \int_0^t [\gamma(s)+\epsilon][e^{P(t)} - e^{P(s)}]ds\right),$$
$$t \in [-r, t_1].$$

Let $U_3(t) = F(t, y_t^2) \cap (f_2(t) + p(t)\|y_t^2 - y_t^1\|_\infty B(0,1))$. Arguing as for U_2, we can prove that U_3 is a measurable multivalued map with nonempty values; so there exists a measurable selection $f_3(t) \in U_3(t)$. This allows us to define

$$y^3(t) = \begin{cases} \phi(t), & t \in [-r, 0], \\ \phi(0) - g(0,\phi) + g(t, y_t) + \int_0^t f_3(s)ds, & t \in [0, t_1], \end{cases}$$

where y^3 is a solution of the problem

$$\begin{cases} \frac{d}{dt}[y(t) - g(t, y_t)] = f_3(t), \quad \text{a.e.} \quad t \in J = [0, t_1], \\ y(t) = \phi(t), \quad t \in [-r, 0]. \end{cases} \tag{8.10}$$

For $t \in [0, t_1]$, we have

$$\begin{aligned} |y^3(t) - y^2(t)| &\leq c\|y_t^3 - y_t^2\|_\infty + \frac{1}{2(1-c)^2}\int_0^t 2p(s) \\ &\quad \times \left(\delta e^{2P(s)} + \int_0^s [\gamma(u)+\epsilon][e^{P(s)} - e^{P(u)}]du\right) ds \\ &\leq c\|y_t^3 - y_t^2\|_\infty + \frac{1}{2(1-c)^2} \\ &\quad \times \left(2\delta e^{2P(t)} + \int_0^t 2p(s)ds \int_0^s [\gamma(u)+\epsilon][e^{P(s)} - e^{P(u)}]du\right) \\ &\leq c\|y_t^3 - y_t^2\|_\infty + \frac{1}{2(1-c)^2} \\ &\quad \times \left(2\delta e^{2P(t)} + \int_0^t [\gamma(u)+\epsilon] \int_u^t \frac{d}{ds}[e^{P(s)} - e^{P(u)}]^2 dsdu\right) \\ &\leq c\|y_t^3 - y_t^2\|_\infty + \frac{1}{2(1-c)^2} \\ &\quad \times \left(2\delta e^{2P(t)} + \int_0^t [\gamma(s)+\epsilon][e^{P(t)} - e^{P(s)}]^2 ds\right). \end{aligned}$$

Then,

$$\begin{aligned} |y^3(t) - y^2(t)| &\leq \frac{1}{2!(1-c)^3} \\ &\quad \times \left(2\delta e^{2P(t)} + \int_0^t [\gamma(s)+\epsilon][e^{(P(t)} - e^{P(s)}]^2 ds\right), \ t \in [-r, t_1]. \end{aligned}$$

Repeating the process for $n = 1, 2, 3, \ldots$, we arrive at the following bound

$$
\begin{aligned}
|y^n(t) - y^{n-1}(t)| \le\ & \frac{1}{(n-1)!(1-c)^n} \\
& \times \int_0^t [\gamma(s) + \epsilon][e^{P(t)} - e^{P(s)}]^{(n-1)} ds \\
& + \frac{1}{(n-1)!(1-c)^n} 2\delta e^{(n-1)P(t)}, \ t \in [-r, t_1].
\end{aligned} \tag{8.11}
$$

By induction, suppose that (8.11) holds for some n and check (8.11) for $n+1$. Let $U_{n+1}(t) = F(t, y^n(t)) \cap (f_n(t) + p(t)\|y_t^n - y_t^{n-1}\|_\infty B(0,1)$. Since U_{n+1} is a nonempty measurable set, there exists a measurable selection $f_{n+1}(t) \in U_{n+1}(t)$, which allows us to define for $n \in \mathbb{N}$

$$
y^{n+1}(t) = \begin{cases} \phi(t), & t \in [-r, 0], \\ \phi(0) - g(0, \phi) + g(t, y_t^{n+1}) + \int_0^t f_{n+1}(s) ds, & t \in [0, t_1]. \end{cases} \tag{8.12}
$$

Therefore, for a.e. $t \in [0, t_1]$, we have

$$
\begin{aligned}
|y^{n+1}(t) - y^n(t)| &\le c\|y_t^{n+1} - y_t^n\|_\infty + \int_0^t |f_{n+1}(s) - f_n(s)| \, ds \\
&\le c\|y_t^{n+1} - y_t^n\|_\infty \\
&\quad + \frac{1}{(n-1)!(1-c)^n} \int_0^t p(s)\|y_s^n - y_s^{n-1}\|_\infty \, ds \\
&\le c\|y_t^{n+1} - y_t^n\|_\infty + \frac{1}{(n-1)!(1-c)^n} \int_0^t p(s) ds \\
&\quad \times \left(2\delta e^{(n-1)P(s)} + \int_0^s [\gamma(u) + \epsilon][e^{P(s)} - e^{P(u)}]^{(n-1)} du \right) \\
&\le c\|y_t^{n+1} - y_t^n\|_\infty + \frac{1}{n!(1-c)^n} \int_0^t 2\delta n p(s) e^{nP(s)} ds \\
&\quad + \frac{1}{n!(1-c)^n} \int_0^t [\gamma(u) + \epsilon] du \int_u^t \frac{d}{ds}[e^{P(s)} - e^{P(u)}]^n ds.
\end{aligned}
$$

Again, an integration by parts leads to

$$
\begin{aligned}
|y^{n+1}(t) - y^n(t)| \le\ & \frac{1}{n!(1-c)^{n+1}} \int_0^t [\gamma(s) + \epsilon][e^{P(t)} - e^{P(s)}]^n ds \\
& + \frac{2}{n!(1-c)^{n+1}} \delta e^{nP(t)}.
\end{aligned}
$$

Consequently, (8.11) holds true for all $n \in \mathbb{N}$. We infer that $\{y^n\}$ is a Cauchy sequence in Ω_1, converging uniformly to a limit function $y \in \Omega_1$, where

$$
\Omega_1 = C([0, t_1], E) \cap \mathcal{D}.
$$

Moreover, from the definition of $\{U_n\}$, we have

$$|f_{n+1}(t) - f_n(t)| \leq p(t)\|y_t^n - y_t^{n-1}\|_\infty, \quad \text{for a.e } t \in [0, t_1].$$

Hence, for almost every $t \in [0, t_1]$, $\{f_n(t)\}$ is also a Cauchy sequence in E and so it converges almost everywhere to some measurable function $f(\cdot)$ in E. In addition, since $f_0 = \bar{g}$, we have

$$\begin{aligned} |f_n(t)| &\leq \sum_{l=1}^{n} p(t)|f_l(t) - f_{l-1}(t)| + |f_0(t)| \\ &\leq \sum_{l=2}^{n} p(t)\|y_t^{l-1} - y_t^{l-2}\|_\infty + \gamma(t) + |\bar{g}(t)| + \epsilon \\ &\leq p(t)\sum_{l=1}^{\infty} \|y_t^{l} - y_t^{l-1}\|_\infty + \gamma(t) + |\bar{g}(t)| + \epsilon \end{aligned}$$

for a.e. $t \in [0, t_1]$. Hence,

$$|f_n(t)| \leq H_0(t)p(t) + \gamma(t) + |\bar{g}(t)| + \epsilon,$$

where

$$H_0(t) := \tfrac{1}{1-c}\left(2\delta \exp\left(\frac{e^{P(t)}}{1-c}\right) + \int_0^t [\gamma(s) + \epsilon] \exp\left(\frac{e^{P(t)} - e^{P(s)}}{1-c}\right) ds\right). \tag{8.13}$$

From the Lebesgue dominated convergence theorem, we deduce that $\{f_n\}$ converges to f in $L^1([0, t_1], E)$. Passing to the limit in (8.12), we find that the function

$$y(t) = \begin{cases} \phi(t), & t \in [-r, 0], \\ \phi(0) - g(0, \phi) + g(t, y_t) + \int_0^t f(s)ds, & t \in [0, t_1], \end{cases}$$

is solution to problem (8.1)–(8.3) on $[0, t_1]$; thus $y \in S_{[0,t_1]}(\phi)$. Moreover, for a.e. $t \in [0, t_1]$, we have

$$\begin{aligned} |x(t) - y(t)| &= \Big|\phi(0) + \int_0^t \bar{g}(s)ds - \psi(0) - \int_0^t f(s)ds + g(t, y_t) \\ &\quad - g(t, x_t) + g(0, \phi) - g(0, \psi)\Big| \\ &\leq 2\|\phi - \psi\|_\infty + \int_0^t |f(s) - f_0(s)|ds + c\|x_t - y_t\|_\infty \\ &\leq 2\|\phi - \psi\|_\infty + \int_0^t |f(s) - f_n(s)|ds + c\|x_t - y_t\|_\infty \\ &\quad + \int_0^t |f_n(s) - f_0(s)|ds \end{aligned}$$

$$\begin{aligned}
&\le 2\|\phi-\psi\|_\infty + c\|x_t-y_t\|_\infty + \int_0^t |f(s)-f_n(s)|ds \\
&\quad + \int_0^t p(s)\sum_{l=1}^{\infty}\|y_s^l - y_s^{l-1}\|_\infty ds + \int_0^t [\gamma(s)+\epsilon]ds \\
&\le 2\|\phi-\psi\|_\infty + c\|x_t-y_t\|_\infty + \|f_n-f\|_{L^1} + \int_0^t \frac{p(s)}{1-c} \\
&\quad \times \left(2\delta\exp\left(\frac{e^{P(s)}}{1-c}\right) + \int_0^s [\gamma(u)+\epsilon]\exp\left(\frac{e^{P(s)}-e^{P(u)}}{1-c}\right)du\right)ds \\
&\le 2\|\phi-\psi\|_{\mathcal{D}} + c\|x_t-y_t\|_\infty + \|f_n-f\|_{L^1} + 2\delta\exp\left(\frac{e^{P(t)}}{1-c}\right) \\
&\quad + \int_0^t [\gamma(s)+\epsilon]\exp\left(\frac{e^{P(s)}-e^{P(u)}}{1-c}\right)ds + \int_0^t [\gamma(s)+\epsilon]ds.
\end{aligned}$$

Thus,

$$\begin{aligned}
|x(t)-y(t)| \le{}& \frac{1}{1-c}\left(2\delta\exp\left(\frac{e^{P(t)}}{1-c}\right) + \int_0^t [\gamma(s)+\epsilon]\exp\left(\frac{e^{P(s)}-e^{P(u)}}{1-c}\right)ds\right) \\
&+ \frac{1}{1-c}\int_0^t [\gamma(s)+\epsilon]ds + \|f_n-f\|_{L^1}.
\end{aligned}$$

Passing to the limit as $n \to \infty$, we obtain

$$|x(t)-y(t)| \le \eta_0(t), \quad \text{a.e. } t \in [-r, t_1] \tag{8.14}$$

with

$$\eta_0(t) := \frac{1}{1-c}\left(2\delta\exp\left(\frac{e^{P(t)}}{1-c}\right) + \int_0^t [\gamma(s)+\epsilon]\left(\exp\left(\frac{e^{P(s)}-e^{P(u)}}{1-c}\right)+1\right)ds\right).$$

Next, we give an estimate for $|\frac{d}{dt}[y(t)-g(t,y_t)] - \bar{g}(t)|$ for $t \in [0, t_1]$. We have

$$\begin{aligned}
|\frac{d}{dt}[y(t)-g(t,y_t)] - \bar{g}(t)| &= |f(t)-f_0(t)| \\
&\le |f_n(t)-f_0(t)| + |f_n(t)-f(t)| \\
&\le p(t)\sum_{k=1}^{\infty}\|y_t^{k+1}-y_t^k\|_\infty + \gamma(t) + \epsilon + |f_n(t)-f(t)|.
\end{aligned}$$

Arguing as in (8.13) and passing to the limit as $n \to +\infty$, we see that

$$|\frac{d}{dt}[y(t)-g(t,y_t)] - \bar{g}(t)| \le H_0(t)p(t) + \gamma(t) + \epsilon, \quad t \in [0, t_1].$$

The resulting solution is denoted by $y_1 := y_{|[-r,t_1]}$.

Step 2. Now consider problem (8.1)–(8.3) on the second interval $[t_1 - r, t_2]$, i.e.,

$$\begin{cases} \frac{d}{dt}[y(t) - g(t, y_t)] \in F(t, y_t), & \text{a.e. } t \in (t_1, t_2], \\ y(t_1^+) = y_1(t_1) + I_1(y_1(t_1)), & \\ y(t) = y_1(t), & t \in [t_1 - r, t_1]. \end{cases} \tag{8.15}$$

Let $f_0 = \bar{g}$ and set

$$y^0(t) = \begin{cases} x(t), & t \in [t_1 - r, t_1], \\ x(t_1) + I_1(x(t_1)) - g(t_1, x_{t_1}) + g(t, x_t) + \int_{t_1}^t f_0(s)ds, & t \in (t_1, t_2]. \end{cases}$$

Notice that (8.14) allows us to use Assumption (H-2). Again applying Lemma 1.4 in [201] and arguing as in Step 1 we can show that the multivalued map U_1: $[t_1, t_2] \to \mathcal{P}(E)$ defined by $U_1(t) = F(t, y_t^0) \cap (\bar{g}(t) + [\gamma(t) + \epsilon] B(0, 1))$ is $U_1(t)$ is measurable. Hence, there exists a function $t \mapsto f_1(t)$ which is a measurable selection for U_1. Define

$$y^1(t) = \begin{cases} y_1(t), & t \in [t_1 - r, t_1], \\ y_1(t_1) + I_1(y_1(t_1)) - g(t_1, (y_1)_{t_1}) + g(t, y_t^1) + \int_{t_1}^t f_1(s)ds, & t \in (t_1, t_2]. \end{cases}$$

Next define the measurable multivalued map $U_2(t) = F(t, y_t^1) \cap (f_1(t) + p(t)\|y_t^1 - y_t^0\|_\infty B(0, 1))$. It has a measurable selection $f_2(t) \in U_2(t)$ by the Kuratowski–Ryll–Nardzewski selection theorem. Repeating the process of selection as in Step 1, we can define by induction a sequence of multivalued maps $U_n(t) = F(t, y_t^{n-1}) \cap (f_{n-1}(t) + p(t)\|y_t^{n-1} - y_t^{n-2}\|_\infty B(0, 1))$ where $\{f_n\} \in U_n$ and $(y^n)_{n \in \mathbb{N}}$ is defined by

$$y^n(t) = \begin{cases} y_1(t), & t \in [t_1 - r, t_1], \\ y_1(t_1) + I_1(y_1(t_1)) - g(t_1, y_{t_1}^n) + g(t, y_t^n) + \int_{t_1}^t f_n(s)ds, & t \in (t_1, t_2]. \end{cases}$$

Let

$$\Omega_2 = \{y : y \in \mathcal{D} \cap C[0, t_1] \cap C(t_1, t_2] \text{ and } y(t_1^+) \text{ exists}\}.$$

As in Step 1, we can prove that the sequence $\{y^n\}$ converges to some $y \in \Omega_2$, a solution to problem (8.15) such that, for a.e. $t \in (t_1, t_2]$, we have

$$|x(t) - y(t)| \leq \frac{1}{1 - c}(|x(t_1) - y_1(t_1)| + c\|x_{t_1} - (y_1)_{t_1}\|_\infty + |I_1(x(t_1)) - I_1(y_1(t_1))| + \eta_1(t))$$

where

$$\eta_1(t) = \frac{1}{1 - c}\left(R_1(t_1)\exp\left(\frac{e^{P(t)}}{1 - c}\right) + \int_{t_1}^t [\gamma(s) + \epsilon]\left(\exp\left(\frac{e^{P(t)} - e^{P(s)}}{1 - c}\right) + 1\right)ds\right),$$

$$\bar{R}_1(t_1) = |x(t_1) - y_1(t_1)| + c\|x_{t_1} - (y_1)_{t_1}\|_\infty + |I_1(x(t_1)) - I_1(y_1(t_1))|,$$
$$\|x_{t_1} - (y_1)_{t_1}\|_\infty = \sup\{|x(t_1+\theta) - y_1(t_1+\theta)| : \theta \in [t_1 - r, t_1]\},$$
$$\left|\frac{d}{dt}[y(t) - g(t, y_t)] - g(t)\right| \leq H_1(t)p(t) + \gamma(t) + \epsilon,$$

and

$$H_1(t) :=$$
$$\frac{1}{1-c}\left(\bar{R}_1(t_1)\exp\left(\frac{e^{P(t)}}{1-c}\right) + \int_{t_1}^{t}[\gamma(s)+\epsilon]\exp\left(\frac{e^{P(t)} - e^{P(s)}}{1-c}\right)ds\right). \quad (8.16)$$

Denote the restriction $y_{|[t_1,t_2]}$ by y_2.

Step 3. We continue this process until we arrive at the function

$$y_{m+1} := y\Big|_{[t_m - r, t_m]\cup(t_m, b]}$$

as a solution of the problem

$$\begin{cases} \frac{d}{dt}[y(t) - g(t, y_t)] \in F(t, y_t), & \text{a.e. } t \in (t_m, b], \\ y(t_m^+) = y_{m-1}(t_m) + I_m(y_{m-1}(t_m)), & \\ y(t) = y_{m-1}(t), & t \in [t_m - r, t_m]. \end{cases}$$

Then, for a.e. $t \in (t_m, b]$, the following estimates are easily derived:

$$|x(t) - y(t)| \leq \frac{1}{1-c}\left[(1+c)\eta_{m-1}(t) + |I_m(x(t_m)) - I_m(y(t_m))| + \eta_m(t)\right]$$

and

$$\left|\frac{d}{dt}[y(t) - g(t, y_t)] - g(t)\right| \leq H_m(t)p(t) + \gamma(t) + \epsilon.$$

Step 4. Summarizing, a solution y of problem (8.1)–(8.3) can be defined as follows

$$y(t) = \begin{cases} y_1(t), & \text{if } t \in [-r, t_1], \\ y_2(t), & \text{if } t \in (t_1, t_2], \\ \dots & \dots \\ y_{m+1}(t), & \text{if } t \in (t_m, b]. \end{cases}$$

From Steps 1 to 3, we have that

$$|x(t) - y(t)| \leq \eta_0(t), \quad \text{for } t \in [-r, t_1],$$

as well as the estimates

$$\begin{aligned}|x(t)-y(t)| &\le \sum_{k=2}^{m+1}|x(t)-y_k(t)|\\ &\le \frac{1}{1-c}\sum_{0<t_k<t}[|x(t_k)-y_k(t_k)|+\|x_{t_k}-(y_k)_{t_k}\|_\infty]\\ &+\frac{1}{1-c}\sum_{0<t_k<t}|I_k(x(t_k))-I_k(y_k(t_k))|+\frac{1}{1-c}\sum_{k=0}^{m}\eta_k(t)\end{aligned}$$

for $t\in(t_1,b]$. Similarly,

$$\left|\frac{d}{dt}[y(t)-g(t,y_t)]-g(t)\right| \le p(t)\sum_{0<t_k<t}H_k(t)+\sum_{0<t_k<t}[\gamma_k(t)+\epsilon],$$

where $\gamma_k := \gamma_{|J_k}$. The proof of Theorem 8.3 is complete. □

8.2 The Relaxed Problem

In this section, we compare trajectories of the problem

$$\frac{d}{dt}[y(t)-g(t,y_t)] \in F(t,y_t),\quad \text{a.e.}\quad t\in J\backslash\{t_1,\dots,t_m\}, \tag{8.17}$$

$$y(t_k^+)-y(t_k)=I_k(y(t_k^-)),\qquad k=1,\dots,m, \tag{8.18}$$

$$y(t)=\phi(t),\qquad t\in[-r,0], \tag{8.19}$$

to those of the convexified impulsive differential inclusion problem

$$\frac{d}{dt}[x(t)-g(t,x_t)] \in \overline{co}\,F(t,x_t),\quad \text{a.e.}\quad t\in J\backslash\{t_1,\dots,t_m\}, \tag{8.20}$$

$$x(t_k^+)-x(t_k)=I_k(x(t_k^-)),\qquad k=1,\dots,m, \tag{8.21}$$

$$x(t)=\phi(t),\qquad t\in[-r,0]. \tag{8.22}$$

With E being a reflexive Banach space, the following hypotheses will be assumed in this section:

(H-4) The function $F\colon J\times\mathcal{D}\to\mathcal{P}_{cl}(E)$ satisfies
 (a) for all $y\in\mathcal{D}$, the map $t\mapsto F(t,y_t)$ is measurable,
 (b) there exists $p_*\in L^1(J,(0,\infty))$ such that

$$F(t,0)\subset p_*(t)B(0,1).$$

 (c) For every $t\in[0,b]$, the multifunction $t\to F(t,y_t)$ has a measurable selection.

(H-5) There exist constants $c_k \geq 0$ such that

$$|I_k(u_1) - I_k(u_2)| \leq c_k|u_1 - u_2| \text{ for each } u_1, u_2 \in E.$$

Then our main contribution is the following.

Theorem 8.4. *Assume that* (H-2)–(H-5) *hold. Then problem* (8.20)–(8.22) *has at least one solution. In addition, for all* $\varepsilon > 0$ *and every solution* x *of problem* (8.20)–(8.22), *problem* (8.17)–(8.19) *has a solution* y *defined on* $[0, b]$ *satisfying*

$$x(t) = y(t),\ t \in [-r, 0] \ \textit{and}\ \|x - y\|_{PC} \leq \varepsilon.$$

In particular, $S^{co}_{[-r,b]}(\phi) = \overline{S_{[-r,b]}(\phi)}$ *where*

$$S^{co}_{[-r,b]} = \{y\colon y \text{ is a solution to (8.20)–(8.22)}\}.$$

Proof. We will separate the proof into two parts.

Part 1. $S^{co}_{[-r,b]} \neq \emptyset$.
For this, we first transform problem (8.20)–(8.22) into a fixed point problem and then make use of Lemma 2.88. Consider the problem on the interval $[-r, t_1]$, that is,

$$\begin{cases} \frac{d}{dt}[y(t) - g(t, y_t)] \in \overline{co}\, F(t, y_t), & \text{a.e. } t \in [0, t_1], \\ y(t) = \phi(t), & t \in [-r, 0]. \end{cases} \tag{8.23}$$

It is clear that all solutions of problem (8.23) are fixed points of the multivalued operator $N : \Omega([-r, t_1]) \to \mathcal{P}(\Omega[-r, t_1])$ defined by

$$N(y) := \left\{ h \in \Omega([-r, t_1]) : h(t) = \begin{cases} \phi(t), & t \in [-r, 0], \\ \phi(0) - g(0, \phi) + g(t, y_t) & \\ \quad + \int_0^t f(s)ds, & t \in [0, t_1], \end{cases} \right\}$$

where

$$g \in S_{\overline{co}F,y} = \left\{ g \in L^1([0, t_1], E) : g(t) \in \overline{co}\, F(t, y_t) \text{ a.e. } t \in (0, t_1] \right\}$$

and $\Omega([-r, t_1]) = \mathcal{D} \cap C([0, t_1], E)$. To show that N satisfies the assumptions of Lemma 2.88, that part of the proof will be given in two steps. Following that, in Steps 3 and 4, we study the problem on the intervals $(t_k, t_{k+1}]$ for $k = 1, \ldots, m-1$.

Step 1. $N(y) \in \mathcal{P}_{cl}(\Omega([-r, t_1])$ for each $y \in \Omega([-r, t_1])$. Indeed, let $\{y_n\} \in N(y)$ be such that $y_n \to \tilde{y}$ in $\Omega([-r, t_1]$, as $n \to +\infty$. Then $\tilde{y} \in \Omega([-r, t_1])$ and there exists a sequence $g_n \in S_{\overline{co}\, F,y}$ such that $y_n(t) = \phi(t)$ for $t \in [-r, 0]$ and

$$y_n(t) = \phi(0) - g(0, \phi) + g(t, y_t) + \int_0^t g_n(s)ds, \quad t \in (0, t_1].$$

The sequence $\{g_n\}$ is integrably bounded. From Lemma 2.76, for every $n \in \mathbb{N}$, there exists $w_n(\cdot) \in F(\cdot, 0)$ such that

$$|g_n(t) - w_n(t)| \leq d(g_n(t), F(t,0)) + p_*(t).$$

Hence, by (H-4) and (H-5), we infer that for a.e. $t \in [0, t_1]$

$$\begin{aligned} |g_n(t)| &\leq H(F(t, y_t), F(t,0)) + p_*(t) \\ &\leq p(t)\|y\|_{PC} + p_*(t) := M(t) \end{aligned}$$

for all $n \in \mathbb{N}$, that is

$$g_n(t) \in M(t)\mathcal{B}(0,1), \quad \text{a.e.} \quad t \in [0, t_1].$$

Since $\mathcal{B}(0,1)$ is weakly compact in the reflexive Banach space E, there exists a subsequence, still denoted $\{g_n\}$, which converges weakly to g by the Dunford–Pettis theorem. By Mazur's Lemma (Lemma 2.66), there exists a second subsequence which converges strongly to g in E, hence almost everywhere (see [186], p. 150). Then the Lebesgue dominated convergence theorem implies that, as $n \to \infty$,

$$\|g_n - g\|_{L^1} \to 0 \text{ and thus } y_n(t) \to \tilde{y}(t),$$

with $\tilde{y}(t) = \phi(t)$, a.e. $t \in [-r, 0]$, and

$$\tilde{y}(t) = \phi(0) - g(0, \phi) + g(t, y_t) + \int_0^t g(s)ds, \; t \in [0, t_1].$$

This proves $\tilde{y} \in N(y)$.

Step 2. There exists $\gamma < 1$ such that $H(N(y), N(\overline{y})) \leq \gamma \|y - \overline{y}\|_{[-r,t_1]}$ for each $y, \overline{y} \in \Omega([-r, t_1])$ where the norm $\|y - \overline{y}\|_{[-r,t_1]}$ will be chosen conveniently. Indeed, let $y, \overline{y} \in \Omega([-r, t_1])$ and $h_1 \in N(y)$. Then there exists $g_1(t) \in \overline{co}\, F(t, y_t)$ such that for each $t \in [0, t_1]$,

$$h_1(t) = \phi(0) - g(0, \phi) + g(t, y_t) + \int_0^t g_1(s)ds.$$

Since, for each $t \in [-r, t_1]$,

$$H(\overline{co}\, F(t, y_t), \overline{co}\, F(t, \overline{y}_t)) \leq p(t)\|y_t - \overline{y}_t\|_\infty,$$

there exists some $w(t) \in \overline{co}\, F(t, \overline{y}_t)$ such that

$$|g_1(t) - w(t)| \leq p(t)\|y_t - \overline{y}_t\|_\infty, \quad t \in [0, t_1].$$

Consider the multivalued mapping $U_1 : [0, t_1] \to \mathcal{P}(E)$ defined by

$$U_1(t) = \{w \in E : |g_1(t) - w| \leq p(t)\|y_t - \overline{y}_t\|_\infty\}.$$

As in the proof of Theorem 8.3, we can show that the multivalued operator $V_1(t) = U_1(t) \cap \overline{co}\, F(t, \overline{y}_t)$ is measurable and takes nonempty values. Then there exists a function $g_2(t)$, which is a measurable selection for V_1. Thus, $g_2(t) \in \overline{co}\, F(t, \overline{y}_t)$ and

$$|g_1(t) - g_2(t)| \leq p(t)\|y - \overline{y}\|_\infty, \quad \text{for a.e. } t \in [0, t_1].$$

For each $t \in [0, t_1]$, let

$$h_2(t) = \phi(0) - g(0, \phi) + g(t, \overline{y}_t) + \int_0^t g_2(s)ds.$$

Therefore, for each $t \in [0, t_1]$, we have

$$\begin{aligned}|h_1(t) - h_2(t)| &\leq c\|y_t - \overline{y}_t\|_\infty + \int_0^t |g_1(s) - g_2(s)|\, ds \\ &\leq c\|y_t - \overline{y}_t\|_\infty + \int_0^t p(s)\|y_s - \bar{y}_s\|_\infty ds \\ &= c\|y_t - \overline{y}_t\|_\infty + \int_0^t p(s)\Big(\sup_{-r\leq\theta\leq 0} |y_s(\theta) - \bar{y}_s(\theta)|\Big)ds \\ &\leq c\|y_t - \overline{y}_t\|_\infty + \int_0^t p(s)e^{\tau\int_0^s p(u)du} \\ &\quad \times \Big(\sup_{-r\leq z\leq t_1} e^{-\tau\int_0^z p(u)du}|y(z) - \bar{y}(z)|\Big)ds \\ &\leq c\|y_t - \overline{y}_t\|_\infty + \frac{1}{\tau}\int_0^t (e^{\tau\int_0^s p(u)du})'\|y - \bar{y}\|_{[-r,t_1]}ds.\end{aligned}$$

Hence,

$$\|h_1 - h_2\|_{[-r,t_1]} \leq \left(c + \frac{1}{\tau}\right)\|y - \bar{y}\|_{[-r,t_1]},$$

where

$$\|y\|_{[-r,t_1]} = \sup\{e^{-\tau\int_0^t p(s)ds}|y(t)| : t \in [-r, t_1],\ \tau \gg \frac{1}{1-c}\}.$$

By an analogous relation, obtained by interchanging the roles of y and $\bar{y}$, we find that

$$H(N(y), N(\bar{y})) \leq \left(c + \frac{1}{\tau}\right)\|y - \bar{y}\|_{[-r,t_1]}.$$

Then N is a contraction and hence, by Lemma 2.88, N has a fixed point y_0 that is a solution to problem (8.23).

For the remaining steps of this part of the proof, we deal with the problem on the intervals, $[t_1, t_2], [t_2, t_3], \ldots, [t_m, b]$.

Step 3. Let $y_2 := y\big|_{[t_1,t_2]}$ be a possible solution to the problem

$$\begin{cases} \frac{d}{dt}[y(t) - g(t, y_t)] \in \overline{co}\, F(t, y_t), & t \in (t_1, t_2], \\ y(t_1^+) = y_0(t_1) + I_1(y_0(t_1^-)), & \\ y(t) = y_0(t), & t \in [t_1 - r, t_1]. \end{cases} \tag{8.24}$$

Then y_2 is a fixed point of the multivalued operator $N : \Omega([t_1 - r, t_2]) \to \mathcal{P}(\Omega([t_1 - r, t_2]))$ defined by

$$N(y) := \left\{ h \in \Omega([t_1 - r, t_2]) : h(t) = \begin{cases} y_0(t), & t \in [t_1 - r, t_1], \\ y_0(t_1) + I_1(y_0(t_1)) - g(t_1, (y_0)_{t_1}) & \\ \quad +g(t, y_t) + \int_{t_1}^t f(s)ds, & t \in (t_1, t_2] \end{cases} \right\}$$

where

$$f \in S_{\overline{co}\, F, y} = \left\{ f \in L^1([t_1, t_2], E) : \ g(t) \in \overline{co}\, F(t, y_t) \ \text{a.e. } t \in [t_1, t_2] \right\}.$$

Again, we show that N satisfies the assumptions of Lemma 2.88. Clearly, $N(y) \in \mathcal{P}_{cl}(\Omega([t_1 - r, t_2]))$ for each $y \in \Omega([t_1 - r, t_2])$. It remains to show that there exists $0 < \gamma < 1$ such that

$$H(N(y), N(\overline{y})) \leq \gamma \|y - \overline{y}\|_{[t_1 - r, t_2]}$$

for each $y, \overline{y} \in \Omega([t_1 - r, t_2])$. For this purpose, let $y, \overline{y} \in \Omega([t_1 - r, t_2])$ and $h_1 \in N(y)$. Then there exists $g_1(t) \in \overline{co}\, F(t, y_t)$ such that, for each $t \in [t_1 - r, t_2]$,

$$h_1(t) = \int_{t_1}^t g_1(s)ds + y_0(t_1) - g(t_1, (y_0)_{t_1}) + g(t, y_t) + I_1(y_0(t_1)).$$

Since from (H-2)

$$H(\overline{co}\, F(t, y_t), \overline{co}\, F(t, \overline{y}_t)) \leq p(t)\|y_t - \overline{y}_t\|_\infty, \ t \in [t_1, t_2],$$

we deduce that there is a $w(\cdot) \in \overline{co}\, F(\cdot, \overline{y}_.)$ such that

$$|g_1(t) - w(t)| \leq p(t)\|y_t - \overline{y}_t\|_\infty, \quad t \in [t_1, t_2].$$

Consider the multivalued map $U_2 : [t_1, t_2] \to \mathcal{P}(E)$ defined by

$$U_2(t) = \{w \in E : |g_1(t) - w| \leq p(t)\|y_t - \overline{y}_t\|_\infty\}.$$

Since the multivalued operator $V_2(t) = U_2(t) \cap \overline{co}\, F(t, \overline{y}_t)$ is measurable with non-empty values, there exists $g_2(t)$ which is a measurable selection for V_2. Then, $g_2(t) \in \overline{co}\, F(t, \overline{y}_t)$ and

$$|g_1(t) - g_2(t)| \leq p(t)\|y_t - \overline{y}_t\|_\infty, \ \text{for a.e. } t \in [t_1, t_2].$$

For a.e. $t \in [t_1, t_2]$, define

$$h_2(t) = \int_{t_1}^t g_2(s)ds + y_0(t_1) + I_1(y_0(t_1)) - g(t_1, (y_0)_{t_1}) + g(t, y_t).$$

For some $\tau > \frac{1}{1-c}$, we have the estimates

$$\begin{aligned}
|h_1(t) - h_2(t)| &\leq c\|y_t - \overline{y}_t\|_\infty + \int_{t_1}^t |g_1(s) - g_2(s)|\, ds \\
&\leq c\|y_t - \overline{y}_t\|_\infty + \int_{t_1}^t p(s)\|(y_1)_s - (y_2)_s\|_\infty ds \\
&\leq c\|y_t - \overline{y}_t\|_\infty + \int_{t_1}^t p(s)\Big(\sup_{-r\leq\theta\leq 0} |y_{1s}(\theta) - y_{2s}(\theta)|\Big) ds \\
&\leq c\|y_t - \overline{y}_t\|_\infty + \frac{1}{\tau}\|y - \overline{y}\|_{[t_1,t_2]}.
\end{aligned}$$

By an analogous relation, obtained by interchanging the roles of y and $\overline{y}$, we obtain

$$H(N(y), N(\overline{y})) \leq \left(c + \frac{1}{\tau}\right)\|y - \overline{y}\|_{[t_1,t_2]},$$

where

$$\|y\|_{[t_1-r,t_2]} = \sup\{e^{-\tau\int_{t_1}^t p(s)ds}|y(t)| : t \in [t_1 - r, t_2]\}.$$

Therefore, N is a contraction and thus, by Lemma 2.88, N has a fixed point y_2 which is a solution of problem (8.24).

Step 4. We continue this process taking into account that $y_m := y\big|_{[t_m,b]}$ is a solution of the following problem,

$$\begin{cases} \frac{d}{dt}[y(t) - g(t, y_t)] \in F(t, y_t), & t \in (t_m, b], \\ y(t_m^+) = y_{m-1}(t_m) + I_m(y_{m-1}(t_m^-)), & \\ y(t) = y_{m-1}(t), & t \in [t_{m-1} - r, t_m]. \end{cases} \tag{8.25}$$

Then a solution y of problem (8.20)–(8.22) may be defined by

$$y(t) = \begin{cases} y_1(t), & \text{if } t \in [-r, t_1] \\ y_2(t), & \text{if } t \in (t_1, t_2], \\ \dots & \dots \\ y_m(t), & \text{if } t \in (t_m, b]. \end{cases}$$

Part 2. Let x be a solution of problem (8.20)–(8.22). Then, there exists $g \in S_{\overline{co}\,F,x}$ such that

$$x(t) = \begin{cases} \phi(t), & t \in [-r, 0], \\ \phi(0) - g(0, \phi) + g(t, x_t) + \int_0^t \bar{g}(s)ds + \sum_{0<t_k<t} I_k(x(t_k)), & t \in [0, b], \end{cases}$$

that is, x is a solution of the problem

$$\begin{cases} \frac{d}{dt}[x(t) - g(t, x_t)] = \bar{g}(t), & \text{a.e. } t \in [0, b]\backslash\{t_1, \dots, t_m\}, \\ x(t_k^+) - x(t_k) = I_k(x(t_k^-)), & k = 1, \dots, m, \\ x(t) = \phi(t) & t \in [-r, 0]. \end{cases} \tag{8.26}$$

Let $\varepsilon > 0$ and $\delta > 0$ be given by the relation $\delta = \varepsilon/(L\sum_{k=1}^{m}\bar{R}_k)$, where L and R_k, for $k = 0, 1, \dots, m$, will be defined later. From Lemma 5.4, there exists a measurable selection f_* of $t \mapsto F(t, x_t)$ such that

$$\sup_{t\in[0,b]}\left|\int_0^t g(s)ds - \int_0^t f_*(s)ds\right| \leq \delta.$$

Let

$$z(t) = \begin{cases} \phi(t), & t \in [-r, 0], \\ \phi(0) - g(0,\phi) + g(t, x_t) + \int_0^t f_*(s)ds & \\ \quad + \sum_{0<t_k<t} I_k(x(t_k)), & t \in (0, b]. \end{cases}$$

Hence,

$$\|x_t - z_t\|_\infty \leq \delta.$$

With Assumption (H-2), we infer that for $u \in \overline{co}\, F(t, z_t)$,

$$\begin{aligned} \gamma(t) := d(g(t), F(t,x_t)) &\leq d(g(t), u) + H(F(t,z_t), F(t,x_t)) \\ &\leq H(\overline{co}\, F(t,x_t), \overline{co}\, F(t,z_t)) + H(F(t,z_t), F(t,x_t)) \\ &\leq 2p(t)\|x_t - z_t\|_\infty \leq 2\delta p(t). \end{aligned}$$

Since, under (H-4)(a) and (H-2), γ is measurable (see [35] or [201, Lemma 1.5], or Remark 8.2), by the above inequality, we deduce that $\gamma \in L^1(J, E)$. From Theorem 5.2, problem (8.17)–(8.19) has a solution y satisfying

$$|y(t) - x(t)| \leq \eta_0(t),\ t \in [0, t_1].$$

Also, for $t \in (t_1, t_2]$, we have the estimates

$$\begin{aligned} |y(t) - x(t)| &\leq \frac{1}{1-c}((1 + c_1 + c)\eta_0(t_1) + \eta_1(t_2)) \\ &\leq L_1[\eta_0(t_1) + \eta_1(t_2)], \end{aligned}$$

where

$$L_1 = \frac{1 + c_1 + c}{1 - c}$$

and

$$\begin{aligned} \eta_0(t_1) &\leq \delta\frac{\exp\left(\frac{e^{P(t_1)}}{1-c}\right)}{1-c}\left(2 + 4\int_0^{t_1} p(s)ds\right) + t_1\epsilon\frac{\exp\left(\frac{e^{P(t_1)}}{1-c}\right)}{1-c} \\ &:= \delta R_0 + t_1\epsilon\frac{\exp\left(\frac{e^{P(t_1)}}{1-c}\right)}{1-c}. \end{aligned}$$

For $t \in (t_2, t_3]$, we have

$$
\begin{aligned}
|y(t) - x(t)| &\leq \frac{1}{1-c}\left(|y(t_2) - x(t_2)| + c\|x_{t_2} - (y_2)_{t_2}\|_{\mathcal{D}}\right. \\
&\qquad \left. + c_2|y(t_2) - x(t_2)| + \eta_3(t_3)\right) \\
&\leq \left(3L_1 + \frac{c_2}{1-c}\right)(\eta_0(t_1) + \eta_1(t_2)) + \frac{\eta_3(t_3)}{1-c} \\
&\leq L_2[\eta_0(t_1) + \eta_2(t_2) + \eta_2(t_3)],
\end{aligned}
$$

where

$$
L_2 = \frac{4 + 3c_1 + 3c + c_2}{1-c}
$$

and

$$
\begin{aligned}
\eta_1(t_2) \leq \delta &\frac{\exp\left(\frac{e^{P(t_2)}}{1-c}\right)}{1-c}\left((1 + c_1 + c)R_0 + 4\int_{t_1}^{t_2} p(s)ds\right) \\
&+ \epsilon(t_2 - t_1)\frac{\exp\left(\frac{e^{P(t_2)}}{1-c}\right)}{1-c} \\
&=: \delta R_1 + \epsilon(t_2 - t_1)\frac{\exp\left(\frac{e^{P(t_2)}}{1-c}\right)}{1-c}.
\end{aligned}
$$

Thus, there exist $L > 0$ and R_k, $k = 0, 1, \ldots, m$ such that, for all $t \in [-r, b]$, we have

$$
\begin{aligned}
|y(t) - x(t)| &\leq L\sum_{i=0}^{m} \eta_i(t_{i+1}) \\
&\leq L\delta\sum_{k=0}^{m} R_k + \frac{\epsilon}{1-c}\sum_{k=1}(t_k - t_{k-1})\exp\left(\frac{e^{P(t_k)}}{1-c}\right).
\end{aligned}
$$

Using the definition of δ, we obtain the upper bound

$$
\|y - x\|_\infty \leq \epsilon(1 + \frac{1}{1-c}\sum_{k=1}(t_k - t_{k-1})\exp\left(\frac{e^{P(t_k)}}{1-c}\right).
$$

Since ε is arbitrary, $\|y-x\|_\infty \leq \varepsilon$, showing the density relation $S^{co}_{[0,b]}(\phi) = \overline{S_{[0,b]}(\phi)}$. □

8.2.1 Existence and Compactness Result: an MNC Approach

Let $F\colon J \times \mathcal{D} \to \mathcal{P}_{cp,cv}(E)$ be a Carathéodory multimap, for each $t \in [0, b]$ the multifunction $t \to F(t, y_t)$ has a measurable selection and let χ denote the Hausdorff measure of noncompactness (MNC). We will make use of the following conditions.

(H-6) There exist a function $p \in L^1(J, \mathbb{R}^+)$ and a continuous nondecreasing function $\rho\colon [0,\infty) \to [0,\infty)$ such that

$$\|F(t,z)\| \le p(t)\rho(\|z\|_\infty) \text{ for a.e. } t \in J \text{ and each } z \in \mathcal{D},$$

with

$$\int_0^b p(s)ds < \int_1^\infty \frac{du}{\rho(u)}.$$

(H-7) Let $D(\theta) := \{\psi(\theta),\ \psi \in \mathcal{D}\}$. There exists $\overline{p} \in L^1([0,b], \mathbb{R}^+)$ such that, for every bounded subset D in $\mathcal{D}$,

$$\chi(F(t,D)) \le \overline{p}(t) \sup\{\chi(D(\theta))\colon \theta \in [-r,0]\},$$

and there exists $L_k > 0,\ k = 0,\dots,m$, such that

$$q_k := 2 \sup_{t \in [t_k, t_{k+1}]} \int_{t_k}^{t_{k+1}} p(s)ds < 1, \quad k = 0,\dots,m.$$

(H-8) There exists $\omega \in (0,1)$ such that, for every bounded subset D in $\mathcal{D}$,

$$\chi(g(t,D)) \le \omega \sup\{\chi(D(\theta))\colon \theta \in [-r,0]\}.$$

(H-9) The function g is continuous and for any bounded set D in Ω, the set $\{t \to g(t, y_t) : y \in D\}$ is equicontinuous in Ω, and there exist constants $0 \le d_1 < 1$, $d_2 \ge 0$ such that

$$|g(t,u)| \le d_1\|u\|_\infty + d_2, \quad t \in [0,b], \quad u \in D.$$

Remark 8.5.

(i) Assumption (H-6) is satisfied if, for example, F satisfies the standard estimation

$$\|F(t,u)\| \le p(t)(1+|u|), \quad p \in L^1(J, \mathbb{R}^+),\ t \in J, \quad u \in \mathcal{D}.$$

(ii) Condition (H-7) holds if the multifunction F can be represented as

$$F(t,u) = F_1(t,u) + F_2(t,u),$$

where the multifunctions $F_1, F_2 : J \times \mathcal{D} \to \mathcal{P}_{cp}(E)$ are such that F_1 is compact in its second argument (i.e., for each bounded $D \subset \mathcal{D}$, the set $F(t,D)$ is relatively compact for a.e. $t \in J$) and F_2 is p–Lipschitz in its second argument.

(iii) (H-8) is satisfied if, for example, the operator $L(y) = g(\cdot, y)$ is compact.

(iv) (H-9) is satisfied if, for example, there exists $d_2 \geq 0$ such that

$$|g(\tau_1, u) - g(\tau_2, \overline{u})| \leq d_2|\tau_1 - \tau_2| \; \tau_1, \tau_2 \in J, \; u, \; \overline{u} \in \mathcal{D},$$

$$\sup\{|g(t, 0)| : t \in J\} < \infty,$$

and for each bounded $D \subset \mathcal{D}$, the set $g(t, D)$ is relatively compact for a.e. $t \in J$. (For more information about this condition see Arino et al. [30]).

Theorem 8.6. *Assume that F satisfies assumptions (H-6)–(H-9). If $\omega + \max\{q_k : k = 0, \ldots, m\} < 1$, then the set of solutions for problem (8.17)–(8.19) is nonempty and compact.*

Proof. **Step 1.** $S_{[-r,b]} \neq \emptyset$.

To see this, let $N_0 : \mathcal{D} \cap C([0, t_1], E) \to \mathcal{P}(\mathcal{D} \cap C([0, t_1], E))$ be defined by

$$N_0(y) = \left\{ h \in \mathcal{D} \cap C([0, t_1], E) : \; h(t) = \begin{cases} \phi(t), & t \in [-r, 0], \\ \phi(0) - g(t, \phi) + g(t, y_t) & \\ \quad + \int_0^t v(s) ds, & t \in [0, t_1], \end{cases} \right.$$

where

$$v \in S_{F,y} = \{v \in L^1([0, t_1], E) : v(t) \in F(t, y_t) \;\; \text{a.e.} \;\; t \in [0, t_1]\}.$$

Set

$$\mathcal{K}_0 := \{y \in \mathcal{D} \cap C([0, t_1], E) \colon \|y_t\|_{\mathcal{D}} \leq \frac{a_0(t)}{1 - d_1}, \; t \in [0, t_1]\},$$

where

$$a_0(t) = \Gamma^{-1}\left(\int_0^t \widehat{M}(s) ds\right), \; \Gamma(z) = \int_{\overline{c}}^z \frac{du}{\rho(u)}$$

and

$$\overline{c} = (1 + d_1)\|\phi\|_{\mathcal{D}} + 2d_2 \;\; \text{and} \; \widehat{M}(t) = \frac{p(t)}{1 - d_1}, \; t \in [0, t_1].$$

It is clear that $\mathcal{K}_0$ is a closed bounded convex set in $\mathcal{D} \cap C([0, t_1], E)$. If $h \in N_0(y)$, then there exists $v \in S_{F,y}$ such that $h(t) = \phi(t)$, $t \in [-r, 0]$, and

$$h(t) = \phi(0) - g(0, \phi) + g(t, y_t) + \int_0^t v(s) ds, \; t \in [0, t_1].$$

Claim 1. $N_0(\mathcal{K}_0) \subset \mathcal{K}_0$. We have

$$\begin{aligned} |h(t)| &\leq (1 + d_1)\|\phi\|_{\mathcal{D}} + 2d_2 + d_1|y_t| + \int_0^t |v(s)| ds \\ &\leq (1 + d_1)\|\phi\|_{\mathcal{D}} + 2d_2 + \frac{d_1}{1 - d_1} a_0(t) + (1 - d_1) \int_0^t \widehat{M}(s)\rho(a_0(s)) ds. \end{aligned}$$

It follows that, for each $t \in [0, t_1]$,

$$|h(t)| \leq \overline{c} + \frac{d_1}{1-d_1} a_0(t) + \int_0^t a_0'(s)ds = \frac{a_0(t)}{1-d_1},$$

proving our claim.

Claim 2. The multivalued map $N_0 : \mathcal{K}_0 \to \mathcal{P}(\mathcal{K}_0)$ has at least one fixed point. Since F is a multifunction with convex values, N_0 has convex values. Moreover, N_0 has a closed graph. Indeed, let $\{y_n : n \in \mathbb{N}\} \subseteq \mathcal{K}_0$ be such that $y_n \to y_*$, $h_n \in N_0(y_n)$ and $h_n \to h_*$, as $n \to +\infty$. We shall prove that $h_* \in N_0(y_*)$. Now $h_n \in N_0(y_n)$ means there exists $v_n \in S_{F,y_n}$ such that

$$h_n(t) = \phi(0) - g(0, \phi) + g(t, y_t) + \int_0^t v_n(s)ds$$

for a.e. $t \in [0, t_1]$. We must prove that there exists $v_* \in S_{F,y_*}$ such that for a.e. $t \in [0, t_1]$, we have

$$h_*(t) = \phi(0) - g(0, \phi) + g(t, y_t) + \int_0^t v_*(s)ds. \tag{8.27}$$

Since $\{y_n : n \in \mathbb{N}\} \subseteq \mathcal{K}_0$ and $\{v_n : n \in \mathbb{N}\} \subseteq F(t, (y_t)_n)$, Assumption (H-6) implies that

$$|v_n(t)| \leq p(t)\rho(a_0(t_1)), \ t \in [0, t_1].$$

In addition, the set $\{v_n(t) : n \in \mathbb{N}\}$ is relatively compact for a.e. $t \in J$ because assumption (H-9) and the convergence of $\{y_n\}_{n \in \mathbb{N}}$ imply

$$\chi(\{v_n(t) : n \in \mathbb{N}\}) \leq \chi(F(t, (y_t)_n) \leq \bar{p}(t)\chi((y_t)_n) = 0.$$

Then the sequence $\{v_n : n \in \mathbb{N}\}$ is semicompact, hence weakly compact in $L^1([0, t_1]; E)$ by Lemma A.39, i.e., there exists $v_* \in L^1$ such that v_n converges weakly to v_*. Finally, (8.27) follows from the Lebesgue dominated convergence theorem

Claim 3. N_0 is a β-condensing operator for a suitable MNC β. For a bounded subset $D \subset \mathcal{K}_0$, let $\operatorname{mod}_C(D)$ be the modulus of quasiequicontinuity of the set of functions D given by

$$\operatorname{mod}_C(D) = \lim_{\delta \to 0} \sup_{x \in D} \max_{|\tau_2 - \tau_1| \leq \delta} |x(\tau_1) - x(\tau_2)|.$$

It is well known (see e.g. Example 2.1.2 in [311]) that $\operatorname{mod}_C(D)$ defines an MNC in PC that satisfies all of the properties in the Definition 2.92 except regularity.

Given the Hausdorff MNC χ, let γ_0 be the real MNC defined on bounded subsets on $\mathcal{K}_0$ by

$$\gamma_0(D) = \sup_{t \in [0, t_1]} e^{-L_0 t} \chi(D(t)).$$

Finally, define the following MNC on bounded subsets of $\mathcal{K}_0$ by

$$\beta_0(B) = \max_{D\in\Delta(\mathcal{K}_0)} (\gamma_0(D), \mathrm{mod}_C(D)),$$

where $\Delta(\mathcal{K}_0)$ is the collection of all countable subsets of B. Then the MNC β is monotone, regular and nonsingular (see Example 2.1.4 in [311]). This measure is also used in [131,223,225] in the discussion of semilinear evolution differential inclusions.

To show that N_0 is $\beta-$condensing, let $B \subset \mathcal{K}_0$ be a bounded set in $\mathcal{K}_0$ such that

$$\beta_0(B) \leq \beta_0(N_0(B)). \tag{8.28}$$

We will show that B is relatively compact. Let $\{y_n : n \in \mathbb{N}\} \subset B$ and let $N_0 = G_0 + \widetilde{L_0} \circ S_F$, where $S_F : \mathcal{D} \cap C([0,t_1], E) \to L^1([0,t_1], E)$ is defined by

$$S_F(y) = S_{F,y} = \left\{v \in L^1([0,t_1], E) : v(t) \in F(t, y_t) \ \text{ a.e } t \in [0,t_1]\right\},$$

and $\widetilde{L_0} : L^1([0,t_1], E) \to \mathcal{D} \cap C([0,t_1], E)$ is defined by

$$\widetilde{L_0}(v)(t) = \int_0^t v(s)ds, \quad t \in [0,t_1].$$

Then,

$$\left|\widetilde{L_0}v_1(t) - \widetilde{L_0}v_2(t)\right| \leq \int_0^t |v_1(s) - v_2(s)|\, ds$$

Moreover, each element h_n in $N_0(y_n)$ can be represented as

$$h_n = \phi(0) - g(0,\phi) + g(\cdot,(y_.)_n) + \widetilde{L_0}(v_n),\ v_n \in S_F(y_n).$$

Using (8.28), we infer that

$$\beta_0(\{h_n : n \in \mathbb{N}\}) \geq \beta_0(\{y_n : n \in \mathbb{N}\}).$$

From assumptions (H-7) and (H-8), it follows that for a.e. $t \in [0,t_1]$,

$$\begin{aligned}\chi(\{v_n(t) : n \in \mathbb{N}\}) &\leq \chi(F(t,\{(y_t)_n)\}_{n=1}^{\infty})\\ &\leq \overline{p}(t) \sup_{-r\leq\theta\leq 0} \chi(\{(y_t)_n(\theta)\}_{n=1}^{\infty})\\ &\leq \overline{p}(t) \sup_{0\leq s\leq t} \chi(\{y_n(s)\}_{n=1}^{\infty})\\ &\leq e^{L_0 t}\overline{p}(t) \sup_{0\leq s\leq t} e^{-L_0 s}\chi(\{y_n(s)\}_{n=1}^{\infty})\\ &\leq e^{L_0 t}\overline{p}(t)\gamma_0(\{y_n\}_{n=1}^{\infty}),\end{aligned}$$

and

$$\chi(g(t,\{(y_n)_t : n \in \mathbb{N}\})) \leq \omega \sup_{0\leq s\leq t} \chi(\{y_n(s)\}_{n=1}^{\infty}).$$

From Lemmas 2.96 and 2.97, we see that

$$e^{-L_0 t}\chi(\{\widetilde{L_0}(v_n)(t)\}_{n=1}^{\infty}) \le \gamma_0(\{y_n\}_{n=1}^{\infty}) \sup_{t\in[0,t_1]} 2\int_0^t e^{-L_0(t-s)}\overline{p}(s)ds,$$

and

$$e^{-L_0 t}\chi(g(t,\{(y_n)_t : n \in \mathbb{N}\})) \le \omega\gamma_0(\{y_n\}_{n=1}^{\infty}).$$

Therefore,

$$\gamma_0(\{y_n\}) \le \gamma_0(\{h_n\}) = \sup_{t\in[0,t_1]} e^{-L_0 t}\chi(\{h_n(t)\} \le (q_0+\omega)\gamma_0(\{y_n\}).$$

Since $0 < \omega + q_0 < 1$, we infer that

$$\gamma_0(\{y_n\}_{n=1}^{\infty}) = 0.$$

Next, we show that $mod_C(B) = 0$, that is, the set B is equicontinuous.

Step 2. *N_0 maps bounded sets into bounded sets in $\mathcal{D} \cap C([0,t_1],E)$.*

Using (H-8), it suffices to show that $\overline{N}_0(B)$ is an equicontinuous set in $\mathcal{D} \cap C([0,t_1],E)$ where $\overline{N}_0$ the operator $\overline{N}_0 : \mathcal{D} \cap C([0,t_1],E) \to \mathcal{D} \cap C([0,t_1],E)$ defined by,

$$\overline{N}_0(y) = \left\{h \in \mathcal{D} \cap C([0,t_1],E) : h(t) = \left\{\begin{array}{ll} \phi(t), & t \in [-r,0] \\ \phi(0) + \int_0^t v(s), & t \in [0,t_1], \end{array}\right\}\right..$$

To see that $\overline{N}_0(B)$ is an equicontinuous set of Ω, let $0 < \tau_1 < \tau_2 \le t_1$, $y \in B$, and $h \in \overline{N}_0(y)$. Then there exists $v \in S_{F,y}$ such that

$$h(t) = \left\{\begin{array}{ll} \phi(t), & t \in [-r,0], \\ \phi(0) + \int_0^t v(s)ds, & t \in [0,t_1]. \end{array}\right.$$

We then have

$$|h(\tau_2) - h(\tau_1)| \le \int_{\tau_1}^{\tau_2} |v(s)|ds \le \int_{\tau_1}^{\tau_2} p(s)\psi(\|y_s\|_{\mathcal{D}})ds \le \psi(s)\int_{\tau_1}^{\tau_2} p(s)ds,$$

where

$$B \subseteq B_q = \{y \in \mathcal{D} \cap C([0,t_1],E) : \sup_{t\in[-r,t_1]} |y(t)| \le q\}.$$

Thus, the right-hand side tends to zero as $\tau_2 - \tau_1 \to 0$. The equicontinuity for the cases $r_1 < r_2 \le 0$ and $r_1 \le 0 \le r_2$ is obvious.

Therefore, $\beta_0(\{h_n\}_{n=1}^{\infty}) = 0$ which implies that $\beta_0(\{y_n\}_{n=1}^{\infty}) = 0$. We have proved that B is relatively compact and so the map N_0 is $\beta-$condensing. From Lemma 2.98, we conclude that N_0 has at least one fixed point, denoted as y_0. Moreover, since $FixN_0$ is bounded, by Lemma 2.99, $FixN_0$ is compact.

Step 3. *$S_{[-r,b]}$ is compact.*

Let

$$C_1 = \{y \in C((t_1,t_2],E) : \ y(t_1^+) \text{ exists}\}, \ C^* = \mathcal{D} \cap C([0,t_1],E) \cap C_1$$

and

$$\mathcal{K}_1 := \left\{ z \in C^* : \|y_t\|_\infty \leq \frac{a_1(t)}{1-d_1}, \ t \in [t_1,t_2] \right\},$$

where

$$a_1(t) = \Gamma^{-1}\left(\int_{t_1}^t \widehat{M}(s)ds \right) \text{ and } \Gamma(z) = \int_c^z \frac{du}{\rho(u)}.$$

Define the operator $N_1 : C^* \to \mathcal{P}(C^*)$ by $N_1(y) = \{h\}$ with

$$h(t) = \begin{cases} y_0(t), & t \in [-r,t_1], \\ \int_{t_1}^t v(s)ds + y_0(t_1) + I_1(y_0(t_1^-)) - g(t_1,(y_0)_{t_1}) + g(t,y_t), & t \in [t_1,t_2], \end{cases}$$

and

$$v \in S_{F,y} = \{v \in L^1([t_1,t_2],E) : v(t) \in F(t,y_t), \text{ a.e. } t \in [t_1,t_2]\}.$$

We can easily check that $N_1(\mathcal{K}_1) \subset (\mathcal{K}_1)$. Thus, we only prove that N_1 is a β-condensing operator. For a bounded subset $B \subset \mathcal{K}_1$, let $\text{mod}_C(B)$ be the modulus of quasiequicontinuity of the set of functions B, γ_1 be the real MNC defined on bounded subsets of $\overline{U}$ by

$$\gamma_1(B) = \sup_{t \in [t_1,t_2]} e^{-L_1 t}\chi(B(t)),$$

and β_1 the MNC defined on $\mathcal{K}_1$ by

$$\beta_1(B) = \max_{\Delta(\mathcal{K}_1)} (\gamma_1(B), \text{mod}_C(B)),$$

where $\Delta(\mathcal{K}_1)$ is the collection of all countable subsets of B. Let $B \subset \mathcal{K}_1$ be a bounded set in $\mathcal{K}_1$ such that

$$\beta_1(B) \leq \beta_1(N_1(B)).$$

We will show that B is relatively compact. It is clear that h_n has the representation:

$$h_n(t) = \widetilde{L_1}(v_n(t)) + y_0(t_1) + I_1(y_0(t_1)) - g(t_1,(y_0)_{t_1}) + g(t,y_t),$$

where $\widetilde{L_1}$ is as defined in Step 1, Claim 3. Then, we have the estimates

$$\begin{aligned} \chi(\{v_n(t) : n \in \mathbb{N}\}) &\leq \chi(F(t,\{(y_n)_n\}_{n=1}^\infty) \\ &\leq \chi(F(t,\{(y_n)_s\}_{n=1}^\infty)) \\ &\leq \overline{p}(t) \sup_{-r \leq \theta \leq 0} \chi(\{(y_n)_t(\theta)\}_{n=1}^\infty)) \\ &\leq \overline{p}(t) \sup_{t_1-r \leq s \leq t_2} \chi(\{(y_t)_n(s)\}_{n=1}^\infty)) \\ &\leq e^{L_1 t}\overline{p}(t) \sup_{t_1-r \leq s \leq t_1} e^{-L_1 s}\chi(\{(y_n)_s\}_{n=1}^\infty)) \\ &\leq e^{L_1 t}\overline{p}(t)\gamma_1(\{y_n\}_{n=1}^\infty), \end{aligned}$$

and

$$\chi(g(t,\{(y_n)_t : n \in \mathbb{N}\})) \leq \omega \sup_{t_1 \leq s \leq t_2} \chi(\{y_n(s)\}_{n=1}^{\infty}).$$

From Lemmas 2.96 and 2.97, we have

$$e^{-L_1 t}\chi(\{\widetilde{L_1}(v_n)(t)\}_{n=1}^{\infty}) \leq \gamma_1(\{y_n\}_{n=1}^{\infty}) \sup_{t\in[t_1,t_2]} 2\int_0^t e^{-L_1(t-s)}\overline{p}(s)ds$$

and

$$e^{-L_1 t}\chi(g(t,\{(y_n)_t : n \in \mathbb{N}\})) \leq \omega\gamma_1(\{y_n\}_{n=1}^{\infty}).$$

Therefore,

$$\gamma_1(\{y_n\}) \leq \gamma_1(\{h_n\}) = \sup_{t\in[t_1,t_2]} e^{-L_1 t}\chi(\{h_n(t)\} \leq (\omega + q_1)\gamma_1(\{y_n\}).$$

Since $0 < \omega + q_1 < 1$, it follows that

$$\gamma_1(\{y_n\}_{n=1}^{\infty}) = 0.$$

By the same argument used in Step 1, we can show that

$$mod_C(\{y_n\}_{n=1}^{\infty}) = 0$$

and so $\beta_1(\{y_n\}_{n=1}^{\infty}) = 0$. Therefore, N_1 is $\beta-$condensing and by Lemma 2.98, N_1 has a fixed point in $\mathcal{K}_1$ denoted by y_1. As in Step 1, we can prove that $FixN_1$ is a compact set.

Step 4. We continue this process taking into account that $y_m := y\big|_{[t_m-r,b]}$ is a solution of the problem

$$\begin{cases} \frac{d}{dt}[y(t) - g(t,y_t)] \in F(t,y_t), & \text{a.e.} \quad t \in (t_m,b], \\ y(t_m^+) = y_{m-1}(t_{m-1}) + I_m(y_{m-1}(t_m^-)), & \\ y(t) = y_{m-1}(t), & t \in [t_m - r, b]. \end{cases}$$

A solution y of problem (8.17)–(8.19) is ultimately defined by

$$y(t) = \begin{cases} y_0(t), & \text{if } t \in [-r,t_1], \\ y_1(t), & \text{if } t \in (t_1,t_2], \\ \dots & \dots \\ y_m(t), & \text{if } t \in (t_m,t_{m+1}]. \end{cases}$$

Step 5. $S(\phi)$ is compact. Let $\{y_n\}_{n\in\mathbb{N}} \subset S(\phi)$; then there exists $v_n \in S_{F,y_n}$ such that

$$y_n(t) = \begin{cases} \phi(t), & t \in [-r,0], \\ \phi(0) - g(0,\phi) + g(t,(y_n)_t)\int_0^t v_n(s)ds & \\ \quad + \sum_{0<t_k<t} I_k(y_n(t_k)), & t \in [0,b]. \end{cases}$$

Hence

$$y_n(t) = \begin{cases} \phi(t), & t \in [-r, 0], \\ \phi(0) - g(0, \phi) + g(t, (y_n)_t) + \int_0^t v_n(s)ds, & t \in [0, t_1], \end{cases}$$

where

$$S_{F,y_n} = \{v \in L^1([0, t_1], E) \mid v(.) \in F(t, y_n(t)) \quad \text{a.e.} \quad t \in [0, t_1]\}$$

Then $\{y_n\} \subset Fix(N_0)$. Since $Fix(N_0)$ is compact, there exists a subsequence of $\{y_n\}_{n \in \mathbb{N}}$ which converges to y_0 in $C([0, t_1], E) \cap \mathcal{D}$ and there exists $v_0 \in L^1([0, t_1], E)$ such that

$$y_0(t) = \begin{cases} \phi(t), & t \in [-r, 0], \\ \phi(0) - g(0, \phi) + g(t, (y_0)_t) + \int_0^t v_0(s)ds, & t \in [0, t_1], \end{cases}$$

where $v_0 \in S_{F,y_0}$. Now

$$y_n(t) = \begin{cases} y_0(t), & t \in [t_1 - r, t_1], \\ y_n(t_1) + I_1(y_n(t_1)) - g(t_1, (y_n)_{t_1}) + g(t, (y_n)_t) & \\ \quad + \int_{t_1}^t v_n(s)ds, & t \in (t_1, t_2]. \end{cases}$$

where

$$S_{F,y_n} = \{v \in L^1([t_1, t_2], E) \mid v(.) \in F(t, y_n(t)) \quad \text{a.e.} \quad t \in [t_1, t_2]\}.$$

We have

$$[y_n(t_1) + I_1(y_n(t_1))] \to [y_0(t_1) + I_1(y_0(t_1))], \text{ as } n \to \infty,$$

and

$$g(t_1, (y_n)_{t_1}) \to g(t_1, (y_0)_{t_1}), \text{ as } n \to \infty.$$

Thus

$$\chi(\{v_n | \, n \in \mathbb{N}\}) = 0, \ \beta_1(\{y_n | \, n \in \mathbb{N}\}) = 0,$$

where χ, β_1 are defined in Step 2. Hence $\{y_n \mid n \in \mathbb{N}\}$ is relatively compact in C_1. Then there exists a subsequence $\{y_n\}$ converging to some $y_1 \in C_1$ and there exists $v_1 \in L^1([t_1, t_2], E)$ such that

$$y_1(t) = \begin{cases} y_0(t), & t \in [t_1 - r, t_1], \\ y_0(t_1) + I_1(y_0(t_1)) + g(t_1, (y_0)_{t_1}) + g(t, (y_1)_t) & \\ \quad + \int_{t_1}^t v_1(s)ds, & t \in (t_1, t_2]. \end{cases}$$

We continue this process taking into account that

$$y_n(t) = \begin{cases} y_{m-1}(t), & t \in [t_m - r, t_m], \\ y_n(t_m) + I_1(y_n(t_m)) - g(t_m, (y_n)_{t_m}) + g(t, (y_n)_t) & \\ \quad + \int_{t_m}^t v_n(s)ds, & t \in (t_m, b]. \end{cases}$$

where

$$S_{F,y_n} = \{v \in L^1([t_m,b],E) \mid v(t) \in F(t,y_n(t)) \text{ a.e. } t \in [t_m,b]\}$$

with

$$y_n(t_m) + I_1(y_n(t_m)) \to y_{m-1}(t_m) + I_m(y_{m-1}(t_1)) \text{ as } n \to \infty,$$

and

$$g(t_m,(y_n)_{t_m}) \to g(t_m,(y_{m-1})_{t_m}), \text{ as } n \to \infty.$$

Then, the sequence

$$\{y_n \mid n \in \mathbb{N}\}$$

is relatively compact in C_m; hence there exists a subsequence of $\{y_n\}_{n\in\mathbb{N}}$ which converges to y_m in C_m. Then $\{y_n \mid n \in \mathbb{N}\}$ has a subsequence in Ω which converges to a limit y defined by

$$y(t) = \begin{cases} y_0(t), & \text{if } t \in [-r,t_1], \\ y_1(t), & \text{if } t \in (t_1,t_2], \\ \dots & \dots \\ y_m(t), & \text{if } t \in (t_m,b]. \end{cases}$$

Hence $S(\phi) \subset \mathcal{P}_{cp}(\Omega)$. This completes the proof of the theorem. □

Chapter 9

Topology and Geometry of Solution Sets

This chapter deals with topological and geometric properties of solution sets for impulsive differential equations and inclusions for periodic and terminal problems. This involves the notions of contractibility, absolute retracts, R_δ-sets, and related concepts.

9.1 Background in Geometric Topology

First, we start with some elementary notions and notations from geometric topology. For details, we recommend [89, 178, 184, 229, 230, 232, 307, 341]. In what follows, (X, d) and (Y, d') stand for two metric spaces.

Definition 9.1. A set $A \in \mathcal{P}(X)$ is called a contractible space provided there exists a continuous homotopy $h : A \times [0, 1] \to A$ and $x_0 \in A$ such that

(i) $h(x, 0) = x$, for every $x \in A$,

(ii) $h(x, 1) = x_0$, for every $x \in A$,

i.e., if the identity map $A \longrightarrow A$ is homotopic to a constant map (A is homotopically equivalent to a point).

Definition 9.2. A subset $A \subset X$ is called *the retract of* X if there exists a retraction $r: X \to A$, i.e. $r(x) = x$, for every $x \in A$.

Observe that A is a retract of X if and only if the identity map Id_A over A possesses a continuous extension onto X.

Similarly, we have the following definition.

Definition 9.3. A is a *neighborhood retract* of X if there exists an open subset $U \subset X$ such that $A \subset U$ and A is a retract of U.

Proposition 9.4. ([178]) *Every retract of a Hausdorff space is a closed subset.*

Note that if $A \in \mathcal{P}_{cv,cl}(X)$, then A is contractible. Also the class of contractible sets is much larger than the class of closed convex sets.

Definition 9.5. A compact nonempty space X is called an R_δ-set provided there exists a decreasing sequence of compact nonempty contractible spaces $\{X_n\}$ such that $X = \bigcap_{n=1}^{\infty} X_n$.

Definition 9.6. A space X is called an absolute retract (in short $X \in AR$) provided that for every space Y, every closed subset $B \subseteq Y$, and any continuous map $f : B \to X$, there exists a continuous extension $\widetilde{f} : Y \to X$ of f over Y, i.e. $\widetilde{f}(x) = f(x)$ for every $x \in B$. In other words, for every space Y and for any embedding $f : X \longrightarrow Y$, the set $f(X)$ is a retract of Y.

Definition 9.7. We say that X is an absolute neighborhood retract ($X \in ANR$) if for any metric space Y, any closed subset B of Y, and any continuous $f : B \to X$, there exist an open neighborhood U of B in Y and an extension $g : U \to X$ of f.

From [19, Proposition 2.15], if $X \in AR$, then X is a contractible space.

Definition 9.8. A space A is closed acyclic if

(i) $H_0(A) = \mathbb{Q}$,

(ii) $H_n(A) = 0$, for every $n > 0$,

where $H_* = \{H_n\}_{n \geq 0}$ is the Čech-homology functor with compact carriers and coefficients in the field of rationals $\mathbb{Q}$. In other words, a space A is acyclic if the map $j : \{p\} \to X,\ j(p) = x_0 \in A$, induces an isomorphism $j_* : H_*(\{p\}) \to H_*(A)$.

Definition 9.9. An u.s.c. map $F : X \to \mathcal{P}(Y)$ is called acyclic if for each $x \in X$, the image set $F(x)$ is compact and acyclic.

From the continuity of Čech-homology functors, we have the following result.

Lemma 9.10 ([229]). *Let X be a compact metric space. Then X is an acyclic space if its structure corresponds to one of the following types:*

(i) *X is convex.*

(ii) *X is contractible.*

(iii) *X is AR.*

(iv) *X is an R_δ set.*

We need the following result owed to Bressan–Cellina–Fryszkowski.

Lemma 9.11 ([103, Theorem 2]). *Let E be a Banach space, $X = L^1(T, E)$, for some measurable space T, and $N : E \to \mathcal{P}(X)$ a contraction map with decomposable values. Then $Fix(N)$ is an absolute retract.*

The next definitions were introduced in [232].

Definition 9.12. A metric space X is called acyclically contractible if there exists an acyclic homotopy $\Pi : X \times [0, 1] \to \mathcal{P}(X)$ such that

(i) $x_0 \in \Pi(x, 1)$, for every $x \in X$ and for some $x_0 \in X$,

(ii) $x \in \Pi(x, 0)$, for every $x \in X$.

Notice that any contractible space and any acyclic, compact metric space are acyclically contractible (see [19, Theorem 19]). Also from [230], any acyclically contractible space is acyclic.

Definition 9.13. A metric space X is called R_δ-contractible if there exists a multivalued homotopy $\Pi : X \times [0,1] \to \mathcal{P}(X)$ which is u.s.c. and satisfies

(i) $x \in \Pi(x, 1)$, for every $x \in X$,

(ii) $\Pi(x, 0) = B$ for every $x \in X$ and for some $B \subset X$,

(iii) $\Pi(x, \alpha)$ is an R_δ−set, for every $\alpha \in [0,1]$ and $x \in X$.

The following Lasota–Yorke Approximation theorem (see [230]) will be needed in this section.

Lemma 9.14. *Let E be a normed space, X a metric space, and $f : X \to E$ be a continuous map. Then, for each $\varepsilon > 0$, there is a locally Lipschitz map $f_\varepsilon : X \to E$ such that*

$$\|f(x) - f_\varepsilon(x)\| < \varepsilon, \quad \textit{for every } x \in X.$$

9.2 Aronszajn Type Results

In 1890, Peano [411] showed that the Cauchy problem

$$y'(t) = f(t, y(t)), \quad \text{a.e.} \quad t \in [0,b], \tag{9.1}$$

$$y(0) = y_0, \tag{9.2}$$

where $f : [0,b] \times \mathbb{R}^n \to \mathbb{R}^n$ is continuous, has local solutions but uniqueness does not hold in general. This observation became a motivation for studying the structure of the set S of solutions to (9.1)–(9.2). Peano himself showed that, in the case $n = 1$, all sections $S(t) = \{x(t) : x \in S\}$ are nonempty, compact, and connected (that is, a continuum) in the standard topology of the real line, for t in some neighborhood of 0. Kneser generalized this result in 1923 [323] for the case of arbitrary n. In 1928, Hukuhara [298] proved that S is a continuum in the Banach space of continuous functions with the sup norm. A more precise characterization of S was found in 1942 by Aronszajn [29], who showed that S is an R_δ−set. Aronzajn's results were improved by several authors [8, 15–18, 20, 22, 46, 48, 89, 91–99, 115, 116, 150, 156, 160, 161, 236, 298, 388, 451–459, 466, 467, 474, 475].

The following result on the topological structure of the set of solutions of nonlinear functional equations is due to N. Aronszajn [29] and further developed by F. Browder and Ch.P. Gupta in [113] (see also [19, Theorem 1.2]).

Definition 9.15. The mapping $f : X \to E$ from a space X to a Banach space $(E, \|\cdot\|)$ is called a proper map if f is continuous and for every compact $K \subset E$ the set $f^{-1}(K)$ is compact.

Lemma 9.16. *Let E be a Banach space, $C \subset E$ be a nonempty closed bounded subset of E and $F : C \to E$ be a completely continuous map. Then $G = Id - F$ is a proper map.*

Proof. Let $K \subset E$ be a compact set of E we show that $G^{-1}(K)$ is compact. Let $\{x_n\}_{n\in\mathbb{N}} \subset G^{-1}(K)$ be a sequence, thus for every $n \in \mathbb{N}$, there exist $y_n \in K$ such that $x_n - F(x_n) = y_n$. Since F is completely continuous map and K is compact then there exist subsequence of $\{x_n\}_{n\in\mathbb{N}}$, $\{y_n\}_{n\in\mathbb{N}}$ converge to $(x, y) \in C \times K$ respectively. Hence

$$x - F(x) = y \Rightarrow x \in G^{-1}(K). \qquad \square$$

Theorem 9.17 ([29]). *Let X be a space, $(E, \|\cdot\|)$ a Banach space, and $f : X \to E$ a proper map. Assume further that, for each $\epsilon > 0$, a proper map $f_\epsilon : X \to E$ is given and the following two conditions are satisfied:*

(i) $\|f_\epsilon(x) - f(x)\| < \epsilon$, *for every $x \in X$;*

(ii) *for every $\epsilon > 0$ an $u \in E$ such that $\|u\| \le \epsilon$, the equation $f_\epsilon(x) = u$ has exactly one solution.*

Then the set $S = f^{-1}(0)$ is R_δ.

The next result is an simple impulsive version of the results in [451]. The proof is exactly analogous to [230, Theorem 69.2] and and [231, Theorem 7.2].

Theorem 9.18 ([29]). *Let $E = PC([0,b], \mathbb{R}^n) = \{y : y$ is a map from J into $\mathbb{R}^n$ such that $y(t)$ is continuous at $t \ne t_k$, left continuous at $t = t_k$, and $y(t_k^+)$ exist, $k = 1, 2, \dots, m\}$ be a Banach space with the usual sup norm, let $X = B(0,r) = \{u \in PC : \|u\|_{PC} \le r\}$ be the closed ball in E, $F : X \to E$ be a compact map, and $f : X \to E$ be a compact vector field associated with F, i.e., $f = u - F(u)$, such that the following conditions are satisfied:*

(iii) *there exists $y_0 \in \mathbb{R}^n$ such that $F(u)(0) = y_0$, for every $u \in B(0,r)$.*

(iv) *for every $\epsilon \in (0,b]$ and every u, $v \in X$, if $u(t) = v(t)$ for each $t \in [0,\epsilon]$, then $F(u)(t) = F(v)(t)$ for each $t \in [0,\epsilon]$.*

(v) $u(t_k^+) - f(u)(t_k^+) = \lim_{t\to t_k^+} F(u)(t)$, $k = 1, 2, \dots, m$, $u \in B(0,r)$.

Then there exists a sequence $f_n : X \to E$ of continuous proper mappings satisfying conditions (i) and (ii) of Theorem 9.17 with respect to f.

Proof. The proof will given in several steps.

Step 1. We consider $E_0 = C([0,t_1], \mathbb{R}^n)$ and $X_0 = \{u \in E_0 : \|u\|_0 \le r\}$, where

$$\|u\|_0 = \sup\{\|u(t)\| : t \in [0,t_1]\}, \text{ for all } u \in E_0.$$

Let $F_0 : X_0 \to E_0$, and $f_0 : X_0 \to E_0$ be the vector field associated to F_0, where $F_0 = F|_{X_0}$ and $f_0 = f|_{X_0}$. For the proof, it is sufficient to define a sequence $F_n : X_0 \to E_0$ of compact maps such that:

$$F(u) = \lim_{n\to\infty} F_n(u), \text{ uniformly in } u \in X_0, \tag{9.3}$$

and

$$f_{0n} : X_0 \to E_0, \quad f_{0n}(u) = u - F_{0n}(u), \text{ is a one-to-one map.} \tag{9.4}$$

To do this, we additionally define the mappings $r_n : [0, t_1] \to [0, t_1]$ by

$$r_n(t) = \begin{cases} 0, & \text{if } t \in [0, \frac{t_1}{n}], \\ t - \frac{t_1}{n}, & \text{if } t \in (\frac{t_1}{n}, t_1]. \end{cases}$$

Now, we are in the position for defining the sequence F_{0n} as follows:

$$F_{0n}(u)(t) = F_0(u)(r_n(t)), \quad \text{for } u \in X_0,\ n = 1, 2, \ldots . \tag{9.5}$$

Clearly, F_{0n} is a continuous and compact mapping for $n = 1, 2, \ldots$. Since $|r_n(t) - t| < \frac{t_1}{n}$, the compactness of F_0 and (9.5) imply that

$$\lim_{n\to\infty} F_{0n}(u) = F(u), \text{ uniformly in } u \in X_0.$$

Now we shall prove that f_{0n} is a one-to-one map. Assume that for some $x, y \in X_0$, we have

$$(f_0)_n(x) = (f_0)_n(y).$$

This implies

$$x - y = (F_0)(x) - (F_0)(y).$$

If $t \in [0, \frac{t_1}{n}]$, then we have

$$x(t) - y(t) = F_0(x)(r_n(t)) - F_0(y)(r_n(t))$$

which implies

$$x(t) - y(t) = F_0(x)(0) - F_0(y)(0).$$

Thus, in view of (iv), we obtain

$$x(t) = y(t), \text{ for every } t \in [0, \frac{t_1}{n}].$$

Finally, successively repeating the above procedure n times implies that

$$x(t) = y(t), \quad \text{for every } t \in [0, t_1].$$

For every $\epsilon \in (0, t_1]$ and $\|u\|_0 \le \epsilon$, the equation $f_\epsilon(x) = u$ has exactly one solution, (for complete details see Szufla [451]).

Step 2. We consider $E_1 = \{u \in C((t_1, t_2], \mathbb{R}^n) : u(t_1^+) \text{ exists}\}$ and $X_1 = \{u \in E_0 : \|u\|_1 \leq r\}$, where

$$\|u\|_1 = \sup\{\|u(t)\| : t \in (t_1, t_2]\}, \text{ for all } u \in E_1.$$

Let $F_1 : X_1 \to E_1$ be defined by

$$F_1(u)(t) = \begin{cases} F(u)(t), & \text{if } t \in (t_1, t_2], \\ u(t_1^+) - f(u)(t_1^+), & \text{if } t = t_1, \end{cases}$$

and let $f_1 : X_1 \to E_1$ be the vector field associated to F_1 given by

$$f_1(u) = u - F_1(u), \text{ for every } u \in X_1.$$

For the proof it is sufficient to define a sequence $(F_1)_n : X_1 \to E_1$ of compact maps such that,

$$F_1(u) = \lim_{n\to\infty} (F_1)_n(u), \text{ uniformly in } u \in X_1, \tag{9.6}$$

and the sequence $f_{1n} : X_1 \to E_1$ given by

$$f_{1n} : X_1 \to E_1, \ f_{1n}(u) = u - F_{1n}(u), \text{ is a one-to-one map.} \tag{9.7}$$

To do this, we also define the mappings $r_n : [t_1, t_2] \to [t_1, t_2]$ by putting:

$$r_n(t) = \begin{cases} t_1, & \text{if } t \in [t_1, \frac{t_2-t_1}{n} + t_1], \\ t - \frac{t_2-t_1}{n}, & \text{if } t \in (\frac{t_2-t_1}{n} + t_1, t_2]. \end{cases}$$

Next, we define the sequence F_{1n} by

$$F_{1n}(u)(t) = F_1(u)(r_n(t)), \text{ for } u \in X_1, \ n = 1, 2, \dots . \tag{9.8}$$

Clearly, F_{1n} is a continuous and compact mapping for $n = 1, 2, \dots$. Since $|r_n(t) - t| < \frac{t_2-t_1}{n}$, the compactness of F_1 and (9.8) imply

$$\lim_{n\to\infty} F_{1n}(u) = F(u), \ \text{ uniformly in } u \in X_1.$$

To show that f_{1n} is one-to-one map. Assume that for some $x, y \in X_1$, we have

$$(f_1)_n(x) = (f_1)_n(y).$$

This implies that

$$x - y = (F_1)(x) - (F_1)(y).$$

If $t \in (t_1, \frac{t_2-t_1}{n} + t_1]$, then in view of (v), we have

$$x(t) - y(t) = F_1(x)(r_n(t)) - F_1(y)(r_n(t))$$

which implies

$$x(t) - y(t) = F_1(x)(t_1^+) - F_1(y)(t_1^+).$$

As in the above step, we obtain

$$x(t) = y(t), \text{ for every } t \in [t_1, \frac{t_2 - t_1}{n} + t_1].$$

Finally, successively repeating the above procedure n times implies that

$$x(t) = y(t), \quad \text{for every } t \in [t_1, t_2].$$

For every $\epsilon \in (t_1, t_2]$ and $\|u\|_1 \le \epsilon$, the equation $f_\epsilon(x) = u$ has exactly one solution, (for the complete proof, see Szufla [451]).

Step 3. We continue this process and consider $E_m = \{u \in C((t_m, b], \mathbb{R}^n) : u(t_m^+) \text{ exists}\}$ and $X_m = \{u \in E_m : \|u\|_m \le r\}$, where

$$\|u\|_m = \sup\{\|u(t)\| : t \in (t_m, b]\}, \text{ for all } u \in E_m.$$

Let $F_m : X_m \to E_m$ and $f_m : X_m \to E_m$ be given by

$$F_m(u)(t) = \begin{cases} F(u)(t), & \text{if } t \in (t_1, t_2], \\ u(t_m^+) - f(u)(t_m^+), & \text{if } t = t_1, \end{cases}$$

and $f_m(u) = u - F_m(u)$ for every $u \in X_1$. For the proof it is sufficient to define a sequence $(F_m)_n : X_m \to E_m$ of compact maps such that

$$F_m(u) = \lim_{n\to\infty} (F_m)_n(u), \text{ uniformly in } u \in X_m, \tag{9.9}$$

and

$$f_{mn} : X_m \to E_m, \ f_{mn}(u) = u - F_{mn}(u), \text{ is a one-to-one map.} \tag{9.10}$$

To do this, we additionally define the mappings $r_n : [t_m, b] \to [t_m, b]$ by putting

$$r_n(t) = \begin{cases} t_m, & \text{if } t \in \left[t_m, \frac{b-t_m}{n} + t_m\right], \\ t - \frac{b-t_m}{n}, & \text{if } t \in \left(\frac{b-t_m}{n} + t_m, b\right]. \end{cases}$$

We define the sequence F_{mn} by

$$(F_m)_n(u)(t) = F_1(u)(r_n(t)), \text{ for } u \in X_m, \ n = 1, 2, \ldots .$$

Now the functions $f_n : X \to E$ defined by

$$f_n(u)(t) = \begin{cases} u(t) - F_0(u)(t), & \text{if } \ t \in [0, t_1], \\ u(t) - F_1(u)(t), & \text{if } \ t \in (t_1, t_2], \\ \ldots \\ u(t) - F_m(u)(t), & \text{if } t \in (t_m, b], \end{cases}$$

satisfy (i) and (ii) of Theorem 9.17. □

From Theorem 9.18, we immediately have the following result.

Corollary 9.19. *Assume that f and F are as in Theorem 9.18. Then $f^{-1}(0) = \mathrm{Fix}(F)$ is an R_δ-set.*

9.2.1 Solution Sets for Impulsive Differential Equations

Consider the first-order impulsive single-valued problem,

$$\begin{cases} \frac{d}{dt}[y(t) - g(t, y_t)] = f(t, y_t), & \text{a.e.} \quad t \in J = [t_0, b] \backslash \{t_1, \dots, t_m\}, \\ \Delta y|_{t=t_k} = I_k(y(t_k^-)), & k = 1, \dots, m, \\ y(t) = \phi(t), & t \in [-r, t_0], \end{cases} \tag{9.11}$$

where $g, f : J \times \mathcal{D} \to E$ are given functions, $t_0 < t_1 < \cdots < t_m < t_{m+1} = b$, $\phi \in \mathcal{D}$, E is a separable Banach space, $\Delta y|_{t=t_k} = y(t_k^+) - y(t_k^-)$, and $y(t_k^+) = \lim_{h \to 0^+} y(t_k + h)$ and $y(t_k^-) = \lim_{h \to 0^+} y(t_k - h)$ represent the right and left hand limits of $y(t)$ at $t = t_k$, respectively.

Let $S(f, \phi)$ denote the set of all solutions to problem (9.11). We wish to prove an Aronszajn-type result for this problem. We need to make the following assumptions:

(I-1) $f : J \times \mathcal{D} \to E$ is an Carathéodory function and for every $t \in [0, b]$ the function $t \to f(t, y_t)$ is strongly measurable.

(I-2) There exist a function $p \in L^1(J, \mathbb{R}^+)$ and a continuous nondecreasing function $\rho : [0, \infty) \to [0, \infty)$ such that

$$\|f(t, x)\| \le p(t)\rho(\|x\|_\infty) \quad \text{for a.e. } t \in J \text{ and each } x \in \mathcal{D}$$

and

$$\int_0^b p(s)ds < \int_{(1+d_1)\|\phi\|_\infty + 2d_2}^\infty \frac{du}{\rho(u)}.$$

(I-3) There exists $\overline{p} \in L^1([0, b], \mathbb{R}^+)$ such that, for every bounded subset D in $\mathcal{D}$,

$$\chi(f(t, D)) \le \overline{p}(t) \sup\{\chi(D(\theta)) : \theta \in [-r, 0]\},$$

and there exist $L_k > 0$, $k = 0, \dots, m$ such that

$$q_k := 2 \sup_{t \in [t_k, t_{k+1}]} \int_{t_k}^{t_{k+1}} e^{-L_k(t-s)} p(s)ds < 1, \quad k = 0, \dots, m.$$

(I-4) There exist constants $\overline{c}_k > 0$ and continuous functions $\phi_k \colon \mathbb{R}^+ \to \mathbb{R}^+$ such that

$$|I_k(x)| \le \overline{c}_k \phi_k(|x|) \quad \text{for each } x \in E, \ k = 1, \dots, m.$$

Our first result in this section is contained in the following theorem.

Theorem 9.20. *Assume that* (I-1)–(I-4) *and* (H-8)–(H-9) *are satisfied. If*

$$\omega + \max\{q_k : k = 0, \dots, m\} < 1,$$

then the set $S(f, \phi)$ *is an* R_δ*–set, and hence is an acyclic space.*

Proof. Let $F : \Omega \to \Omega$ be defined by

$$F(y)(t) = \begin{cases} \phi(t), & t \in [-r, t_0], \\ \phi(t_0) - g(t_0, \phi) + g(t, y_t) & \\ \quad + \int_{t_0}^t f(s, y_s)ds + \sum_{t_0 < t_k < t} I_k(y(t_k)), & t \in [t_0, b]. \end{cases}$$

Thus, $Fix\ F = S(f, \phi)$. From Theorem 8.6, we know that $S(f, \phi) \neq \emptyset$, and there exists $\overline{M} > 0$ such that

$$\|y\|_\Omega \leq \overline{M}, \text{ for every } y \in S(f, \phi).$$

Define

$$\widetilde{f}(t, y_t) = \begin{cases} f(t, y_t), & \text{if } \|y_t\|_\infty \leq \overline{M}, \\ f\left(t, \frac{\overline{M} y_t}{\|y_t\|_\infty}\right), & \text{if } \|y_t\|_\infty \geq \overline{M}. \end{cases}$$

Since f is L^1-Carathéodory, the function $\widetilde{f}$ is Carathéodory and is integrably bounded by (I-2), so there exists $h \in L^1(J, \mathbb{R}^+)$ such that

$$\|\widetilde{f}(t, x)\| \leq h(t), \quad \text{a.e. } t \text{ and all } x \in \mathcal{D}. \tag{9.12}$$

Consider the modified problem,

$$\begin{cases} \frac{d}{dt}[y(t) - g(t, y_t)] = \widetilde{f}(t, y_t), & \text{a.e. } t \in J \backslash \{t_1, \dots, t_m\}, \\ \Delta y|_{t=t_k} = I_k(y(t_k^-)), \ k = 1, \dots, m, & \\ y(t) = \phi(t), & t \in [-r, t_0]. \end{cases}$$

We can easily prove that $S(f, \phi) = S(\widetilde{f}, \phi) = Fix\widetilde{F}$, where $\widetilde{F} : \Omega \to \Omega$ is defined by

$$\widetilde{F}(y)(t) = \begin{cases} \phi(t), & t \in [-r, t_0], \\ \phi(t_0) - g(t_0, y_{t_0}) + g(t, y_t) + \int_{t_0}^t \widetilde{f}(s, y_s)ds & \\ \quad + \sum_{0 < t_k < t} I_k(y(t_k^-)), & t \in [t_0, b]. \end{cases}$$

By inequality (9.12) and the continuity of I_k, we deduce that

$$\|\widetilde{F}(y)\|_\Omega \leq (1 + d_1)\|\phi\|_\infty + d_1 M + 2d_2 + |h\|_{L^1} + \sum_{k=1}^m \overline{c}_k \phi_k(\overline{M}) := R.$$

Then $\widetilde{F}$ is uniformly bounded. As in Theorem 8.6, we can prove that $\widetilde{F} : \Omega \to \Omega$ is compact which allows us to define the compact perturbation of the identity $\widetilde{G}(y) = y - \widetilde{F}(y)$ which is a proper map. From the compactness of $\widetilde{F}$ and the Lasota–Yorke approximation theorem, we can easily prove that all conditions of Theorem 9.17 are met. Therefore the solution set $S(\widetilde{f}, \phi) = \widetilde{G}^{-1}(0)$ is an R_δ-set and hence an acyclic space by Lemma 9.10. □

9.3 Solution Sets of Differential Inclusions

In this section, we present some geometrical properties of solution sets for some classes of impulsive functional differential inclusions in Banach spaces. In the case where $I_k \equiv 0$ and F is a scaler multimap (i.e., $F : [0,b] \times \mathbb{R}^n \to \mathcal{P}(\mathbb{R}^n)$), Lasry and Robert [341] proved that the set $S(F)$ of all solutions of

$$y'(t) \in F(t, y(t)), \quad \text{a.e.} \quad t \in J = [0,b], \tag{9.13}$$

$$y(0) = 0, \tag{9.14}$$

where F is a compact, convex valued upper semicontinuous and bounded map, is an acyclic and compact set. Himmelberg and Vleck [291], using an idea given by Yorke [490] and Hyman's characterization of R_δ−sets [307], proved that under the same assumptions, the set $S(F)$ is an R_δ−set in $C([0,b], \mathbb{R}^n)$. For the information about R_δ sets of the problem (9.13)–(9.14), see De Blasi and Myjak [171], Papageorgiou [403] and Lasry and Robert [341]. By using the σ-selection and measurable locally Lipschitzian properties, we can prove that the solution set of the problem (9.13)–(9.14) is an R_δ−set. For more information about the regulation of solution sets, see [99,164, 168,183,232,233,303] and the references therein.

Very recently, the topological and geometric structure of solution sets for impulsive differential inclusions on compact and noncompact intervals were investigated in [77,78,174–176,244,245,274,285] where contractibility, AR, acyclicity, R_δ−sets properties are obtained.

9.4 σ-selectionable Multivalued Maps

The following definitions and the next result can be found in [230,264] (see also [34, p. 86]). Let (X,d) and (Y,d') be two metric spaces.

Definition 9.21. We say that a map $F : X \to \mathcal{P}(Y)$ is σ-Ca-selectionable if there exists a decreasing sequence of compact valued u.s.c. maps $F_n : X \to Y$ satisfying:

(a) F_n has a Carathéodory selection for all $n \geq 0$ (the F_n are called Ca-selectionable),

(b) $F(x) = \bigcap_{n\geq 0} F_n(x)$ for all $x \in X$.

Definition 9.22.

(a) A single-valued map $f : [0,a] \times X \to Y$ is said to be measurable-locally-Lipschitz (mLL), if $f(\cdot, x)$ is measurable for every $x \in X$ and for every $x \in X$, there exists a neighborhood V_x of x and an integrable function $L_x : [0,a] \to [0,\infty)$ such that, for all $t \in [0,a]$ and $x_1, x_2 \in V_x$,

$$d'(f(t,x_1), f(t,x_2)) \leq L_x(t) d(x_1,x_2).$$

(b) A multivalued mapping $F : [0,a] \times X \to \mathcal{P}(Y)$ is mLL-selectionable if it has an mLL-selection.

Definition 9.23. We say that a multivalued map $\phi : [0,a] \times E \to \mathcal{P}(E)$ with closed values is upper-Scorza–Dragoni if, given $\delta > 0$, there exists a closed subset $A_\delta \subset [0,1]$ such that the measure $\mu([0,a] \setminus A_\delta) \leq \delta$ and the restriction ϕ_δ of ϕ to $A_\delta \times E$ is u.s.c.

Theorem 9.24 ([230, Theorem 19.19]). *Let E and E_1 be two separable Banach spaces and let $F : [a,b] \times E \to \mathcal{P}_{cp,cv}(E_1)$ be an upper-Scorza–Dragoni map. Then F is $\sigma-$Ca-selectionable, the maps $F_n : [a,b] \times E \to \mathcal{P}(E_1)$ $(n \in \mathbb{N})$ are almost upper semicontinuous, and have*

$$F_n(t,x) \subset \overline{co}\left(\cup_{x\in E} F_n(t,x)\right).$$

Moreover, if F is integrably bounded, then F is σ-mLL-selectionable.

Proof. Consider the family $\{B(y,r_k)\}_{y\in E}$, where $r_k = (1/3)^k$, $k = 1,2,\ldots$ Using Stone's theorem for every $k = 1,2,\ldots$, we get locally finite subcovering $\{U_i^k\}_{i\in I_k}$ of $\{B(y,r_k)\}_{y\in E}$. For every $i \in I_k$, $k = 1,2,\ldots$, we fix the center $y_i^k \in E$ such that $U_i^k \subset B(y_i^k, r_k)$. Now, let $\eta_i^k : E \to [0,1]$ be a locally Lipschitz partition of unity subordinated to $\{U_i^k\}_{i\in I^k}$.

Define $\psi_i^k : [0,a] \to \mathcal{P}(E)$ and $f_i^k : [0,a] \to E$ as follows:

$$\psi_i^k(t) = \overline{co}\Bigg(\bigcup_{y\in B(y_i^k, 2r_k)} \varphi(t,y)\Bigg),$$

and let f_i^k be a measurable selection of ψ_i^k which exists in view of the Kuratowski–Ryll–Nardzewski theorem.

Finally, define $\varphi_k : [a,b] \times E \to \mathcal{P}(E_1)$ and $f_k : [a,b] \times E \to E_1$ as follows:

$$\varphi_k(t,z) = \sum_{i\in I_k} \eta_i^k(z) \cdot \psi_i^k(t), \quad \text{and} \quad f_k(t,z) = \sum_{i\in I_k} \eta_i^k(z) \cdot f_i^k(t).$$

Then $f_k \subset \varphi_k$. Fix $t \in [a,b]$. If $\varphi(t,\cdot)$ is *u.s.c.* then $\varphi(t,z) = \bigcap_{k=1}^{\infty} \varphi_k(t,z)$ and $\varphi_{k+1}(t,z) \subset \varphi_k(t,z)$, for every $z \in E$. By the assumptions on φ, the map $\varphi(t,\cdot)$ is *u.s.c.* for almost all $t \in [0,a]$, and the first part of Theorem 9.24 is proved. The second claim is an immediate consequence of the first one. □

The following σ-selectionability lemma can be proved as in Proposition 4.1 of De Belasi [93] or the Lemma of Papageorgiou [406].

Lemma 9.25. *Assume that X is a Polish space (a complete separable metrizable space), Y is a separable Banach space, and $F : [0,b] \times X \to \mathcal{P}_{cl,cv}(Y)$ satisfies*

(i) $t \to F(t,x)$ *is measurable;*

(ii) $x \to F(t,x)$ *is* H*-u.s.c. (i.e.* $\varepsilon - \delta$*-u.s.c);*

(iii) $\|F(t,x)\|_{\mathcal{P}} \leq l(t)$ *a.e. on* $[0,b]$ *with* $l \in L^p([0,b],\mathbb{R}^+)$, $1 \leq p < \infty$.

Then, there exists a sequence of $F_n : [0,b] \times X \to \mathcal{P}_{cl,cv}(Y)$ *such that for every* $n \in \mathbb{N}$ *and* $x \in X$, *there exists* $l_n(x) > 0$ *and* $\varepsilon_n > 0$ *such that if* $x_1, x_2 \in B(x, \varepsilon_n)$, *then*

$$\begin{aligned} &H(F_n(t,x_1), F_n(t,x_2)) \leq l_n(x)p(t)|x_1 - x_2|, \quad \text{a.e. on } [0,b], \\ &F(t,x) \subset \cdots F_n(t,x) \subset F_{n-1}(t,x) \cdots, \ \|F_n(t,x)\|_{\mathcal{P}} \leq p(t), \quad \text{a.e. on } [0,b], \\ &H(F_n(t,x), F(t,x)) \to 0, \quad \text{as } n \to \infty \text{ for all } (t,x) \in [0,b] \times X, \end{aligned}$$

and there exists $u_n : [0,b] \times X \to Y$ *measurable in* $t \in [0,b]$, *locally Lipschitz in* $x \in X$ *and* $u_n(t,x) \in F_n(t,x)$ *for* $(t,x) \in [0,b] \times X$ *and* $n \in \mathbb{N}$.

For *u.s.c* multivalued maps, there exists a useful approximation result similar to the Lasota–Yorke Lemma [343].

Lemma 9.26 ([167, Lemma 2.2]). *Let* Ω *be a subset of a Banach space,* E *be a Banach space and* $F : \Omega \to \mathcal{P}_{cl,cv}(E)$ *be a multivalued map. Let* $r_n = 3^{-n}$, $\{U_\lambda\}_{\lambda\Lambda}$ *be a locally finite refinement of* $\Omega = \bigcup_{\omega\in\Omega} B(\omega, r_n)$, $\{\phi_\lambda\}_{\lambda\in\Lambda}$ *be a locally Lipschitz partition of unity subordinate to* $\{U_\lambda\}_{\lambda\in\Lambda}$, *pick* $\omega_\lambda \in U_\lambda \subset B(\omega_\lambda, r_n) \cap \Omega$ *and let*

$$F_n(\omega) = \sum_{\lambda\in\Lambda} \phi_\lambda(\omega) C_\lambda,$$

where $C_\lambda = \overline{co}\, F(B(\omega_\lambda, 2r_n) \cap \Omega)$. *Then*

(i) $F(\omega) \subset F_{n+1}(\omega) \subset F_n(\omega) \subset \overline{co}\, F(B(\omega, 3r_n) \cap \Omega)$ *on* Ω.

(ii) *If* F *is uniformly locally bounded, then* F_n *is locally Lipschitz for large* n.

(iii) *If* F *is* $\varepsilon - \delta$*-u.s.c., then* $H(F_n(\omega), F(\omega)) \to 0$ *on* Ω, *as* $n \to \infty$.

The following result is in part the multivalued version of a well-know theorem of Scorza–Dragoni and Lusin's Theorem which states, roughly speaking, that every measurable function is continuous almost everywhere.

Theorem 9.27 ([433]). *Let* X, Y *be a sparable Banach spaces and* $F : J \times X \to \mathcal{P}_{cl}(Y)$ *with* J *a measurable subset of* $\mathbb{R}$. *Assume that* $Graph(F(t,.))$ *is a closed in* $X \times Y$ *for almost* $t \in J$. *Then there exists* $\widehat{F} : J \times X \to \mathcal{P}(Y)$ *such that*

(i) *for almost all* $t \in J$, $\widehat{F}(t,x) \subset F(t,x)$ *for all* $x \in X$;

(ii) *if* $\Delta \subset J$ *is measurable and* $u : \Delta \to X$, $v : \Delta \to Y$ *are measurable functions with* $v(t) \in F(t,u(t))$ *a.e in* Δ, *then* $v(t) \in \widehat{F}(t,u(t))$ *a.e. in* Δ;

(iii) *for every* $\varepsilon > 0$, *there exists a closed* $J_\varepsilon \subset J$ *with* $\mu(J \backslash J_\varepsilon) < \varepsilon$ *and the graph of* $\widehat{F}|_{J_\varepsilon \times X}$ *is closed in* $J \times X \times Y$.

If E is a separable Banach space, we present the following result of Scorza–Dragoni type (essentially due to Rzézuchowski [433]):

Theorem 9.28 ([433]). *Let $F : [0,b] \times D \to \mathcal{P}_{cl}(E)$ with D is a closed convex subset of E. Assume that F satisfies the following conditions:*

(i) *F is an upper-Carathédory map;*

(ii) *F maps compact subsets of $J \times D$ into compact ones.*

Then there is a map $F_0 : [0,b] \times D \to \mathcal{P}(E) \cup \{\emptyset\}$ such that:

(a) *for all $t \in [0,b]$ and $x \in D$, $F_0(t,x) \subset F(t,x)$;*

(b) *if $\Delta \subset [0,b]$ is measurable, $u, v : \Delta \to D$, then $v(t) \in F_0(t,u(t))$ a.e. in Δ;*

(c) *for any $\varepsilon > 0$, there is a closed $\Delta_\varepsilon \subset [0,b]$ such that F_0 restricted to $\Delta_\varepsilon \times D$ has nonempty values and is (jointly) upper semicontinuous.*

Finally, we present some nice approximations of upper semicontinuous multivalued maps.

Theorem 9.29. *Let Ω be a complete separable metric space. Let μ be a nonnegative finite measure on the completion $\mathcal{L}$ of the Borel σ-algebra Σ. Let X be a complete separable metric space with Borel σ-algebra $\mathcal{B}(X)$, and let E be a separable Banach space. Suppose that $F : \Omega \times X \to \mathcal{P}_{cl,cv,b}(E)$ is a bounded multivalued map such that:*

(i) *F is $\mathcal{L} \otimes \mathcal{B}(X)$ weakly measurable;*

(ii) *for each $\omega \in \Omega$, $x \to F(\omega, x)$ is H_d-u.s.c..*

Then there exists a sequence $\{F_n \mid n \in \mathbb{N}\}$ of multivalued maps $F_n : \Omega \times X \to \mathcal{P}_{cv,cl,b}(E)$ and a sequence $\{f_n \mid n \in \mathbb{N}\}$ of single functions $f_n : \Omega \times X \to E$ satisfying, for every $n \in \mathbb{N}$, the following conditions:

(a_1) *for each $x \in X$, $\omega \to F_n(\omega, x)$ is $\mathcal{L}$-weakly measurable;*

(a_2) *for each $\omega \in \Omega$, $x \to F_n(\omega, x)$ is locally Lipschitz;*

(a_3) *for each $(\omega, x) \in \Omega \times X$ and for all $n \in \mathbb{N}$,*

$$F(t,x) \subset F_{n+1}(t,x) \subset F_n(t,x) \subset \overline{co}\left(\bigcup_{x \in X} (F(t,x)\right);$$

(a_4) *for each $(t,x) \in \Omega \times X$, $\lim_{n\to\infty} H(F_n(t,x), F(t,x)) = 0$;*

(a_5) *f_n is a Carathéodory–Lipschitz selector of F_n.*

9.4.1 Contractible and R_δ-contractible

Let $S_{[-r,b]}(\phi)$ denote the set of all solutions of problem (8.1)–(8.3). Now, we are in position to state and prove another characterization of the geometric structure of $S_{[-r,b]}(\phi)$.

Theorem 9.30. *Let $F\colon J \times \mathcal{D} \to \mathcal{P}_{cp,cv}(E)$ be a Carathéodory and mLL-selectionable multivalued map that satisfies conditions* (H-6), (H-7), (H-5) *and* (H-3).

Then, for every $\phi \in \mathcal{D}$, the set $S_{[-r,b]}(\phi)$ is contractible.

Proof. Let $f \subset F$ be a measurable, locally Lipschitz selection and consider the single-valued problem

$$\begin{cases} \frac{d}{dt}[y(t) - g(t, y_t)] = f(t, y_t), & \text{a.e. } t \in J\backslash\{t_1, t_2, \dots, t_m\}, \\ y(t_k^+) - y(t_k) = I_k(y(t_k^-)), & k = 1, \dots, t_m, \\ y(t) = \phi(t), & t \in [-r, 0]. \end{cases} \tag{9.15}$$

As in [82], [397, Theorems 3.3], or [398, Theorem 3.5], we can prove that problem (9.15) has exactly one solution for every $\phi \in \mathcal{D}$. Define the homotopy $h : S_{[-r,b]}(\phi) \times [0, 1] \to S_{[-r,b]}(\phi)$ by

$$h(y, \alpha)(t) = \begin{cases} y(t), & \text{for } -r \leq t \leq \alpha b, \\ \overline{x}(t), & \text{for } \alpha b < t \leq b, \end{cases}$$

where $\overline{x} = S_{[-r,b]}(f, \phi)$ is the unique solution of problem (9.15). In particular,

$$h(y, \alpha) = \begin{cases} y, & \text{for } \alpha = 1, \\ \overline{x}, & \text{for } \alpha = 0. \end{cases}$$

We prove that h is a continuous homotopy. Let $(y_n, \alpha_n) \in S_{[-r,b]}(\phi) \times [0, 1]$ be such that $(y_n, \alpha_n) \to (y, \alpha)$, as $n \to \infty$. We shall prove that $h(y_n, \alpha_n) \to h(y, \alpha)$. We have

$$h(y_n, \alpha_n)(t) = \begin{cases} y_n(t), & \text{for } t \in [-r, \alpha_n b], \\ \overline{x}(t), & \text{for } t \in (\alpha_n b, b]. \end{cases}$$

(a) If $\lim_{n\to\infty} \alpha_n = 0$, then

$$h(y, 0)(t) = \begin{cases} \phi(t), & \text{for } t \in [-r, 0], \\ \overline{x}(t), & \text{for } t \in (0, b]. \end{cases}$$

Hence,

$$\|h(y_n, \alpha_n) - h(y, \alpha)\|_\Omega \leq \|y_n - \phi\|_\Omega + \|y_n - \bar{x}\|_{[0,\alpha_n b]},$$

which tends to 0 as $n \to +\infty$ for $y_n \equiv \phi$ on $[-r, 0]$. The case where $\lim_{n\to\infty} \alpha_n = 1$ is treated similarly.

(b) If $\alpha_n \neq 0$ and $0 < \lim_{n\to\infty} \alpha_n = \alpha < 1$, then we may distinguish between two sub-cases:

(i) If $t \in [-r, \alpha b]$, then

$$y_n(t) = \phi(t) \text{ implies } h(y_n, \alpha_n)(t) = h(\phi, \alpha_n)(t) \text{ for all } t \in [-r, 0].$$

Furthermore, $y_n \in S_{[-r,b]}(\phi)$ implies the existence of $v_n \in S_{F,y_n}$ such that, for $t \in [0, \alpha_n b]$,

$$y_n(t) = \phi(t) - g(0,\phi) + g(t,(y_n)_t) + \int_0^t v_n(s)ds + \sum_{0<t_k<t} I_k(y_n(t_k)).$$

It remains to prove that $v \in F(t, y_t)$, for a.e. $t \in J$. Lemma 2.66 yields the existence of $\alpha_i^n \geq 0$, $i = n, \ldots, k(n)$, such that $\sum_{i=1}^{k(n)} \alpha_i^n = 1$ and the sequence of convex combinations $g_n(\cdot) = \sum_{i=1}^{k(n)} \alpha_i^n v_i(\cdot)$ converges strongly to v in L^1. Since F takes convex values, using Lemma 2.64, we obtain that, for a.e. $t \in J$,

$$\begin{aligned} v(t) &\in \bigcap_{n\geq 1} \overline{\{g_k(t) : k \geq n\}}, \\ &\subset \bigcap_{n\geq 1} \overline{co}\{v_k(t),\ k \geq n\} \\ &\subset \bigcap_{n\geq 1} \overline{co}\{\bigcup_{k\geq n} F(t, y_k(t))\} \\ &= \overline{co}(\limsup_{k\to\infty} F(t, y_k(t))). \end{aligned} \tag{9.16}$$

Since F is u.s.c. with compact values and (9.16) we have that $v(t) \in \overline{co}\, F(t, y(t))$. Since $F(\cdot,\cdot)$ has closed, convex values, we deduce that $v(t) \in F(t, y(t))$, for a.e. $t \in J$. Now $\{y_n\}$ converges to y so

$$\|y_n\|_\Omega \leq R,$$

for some $R > 0$. Then (H-6) implies that

$$|v_n(t)| \leq p(t)\rho(R), \text{ a.e. } t \in [0, b].$$

By the Lebesgue dominated convergence theorem, it follows that

$$v \in L^1([0,b], E) \text{ implies } v \in S_{F,y}.$$

Using the continuity of I_k, we find that for $t \in [0, b]$,

$$y(t) = \phi(t) - g(0,\phi) + g(t, y_t) + \int_0^t v(s)ds + \sum_{0<t_k<t} I_k(y(t_k)).$$

(ii) If $t \in (\alpha_n b, b]$, then

$$h(y_n, \alpha_n)(t) = h(y, \alpha)(t) = \bar{x}(t).$$

Thus,

$$\|h(y_n, \alpha_n) - h(y, \alpha)\|_\Omega \to 0, \text{ as } n \to \infty.$$

Therefore, h is a continuous function, proving that $S_{[-r,b]}(\phi)$ is contractible to the point $\overline{x} = S_{[-r,b]}(f, \phi)$. □

A further precise result is given in the following theorem.

Theorem 9.31. *Let $F\colon J \times \mathcal{D} \to \mathcal{P}_{cp,cv}(E)$ be a Carathéodory and a Ca-selectionable multivalued map. Assume that conditions* (H-6)–(H-8) *hold. In addition, assume that*

(I-5) *There exist $d \in [0,1)$, $\bar{d} > 0$ such that*

$$|g(\tau_1, u) - g(\tau_2, \overline{u})| \le d\,\|u - \overline{u}\|_\infty + \bar{d}\,|\tau_1 - \tau_2| \ \tau_1, \tau_2 \in J,\ u,\ \overline{u} \in \mathcal{D}.$$

Then the solution set $S_{[-r,b]}(\phi)$ is R_δ−contractible and acyclic.

Proof. Replace the single-valued homotopy $h : S_{[-r,b]}(\phi) \times [0,1] \to S_{[-r,b]}(\phi)$ in Theorem 9.30 by the multivalued homotopy $\Pi : S_{[-r,b]}(\phi) \times [0,1] \to \mathcal{P}(S_{[-r,b]}(\phi))$ defined by

$$\Pi(x, \alpha) = \left\{ y \in S_{[-r,b]}(\phi) : y(t) = \begin{cases} x(t), & \text{if } -r \le t \le \alpha b, \\ z(t), & \text{if } \alpha b \le t \le b, \end{cases} \right\},$$

where $f \subset F$ and $z \in S(f, \alpha b, x)$ is the solution set of the problem

$$\begin{cases} \frac{d}{dt}[y(t) - g(t, y_t)] = f(t, y_t), & \text{a.e.} \quad t \in [\alpha b, b] \backslash \{t_1, t_2, \dots t_m\}, \\ y(t_k^+) - y(t_k) = I_k(y(t_k^-)), & \qquad k = 1, \dots, m, \\ y(t) = x(t), & \qquad t \in [-r, \alpha b]. \end{cases}$$

From the definition of Π, $\Pi(x, 0) = S(f, 0, x)$ and $x \in \Pi(x, 1)$ for every $x \in S_{[-r,b](\phi)}$. It remains to prove that $\Pi(\cdot,\cdot)$ is u.s.c. Since $\Pi(\cdot,\cdot)$ has nonempty compact values, we only check (see Lemma 2.44) that Π is locally compact and has a closed graph. Finally, we show that $\Pi(x, \alpha)$ is an R_δ set for each x, α. This will be performed in three steps.

Step 1. Π is locally compact. We argue in two sub-steps.

(a) The multivalued map $\widetilde{S} : [0, b] \times \mathcal{D} \to \mathcal{P}(\Omega)$ defined by

$$\widetilde{S}(t, \phi) = S(f, t, \phi)$$

is u.s.c. Here, $S(f,u,\phi)$ refers to the solution set of the problem

$$\begin{cases} \frac{d}{dt}[y(t)-g(t,y_t)]) = f(t,y_t), & \text{a.e.} \quad t\in[u,b]\backslash\{t_1,t_2,\dots t_m\},\\ y(t_k^+)-y(t_k) = I_k(y(t_k^-)), & k=1,\dots,m,\\ y(t) = x(t), & t\in[-r,u]. \end{cases}$$

To the contrary, assume that $\widetilde{S}$ is not u.s.c. at some point (t_0,ϕ_0). Then there exists an open neighborhood U of $\widetilde{S}(t_0,\phi_0)$ in Ω such that, for every open neighborhood V at (t_0,ϕ_0) in the metric space $[0,b]\times\mathfrak{D}$, there exists $(t_1,\phi_1)\in V$ such that $\widetilde{S}(t_1,\phi_1)\not\subset U$. Let $V_n=\{(t,\phi)\in[0,b]\times\mathfrak{D} : d((t,\phi),(t_0,\phi_0))<1/n\}$, for each $n=1,2,\dots$, where d denotes the product metric in $[0,b]\times\mathfrak{D}$. Then for each $n=1,2,\dots$, there is some $(t_n,\phi_n)\in V_n$ and $y_n\in\widetilde{S}(t_n,\phi_n)$ such that $y_n\notin U$. Define the maps

$$G_{t_0,\phi_0},\ F_{t_0,\phi_0}:\Omega\to\Omega$$

by

$$F_{t_0,\phi_0}(x)(t)=\begin{cases}\phi_0(t), & t\in[-r,t_0],\\ \phi_0(t_0)-g(t_0,\phi_0)+g(t,x_t)+\int_{t_0}^t f(s,x_s)ds \\ \quad +\sum_{t_0<t_k<t} I_k(x(t_k)), & t\in[t_0,b],\end{cases}$$

and the compact perturbation of the identity

$$G_{t_0,\phi_0}(x)=x-F_{t_0,\phi_0}(x),\quad \text{for } t\in[0,b] \text{ and } x\in\Omega.$$

By a simple calculation, for $x\in\Omega$ $t\in[0,b]$, $t_0\in[0,b]\backslash\{t_1,\dots,t_m\}$, and $\phi_0\in\mathfrak{D}$, we have

$$F_{t_0,\phi_0}(x)(t)=\phi_0(t_0)-g(t_0,\phi_0)+g(t_0,0)-F_{t_0,0}(x)(t_0)+F_{t_0,0}(x)(t).$$

Consequently,

$$G_{t_0,\phi_0}(x)(t)=-\phi_0(t_0)+g(t_0,\phi_0)-g(t_0,0)+F_{t_0,0}(x)(t_0)+G_{t_0,0}(x)(t).$$

From the definition above, we have that

$$\widetilde{S}(t_0,\phi_0)=G^{-1}_{t_0,\phi_0}(0),\quad \text{for each } (t_0,\phi_0)\in[0,b]\times\mathfrak{D}.$$

Since F_{t_0,ϕ_0} is a compact map (see Theorem 3.3 in [398]), the compact perturbation of the identity G_{t_0,ϕ_0} is proper. Moreover, $y_n\in\widetilde{S}(t_n,\phi_n)$. Then,

$$y_n(t)=\begin{cases}\phi_n(t), & t\in[-r,t_n],\\ \phi_n(t_n)-g(t_n,\phi_n)+g(t,(y_n)_t) \\ \quad +\int_{t_n}^t f(s,(y_s)_n)ds+\sum_{t_n<t_k<t} I_k(y_n(t_k)), & t\in[t_n,b].\end{cases}$$

It follows that

$$\begin{aligned}0&=G_{t_n,\phi_n}(y_n)(t)\\ &=-\phi_n(t_n)+g(t_n,\phi_n)-g(t_n,0)+F_{t_n,0}(y_n)(t_n)+G_{t_n,0}(y_n)(t)\end{aligned}$$

and

$$G_{t_0,\phi_0}(y_n)(t) = -\phi_0(t_0) + g(t_0,\phi_0) - g(t_0,0) + F_{t_0,0}(y_n)(t_0) + G_{t_0,0}(y_n)(t).$$

We then have the estimate

$$\begin{aligned}
|G_{t_0,\phi_0} y_n(t)| &\le |\phi_n(t_n) - \phi_0(t_0)| + |g(t_0,0) - g(t_n,0)| \\
&\quad + |g(t_0,\phi_0) - g(t_n,\phi_n)| \\
&\quad + |F_{t_n,0}(y_n)(t_n) - F_{t_0,0}(y_n)(t_0)| \\
&\quad + |G_{t_n,0}(y_n)(t) - G_{t_0,0}(y_n)(t)| \\
&\le |\phi_n(t_n) - \phi_0(t_0)| + |g(t_0,0) - g(t_n,0)| \\
&\quad + |g(t_0,\phi_0) - g(t_n,\phi_n)| \\
&\quad + \left| \int_0^{t_n} f(s,(y_s)_n)ds - \int_0^{t_0} f(s,(y_n))ds \right| \\
&\quad + \left| \sum_{0<t_k<t_0} I_k(y_n(t_k)) - \sum_{0<t_k<t_n} I_k(y_n(t_k)) \right| \\
&\le |\phi_n(t_n) - \phi_0(t_0)| + |g(t_0,0) - g(t_n,0)| \\
&\quad + |g(t_0,\phi_0) - g(t_n,\phi_n)| \\
&\quad + \int_{t_0}^{t_n} \|f(s,(y_s)_n)\| ds + \left| \sum_{t_n<t_k<t_0} I_k(y_n(t_k)) \right| \\
&\le \|\phi_n(t_n) - \phi_n(t_0)\| + \|\phi_n(t_0) - \phi_0(t_0)\| \\
&\quad + (d+\bar{d})|t_n - t_0| + \|\phi_n - \phi_0\|_\infty \\
&\quad + \int_{t_0}^{t_n} \|f(s,(y_s)_n)\| ds + \left| \sum_{t_n<t_k<t_0} I_k(y_n(t_k)) \right|.
\end{aligned}$$

In addition, condition (H-6) implies that

$$\begin{aligned}
\|G_{t_0,\phi_0} y_n(t)\| &\le 2\|\phi_n - \phi_0\|_\infty + \|\phi_n(t_n) - \phi_n(t_0)\| + (d+\bar{d})|t_n - t_0| \\
&\quad + \int_{t_0}^{t_n} p(s)\rho(R) ds + \sum_{t_n<t_k<t_0} c_k \overline{\phi}(R).
\end{aligned}$$

Also, $\lim_{n\to\infty} \phi_n = \phi_0$ and $\lim_{n\to\infty} t_n = t_0$ imply that $\lim_{n\to\infty} G_{t_0,\phi_0}(y_n) = 0$. The set $A = \overline{\{G_{t_0,\phi_0}(y_n)\}}$ is a compact set and so is $G^{-1}_{t_0,\phi_0}(A)$. It is clear that $\{y_n\} \subset A$. Without loss of generality, we may assume that $\lim_{n\to\infty} y_n = y_0$, so $y_0 \in \widetilde{S}(t_0,\phi_0) \subset U$. But this is a contradiction to the assumption that $y_n \notin U$ for each n.

(b) Π is locally compact. For some $r > 0$, let

$$B \times I = \{(x,\alpha) \in S_{[-r,b]}(\phi) \times [0,1] : \|x\|_\Omega \le r\}$$

and $\{y_n\} \in \Pi(B \times I)$; then there exists $(\phi_n, \alpha_n) \in B \times I$ such that

$$y_n(t) = \begin{cases} x_n(t), & \text{for } -r \le t \le \alpha_n b, \\ z_n(t), & \text{for } \alpha b < t \le b,\ z_n \in S(f,\alpha_n b,\phi_n). \end{cases}$$

Since $S_{[-r,b]}(\phi)$ is compact, there exist subsequences of $\{x_n\}$ and $\{\alpha_n\}$ which converge to x and α respectively. Now $\widetilde{S}$ is u.s.c. so $\widetilde{S}([0,b] \times \{x_n\} \cup \{x\})$ is compact. Hence, there exists a subsequence of $\{z_n\} \in S(f, \alpha b, x)$. By the compactness of $S(f, \alpha b, x)$, there exists z such that thc subsequence $\{z_n\}$ converges to $z \in S(f, \alpha b, x)$. Therefore Π is locally compact.

Step 2. Π has a closed graph. Let $(x_n, \alpha_n) \longrightarrow (x_*, \alpha)$, $h_n \in \Pi(x_n, \alpha_n)$, and $h_n \longrightarrow h_*$ as $n \to +\infty$. We shall prove that $h_* \in \Pi(x_*, \alpha)$. Now $h_n \in \Pi(y_n, \alpha_n)$ means that there exists $z_n \in S(f, \alpha_n b, \phi_n)$ such that, for each $t \in J$,

$$h_n(t) = \begin{cases} \phi_n(t), & \text{for } -r \le t \le \alpha_n b, \\ z_n(t), & \text{for } \alpha_n b < t \le b. \end{cases}$$

We must prove that there exists $z_* \in S(f, \alpha b, x_*)$ such that, for each $t \in J$,

$$h_*(t) = \begin{cases} x_*(t), & \text{for } -r \le t \le \alpha b, \\ z_*(t), & \text{for } \alpha b < t \le b. \end{cases}$$

Clearly $(\alpha_n b, \phi_n) \to (\alpha, x_*)$ as $n \to \infty$ and we can easily show that there exists a subsequence $\{z_n\}$ converging to some limit z_*. The cases $\alpha = 0$ or $\alpha = 1$ can be treated as in the proof of Theorem 9.30. From the above arguments, we find that $z_* \in S(f, \alpha b, x)$, proving our claim.

Step 3. We claim that $\Pi(x, \alpha)$ is an R_δ-set for each fixed $\alpha \in [0,1]$ and $x \in S_{[-r,b]}$. Clearly $\Pi(x, \alpha) = S_{J_0 \cup J}(x)$ where $J_0 = [-r, \alpha b]$ and $J = [\alpha b, b]$. Since F is σ-Ca-selectionable, there exists a decreasing sequence of multivalued maps $F_k : [0,b] \times \mathcal{D} \to \mathcal{P}(E)$ $(k \in \mathbb{N})$ which have Carathéodory selections and satisfy

$$F_{k+1}(t,u) \subset F_k(t,u) \text{ for almost all } t \in [0,b],\ u \in \mathcal{D},$$

and

$$F(t,u) = \bigcap_{k=0}^{\infty} F_k(t,u),\ u \in \mathcal{D}.$$

Then,

$$\Pi(x, \alpha) = \bigcap_{k=0}^{\infty} S_{[-r,b]}(F_k, x).$$

When either condition (H-8) or (H-9) is satisfied, Theorem 8.6 implies that $\Pi(x, \alpha)$ and $S_{[-r,b]}(F_k, x)$ are compact sets. Moreover from Theorem 9.30, the sets $S_{[-r,b]}(F_k, x)$ are contractible sets. Therefore, $\Pi(x, \alpha)$ is an R_δ set.

As a consequence, all properties in Definition 9.12 are met. Therefore, the set $S_{[-r,b]}(\phi)$ is R_δ–contractible, completing the proof of the theorem. □

9.4.2 R_δ-sets

In this subsection, a few more results regarding the topological structure of the solution sets are derived.

Theorem 9.32. *Let $F : J \times \mathcal{D} \to \mathcal{P}_{cp,cv}(E)$ be a Carathéodory and σ-Ca-selectionable multivalued map and assume that conditions (H-6), (H-7), and (H-3) hold. Then the solution set $S_{[-r,b]}(\phi)$ is an R_δ-set.*

Proof. Since F is σ-Ca-selectionable, there exists a decreasing sequence of multivalued maps $F_k : [0,b] \times \mathcal{D} \to \mathcal{P}(E)$ $(k \in \mathbb{N})$ that have Carathéodory selections and satisfy

$$F_{k+1}(t,u) \subset F_k(t,x) \text{ for almost all } t \in [0,b],\ x \in \mathcal{D},$$

and

$$F(t,x) = \bigcap_{k=0}^{\infty} F_k(t,x),\ x \in \mathcal{D}.$$

Then,

$$S_{[-r,b]}(F,\phi) = \bigcap_{k=0}^{\infty} S_{[-r,b]}(F_k,\phi).$$

From Theorem 9.31, the set $S_{[-r,b]}(F_k,\phi)$ is contractible for each $k \in \mathbb{N}$. Hence, $S_{[-r,b]}(F,\phi)$ is an R_δ-set. □

Theorem 9.33. *Let F: $J \times \mathcal{D} \to \mathcal{P}_{cp,cv}(E)$ be a Carathéodory and a σ-mLL-selectionable map and assume that conditions* (H-6), (H-7), *and* (H-3) *hold. Then the solution set $S_{[-r,b]}(\phi)$ is an R_δ-set.*

Proof. It suffices to prove that F is $\sigma - mLL$-selectionable and then apply Theorem 9.30. □

Theorem 9.34. *Let F: $J \times \mathcal{D} \to \mathcal{P}_{cp,cv}(E)$ be upper-Scorza–Dragoni and assume that conditions* (H-6), (H-7), *and* (H-5) *hold with $\sum_{k=1}^{k=m} c_k < 1$. Then the solution set $S_{[-r,b]}(\phi)$ is an R_δ.*

Proof. Since F is upper-Scorza–Dragoni, by Theorem 9.24, F is a σ-Ca-selectionable map. Therefore, $S_{[-r,b]}(F,\phi)$ is an R_δ-set. □

Proposition 9.35. ([431]) *Let C be a closed, convex subset of a Banach space E and let $N : C \to \mathcal{P}_{cp,cv}(C)$ be a contraction multivalued map. Then* Fix(N) *is a nonempty, compact AR-space.*

Our new result in this direction is the following.

Theorem 9.36. *Let* $F: J \times \mathcal{D} \to \mathcal{P}_{cp,cv}(E)$ *be a multivalued map. Assume that* (H-3) *and*

(I-6) *The function* $F: J \times \mathcal{D} \to \mathcal{P}_{cp}(E)$ *satisfies that for fixed* y*, the multifunction* $t \mapsto F(t, y_t)$ *is measurable.*

(I-7) *There exists* $p \in L^1([0,b],(0,\infty))$ *such that*

$$H(F(t,z_1),F(t,z_2)) \leq p(t)\|z_1 - z_2\|_\infty \text{ for all } z_1, z_2 \in \mathcal{D}$$

$$\text{and } F(t,0) \subset p(t)B(0,1) \text{ for a.e. } t \in J.$$

Then the solution set $S_{[-r,b]}(\phi) \in AR$.

Proof. Consider the operator $N: \Omega \to \mathcal{P}(\Omega)$ defined for $y \in \Omega$ by

$$N(y) = \left\{ h \in \Omega : h(t) = \begin{cases} \phi(t), & t \in [-r,0], \\ \phi(0) - g(t,\phi) + g(t,y_t) + \int_0^t v(s)ds & \\ \quad + \sum_{0<t_k<t} I_k(y(t_k)), & t \in J, \end{cases} \right\}$$

where $v \in S_{F,y} = \{v \in L^1(J,E) : v(t) \in F(t,y(t)), \text{ a.e. } t \in J\}$. Using the fact that $F(\cdot,\cdot)$ has a convex and compact values and by (H-3), (I-6), and (I-7), for every $y \in \Omega$ we have $N(y) \in \mathcal{P}_{cv,cp}(\Omega)$. With a Bielecki-type norm on Ω, we can prove that N is a contraction. Hence, from Proposition 9.35, the solution set $S_{[-r,b]}(\phi) = Fix(N)$ is a nonempty, compact AR-space. □

9.5 Impulsive DIs on Proximate Retracts

Let K be a compact subset of $\mathbb{R}^n$. For a point $x \in K$, by

$$T_K(x) = \{v \in \mathbb{R}^n : \liminf_{t \to 0+} \frac{d(x + hv, K)}{h} = 0\},$$

we shall denote the (Bouligand) tangent cone to K at x. A compact subset $K \subset \mathbb{R}^n$ is called a *proximate retract* provided there exist an open neighborhood U of K in $\mathbb{R}^n$ and a retraction $r : U \to \mathbb{R}^n$ (called a metric retraction) such that

$$\|x - r(x)\| = \text{dist}(x, K), \text{ for every } x \in U.$$

It is well known that the class of all proximate retracts is quite rich; in particular, it contains convex sets and C^2–manifolds. It is easy to see that, for a given K, if $r : U \to K$ exists, then it is unique. Since one can take a sufficiently small U, for example by restricting U to $U \cap \{x \in \mathbb{R}^n : \text{dist}(x,K) < \delta\}$, $\delta > 0$, we may assume that $\|r(x) - x\| \leq \delta$, for a given $\delta > 0$ and $x \in U$. The following lemmas play an important role in our considerations

Lemma 9.37. ([79]) *Let K be a proximate retract. Then*

$$T_K(r(x)) \subseteq \{y \in \mathbb{R}^n : \langle y, y - r(x)\rangle \leq 0\}, \text{ for any } x \in U,$$

where $\langle \cdot, \cdot \rangle$ denotes the inner product in $\mathbb{R}^n$.

Lemma 9.38 ([79]). *Let K be a proximate retract, $r : U \to K$ a metric retraction, and $s > 0$ be such that $K \cap \overline{B(0,s)} \neq \emptyset$, where $\overline{B(0,s)}$ is the closure of $B(0,s)$ in $\mathbb{R}^n$. Then there exists $\epsilon_0 > 0$ such that, for any $0 < \epsilon \leq \epsilon_0$, there exist subsets $K \subset K_\epsilon \subset U_\epsilon$ of $\mathbb{R}^n$, K_ϵ closed and U_ϵ open, and a continuous retraction $r_\epsilon : U_\epsilon \to K_\epsilon$ such that the following conditions are satisfied:*

(i) $\bigcap_{0<\epsilon\leq\epsilon_0} K_\epsilon = K$;

(ii) $\|r_\epsilon(x) - x\| = \mathrm{dist}(x, K_\epsilon)$, *for all* $x \in U_\epsilon \cap \overline{B(O,s)}$;

(iii) $T_{K_\epsilon}(x) \subseteq \{y \in \mathbb{R}^n : \langle y, y - r(x)\rangle \leq 0\}$, *for any* $x \in K_\epsilon \cap \overline{B(0,s)}$.

Lemma 9.39. ([416, 417]) *Let K be a proximate retract, let U be an open neighborhood of K in $\mathbb{R}^n$, and let $r : U \to K$ be a metric retraction. Assume further that $\epsilon > 0$ is chosen in such a way that $\overline{O}_{2\epsilon}(K) \subset U$. Then we have:*

(i) $\overline{O}_{2\epsilon}(K)$ *is a approximate retract;*

(ii) $\{y \in \mathbb{R}^n : \langle y, x - r(x)\rangle \leq 0\} \subset T_{\overline{O}_\epsilon(K)}(x)$, *for all* $x \in \overline{O}_\epsilon(K)$;

(iii) $T_K(r(x)) \subset \{y \in \mathbb{R}^n : \langle y, x - r(x)\rangle \leq 0\}$, *for all* $x \in \overline{O}_\epsilon(K)$,

where

$$O_{2\epsilon}(K) = \{x \in \mathbb{R}^n : d(x,a) < \epsilon \text{ for some } a \in K\}.$$

9.5.1 Viable Solution

The concept of viability was considered in 1942 by Nagumo [386]. Let f be a bounded continuous function from a closed set K in $\mathbb{R}^n$ to $\mathbb{R}^n$. Nagumo's theorem states that a necessary and sufficient condition for the differential equation

$$y'(t) = f(y(t)),\ y(0) = y_0,$$

to have viable trajectories for all initial states $y_0 \in K$ is that

$$\text{for all } y \in K,\ f(y) \text{ belongs to the tangent cone to } K \text{ at } y.$$

The set-valued (upper semicontinuous case) is due to Haddad [262, 263]. The possibility to replace the tangency condition, i.e.,

$$F(t,x) \cap T_K(x) \neq \emptyset \text{ for all } J \times K,$$

by the convexified tangency conditions, $F(t,x) \cap \mathrm{conv}\, T_K(x) \neq \emptyset$ on $J \times K$, was noted by Guseinov, Subbotin and Ushakov [260, 261]. Let us also note that in the autonomous case (i.e., does not depend on t), the tangency condition is also necessary for the existence of solutions to the problem,

$$y'(t) \in F(t, y(t)), \quad \text{a.e.} \quad t \in J = [0, b], \tag{9.17}$$

$$y(0) = y_0 \in K. \tag{9.18}$$

Problem (9.17)–(9.18), in the case of a time-dependent constraint (i.e., the existence of a solution $y'(t) \in F(t, y(t))$ such that $y(0) = y_0 \in K(0)$, and $y(t) \in K(t)$, where the set-valued map $K : J \times \mathbb{R}^n$ has closed values), has been studied thoroughly in [204] for a closed convex-valued $\mathcal{L} \otimes B$-measurable $F(t, \cdot)$. Having a linear growth with upper semicontinuous $F(t, \cdot)$ and absolutely continuous $K(\cdot)$ (respectively in [203] for $F(\cdot, x)$ measurable, $F(t, \cdot)$ continuous and left absolutely continuous $K(\cdot)$), the appropriate tangency conditions were stated in terms of the convex hull of the Bouligand tangent cone to the graph of $K(\cdot)$; see Theorems 4.2 and 4.7 in [203] and Theorem 3.1 in [204]. A related result (under stronger assumptions) in the context of differential inclusions with delays has been established in [302, Theorems 2 and 3]. Some other issues of viability theory are studied in detail in [34, 35]. Very recently, Aubin et al. [36] considered the viable solution for hybrid systems of differential inclusions. In this subsection, we discuss the Cauchy problem for impulsive differential inclusions on proximate retracts. Our considerations are based on [34, 231, 234]. We give here natural generalizations of some results contained therein.

More precisely, we will consider problems of form,

$$y'(t) \in F(t, y(t)), \quad \text{a.e.} \quad t \in J = [0, b] \backslash \{t_1, \dots, t_m\}, \tag{9.19}$$

$$y(t_k^+) = I_k(y(t_k^-)), \quad k = 1, \dots, m, \tag{9.20}$$

$$y(0) = y_0 \in K, \tag{9.21}$$

where K is a proximate retract, $F : [0, b] \times K \to \mathcal{P}(\mathbb{R}^n)$, and $I_k \in C(K, K)$. Our method to solve (9.19)–(9.21) consists in solving the following associated problem:

$$y'(t) \in \widetilde{F}(t, y(t)), \quad \text{a.e.} \quad t \in J = [0, b] \backslash \{t_1, \dots, t_m\}, \tag{9.22}$$

$$y(t_k^+) = \widetilde{I}_k(y(t_k^-)), \quad k = 1, \dots, m, \tag{9.23}$$

$$y(0) = y_0 \in K, \tag{9.24}$$

where $\widetilde{F} : [0, b] \times \mathbb{R}^n \to \mathcal{P}(\mathbb{R}^n)$ and $\widetilde{I}_k : \mathbb{R}^n \to \mathbb{R}^n$ are given by

$$\widetilde{F}(t, y(t)) = \begin{cases} \alpha(y) F(t, r(y(t))), & \text{if } y \in U, \\ 0, & \text{if } y \notin U, \end{cases}$$

and

$$\widetilde{I}_k(y(t)) = \begin{cases} \alpha(y) I_k(r(y(t))), & \text{if } y \in U, \\ 0, & \text{if } y \notin U, \end{cases}$$

where $r : U \to K$ is the metric retraction and $\alpha : \mathbb{R}^n \to [0,1]$ is a continuous Uryshon function such that $\alpha|_K \equiv 1$ and $\alpha|_{\mathbb{R}^n \setminus K} \equiv 0$. Obviously, $\widetilde{F}$ is unique up to the choice of the Uryshon function α.

Definition 9.40. A multivalued $F : [0,b] \times \mathbb{R}^n \to \mathcal{P}(\mathbb{R}^n)$ is called a strong Carathéodory map on K provided that the following conditions are satisfied:

(i) $F(t,\cdot) : K \to \mathcal{P}_{cp,cv}(\mathbb{R}^n)$ is upper u.s.c., for a.e. $t \in [0,b]$;

(ii) $F(\cdot,x) \to \mathbb{R}^n$ has a measurable selector, for all $x \in K$;

(iii) for every bounded set $A \subset K$ and a.e. $t \in [0,b]$, the set $F(t,A)$ is relatively compact;

(iv) There exist a continuous function $\psi : [0,\infty) \to (0,\infty)$ and $p \in L^1([0,b],\mathbb{R}^n)$ such that

$$\|F(t,x)\| \leq p(t)\psi(\|x\|), \text{ for a.e. } t \in [0,b] \text{ and each } x \in K$$

with

$$\int_0^b p(s)ds < \int_{N_0}^{\infty} \frac{du}{\psi(u)}, k = 1,\dots,m,$$

and

$$N_0 = \|y_0\|.$$

Remark 9.41. If $F : [0,b] \times K \to \mathcal{P}(\mathbb{R}^n)$ is a strong Carathéodory map in K, then $\widetilde{F} : [0,b] \times \mathbb{R}^n \to \mathcal{P}_{cp,cv}(\mathbb{R}^n)$ is a strong Carathéodory map on $\mathbb{R}^n$.

The next lemma is crucial in what follows.

Lemma 9.42. *Let $F : [0,b] \times K \to \mathcal{P}(\mathbb{R}^n)$ be tangent to the proximate retract K in $\mathbb{R}^n$. If $y \in PC \cap \{AC(J_k,\mathbb{R}^n) : k = 1,\dots,m\}$ is a solution of the problem (9.22)–(9.23), then $y(t) \in K$, for all $t \in [0,b]$.*

Proof. Let $y \in PC \cap \{AC(J_k,\mathbb{R}^n) : k = 1,\dots,m\}$ and $d : [0,t_1] \to \mathbb{R}^n$ be defined by $d(t) = \operatorname{dist}(y(t),K)$. We want to show that $d(t) = 0$, for all $t \in [0,t_1]$. Since $y(0) = y_0 \in K$, $d(0) = 0$. We can easily prove that $d \in AC([0,t_1],\mathbb{R}^n)$. Let $t_0 \in (0,t_1]$ such that $y'(t_0) \neq 0$; then there exists $h > 0$ such that $y(t) \neq 0$, for every $t \in (0,t_1 - h)$,

$$\begin{aligned} d(t_0 + h) &= \operatorname{dist}(y(t_0) + hy'(t_0) + \circ(h), K) \\ &\leq \operatorname{dist}(r(y(t_0)) + hy'(t_0), K) + \|y(t_0) - r(y(t_0))\| + \circ(h) \\ &\leq \operatorname{dist}(r(y(t_0)) + hy'(t_0), K) + \operatorname{dist}(y(t_0), K) + \circ(h). \end{aligned}$$

Hence,

$$\frac{d(t_0+h) - d(t_0)}{h} \leq \frac{\operatorname{dist}(r(y(t_0)) + hy'(t_0), K) + \circ(h)}{h},$$

so

$$\lim_{h\to 0+} \frac{d(t_0+h)-d(t_0)}{h} \leq \lim_{h\to 0+} \frac{\text{dist}\,(r(y(t_0)) + hy'(t_0), K) + \circ(h)}{h} = 0,$$

so that $d'(t_0) \leq 0$. Hence,

$$d(t) = 0 \text{ for all } \ t \in [0, t_1] \text{ implies } y(t) \in K, \quad \text{for all } t \in [0, t_1].$$

- Let $t \in (t_1, t_2]$. If $t = t_1$, then $y(t_1^+) = \alpha(y(t_1))I_1(r(y(t_1)))$. It is clear that

$$I_1 \circ r : U \to K \to K.$$

Since $y(t_1) \in K$ and from the definition of α, we have

$$y(t_1^+) = I_1(y(t_1)) \ \text{ implies } \ d(t_1^+) = dist(I_1(r(y(t_1))), K) = 0.$$

In the case where $t \in (t_1, t_2]$, we can easily prove that $y(t) \in K$.

- We continue this process and take into account that y is the solution of the problem (9.22)–(9.24) on $(t_m, b]$ such that

$$y(t) \in K, \text{ for all } t \in (t_m, b].$$

Then

$$y(t) \in K, \text{ for every } t \in [0, b]. \qquad \square$$

Theorem 9.43. *Let $F : [0, b] \times K \to \mathcal{P}(\mathbb{R}^n)$ be tangent to a proximate retract K in $\mathbb{R}^n$ and assume that F is a strong Carathédory map. Then the IVP (9.19)–(9.21) has a viable solution and the solution is R_δ.*

Proof. From part (iv) of Definition 9.40 and the continuity of the I_k, we can prove that there exists $M > 0$ such that, for every solution y of the problem (9.19)–(9.21), we have

$$\sup\{\|y(t)\| : t \in [0, b]\} < M.$$

Notice that any solution of (9.22)–(9.24) is a viable solution of (9.19)–(9.21). To see this, observe that if y is a solution of (9.22)–(9.24), then $y(t) \in K$ for all $t \in [0, b]$ by Lemma 9.42. So $\widetilde{F}(t, y(t)) = \alpha(y(t))F(t, r(y(t)) = F(t, y(t))$. Conversely, if y is a viable solution of (9.19)–(9.21), then y is a solution of (9.22)–(9.24). Let $\epsilon > 0$ be given and let $\tau_\epsilon : \mathbb{R}^n \to [0, 1]$ be the Urysohn function for

$$(\overline{B}(0, M), \mathbb{R}^n \backslash B(0, M))$$

such that $\tau_\epsilon(y) = 1$ if $\|y\| \le M$, and $\tau_\epsilon(y) = 1$ if $\|y\| \ge M + \epsilon$. Let $\widetilde{F}_\epsilon(t, y) = \alpha_\epsilon(y)\widetilde{F}(t, y)$ and consider the differential inclusion

$$y'(t) \in \widetilde{F}_\epsilon(t, y(t)), \quad \text{a.e.} \quad t \in J = [0, b]\backslash\{t_1, \dots, t_m\}, \tag{9.25}$$

$$y(t_k^+) = \widetilde{I}_{\epsilon k}(y(t_k^-)), \qquad k = 1, \dots, m, \tag{9.26}$$

$$y(0) = y_0 \in K. \tag{9.27}$$

The proof will be given in several steps.

Step 1. Consider the problem,

$$y'(t) \in \widetilde{F}_\epsilon(t, y(t)), \quad \text{a.e.} \quad t \in [0, t_1], \tag{9.28}$$

$$y(0) = y_0. \tag{9.29}$$

We transform the problem (9.28)–(9.29) into a fixed point problem by defining the operator $N : C([0, t_1], \mathbb{R}^n) \longrightarrow \mathcal{P}_{cp,cv}(C([0, t_1], \mathbb{R}^n))$ by

$$N(y) = \left\{h \in C([0, t_1], \mathbb{R}^n) : h(t) = y_0 + \int_0^t v(s)ds, \ v \in S_{\widetilde{F}_\epsilon, y}\right\}.$$

Analogous to the proof of Theorem 6.2, we can prove that there exists $y_1 \in N(y_1)$ that is a solution of the problem (9.28)–(9.29). From Lemma 9.42, $y_1(t) \in K$ for all $t \in [0, t_1]$, so $\widetilde{I}_{\epsilon 1}(y_1(t_1)) = I_k(y_1(t_1))$.

Step 2. Next, consider the problem

$$y'(t) \in \widetilde{F}_\epsilon(t, y(t)), \quad \text{a.e.} \quad t \in (t_1, t_2], \tag{9.30}$$

$$y(t_1^+) = I_1(y_0(t_1)). \tag{9.31}$$

Let

$$C_1 = \{y \in C((t_1, t_2], \mathbb{R}^n) : y(t_1^+) \text{ exists}\}.$$

Set $C_* = C([0, t_1], \mathbb{R}^n) \cap C_1$. Let the operator $N_1 : C_* \to \mathcal{P}_{cp,cv}(C_*)$ be defined by:

$$N_1(y) = \left\{y \in C_* : h(t) = I_1(y_1(t_1^-)) + \int_{t_1}^t v(s)ds, \ v \in S_{\widetilde{F}, y}\right\}.$$

Again as in the proof of Theorem 6.2, we can show that N_1 has at least one fixed point that is a solution of the problem (9.30)–(9.31). Denote this solution by y_2.

Step 3. We continue this process and take into account that $y_m := y\big|_{[t_{m-1}, t_m]}$ is a solution to the problem

$$y'(t) \in \widetilde{F}(t, y(t)), \quad \text{a.e.} \quad t \in (t_m, b], \tag{9.32}$$

$$y(t_m^+) = I_m(y_{m-1}(t_m^-)). \tag{9.33}$$

The solution y of the problem (9.25)–(9.27) is then defined by

$$y(t) = \begin{cases} y_1(t), & \text{if } t \in [0, t_1], \\ y_2(t), & \text{if } t \in (t_1, t_2], \\ \dots \\ y_{m+1}(t), & \text{if } t \in (t_m, b]. \end{cases}$$

Now, we show that $S(\widetilde{F})$ is an R_δ–set. From Lemma 9.42, there exists $M > 0$ such that, for every y solution of the problem (9.25)–(9.27), we have $\|y\|_\infty < M$. We set

$$\widetilde{F}_M(t, y) = \begin{cases} \widetilde{F}(t, y), & \text{if } \|y\| \leq M \text{ and } t \in [0, b], \\ \widetilde{F}\left(t, \dfrac{My}{\|y\|}\right), & \text{if } \|y\| \geq M \text{ and } t \in [0, b]. \end{cases}$$

It is clear that $\widetilde{F}_M$ is an integrably bounded Carathédory map and $S(\widetilde{F}) = S(\widetilde{F}_M)$. As in [171] (see Lemma 1), there exists an almost upper semicontinuous, integrably bounded map $G : [0, b] \times \mathbb{R}^n \to \mathcal{P}_{cp,cv}(\mathbb{R}^n)$ such the $S(G) = S(\widetilde{F}_M)$. From results in [230], G is σ-mLL-selectionable, so there exists a sequence of multivalued maps $\{G_k\}_{k=1}^\infty$, such that

$$G_{k+1}(t, u) \subset G_k(t, u) \text{ for almost every } t \in [0, b],\ u \in \mathbb{R}^n,$$

and $G(t, u) = \cap_{k=1}^\infty G_k(t, u)$. By Theorems 8.6 and 9.30, $S(G_k)$ is compact and contractible, and so $S(F)$ is an R_δ–set. □

Theorem 9.44. *Let $F : [0, b] \times K \to \mathcal{P}_{cp,cv}(\mathbb{R}^n)$ be a Carathédory multifunction, weakly tangent to an approximate retract K to $\mathbb{R}^n$, and assume that all the conditions of Theorem 9.43 hold. Then the solution set of problem* (9.19)–(9.21) *is nonempty and an R_δ–set.*

Proof. Let $r : U \to K$ be the metric retraction. According to Lemma 9.38(i), we choose $\epsilon > 0$ such that $O_{2\epsilon}(K) \subset U$ and $\overline{O_{2\epsilon}(K)}$ is a proximate neighborhood retract. We consider the multivalued map $T : \overline{O_\epsilon(K)} \to \mathcal{P}_{cl}(\mathbb{R}^n)$ defined by

$$T(x) = \{y \in \mathbb{R}^n : \langle y, x - r(x) \rangle \leq 0\}.$$

We easily prove that T has a closed graph in $\overline{O_\epsilon(K)} \times \mathbb{R}^n$. The multivalued mapping

$$F_\epsilon : [0, b] \times \overline{O_\epsilon(K)} \to \mathcal{P}_{cl}(\mathbb{R}^n),$$

defined by

$$F_\epsilon(t, y) = F(t, r(y)) \cap T(r(y)),$$

is Carathéodory integrably bounded. From Lemma 9.39, F_ϵ satisfies the tangent to K condition. From Theorem 9.43, $S(F_\epsilon)$ is R_δ. Finally, we can observe that, for every $y_0 \in K$, we have

$$S(F) = \cap_{n=1}^\infty S(F_{\frac{1}{n}}).$$

Hence, $S(F)$ is an R_δ–set. □

9.6 Periodic Problems

In this section, we consider the impulsive periodic problem,

$$y'(t) \in \varphi(t, y(t)), \qquad \text{a.e. } t \in J\backslash\{t_1, \dots, t_m\}, \tag{9.34}$$

$$y(t_k^+) - y(t_k^-) = I_k(y(t_k^-)), \qquad k = 1, \dots, m, \tag{9.35}$$

$$y(0) = y(b), \tag{9.36}$$

where $\varphi : J \times \mathbb{R}^n \to \mathcal{P}(\mathbb{R}^n)$ is a multifunction. Some basic results in the theory of periodic boundary value problems for first order impulsive differential equations and inclusions may be found in [240, 389–391, 396] and the references therein. Our goal in this section is to give an existence result for the above problem by using topological degree theory combined with a Poincaré operator.

9.6.1 Poincaré Translation Operator

By Poincaré operators we mean the translation operator along the trajectories of the associated differential system, and the first return (or section) map defined on the cross section of the torus by means of the flow generated by the vector field. The translation operator is sometimes also called the Poincaré–Andronov, or Levinson, or simply the T-operator. In the classical theory (see [326, 496] and the references therein), both these operators are defined to be single-valued, when assuming, among other things, the uniqueness of solutions of initial value problems. In the absence of uniqueness, it is often possible to approximate the right-hand sides of the given systems by locally Lipschitzian ones (implying uniqueness already), and then apply a standard limiting argument. This might be, however, rather complicated and is impossible for discontinuous right-hand sides. On the other hand, set-valued analysis allows us to effectively handle such classically troublesome situations. For additional background details, see [15, 230].

Let $\varphi : J \times \mathbb{R}^n \to \mathcal{P}(\mathbb{R}^n)$ be a Carathédory map. We define a multivalued map

$$S_\varphi : \mathbb{R}^n \to \mathcal{P}(PC)$$

by

$$S_\varphi(x) = \{y \mid y(\cdot, x) \text{ is a solution of the problem satisfying } y(0, x) = x\}.$$

Consider the operator P_t defined by $P_t = \Psi \circ S_\varphi$, where

$$P_t : \mathbb{R}^n \xrightarrow{\;S_\varphi\;} \mathcal{P}(PC) \xrightarrow{\;\Psi_t\;} \mathcal{P}(\mathbb{R}^n)$$

and

$$\Psi_t(y) = y(0) - y(t).$$

Here, P_t is called the Poincaré translation map associated with the Cauchy problem

$$y'(t) \in \varphi(t, y(t)), \quad \text{a.e. } t \in J\backslash\{t_1, \ldots, t_m\}, \tag{9.37}$$
$$y(t_k^+) - y(t_k^-) = I_k(y(t_k^-)), \quad k = 1, \ldots, m, \tag{9.38}$$
$$y(0) = y_0 \in \mathbb{R}^n. \tag{9.39}$$

The following lemma is easily proved.

Lemma 9.45. *Let $\varphi : J \times \mathbb{R}^n \to \mathcal{P}_{cv,cp}(\mathbb{R}^n)$ be a Carathédory multfunction. Then the periodic problem* (9.34)–(9.36) *has a solution if and only if, for some $y_0 \in \mathbb{R}^n$, we have $0 \in P_b(y_0)$, where P_b is the Poincaré map associated with* (9.37)–(9.39).

9.6.2 Existence Result

Set

$$K^n(r) = K^n(x, r), \quad S^{n-1}(r) = \partial K^n(r), \quad \text{and} \quad P^n = \mathbb{R}^n\backslash\{0\},$$

where $K^n(x, r)$ is a closed ball in $\mathbb{R}^n$ with center x and radius r, and $\partial K^n(r)$ stands for the boundary of $K^n(r)$ in $\mathbb{R}^n$. For any $X \in ANR$-space, we set

$$J(K^n(r), X) = \big\{F : X \to \mathcal{P}(X) \mid F \text{ u.s.c with } R_\delta\text{-values}\big\}.$$

Moreover, for any continuous $f : X \to \mathbb{R}^n$, where $X \in ANR$, we set

$$J_f(K^n(r), X) = \big\{\varphi : K^n(r) \to \mathcal{P}(X) \mid \varphi = f \circ F \text{ for some } F \in J(K^n(r), X) \text{ and } \varphi(S^{n-1}(r)) \subset P^n\big\}.$$

Finally, we define

$$CJ(K^n(r), \mathbb{R}^n) = \cup\big\{J_f(K^n(r), \mathbb{R}^n) \mid f : X \to \mathbb{R}^n \text{ is continuous and } X \in ANR\big\}.$$

It is well-known (see [230]) that, for the multivalued maps in this class, the notion of topological degree is available. To define it, we need an appropriate concept of homotopy in $CJ(K^n(r), \mathbb{R}^n)$).

Definition 9.46. Let $\phi_1, \phi_2 \in CJ(K^n(r), \mathbb{R}^n)$ be two maps of the form

$$\phi_1 = f_1 \circ F_1 : K^n(r) \xrightarrow{F_1} \mathcal{P}(X) \xrightarrow{f_1} \mathbb{R}^n$$
$$\phi_2 = f_2 \circ F_2 : K^n(r) \xrightarrow{F_2} \mathcal{P}(X) \xrightarrow{f_2} \mathbb{R}^n.$$

We say that ϕ_1 and ϕ_2 are homotopic in $CJ(K^n(r), \mathbb{R}^n)$ if there exist an u.s.c. R_δ-valued homotopy $\mathcal{H} : [0, 1] \times K^n(r) \to \mathcal{P}(X)$ and a continuous homotopy $h : [0, 1] \times X \to \mathbb{R}^n$ satisfying

(i) $\mathcal{H}(0,u) = F_1(u)$, $\mathcal{H}(1,u) = F_2(u)$ for every $u \in K^n(r)$;

(ii) $h(0,x) = f_1(x)$, $h(1,x) = f_2(x)$ for every $x \in X$;

(iii) for every $(\lambda, u) \in [0,1] \times S^{n-1}(r)$ and $x \in \mathcal{H}(\lambda, u)$, we have $h(\lambda, x) \neq 0$.

The map $H : [0,1] \times K^n(r) \to \mathcal{P}(\mathbb{R}^n)$ given by

$$H(\lambda, u) = h(\lambda, \mathcal{H}(\lambda, u))$$

is called a homotopy in $CJ(K^n(r), \mathbb{R}^n)$ between ϕ_1 and ϕ_2.

Theorem 9.47 ([230]). *There exists a map* $\mathrm{Deg} : CJ(K^n(r), \mathbb{R}^n) \to \mathbb{Z}$, *called the topological degree function, satisfying the following properties:*

(I-8) *If* $\varphi \in CJ(K^n(r), \mathbb{R}^n)$ *is of the form* $\varphi = f \circ F$ *with* F *single valued and continuous, then* $\mathrm{Deg}(\varphi) = \deg(\varphi)$, *where* $\deg(\varphi)$ *stands for the ordinary Brouwer degree of the single valued continuous map* $\varphi : K^n(r) \to \mathbb{R}^n$.

(I-9) *If* $\mathrm{Deg}(\varphi) = 0$, *where* $\varphi \in CJ(K^n(r), \mathbb{R}^n)$, *then there exists* $u \in K^n(r)$ *such that* $0 \in \varphi(u)$.

(I-10) *If* $\varphi \in CJ(K^n(r), \mathbb{R}^n)$ *and* $\{u \in K^n(r) \mid 0 \in \varphi(u)\} \subset \mathrm{Int}K^n(r_0)$ *for some* $0 < r_0 < r$, *then the restriction* φ_0 *of* φ *to* $K^n(r_0)$ *is in* $CJ(K^n(r), \mathbb{R}^n)$ *and* $\mathrm{Deg}(\varphi_0) = \mathrm{Deg}(\varphi)$.

Let $A \subset \mathbb{R}^n$ and $B \subset \mathbb{R}^m$; $CJ_0(A, B)$ will denote the class of mappings

$$CJ_0(K^n(r), \mathbb{R}^n) = \big\{\varphi : A \to \mathcal{P}(B) \mid \varphi = f \circ F,\ \ F : A \to \mathcal{P}(X),\\ F \text{ is u.s.c. with } R_\delta\text{-values and } f : X \to B \text{ is continuous}\big\},$$

where $X \in ANR$.

The following theorem due to Gorniewicz [230] is essential in the proof of the main result in this section.

Theorem 9.48 (Nonlinear Alternative). *Assume that* $\varphi \in CJ_0(K^n(r), \mathbb{R}^n)$. *Then* φ *has at least one of the following properties:*

(i) $Fix(\varphi) \neq \emptyset$,

(ii) *there is an* $x \in S^{n-1}(r)$ *with* $x \in \lambda\varphi(x)$ *for some* $0 < \lambda < 1$.

The following lemma can be found in [230, 264].

Lemma 9.49. *If* $\varphi : X \to \mathcal{P}_{cp,cv}(\mathbb{R}^n)$ *is an u.s.c. multivalued map, then* φ *is* σ-LL-*selectionable.*

Let $a \in \mathbb{R}^n$. Set

$$S_\varphi = \big\{y \in PC :\ y \text{ is solution of the problem (9.34)–(9.35) with } y(0) = a\big\}.$$

Theorem 9.50. *Let* $\varphi\colon \mathbb{R}^n \to \mathcal{P}_{cp,cv}(\mathbb{R}^n)$ *be an u.s.c. multifunction. Assume that*

(I-11) *There exist functions* $p \in L^1(J, \mathbb{R}_+)$ *and* $\psi : \mathbb{R}^n \to \mathbb{R}_+$ *such that*

$$\|\varphi(t, y)\|_{\mathcal{P}} \leq p(t)\psi(|y|) \quad \textit{for each} \quad (t, y) \in J \times \mathbb{R}^n.$$

(I-12) *There exist constants* c_k^*, $b_k^* \in \mathbb{R}_+$ *and* $\alpha_k \in [0, 1)$ *such that*

$$|I_k(y)| \leq c_k^* + b_k^*|y|^{\alpha_k} \quad y \in \mathbb{R}^n.$$

(I-13) *There exists* $r > 0$ *such that*

$$\frac{r}{\psi(r)\|p\|_{L^1} + \sum\limits_{k=1}^{m} [c_k^* + b_k^* r^{\alpha_k}]} > 1.$$

Then the problem (9.34)–(9.36) *has at least one solution.*

Proof. From Lemma 9.49, φ is $\sigma-$LL-selectionable, and so from Theorem 9.30, S_φ is $R_\delta-$contractible. Set $A = B = \mathbb{R}^n$ and $X = PC \in ANR$. We will prove that

$$\Psi : PC \to \mathbb{R}^n \quad \text{defined by} \quad y \to \Psi(y) = y(0) - y(\cdot)$$

is a continuous map. Let $\{y_n\}$ be a sequence such that $y_n \to y$ in PC. Then,

$$|\Psi(y_n)(t) - \Psi(y)(t)| \leq 2\|y_n - y\|_{PC} \to 0 \text{ as } n \to \infty.$$

Hence,

$$P_b \in CJ_0(K^n(r), \mathbb{R}^n).$$

Suppose $a \in P_t(a) = \lambda(\Psi_t \circ S_\varphi)(a)$ for some $\lambda \in (0, 1)$. Then, there exists $y \in PC$ such that $y \in S_\varphi(a)$. This implies $y(0) = a$ and $a = \lambda(a - y(t))$, $a \in S^{n-1}(r)$. For $t \in J$, we have

$$\begin{aligned}
|a| &\leq \|y(t)\| \\
&\leq \int_0^t p(s)\psi(|y(s)|)ds + \sum_{k=1}^{m}[c_k^* + b_k^*|y(t_k)|^{\alpha_k}] \\
&\leq \psi(r)\int_0^b p(s)ds + \sum_{k=1}^{m}[c_k^* + b_k^* r^{\alpha_k}].
\end{aligned}$$

Hence,

$$\frac{|a|}{\psi(r)\|p\|_{L^1} + \sum\limits_{k=1}^{m} [c_k^* + b_k^* r^{\alpha_k}]} \leq 1.$$

This contradicts condition (I-13).

Next, we will show that S_φ is u.s.c. by proving that the graph

$$\Gamma_\varphi := \{(x, y) \mid y \in S_\varphi(x)\}$$

of S_φ is closed. Let $(x_n, y_n) \in \Gamma_\varphi$, i.e., $y_n \in S_\varphi(x_n)$, and let $(x_n, y_n) \to (x, y)$ as $n \to \infty$. Since $y_n \in S_\varphi(x_n)$, there exists $v_n \in L^1(J, \mathbb{R}^n)$ such that

$$y_n(t) = x_n + \int_0^t v_n(s)ds + \sum_{0<t_k<t} I_k(y_n(t_k)),\ t \in J.$$

Since (x_n, y_n) converges to (x, y), there exists $M > 0$ such that

$$|x_n| \leq M \quad \text{for all } n \in \mathbb{N}.$$

By using (I-11)–(I-12), we can easily prove that there exist $\overline{M} > 0$ such that

$$\|y_n\|_{PC} \leq \overline{M} \quad \text{for all } n \in \mathbb{N}.$$

From the definition of y_n, we have $y_n'(t) = v_n(t)$ a.e. $t \in J$, so

$$|v_n(t)| \leq p(t)\psi(M),\ t \in J.$$

Thus, $v_n(t) \in p(t)\psi(M)\overline{B}(0,1) := \mathcal{H}(t)$ a.e. $t \in J$. It is clear that $\mathcal{H} : J \to \mathcal{P}_{cp,cv}(\mathbb{R}^n)$ is a multivalued map that is integrably bounded. Since $\{v_n(\cdot) : n \geq 1\} \in \mathcal{H}(\cdot)$, we may pass to a subsequence if necessary to obtain that v_n converges weakly to v in $L^1_w(J, \mathbb{R}^n)$. Lemma 2.66 yields the existence of $\alpha_i^n \geq 0$, $i = n, \ldots, k(n)$ such that $\sum_{i=1}^{k(n)} \alpha_i^n = 1$ and the sequence of convex combinations $g_n(\cdot) = \sum_{i=1}^{k(n)} \alpha_i^n v_i(\cdot)$ converges strongly to v in L^1. Since φ takes convex values, using Lemma 2.64, we obtain

$$\begin{aligned}
v(t) &\in \bigcap_{n\geq 1} \overline{\{g_k(t) : k \geq n\}} \\
&\subset \bigcap_{n\geq 1} \overline{co}\{v_k(t),\ k \geq n\} \\
&\subset \bigcap_{n\geq 1} \overline{co}\Big\{\bigcup_{k\geq n} \varphi(t, y_k(t))\Big\} \\
&= \overline{co}(\limsup_{k\to\infty} \varphi(t, y_k(t))).
\end{aligned}$$

for a.e. $t \in J$. Since φ is u.s.c. with compact values and (9.16), then we have that $v(t) \in \overline{co}\,\varphi(t, y(t))$. Since $\varphi(\cdot,\cdot)$ has closed, convex values, we deduce that $v(t) \in \varphi(t, y(t))$, for a.e. $t \in J$. Let

$$z(t) = x + \int_0^t v(s)ds + \sum_{0<t_k<t} I_k(y(t_k)),\ t \in J.$$

Since the functions I_k, $k = 1, \dots, m$, are continuous, we obtain the estimates

$$\begin{aligned}\|y_n - z\|_{PC} \le |x_n - x| + \int_0^b |\bar{v}_n(s) - v(s)|ds \\ + \sum_{k=1}^{m} |I_k(y_n(t_k)) - I_k(y(t_k))|.\end{aligned}$$

The right-hand side of the above expression tends to 0 as $n \to +\infty$. Hence,

$$y(t) = x + \int_0^t v(s)ds + \sum_{0<t_k<t} I_k(y(t_k)), \; t \in J.$$

Thus, $y \in S_\varphi(x)$. Now, we show that S_φ maps bounded sets into relatively compact sets of PC. Let B be a bounded set in $\mathbb{R}^n$ and let $\{y_n\} \subset S_\varphi(B)$. Then there exists $\{x_n\} \subset B$ such that

$$y_n(t) = x_n + \int_0^t v_n(s)ds + \sum_{0<t_k<t} I_k(y_n(t_k)), \; t \in J,$$

where $v_n \in S_{\varphi, y_n}$, $n \in \mathbb{N}$. Since $\{x_n\}$ is a bounded sequence, there exists a subsequence of $\{x_n\}$ converging to x, so from (I-11)–(I-12), there exists $M_* > 0$ such that

$$\|y_n\|_{PC} \le M_*, \; n \in \mathbb{N}.$$

As in the proof of Theorem 6.2, we can show that $\{y_n : n \in \mathbb{N}\}$ is equicontinuous in PC. As a consequence of the Arzelá–Ascoli Theorem, we conclude that there exists a subsequence of $\{y_n\}$ converging to y in PC. By a similar argument to the one above, we can prove that

$$y(t) = x + \int_0^t v(s)ds + \sum_{0<t_k<t} I_k(y(t_k)), \; t \in J,$$

where $v \in S_{F,y}$. Thus, $y \in S_\varphi(x)$. This implies that S_φ is u.s.c.

As a consequence of the nonlinear alternative of Leray–Schauder type Theorem 9.48, we conclude that $Fix\ P_b \neq \emptyset$. This completes the proof of the theorem. □

9.7 Solution Set for Nonconvex Case

In this section, we study the topological structure of solution sets for a first-order impulsive evolution inclusion with initial conditions, where the right-hand side is not necessarily convex valued.

Given a real separable Banach space E with norm $\|\cdot\|$, consider the following problem

$$y'(t) - Ay(t) \in F(t, y(t)), \quad \text{a.e.} \quad t \in J\backslash\{t_1, \dots, t_m\}, \tag{9.40}$$
$$\Delta y_{t=t_k} = I_k(y(t_k^-)), \quad k = 1, \dots, m \tag{9.41}$$
$$y(0) = a, \tag{9.42}$$

where $0 = t_0 < t_1 < \cdots < t_m < t_{m+1} = b$, $J = [0, b]$, $F\colon J \times E \to \mathcal{P}(E)$ is a multifunction, and the operator A is the infinitesimal generator of a C_0 – semigroup $\{T(t)\}_{t\geq 0}$, $I_k \in C(E, E)\,(k = 1, \dots, m)$, and $\Delta y|_{t=t_k} = y(t_k^+) - y(t_k^-)$. The notations $y(t_k^+) = \lim_{h\to 0^+} y(t_k + h)$ and $y(t_k^-) = \lim_{h\to 0^+} y(t_k - h)$ stand for the right and the left limits of the function y at $t = t_k$, respectively.

9.7.1 Continuous Selection and AR of Solution Sets

We now present some proprieties of the solution set of the problem (9.40)–(9.42) where the right-hand side is not necessarily convex. For more information about this section, we cite [78]. Throughout this section, we assume that there exist $M > 0$ and $\omega > 0$ such that

$$\|T(t)\| \leq Me^{\omega t},\ t \geq 0.$$

First, we give some auxiliary lemmas.

Let $f \in L^1(J, E)$, and assume that the operator A generates a semigroup $T(t)$. We consider the following impulsive problem

$$y'(t) - Ay(t) = f(t), \quad \text{a.e.} \quad t \in J, \tag{9.43}$$
$$\Delta y_{t=t_k} = I_k(y(t_k^-)), \quad k = 1, \dots, \tag{9.44}$$
$$y(0) = a, \tag{9.45}$$

where A is the infinitesimal generator of a C_0–semigroup of the linear operator $T(t)$ on a separable Banach space E.

Denote the solution of the problem (9.43)–(9.45) by $y(a, f)$.

If

$$PC := \big\{y : J \to E : y \in AC((t_k, t_{k+1}), E),\ k = 0, \dots m-1\big\},$$

is endowed with

$$\|y\|_{PC} = \|y\|_\infty + \|y'\|_{L^1},$$

then $(PC, \|\cdot\|_{PC})$ is a Banach space.

Definition 9.51. A function $y(a, f) \in PC$ is called a mild solution of problem (9.43)–(9.45) if

$$y(a, f) = T(t)a + \int_0^t T(t-s)f(s)ds + \sum_{0<t_k<t} T(t-t_k)I_k(y(a, f)(t_k)), \quad t \in J.$$

Definition 9.52. A function $y \in PC$ is called a mild solution of problem (9.40) if there exists $f \in L^1(J, E)$ such that

$$y(t) = T(t)a + \int_0^t T(t-s)f(s)ds + \sum_{0<t_k<t} T(t-t_k)I_k(y(t_k)), \quad t \in J,$$

and

$$f(t) \in F(t, y(t)), \quad \text{a.e.} \quad t \in J.$$

Note that $\Omega_* = \{y(a, f) : (a, f) \in E \times L^1(J, E)\}$ is the appropriate space for mild solutions of the problem (9.43)–(9.45). We define the following operator $\overline{L}$: $E \times L^1(J, E) \to \Omega_*$ by

$$\overline{L}(v)(t) = \begin{cases} L_0(a,v)(t), & \text{if } t \in [0, t_1], \\ L_1(a,v)(t), & \text{if } t \in (t_1, t_2], \\ \dots & \dots \\ L_{m-1}(a,v)(t), & \text{if } t \in (t_{m-1}, t_m], \\ L_m(a,v)(t), & \text{if } t \in (t_m, b], \end{cases} \tag{9.46}$$

where

$$\begin{aligned} L_0(a,v)(t) &= T(t)a + \int_0^t T(t-s)v(s)ds, \quad t \in [0, t_1] \\ L_1(a,v)(t) &= T(t-t_1)[L_0(v)(t_1) + I_1(L_0(v)(t_1))] \\ &\quad + \int_{t_1}^t T(t-s)v(s)ds, \quad t \in (t_1, t_2] \\ L_2(a,v)(t) &= T(t-t_2)[L_1(v)(t_2) + I_2(L_1(v)(t_2))] \\ &\quad + \int_{t_2}^t T(t-s)v(s)ds, \quad t \in (t_2, t_3] \\ &\dots \\ L_{m-1}(a,v)(t) &= T(t-t_m)[L_{m-2}(v)(t_{m-1}) + I_{m-1}(L_{m-2}(v)(t_{m-1}))] \\ &\quad + \int_{t_m}^t (T(t-s)v(s)ds, \quad t \in (t_m, b]. \end{aligned}$$

From (9.46), we can easily check that

$$\begin{aligned} \overline{L}(a,v)(t) &= T(t)a + \sum_{0<t_k<t} T(t-s)I_k(L_{k-1}(v)(t_k)) \\ &\quad + \int_0^t T(t-s)v(s)ds, \quad \text{a.e.} \quad t \in J. \end{aligned}$$

Lemma 9.53. *The map $\overline{L} : E \times L^1(J, E) \to \Omega$ defined by $\overline{L}(a, v) = y(a, v)$ is one-to-one, where $y(.,.)$ is solution of the problem* (9.43)–(9.45).

Proof. For the construction of the operator $\overline{L}$ we have $\overline{L}(a,v) = y(a,v)$. Now, we prove that $\overline{L}$ is one-to-one.

Let $v_1, v_2 \in L^1(J,E)$ be such that $\overline{L}(v_1) = \overline{L}(v_2)$. For $t \in [0,t_1]$, we have

$$\int_0^t T(t-s)v_1(s)ds = \int_0^t T(t-s)v_2(s)ds.$$

For $t \in (t_1,t_2]$ we have

$$\begin{aligned} &T(t-t_1)[L_0(v_1)(t_1) + I_1(L_0(v_1)(t_1))] + \int_{t_1}^t T(t-s)v_1(s)ds \\ &\quad = T(t-t_2)[L_0(v_2)(t_1) + I_1(L_0(v_2)(t_1))] + \int_{t_1}^t T(t-s)v_2(s)ds. \end{aligned}$$

For $t \in (t_2,t_3]$ we have

$$\begin{aligned} &T(t-t_2)[L_1(v_1)(t_2) + I_2(L_1(v_1)(t_2))] + \int_{t_2}^t T(t-s)v_1(s)ds \\ &\quad = T(t-t_2)[L_1(v_2)(t_2) + I_2(L_1(v_2)(t_2))] + \int_{t_2}^t T(t-s)v_2(s)ds. \end{aligned}$$

We continue this process and get for $t \in (t_m, b]$

$$\begin{aligned} &T(t-t_m)[L_{m-1}(v_1)(t_{m-1}) + I_{m-1}(L_{m-1}(v_1)(t_{m-1}))] + \int_{t_{m-1}}^t T(t-s)v_1(s)ds \\ &\quad = T(t-t_m)[L_{m-1}(v_2)(t_{m-1}) + I_{m-1}(L_{m-1}(v_2)(t_{m-1}))] \\ &\qquad + \int_{t_{m-1}}^t T(t-s)v_2(s)ds. \end{aligned}$$

Let $t \in J_1 = [0,t_1]$, then

$$\int_0^t T(t-s)[v_1(s) - v_2(s)]ds = 0. \tag{9.47}$$

Thus, for each $h > 0$, we have

$$\int_0^t T(t+h-s)[v_1(s) - v_2(s)]ds = 0. \tag{9.48}$$

From (9.47) and (9.48) we obtain

$$\begin{aligned} &\int_t^{t+h} T(t+h-s)[v_1(s) - v_2(s)]ds = \\ &\int_0^t T(t+h-s)[v_1(s) - v_2(s)]ds + \int_t^{t+h} T(t+h-s)[v_1(s) - v_2(s)]ds = \\ &\qquad \int_0^{t+h} T(t+h-s)[v_1(s) - v_2(s)]ds = 0. \end{aligned}$$

Hence

$$\frac{1}{h}\int_t^{t+h} T(t+h-s)[v_1(s)-v_2(s)]ds = 0. \tag{9.49}$$

Let J_1^* be the set of all $t \in (0, t_1)$ such that

$$\lim_{h\to 0}\frac{1}{2h}\int_{t-h}^{t+h} \|v_1(t)-v_2(t)-v_1(s)+v_2(s)\|ds = 0, \tag{9.50}$$

and we observe that $J_1 \backslash J_1^*$ has Lebesgue measure $m(J_1\backslash J_1^*) = 0$ (see [186] p. 217). Let $t \in J_1^*$ be arbitrary. We have

$$\begin{aligned}\frac{1}{h}\int_t^{t+h} & T(t+h-s)[v_1(s)-v_2(s)]ds\\ &= \frac{1}{h}\int_t^{t+h} T(t+h-s)[v_1(s)-v_2(s)-v_1(t)+v_2(t)]ds\\ &\qquad + \frac{1}{h}\int_t^{t+h} T(t+h-s)[v_1(t)-v_2(t)]ds.\end{aligned}$$

By (9.49), we have

$$\begin{aligned}\frac{1}{h}\int_t^{t+h} & T(t+h-s)[v_1(t)-v_2(t)-v_1(s)+v_2(s)]ds\\ &= \frac{1}{h}\int_t^{t+h} T(t+h-s)[v_1(t)-v_2(t)]ds.\end{aligned}$$

Set

$$\mu(h) = \frac{1}{h}\int_0^h T(s)[v_1(t)-v_2(t)]ds.$$

Then

$$\begin{aligned}\mu(h) &\le \frac{1}{h}\int_t^{t+h} \|T(t+h-s)\|_{B(E)}\|v_1(t)-v_2(t)-v_1(s)+v_2(s)\|ds\\ &\le \frac{1}{h}\int_t^{t+h} \|T(t+h-s)\|_{B(E)}\|v_1(t)-v_2(t)-v_1(s)+v_2(s)\|ds\\ &\le \frac{Me^{2\omega h}}{h}\int_{t-h}^{t+h} \|v_1(t)-v_2(t)-v_1(s)+v_2(s)\|ds.\end{aligned}$$

Hence

$$\mu(h) \le \frac{2Me^{2\omega h}}{2h}\int_{t-h}^{t+h} \|v_1(t)-v_2(t)-v_1(s)+v_2(s)\|ds.$$

In view of (9.50) we conclude that

$$\lim_{h\to 0}\mu(h) = 0 \Rightarrow v_1(t)-v_2(t) = 0.$$

Since t is arbitrary and $m(J_1 \backslash J_1^*) = 0$, we have

$$v_1 - v_2 = 0 \Rightarrow v_1 = v_2 \text{ on } J_1.$$

For $t \in (t_1, t_2]$ we have

$$\int_{t_1}^{t} T(t-s)[v_1(s) - v_2(s)]ds, \qquad t \in (t_1, t_2] = J_2. \tag{9.51}$$

Thus, for each $h > 0$, we have

$$\int_{t_1}^{t} T(t+h-s+t_1)[v_1(s) - v_2(s)]ds = 0. \tag{9.52}$$

From (9.51) and (9.52) we obtain

$$\begin{aligned}
&\int_{t}^{t+h} T(t+h-s+t_1)[v_1(s) - v_2(s)]ds \\
&= \int_{t_1}^{t} T(t+h-s+t_1)[v_1(s) - v_2(s)]ds + \int_{t}^{t+h} T(t+h-s+t_1)[v_1(s) - v_2(s)]ds \\
&= \int_{t_1}^{t+h} T(t+h-s+t_1)[v_1(s) - v_2(s)]ds = 0.
\end{aligned}$$

Hence

$$\frac{1}{h}\int_{t}^{t+h} T(t+h-s+t_1)[v_1(s) - v_2(s)]ds = 0. \tag{9.53}$$

Let J_2^* be the set of all $t \in (t_1, t_2)$ such that

$$\lim_{h \to 0} \frac{1}{2h} \int_{t-h}^{t+h} \|v_1(t) - v_2(t) - v_1(s) + v_2(s)\| ds = 0, \tag{9.54}$$

and we observe that $J_2 \backslash J_2^*$ has Lebesgue measure $m(J_2 \backslash J_2^*) = 0$ (see [186] p. 217). Let $t \in J_2^*$ be arbitrary. We have

$$\begin{aligned}
&\frac{1}{h}\int_{t}^{t+h} T(t+h-s+t_1)[v_1(s) - v_2(s)]ds \\
&\qquad = \frac{1}{h}\int_{t}^{t+h} T(t+h-s)[v_1(s) - v_2(s) - v_1(t) + v_2(t)]ds \\
&\qquad\qquad + \frac{1}{h}\int_{t}^{t+h} T(t+h-s+t_1)[v_1(t) - v_2(t)]ds.
\end{aligned}$$

By (9.53), we have

$$\begin{aligned}
&\frac{1}{h}\int_{t}^{t+h} T(t+h-s+t_1)[v_1(t) - v_2(t) - v_1(s) + v_2(s)]ds = \\
&\qquad\qquad \frac{1}{h}\int_{t}^{t+h} T(t+h-s+t_1)[v_1(t) - v_2(t)]ds.
\end{aligned}$$

Set

$$\mu_1(h) = \frac{1}{h}\int_{t_1}^{t_1+h} T(s)[v_1(t) - v_2(t)]ds.$$

$$\mu_1(h) \le \frac{1}{h}\int_{t}^{t+h} \|T(t+h-s+t_1)\|_{B(E)}\|v_1(t) - v_2(t) - v_1(s) + v_2(s)\|ds$$
$$\le \frac{1}{h}\int_{t}^{t+h} \|T(t+h-s+t_1)\|_{B(E)}\|v_1(t) - v_2(t) - v_1(s) + v_2(s)\|ds$$
$$\le \frac{Me^{2\omega h}}{h}\int_{t-h}^{t+h} \|v_1(t) - v_2(t) - v_1(s) + v_2(s)\|ds.$$

Hence

$$\mu_1(h) \le \frac{2Me^{2\omega(h+t_1)}}{2h}\int_{t-h}^{t+h} \|v_1(t) - v_2(t) - v_1(s) + v_2(s)\|ds.$$

In view of (9.54) we conclude that

$$\lim_{h\to 0} \mu(h) = 0 \Rightarrow v_1(t) - v_2(t) = 0.$$

Since t is arbitrary and $m(J_2 \backslash J_2^*) = 0$, we have

$$v_1 - v_2 = 0 \Rightarrow v_1 = v_2 \text{ on } J_2.$$

We continue this process until we arrive at

$$v_1 = v_2 \text{ on } J_m.$$

This implies that

$$v_1 = v_2 \text{ on } J.$$ □

For each $(a, f) \in E \times L^1(J, E)$ we denote by $y(a, f) \in \Omega_*$ the solution of the problem (9.43)–(9.45). Then (Ω_*, d) is a metric space with distance

$$d(y(a_1, f_1), y(a_2, f_2)) = \|y(a_1, f_1) - y(a_2, f_2)\|_{PC} + \|f_1 - f_2\|_{L^1}. \quad (9.55)$$

Lemma 9.54. *Ω_* is a complete metric space.*

Proof. Since $\overline{L}$ is one-to-one, the distance (9.55) is well-defined. It remains to show that Ω_* with the distance d, is complete. Let $\{y_n(a_n, f_n)\}$ be a Cauchy sequence in Ω_*. For $n, m \in \mathbb{N}$, we have

$$d(y_n(a_n, f_n), y_m(a_m, f_m)) = \|y_n(a_n, f_n) - y_m(a_m, f_m)\|_{PC} + \|f_n - f_m\|_{L^1}.$$

Since L^1 is a Banach space, there exists $f \in L^1$ such that

$$\|f_n - f\|_{L^1} \to 0 \text{ as } n \to \infty.$$

From the definition of y_n, we have

$$\|a_n - a_m\| = \|T(0)a_n - T(0)a_m\| \le \|y_n(a_n, f_n) - y_m(a_m, f_m)\|.$$

Thus $\{a_n\}_{n\ge 1}$ is a Cauchy sequence in E. Then there exists $a \in E$ such that

$$\|a_n - a\| \to 0 \text{ as } n \to \infty.$$

We consider the following impulsive problem

$$y'(t) - Ay(t) = f(t), \quad \text{a.e.} \quad t \in J, \tag{9.56}$$
$$\Delta y_{t=t_k} = I_k(y(t_k^-)), \qquad k = 1, \dots, \tag{9.57}$$
$$y(0) = a. \tag{9.58}$$

The solution of problem (9.56)–(9.56) is defined by

$$y(a, f) = T(t)a + \int_0^t T(t-s)f(s)ds + \sum_{0<t_k<t} T(t-t_k)I_k(y(a, f)(t_k)), \; t \in J.$$

Now we prove that the sequence $\{y_n\}_{n\ge 1}$ converges to $y(a, f)$. Let $t \in [0, t_1]$. We have

$$\|y_n(a_n, f_n)(t) - y(a, f)(t)\| \le Me^{\omega b}\|a_n - a\| + Me^{\omega b}\int_0^b \|f_2(s) - f_1(s)\|ds.$$

Then

$$\sup_{t\in[0,t_1]} \|y_n(a_n, f_n)(t) - y(a, f)(t)\| \le \\ Me^{\omega b}\|a_n - a\| + Me^{\omega b}\int_0^b \|f_n(s) - f(s)\|ds \to 0 \text{ as } n \to \infty.$$

This implies that

$$\|y_n(a_n, f_n)(t_1) - y(a, f)(t_1)\| \to 0 \text{ as } n \to \infty,$$

and by the continuity of I_1, we have

$$\|y_n(a_n, f_n)(t_1^+) - y(a, f)(t_1^+)\| \le \|y_n(a_n, f_n)(t_1) - y(a, f)(t_1)\| \\ + \|I_1(y_n(a_n, f_n)(t_1)) - I_1(y(a, f)(t_1))\| \to 0 \text{ as } n \to \infty.$$

For $t \in (t_1, t_2]$ we have

$$\sup_{t\in[t_1,t_2]} \|y_n(a_n, f_n)(t) - y(a, f)(t)\| \le \\ Me^{\omega b}\|y_n(a_n, f_n)(t_1) - y(a, f)(t_1)\| + Me^{\omega b}\|I_1(y_n(a_n, f_n)(t_1)) \\ - I_1(y(a, f)(t_1))\| + Me^{\omega b}\int_0^b \|f_n(s) - f(s)\|ds \to 0 \text{ as } n \to \infty.$$

We continue this process until we arrive

$$\sup_{t\in[t_m,b]} \|y_n(a_n, f_n)(t) - y(a, f)(t)\| \leq$$
$$Me^{\omega b}\|y_n(a_n, f_n)(t_m) - y(a, f)(t_m)\| + Me^{\omega b}\|I_m(y_n(a_n, f_n)(t_m))$$
$$- I_m(y(a, f)(t_m))\| + Me^{\omega b}\int_0^b \|f_n(s) - f(s)\|ds \to 0 \text{ as } n \to \infty.$$

Hence $\{y_n(a_n, f_n)\}_{n\geq 1}$ converges to $y(a, f)$ in Ω. □

Let

$$G(a, f) = \{v \in L^1(J, E) : v(t) \in F(t, y(a, f)(t)) \text{ a.e. } t \in J\},$$

and

$$G_*(a) = \{f \in L^1(J, E) : f \in G(a, f)\}.$$

Lemma 9.55. *Assume that the multivalued map $F : J \times E \to \mathcal{P}_{cp}(E)$ is such that $t \to F(t, \cdot)$ is measurable and*

($\overline{\mathcal{H}}_1$) *There exists $p \in L^1(J, \mathbb{R}^+)$ such that*

$$H(F(t,x), F(t,y)) \leq p(t)\|x - y\|, \text{ for } x, y \in E \text{ and a.e. } t \in J_k$$

and

$$F(t,0) \subset p(t)\overline{B}(0,1), \text{ for a.e. } t \in J,$$

hold. Then F is of lower semicontinuous type and $G : E \times L^1(J,E) \to \mathcal{P}(DL^1(J,E))$.

Remark 9.56. ($\overline{\mathcal{H}}_1$) implies that the multifunction F has at most the linear growth

$$\|F(t,x)\|_{\mathcal{P}} \leq p(t)(1 + \|x\|), \ p \in L^1(J, \mathbb{R}^+), \ \text{a.e. } t \in J, \ x \in E.$$

Proof of Lemma 9.55. The multivalued map $t \to F(t,\cdot)$ is measurable and $x \to F(\cdot, x)$ is H-continuous. In addition $F(\cdot,\cdot)$ has compact values; hence $G(\cdot,\cdot)$ is measurable, continuous. Since the measurable multifunction F is integrably bounded, Lemma 2.84 implies that the Nemyts'kiĭ operator $\mathcal{F}$ has decomposable values. Let $(a, f) \in E \times L^1(J,E)$, then there exists a unique $y(a, f) \in PC$ such that $y(a, f)$ is a solution of the problem (9.43)–(9.45). By the definition of $\mathcal{F}$ and G we have $\mathcal{F}(y(a, f)) = G(a, f)$. Then for every $(a, f) \in E \times L^1(J,E)$, we obtain that $G(a, f) \in \mathcal{P}(DL^1(J,E))$. □

Theorem 9.57. *$F : J \times E \to \mathcal{P}_{cp}(E)$; $t \to F(t,x)$ is measurable, for each $x \in E$. Assume that ($\overline{\mathcal{H}}_1$) and the following condition*

($\overline{\mathcal{H}}_2$) *there exist* $c_k > 0$, $k = 1, \ldots, m$ *such that*

$$\|I_k(x) - I_k(y)\| \leq c_k \|x - y\| \text{ for every } x, y \in E$$

hold. Then there exists a continuous function $L_* : \overline{\Omega} \to PC$ *satisfying, for each* $a \in E$*, the following properties:*

$$L_*(a, y) \in S(a), \quad \text{for each } y \in \tilde{\Omega}, \quad \text{and}$$
$$L_*(a, f) = y, \qquad \text{for each } y \in S(a),$$

where

$$\overline{\Omega} = \{(a, y) \in E \times \Omega_* : a \in E, \ y \in \tilde{\Omega}\}$$

and

$$\tilde{\Omega} = \{y \in \Omega_* : y(0) = a\}.$$

Remark 9.58. By conditions of Theorem 9.57 the solution set of problem (9.40)–(9.42) is not empty. For the proof, we can see [175, 279].

Proof. From Lemma 9.55, we have that, for every $(a, f) \in E \times L^1(J, E)$, $G(a, f) \in \mathcal{P}(DL^1(J, E))$ with decomposable values. Now we prove that G is Hausdorff continuous. Let (a_n, f_n) be a sequence in $E \times L^1(J, E)$ converging to (a, f). Let $\epsilon > 0$ and $g_n \in G(a_n, f_n)$. ($\overline{\mathcal{H}}_1$) tells us that

$$H(F(t, y(a, f)(t)), F(t, y_n(a_n, f_n)(t))) \leq p(t)\|y_n(a_n, f_n)(t) - y(a, f)(t)\| + \frac{\epsilon}{2}, \quad \text{a.e.} \quad t \in J.$$

Hence there is $w \in F(t, y(a, f)(t))$ such that

$$\|g_n(t) - w\| \leq p(t)\|y(a, f)(t) - y(a_n, f_n)(t)\| + \frac{\epsilon}{2}, \quad t \in J.$$

Consider the mapping $U_n : J \to \mathcal{P}(E)$ defined by

$$U_n(t) = \left\{w \in E : \|f(t) - w\| \leq p(t)\|y(a_n, f_n)(t) - y(a, f)(t)\| + \frac{\epsilon}{2}\right\}, \quad t \in J,$$

that is, $U_n(t) = \overline{\mathcal{B}}(g_n(t), p(t)|y(a_n, f_n)(t) - y(a, f)(t)\| + \frac{\epsilon}{2})$. Since $g, l, y, \bar{y}$ are measurable, Theorem III.4.1 in [136] tells us that the closed ball U is measurable.

Finally the set $V(t) = U(t) \cap F(t, \overline{y}(t))$ is nonempty since it contains w. Therefore the intersection multivalued operator V is measurable with nonempty, closed values. By Lemma 2.73, there exists a function $\overline{g}_n(t)$, which is a measurable selection for V. Thus $\overline{g}_n(t) \in F(t, y(a, f)(t))$ and

$$\|g_n(t) - \overline{g}_n(t)\| \leq p(t)\|y(a_n, f_n)(t) - y(a, f)(t)\| + \frac{\epsilon}{2}, \quad \text{for a.e. } t \in J.$$

Thus for a.e. $t \in J$,

$$\begin{aligned}\|g_n(t) - \overline{g}_n(t)\| &\le p(t)\|y(a_n, f_n)(t) - y(a, f)(t)\|ds + \frac{\epsilon}{2}\\ &\le Me^{\omega b}p(t)\|a_n - a\| + Me^{\omega b}p(t)\|f_n - f\|_{L^1}\\ &\quad + Me^{\omega b}p(t)\sum_{k=1}^{m}\|I_k(y(a_n, f_n))(t_k) - I_k(y(a, f))(t_k)\| + \frac{\epsilon}{2}.\end{aligned}$$

Hence

$$\begin{aligned}\|g_n - \overline{g}_n\|_{L^1} &\le Me^{\omega b}\|p\|_{L^1}\|a_n - a\| + Me^{\omega b}\|p\|_{L^1}\|f_n - f\|_{L^1}\\ &\quad + Me^{\omega b}\|p\|_{L^1}\sum_{k=1}^{m}\|I_k(y(a_n, f_n))(t_k) - I_k(y(a, f))(t_k)\| + b\frac{\epsilon}{2}.\end{aligned}$$

By an analogous relation, obtained by interchanging the roles of $y(a_n, f_n)$ and $y(a, f)$, we finally arrive at

$$\begin{aligned}H(G(a_n, f_n), G(a, f)) &\le Me^{\omega b}\|p\|_{L^1}\|a_n - a\| + Me^{\omega b}\|p\|_{L^1}\|f_n - f\|_{L^1}\\ &\quad + Me^{\omega b}\|p\|_{L^1}\sum_{k=1}^{m}\|I_k(y(a_n, f_n))(t_k)\\ &\quad - I_k(y(a, f))(t_k)\| + b\frac{\epsilon}{2}.\end{aligned}$$

Using the fact that I_k, $k = 1, \ldots, m$, are continuous functions, we have

$$H(G(a_n, f_n), G(a, f)) \to 0 \text{ as } n \to \infty.$$

Now, we show that $a \to G(a, .)$ is $H-$ Lipschitz. Let $f_1, f_2 \in L^1(J, E)$. Denote by $y(a, f_1)$ and $y(a, f_2)$ the mild solutions of the problem (9.43)-(9.45), respectively. Thus for $t \in [0, t_1]$, we have

$$\|y(a, f_1)(t) - y(a, f_2)(t)\| \le Me^{\omega t_1}\int_0^t \|f_1(s) - f_2(s)\|ds. \tag{9.59}$$

From $(\overline{\mathcal{H}}_2)$ we obtain

$$\|I_1(y(a, f_1)(t_1)) - I_1(y(a, f_2)(t_1))\| \le c_1 Me^{\omega t_1}\int_0^{t_1} \|f_1(s) - f_2(s)\|ds. \tag{9.60}$$

For $t \in (t_1, t_2]$, we have

$$\begin{aligned}\|y(a_1, f_1)(t) - y(a_2, f_2)(t)\| &\le \|y(a, f_1)(t_1) - y(a, f_2)(t_1)\|\\ &\quad + \|I_1(y(a, f_1)(t_1)) - I_1(y(a, f_2)(t_1))\|\\ &\quad + \int_{t_1}^t \|T(t-s)\|_{B(E)}\|f_1(s) - f_2(s)\|ds\end{aligned}$$

$$\leq Me^{\omega t}\int_0^t \|f_1(s)-f_2(s)\|ds$$
$$+ c_1 Me^{\omega t_1}\int_0^{t_1}\|f_1(s)-f_2(s)\|ds.$$

Hence, for $t\in(t_{p-1},t_p],\ p=1,\dots,m$, we have

$$\|y(a_1,f_1)(t)-y(a_2,f_2)(t)\|\leq Me^{\omega t}\int_0^t \|f_1(s)-f_2(s)\|ds+\sum_{k=1}^{p-1}c_k Me^{\omega t_k}\times\int_{t_{k-1}}^{t_k}\|f_1(s)-f_2(s)\|ds. \tag{9.61}$$

Let $v_1\in G(a,f_1)$ and $\epsilon>0$. The function

$$F(t,y(a,f_2))\cap B(v_1(t),p(t)\|y(a,f_1)-y(a,f_2)\|+\frac{\epsilon}{b})$$

is measurable. Then there exists $v_2\in F(t,y(a,f_2))$ such that

$$\|v_1(t)-v_2(t)\|\leq p(t)\|y(a,f_1)-y(a,f_2)\|+\frac{\epsilon}{b}.$$

From (9.61), we have

$$\|v_1(t)-v_2(t)\|\leq p(t)Me^{\omega t}\int_0^t\|f_1(s)-f_2(s)\|ds$$
$$+\,p(t)\sum_{k=1}^{p-1}c_k Me^{\omega t_k}\int_{t_{k-1}}^{t_k}\|f_1(s)-f_2(s)\|ds+\frac{\epsilon}{b}.$$

Since $t_{p-1}<t$, we have

$$\|v_1(t)-v_2(t)\|\leq p(t)\left(Me^{\omega b}+\sum_{k=1}^{m}c_k e^{\omega t_k}\right)\int_0^t\|f_1(s)-f_2(s)\|ds+\frac{\epsilon}{b}. \tag{9.62}$$

Let $\tau>Me^{\omega b}+\sum_{k=1}^m c_k e^{\omega t_k}=D$, $m(t)=\int_0^t p(s)ds$, and let $L^1(J,E)$ be the weighted space of Lebesgue measurable functions such that

$$\int_0^b\|v(t)\|e^{-\tau m(t)}dt<\infty.$$

Endowed with the norm,

$$\|v\|_1=\int_0^b\|v(t)\|e^{-\tau m(t)}dt,\ \text{ for } v\in L^1(J,E),$$

$L^1(J,E)$ becomes a Banach space. From (9.62), we have

$$\begin{aligned}
\|v_1 - v_2\|_1 &= \int_0^b e^{-\tau m(t)} \|v_1(t) - v_2(t)\| dt \\
&\le \int_0^b p(t) e^{-\tau m(t)} D dt \int_0^t \|f_1(s) - f_2(s)\| ds \\
&= \frac{-D}{\tau} \int_0^b [e^{-\tau m(t)}]' dt \int_0^t \|f_1(s) - f_2(s)\| ds \\
&= \frac{-D}{\tau} \left\{ e^{-\tau m(b)} \int_0^b \|f_1(t) - f_2(t)\| dt - \int_0^b \|f_1(t) - f_2(t)\| dt \right\} + \epsilon. \\
&\le \frac{D}{\tau} \int_0^b \|f_1(t) - f_2(t)\| dt + \epsilon.
\end{aligned}$$

Since ϵ is arbitrary, we get

$$\|v_1 - v_2\|_1 \le \frac{D}{\tau} \|f_1 - f_2\|_1.$$

By an analogous relation obtained by interchanging the roles of v_1 and v_2, we finally obtain that, for every $f_1, f_2 \in L^1(J,E)$,

$$H(G(a, f_1), G(a, f_2)) \le \frac{D}{\tau} \|f_1 - f_2\|_1.$$

Hence there exists a continuous single-valued map $\phi : E \times L^1(J,E) \to L^1(J,E)$ such that

$$\phi(a, f) \in G(a, f) \text{ for each } f \in L^1(J,E)$$

and

$$\phi(a, f) = f \text{ for every } f \in G_*(a, f).$$

Let $(a, y) \in \overline{\Omega}$ be arbitrary. There exists a unique $y \in \tilde{\Omega}$ for some $f \in L^1(J,E)$, where $y(a, f)$ denotes the mild solution of (9.43)–(9.45). Hence $(a, y) = (a, y(a, f))$. We consider the following function

$$\begin{aligned}
L_*(a, y(a, f))(t) = T(t)a &+ \int_0^t T(t-s)\phi(a, f)(s) ds \\
&+ \sum_{0<t_k<t} T(t - t_k) I_k(y(a, f)(t_k)), \quad t \in J.
\end{aligned}$$

Now we prove that the function L_* is continuous. Let $(a, y(a, f)), (a_0, y(a_0, f_0)) \in E \times \overline{\Omega}$. Then

$$\begin{aligned}
&\|L_*(a, y(a, f))(t) - L_*(a_0, y(a_0, f_0))(t)\| \le \\
&\qquad M e^{\omega b} \|a - a_0\| + M e^{\omega b} \int_0^t \|\phi(a, f)(s) - \phi(a_0, f_0)(s) ds \\
&\qquad\qquad + M e^{\omega b} \sum_{0<t_k<t} \|I_k(y(a, f)(t_k)) - I_k(y(a_0, f_0)(t_k)\|.
\end{aligned}$$

Thus

$$\|L_*(a, y(a, f)) - L_*(a_0, y(a_0, f_0))\|_{PC} \le$$
$$Me^{\omega b}\|a - a_0\| + Me^{\omega b}\int_0^b \|\phi(a, f)(s) - \phi(a_0, f_0)(s)ds$$
$$+ Me^{\omega b}\sum_{k=1}^{m} c_k\|y(a, f)(t_k)) - y(a_0, f_0)(t_k)\|.$$

This implies that

$$d(L_*(a, y(a, f)), L_*(a_0, y(a_0, f_0)) \le$$
$$Me^{\omega b}\|a - a_0\| + Me^{\omega b}\int_0^b \|\phi(a, f)(s) - \phi(a_0, f_0)(s)ds$$
$$+ Me^{\omega b}\sum_{k=1}^{m} c_k\|y(a, f)(t_k)) - y(a_0, f_0)(t_k)\| + \|f - f_0\|_{L^1}$$
$$\to 0 \text{ as } (a, y(a, f)) \to (a_0, y(a_0, f_0)).$$

Hence L_* is continuous. □

Now, we prove that the solutions set is an absolute retract.

Theorem 9.59. *Let $F : J \times E \to \mathcal{P}_{cp}(E)$ be a multivalued map. Assume that $t \to F(t, .)$ is measurable, $(\overline{\mathcal{H}}_1)$ and*

$(\overline{\mathcal{H}}_3)$ *there exist $c_k > 0$, $k = 1, \dots, m$, such that*

$$I_k(x) = c_k x, \text{ for each } x \in E.$$

Then the solution set of problem (9.40)–(9.42) *is an absolute retract.*

Proof. Using the fact that I_k, $k = 1, \dots, m$, are linear functions, then $\overline{L}$ is linear. Also, it is clear that $\overline{L}(E \times L^1(J, E)) = \Omega_*$. Moreover since $\overline{L}$ is one-to-one then from Lemma 9.54, Ω_* is a Banach space with norm

$$\|y(a, f)\|_{\Omega_*} = \|y(a, f)\|_{PC} + \|f\|_{L^1}, \text{ for each } (a, f) \in E \times L^1(J, E).$$

By using Theorem 9.57, we can prove that there exist $L_* : \Omega_* \to PC$ such that

$$L_*(a, y) \in S(a) \quad \text{for each } y \in \tilde{\Omega} \quad \text{and}$$
$$L_*(a, f) = y, \quad \text{for each } y \in S(a),$$

where

$$\overline{\Omega} = \{(a, y) \in E \times \Omega_* : a \in E,\ y \in \tilde{\Omega}\},$$

and

$$\tilde{\Omega} = \{y \in \Omega_* : y(0) = a\}.$$

It is clear that $S(a) \subset \tilde{\Omega}$. Since $\overline{L}$ is linear, continuous and bijective, $\overline{L}$ is an homeomorphism, and

$$S(a) = (\overline{L} \circ G_*)(a).$$

By the same method as used in Theorem 9.57, we can easily prove that G_* is a contraction. Hence from Lemma 9.11, $S(a)$ is an absolute retract of $\tilde{\Omega}$ since $\tilde{\Omega}$ is a closed convex subset of a Banach space Ω_*. Then $\tilde{\Omega}$ is retract of Ω_* (see, [114]). This implies that $S(a)$ is an absolute retract. □

9.8 The Terminal Problem

Our final existence theorem is concerned with the terminal (or target) problem. We also prove compactness and acyclicity of the solution set extending similar results obtained in [17] for differential inclusions. Consider the problem

$$\begin{cases} y'(t) \in F(t, y(t)), & \text{a.e.} \quad t \in J, \\ \Delta y|_{t=t_k} = I_k(y(t_k^-)), & \quad k = 1, \ldots, \\ \lim\limits_{t\to\infty} y(t) = y_\infty \in \mathbb{R}^n. \end{cases} \tag{9.63}$$

9.8.1 Existence and Solution Set

Consider the set

$$PC_b = \big\{y \in PC(\mathbb{R}_+, \mathbb{R}^n) : y \text{ is bounded}\big\},$$

where

$$\begin{aligned} PC(\mathbb{R}_+, \mathbb{R}^n) = \big\{y\colon [0,\infty) \to \mathbb{R}^n \mid\ & y_k \in C((t_k, t_{k+1}), \mathbb{R}^n), \\ & k = 0, \ldots,\ y(t_k^-) \text{ and } y(t_k^+) \text{ exist and} \\ & \text{satisfy } y(t_k) = y(t_k^-) \text{ for } k = 1, \ldots\big\} \end{aligned}$$

and $y_k := y_{|(t_k, t_{k+1}]}$. Endowed with the norm

$$\|y\|_b = \sup\{\|y(t)\| : t \in [0, \infty)\},$$

PC_b is a Banach space.

The following compactness criterion on unbounded domains is a simple extension of a compactness criterion in $C_b(\mathbb{R}_+, \mathbb{R}^n)$ (see [153, p. 62]).

Lemma 9.60. *Let $M \subset PC_\ell = \{x \in PC(\mathbb{R}_+, \mathbb{R}^n) : \lim_{t\to+\infty} x(t)$ exists$\}$. Then M is relatively compact if it satisfies the following conditions:*

(a) *M is uniformly bounded in $PC_\ell(\mathbb{R}_+, \mathbb{R}^n)$.*

(b) *The functions belonging to M are almost equicontinuous on $\mathbb{R}_+$, i.e., equicontinuous on every compact interval of $\mathbb{R}_+$.*

(c) *The functions from M are equiconvergent, that is, given $\varepsilon > 0$, there corresponds $T(\varepsilon) > 0$ such that $|x(t) - x(+\infty)| < \varepsilon$ for any $t \geq T(\varepsilon)$ and $x \in M$.*

We are now in position to state our final existence result in this chapter.

Theorem 9.61. *Assume that the multivalued map $F : J \times \mathbb{R}^n \to \mathcal{P}_{cp}(\mathbb{R}^n)$ is such that $t \to F(t, \cdot)$ is measurable and the following conditions hold:*

(I-14) *there exists $l \in L^1(J, \mathbb{R}_+)$ such that $F(t,0) \subset l(t)\overline{B}(0,1)$, a.e. $t \in J$, and*

$$H(F(t,x), F(t,y)) \leq l(t)\|x - y\|, \ x, y \in \mathbb{R}^n \text{ and a.e. } t \in J.$$

(I-15) *$\sum_{k=1}^{\infty} \|I_k(0)\| < \infty$ and there exist constants $c_k \geq 0$ such that $\sum_{k=1}^{\infty} c_k < 1$, and*

$$\|I_k(x) - I_k(y)\| \leq c_k\|x - y\|, \text{ for each } x, y \in \mathbb{R}^n.$$

If

$$\|l\|_{L^1} + \sum_{k=1}^{\infty} c_k < 1, \tag{9.64}$$

then the boundary value problem (9.63) has at least one solution. If, in addition, F is a Carathéodory multivalued map with compact convex values, then the solution set is a contractible compact set, and hence is acyclic.

Proof.

Part 1. Existence of a solution

Consider the operator $N : PC_b \to \mathcal{P}(PC_b)$ defined by $N(y) = \{h \in PC_b\}$ such that, for $t \in J$,

$$h(t) = y_\infty - \int_t^\infty v(s)ds - \sum_{t_k \geq t} I_k(y(t_k)),$$

where $v \in S_{F,y} = \{v \in L^1(J, \mathbb{R}^n) : v(t) \in F(t, y(t)), \text{ a.e. } t \in J\}$. Clearly, the fixed points of the operator N are solutions of the problem (9.63).

Step 1. $N(y) \in \mathcal{P}_{cl}(PC)$ for each $y \in PC_b$: Indeed, let $(y_n)_{n\geq 0} \in N(y)$ be such that $y_n \longrightarrow \widetilde{y}$ in PC_b. Then there exists $v_n \in S_{F,y}$ such that for a.e. $t \in J$,

$$y_n(t) = y_\infty - \int_t^\infty v_n(s)ds - \sum_{t_k \geq t} I_k(y(t_k)). \tag{9.65}$$

Since F has compact values, then there exists a subsequence $(v_{n_m}(\cdot))$ which converges to $v(\cdot)$ in $\mathbb{R}^n$. From (I-14), we have

$$\|v_{n_m}(t)\| \leq l(t)\|y\|_{PC} + l(t), \quad \text{a.e.} \quad t \in J.$$

By the Lebesgue dominated convergence theorem, $\int_t^\infty v_n(s)ds$ converges to $\int_t^\infty v(s)ds$. Passing to the limit in (9.65) yields that $\widetilde{y} \in N(y)$, as claimed.

Step 2. N is a contraction: Let $y, \overline{y} \in PC_b$ and $h \in N(y)$. Then there exists $v(t) \in F(t, \overline{y}(t))$ such that for a.e. $t \in J$,

$$h(t) = y_\infty - \int_t^\infty v(s))ds - \sum_{t_k \geq t} I_k(y(t_k)).$$

Condition (I-14) implies that there is $u \in F(t, \overline{y}(t))$ such that

$$\|v(t) - u\| \leq l(t)\|y(t) - \overline{y}(t)\|, \quad \text{a.e.} \quad t \in J.$$

Consider the multivalued map defined by $V(t) = U(t) \cap F(t, \overline{y}(t))$ where $U : [0, \infty) \to \mathcal{P}_{cl}(\mathbb{R}^n)$ is given by

$$U(t) = \{u \in \mathbb{R}^n : \|v(t) - u\| \leq l(t)\|y(t) - \overline{y}(t)\|\}.$$

Arguing as in the proof of Theorem 6.4, we can find a function $\overline{v}(t)$, which is a measurable selection for V. Thus, $\overline{v}(t) \in F(t, \overline{y}(t))$ and

$$\|v(t) - \overline{v}(t)\| \leq l(t)\|y(t) - \overline{y}(t)\|, \quad \text{a.e.} \quad t \in J.$$

Let us define, for a.e. $t \in J$,

$$\overline{h}(t) = y_\infty - \int_t^\infty \overline{v}(s)ds - \sum_{t_k \geq t} I_k(y(t_k)).$$

Then, $\overline{h} \in N(\overline{y})$ and

$$\begin{aligned}\|h(t) - \overline{h}(t)\| &\leq \int_0^\infty l(s)\|\overline{y}(s) - y(s)\|ds + \sum_{k=1}^\infty c_k\|\overline{y}(t_k) - y(t_k)\| \\ &\leq \int_0^\infty l(s)\|\overline{y} - y\|_{PC_b}ds + \sum_{k=1}^\infty c_k\|\overline{y} - y\|_{PC_b}.\end{aligned}$$

Hence,

$$\|h - \overline{h}\|_{PC_b} \leq \left(\|l\|_{L^1} + \sum_{k=1}^\infty c_k\right)\|y - \overline{y}\|_{PC_b}.$$

Since $y \in PC_b$ and $h \in N(y)$ are arbitrary, it follows that

$$\sup_{h \in N(y)} d(h, N(\overline{y})) \leq \left(\|l\|_{L^1} + \sum_{k=1}^\infty c_k\right)\|y - \overline{y}\|_{PC_b}.$$

By an analogous relation, obtained by interchanging the roles of y and $\overline{y}$, we obtain that

$$H(N(y), N(\overline{y})) \leq \Big(\|l\|_{L^1} + \sum_{k=1}^{\infty} c_k\Big)\|y - \overline{y}\|_{PC_b}.$$

In view of (9.64), we have a contraction and so Lemma 2.88 implies that N has a fixed point in PC_b that is a solution of problem (9.63). For each $y_\infty \in \mathbb{R}^n$, let $S(y_\infty)$ be the solution set of problem (9.63).

Part 2. Structure of the solution set

First, we prove that for each $y_\infty \in \mathbb{R}^n$, $S(y_\infty)$ is acyclic. Let $y \in S(y_\infty)$; then there exists $v \in L^1(J, \mathbb{R}^+)$ such that $v(t) \in F(t, y(t))$ for a.e. $t \in J$, and

$$y(t) = y_\infty - \int_t^\infty v(s)ds - \sum_{t_k \geq t} I_k(y(t_k)).$$

Consequently,

$$\|y(t)\| \leq \|y_\infty\| + \int_0^\infty l(s)\|y(s)\|ds + \|l\|_{L^1} + \sum_{k=1}^{\infty} c_k\|y(t_k)\| + \sum_{k=1}^{\infty}\|I_k(0)\|, \quad \text{a.e.} \quad t \in J.$$

Then

$$\|y\|_{PC_b} \leq \|y_\infty\| + \int_0^\infty l(s)\|y\|_{PC_b}ds + \|l\|_{L^1} + \sum_{k=1}^{\infty} c_k\|y\|_{PC_b} + \sum_{k=1}^{\infty}\|I_k(0)\|.$$

Hence

$$\|y\|_{PC_b} \leq M := \frac{1}{1 - \|l\|_{L^1} - \sum_{k=1}^{\infty} c_k}\Big(\|y_\infty\| + \|l\|_{L^1} + \sum_{k=1}^{\infty}\|I_k(0)\|\Big). \tag{9.66}$$

Step 1. *Definition of an homotopy:* Consider the modified problem

$$\begin{cases} y'(t) \in G(t, y(t)), & \text{a.e.} \quad t \in J, \\ \Delta y|_{t=t_k} = I_k(y(t_k^-)), & k = 1, \dots, \\ \lim\limits_{t\to\infty} y(t) = y_\infty \in \mathbb{R}^n, \end{cases} \tag{9.67}$$

where $G : \mathbb{R}^+ \times \mathbb{R}^n \to \mathcal{P}_{cp,cv}(\mathbb{R}^n)$ is the multivalued map defined by

$$G(t, y)) = \begin{cases} F(t, y), & \|y\| \leq M_*, \\ F\left(t, \dfrac{yM_*}{\|y\|}\right), & \|y\| > M_*, \end{cases}$$

where $M_* > M$. Now, we show that $S(y_\infty) = G(y_\infty)$ where $G(y_\infty)$ is the solution set of problem (9.67). Let $y \in S(y_\infty)$. Then $y(t_k^+) - y(t_k^-) = I_k(y(t_k))$ for every $k \in \{1, 2, \ldots\}$, $\lim_{t\to\infty} y(t) = y_\infty$, and $y'(t) \in F(t, y(t))$, for a.e. $t \in J$. Thus,

$$\|y(t)\| \le M, \text{ a.e. } t \in J \text{ implies } F(t, y(t)) = G(t, y(t)), \quad \text{a.e.} \quad t \in J.$$

This implies that $y \in G(y_\infty)$, so $S(y_\infty) \subset G(y_\infty)$.

Conversely, let $y \in G(y_\infty)$; then

$$y(t) = y_\infty - \int_t^\infty v(s)ds - \sum_{t_k \ge t} I_k(y(t_k)), \text{ a.e. } t \in J,$$

where $v \in S_{G,y}$. Set $W = \{t \in [0, \infty) : \|y(t)\| > M_*\}$. We show that $mes(W) = 0$ where mes is the Lebesgue measure. Assume that $mes(W) \ne 0$. Then,

$$y(t) = y_\infty - \int_W^\infty v(s)ds - \int_{[t,\infty)\setminus W}^\infty v(s)ds - \sum_{t_k \ge t} I_k(y(t_k)).$$

As a consequence,

$$\begin{aligned}\|y(t)\| \le \|y_\infty\| &+ \int_W \|v(s)\|ds + \int_{[0,\infty)\setminus W} \|v(s)\|ds \\ &+ \sum_{k=1}^\infty c_k\|y(t_k)\| + \sum_{k=1}^\infty \|I_k(0)\|.\end{aligned}$$

From the definition of G, (I-14), and (I-15), we obtain the estimate

$$\begin{aligned}\|y(t)\| &\le \|y_\infty\| + \int_W \|v(s)\|ds + \int_{[0,\infty)\setminus W} \|v(s)\|ds \\ &\quad + \sum_{k=1}^\infty c_k\|y(t_k)\| + \sum_{k=1}^\infty \|I_k(0)\| \\ &\le \|y_\infty\| + \int_W l(s)ds + \int_W M_* l(s)ds + \int_{[0,\infty)\setminus W} l(s)\|y(s)\|ds \\ &\quad + \int_{[0,\infty)\setminus W} l(s)ds + \sum_{k=1}^\infty c_k\|y(t_k)\| + \sum_{k=1}^\infty \|I_k(0)\| \\ &\le \|y_\infty\| + \int_0^\infty l(s)ds + \int_0^\infty l(s)\|y(s)\|ds \\ &\quad + \sum_{k=1}^\infty c_k\|y(t_k)\| + \sum_{k=1}^\infty \|I_k(0)\|.\end{aligned}$$

Hence,

$$\|y\|_{PC_b} \le \|y_\infty\| + \|l\|_{L^1} + \|l\|_{L^1}\|y\|_{PC_b} + \sum_{k=1}^\infty c_k\|y\|_{PC_b} + \sum_{k=1}^\infty \|I_k(0)\|.$$

Therefore,

$$\|y\|_{PC_b} \leq \frac{1}{1-\|l\|_{L^1}-\sum\limits_{k=1}^{\infty} c_k}\left(\|y_\infty\| + \|l\|_{L^1} + \sum_{k=1}^{\infty}\|I_k(0)\|\right) = M.$$

This is a contradiction to $M < M_*$. Thus, $meas(W) = 0$, and this implies

$$G(t, y(t)) = F(t, y(t)), \ \text{a.e. } t \in J.$$

So, $S(y_\infty) = G(y_\infty)$. Therefore, $G(\cdot,\cdot) \in \mathcal{P}_{cp,cv}(\mathbb{R}^n)$, and since $F(\cdot,\cdot) \in \mathcal{P}_{cp,cv}(\mathbb{R}^n)$ and is u.s.c, there exists a Carathéodory Lipschitzian selection (see [34]) $g : \mathbb{R}^+ \times \mathbb{R}^n \to \mathbb{R}^n$ such that

$$g(t,x) \in G(t,x), \ \text{for a.e. } t \in J \ \text{and} \ x \in \mathbb{R}^n.$$

Since $\|l\|_{L^1} + \sum_{k=1}^{\infty} c_k < 1$, we can choose a Lipschitzian function $\gamma \in L^1(J,\mathbb{R}^+)$ such that $\int_0^\infty \gamma(s)ds + \sum_{k=1}^{\infty} c_k < 1$ (see [34]) and

$$\|g(t,x) - g(t,y)\| \leq \gamma(t)\|x-y\|, \ \text{for a.e. } t \in J \ \text{and each } x,y \in \mathbb{R}^n.$$

Consider the single-valued problem,

$$\begin{cases} x'(t) = g(t,x(t)), & \text{a.e.} \quad t \in J, \\ \Delta x|_{t=t_k} = I_k(x(t_k^-)), & \qquad k = 1,\ldots, \\ \lim\limits_{t\to\infty} x(t) = y_\infty \in \mathbb{R}^n. & \end{cases} \tag{9.68}$$

Since g is a Carathéodory Lipschitz function, by Banach's fixed point theorem, we can prove that the problem (9.68) has a unique global solution, which we denote by $\overline{x}$. Furthermore, $\overline{x} \in PC$ and $\lim\limits_{t\to\infty} \overline{x}(t) = y_\infty$. Since $g(t,x) \in G(t,x)$,

$$\|\overline{x}\|_{PC_b} \leq M_*,$$

which implies that $\overline{x} \in PC_b$. Define the homotopy $h : S(y_\infty) \times [0,1] \to S(y_\infty)$ by

$$h(y,\alpha)(t) = \begin{cases} y(t), & \text{for } t \geq \frac{1}{\alpha} - \alpha, \ \alpha \neq 0, \\ \overline{z}(t), & \text{for } 0 \leq t < \frac{1}{\alpha} - \alpha, \ \alpha \neq 0, \\ \overline{x}(t), & \text{for } \alpha = 0, \end{cases}$$

where, for each $\alpha \in (0,1)$, $\overline{z}$ is the unique solution of the backward Cauchy problem

$$\begin{cases} z'(t) = g(t,z(t)), & \text{a.e.} \quad t \in [0, \frac{1}{\alpha} - \alpha]\backslash\{t_1, t_2, \ldots\}, \\ \Delta z|_{t=t_k} = I_k(z(t_k^-)), & \qquad k = 1,\ldots, \\ z\left(\frac{1}{\alpha} - \alpha\right) = \overline{x}\left(\frac{1}{\alpha} - \alpha\right). & \end{cases} \tag{9.69}$$

Step 2. *h is a continuous function:* Let $(y_n, \alpha_n) \in S(y_\infty) \times [0,1]$ be such that $(y_n, \alpha_n) \to (y, \alpha)$, as $n \to +\infty$. We shall prove that $h(y_n, \alpha_n) \to h(y, \alpha)$. We have

$$h(y_n, \alpha_n)(t) = \begin{cases} y_n(t), & \text{for } t \geq \frac{1}{\alpha_n} - \alpha_n,\ \alpha_n \neq 0, \\ \overline{z}(t), & \text{for } 0 \leq t < \frac{1}{\alpha_n} - \alpha_n,\ \alpha_n \neq 0, \\ \overline{x}(t), & \text{for } \alpha_n = 0. \end{cases}$$

We distinguish three different cases:

(1) If $\lim_{n\to\infty} \alpha_n = 0$, then $\|h(y_n, \alpha_n)(t) - h(y, 0)(t)\| = \|\overline{z}(t) - \overline{x}(t)\|$, for n large enough. Indeed when $\lim_{n\to\infty} \alpha_n = 0$, for every $t \in [0, \frac{1}{\alpha} - \alpha]$, there exists $n_0 \in \mathbb{N}$ such that $n \geq n_0$ implies $\frac{1}{\alpha_n} - \alpha_n > t$ and thus $h(y_n, \alpha_n) = \overline{x}(t)$. Moreover,

$$\lim_{n\to+\infty} \overline{z}\left(\frac{1}{\alpha_n} - \alpha_n\right) = \lim_{n\to+\infty} \overline{x}\left(\frac{1}{\alpha_n} - \alpha_n\right) = y_\infty.$$

Hence, $\lim_{n\to\infty} \|h(y, \alpha_n)(t) - h(y, 0)(t)\| = 0$.

(2) Let $\alpha_n \neq 0$ and $0 < \lim_{n\to\infty} \alpha_n = \alpha \leq 1$. Since $\overline{z}_n$ is a solution of problem (9.69), for every $n \geq 1$ and for a.e. $t \in [0, \frac{1}{\alpha_n} - \alpha_n]$, we have

$$\overline{z}(t) = \overline{x}\left(\frac{1}{\alpha_n} - \alpha_n\right) - \int_t^{1/\alpha_n - \alpha_n} g(s, \overline{z}(s))ds - \sum_{t < t_k < 1/\alpha_n - \alpha_n} I_k(\overline{z}(t_k)).$$

Without loss of generality, we may assume $\alpha_n \geq \alpha$; hence $\frac{1}{\alpha_n} - \alpha_n \leq \frac{1}{\alpha} - \alpha$. Consider the function

$$\widetilde{z}_n(t) = \begin{cases} \overline{z}(t), & \text{for } t \in [0, \frac{1}{\alpha_n} - \alpha_n], \\ \overline{x}(\frac{1}{\alpha_n} - \alpha_n), & \text{for } \frac{1}{\alpha_n} - \alpha_n \leq t \leq \frac{1}{\alpha} - \alpha. \end{cases}$$

Since g and I_k, for $k = 1, \ldots,$ are Lipschitzian functions and $\overline{x}$ is bounded, then there exists a positive real number M independent of $n \in \mathbb{N}$ such that

$$\sup\{\|\widetilde{z}_n(t)\| : t \in [0, \frac{1}{\alpha} - \alpha]\} \leq M, \quad \text{for every } n \in \mathbb{N}.$$

Moreover, the set $\{\widetilde{z}_n(t) : t \in [0, \frac{1}{\alpha} - \alpha]\}$ is equicontinuous in $PC_b([0, \frac{1}{\alpha} - \alpha], \mathbb{R}^n)$. If $\tau_1, \tau_2 \in [0, \frac{1}{\alpha_n} - \alpha_n]$, then

$$\|z_n(\tau_1) - z_n(\tau_2)\| \leq \int_{\tau_1}^{\tau_1} \|g(s, \overline{z}(s))\|ds + \sum_{\tau_1 < t_k < \tau_2} \|I_k(\overline{z}(t_k)\|.$$

Using the fact that $\overline{x}$ is bounded, we can easily prove that there exists $K_* > 0$ such that $\|\overline{z}\|_{PC} \leq K_*$. Then,

$$\|\widetilde{z}_n(\tau_1) - \widetilde{z}_n(\tau_2)\| \leq \int_{\tau_1}^{\tau_2} K_* \gamma(s)ds + \int_{\tau_1}^{\tau_2} l(s)ds + \sum_{\tau_1 < t_k < \tau_2} \sup\{\|I_k(x)\| : x \in \overline{B}(0, K_*)\}.$$

If $\tau_1, \tau_2 \in [\frac{1}{\alpha_n} - \alpha_n, \frac{1}{\alpha} - \alpha]$, then

$$\|\widetilde{z}_n(\tau_1) - \widetilde{z}_n(\tau_2)\| = 0.$$

In the case $\tau_1 \in [0, \frac{1}{\alpha_n} - \alpha_n]$ and $\tau_2 \in [\frac{1}{\alpha_n} - \alpha_n, \frac{1}{\alpha} - \alpha]$, we obtain

$$\begin{aligned}\|\widetilde{z}_n(\tau_1) - \widetilde{z}_n(\tau_2)\| &\leq \int_{\tau_1}^{\frac{1}{\alpha_n} - \alpha_n} K_* \gamma(s)ds + \int_{\tau_1}^{\tau_2} l(s)ds \\ &\quad + \sum_{\tau_1 < t_k < \tau_2} \sup\{\|I_k(x)\| : x \in \overline{B}(0, K_*)\} \\ &\leq \int_{\tau_1}^{\tau_2} K_* \gamma(s)ds + \int_{\tau_1}^{\tau_2} l(s)ds \\ &\quad + \sum_{\tau_1 < t_k < \tau_2} \sup\{\|I_k(x)\| : x \in \overline{B}(0, K_*)\}.\end{aligned}$$

By the Ascoli–Arzelà Lemma, there exists $\widetilde{z} \in PC_b([0, \frac{1}{\alpha} - \alpha], \mathbb{R}^n)$ such that $(\widetilde{z}_n)_{n \in \mathbb{N}}$ converges to $\widetilde{z}$ and

$$\widetilde{z}(t) = \overline{x}\left(\frac{1}{\alpha} - \alpha\right) - \int_t^{1/\alpha - \alpha} g(s, \widetilde{z}(s))ds - \sum_{t < t_k < \frac{1}{\alpha} - \alpha} I_k(\widetilde{z}(t_k)).$$

Hence, $\widetilde{z}$ is solution of problem (9.69). By uniqueness, $\widetilde{z} = \overline{z}$. Thus,

$$\sup\{\|\widetilde{z}_n(t) - \overline{z}(t)\| : t \in [0, \frac{1}{\alpha} - \alpha]\} \to 0, \text{ as } n \to \infty.$$

(3) Finally, let $t \geq \frac{1}{\alpha_n} - \alpha_n$ and $\alpha_n \neq 0$. Since (y_n) converges to y in PC_b, we have

$$\sup\{\|y_n(t) - y(t)\| : t \in [0, \infty)\} \to 0, \text{ as } n \to \infty.$$

In summary,

$$\sup\{\|h(y_n, \alpha_n)(t) - h(y, \alpha)(t)\| : t \in [0, \infty)\} \to 0, \text{ as } n \to \infty,$$

proving our claim.

Step 3. *$S(y_\infty)$ is closed in PC_b:* Let $\{y_n : n \in \mathbb{N}\} \subset S(y_\infty)$ be such that (y_n) converges to y in PC_b, as $n \to +\infty$. Then there exists $v_n \in S_{G, y_n}$ such that

$$y_n(t) = y_\infty - \int_t^\infty v_n(s)ds - \sum_{t_k \geq t} I_k(y_n(t_k)), \text{ a.e. } t \in J.$$

From (9.66), there exists $M > 0$ such that

$$\sup\{\|y_n(t)\| : t \in \mathbb{R}^+\} \leq M \tag{9.70}$$

and

$$\|v_n(t)\| \leq l(t)M + l(t), \text{ a.e. } t \in J.$$

Since $\overline{B}(0,1)$ is compact in $\mathbb{R}^n$, there exists subsequence $(v_{n_m}(\cdot))$ which converges to some limit $v(\cdot)$. By the Lebesgue dominated convergence theorem, we conclude that $v \in L^1(\mathbb{R}^+, \mathbb{R}^n)$. Using the fact that $G(\cdot,\cdot) \in \mathcal{P}_{cp,cv}(\mathbb{R}^n)$ and that $G(t,\cdot)$ is u.s.c., the map $G(t,\cdot)$ has a closed graph, hence $v \in S_{G,y}$. Therefore,

$$y(t) = y_\infty - \int_t^\infty v(s)ds - \sum_{t_k \geq t} I_k(y(t_k)), \text{ a.e. } t \in J,$$

proving that $y \in S(y_\infty)$. From Steps 1–3, h is a continuous homotopy with $h(y,0) = \overline{x}$ and $h(y,1) = y$ (see also [218] for the case of differential inclusions). Therefore, the set $S(\cdot)$ is contractible. Let $(y_n)_{n\in\mathbb{N}} \subset S(y_\infty)$. Now (9.66) shows that $(y_n)_{n\in\mathbb{N}}$ is uniformly bounded. Also, we can easily prove that, for every compact interval $J_* \subset [0,\infty)$, $\{y_n : n \in \mathbb{N}\}$ is also equicontinuous. It remains to only prove that $\{y_n : n \in \mathbb{N}\}$ is equiconvergent at ∞, i.e., for every $\varepsilon > 0$, there exists $T(\varepsilon) > 0$ such that $\|y_n(t) - y_n(\infty)\| \leq \varepsilon$ for every $t \geq T$ and each $n \in \mathbb{N}$. Let $y_n \in \{y_n : n \in \mathbb{N}\}$; then there exists $v_n \in S_{F,y_n}$ such that

$$y_n(t) = y_\infty - \int_t^\infty v_n(s)ds - \sum_{t_k \geq t} I_k(y_n(t_k)), \text{ a.e. } t \in J.$$

Thus,

$$\begin{aligned}\|y_n(t) - y_n(\infty)\| &\leq \int_t^\infty \|v_n(s)\|ds + \sum_{t \leq t_k < \infty} \|I_k(y_n(t_k))\| \\ &\leq (1+M)\int_t^\infty l(s)ds + \sum_{t \leq t_k < \infty} (c_k M + \|I_k(0)\|).\end{aligned}$$

Since $\sum_{k=1}^\infty c_k < \infty$, $\sum_{k=1}^\infty \|I_k(0)\| < \infty$, and $p \in L^1(J,\mathbb{R}^+)$, there exist k_0 and $T(\varepsilon) > 0$ such that

$$\sum_{k=k_0}^\infty (c_k M + \|I_k(0)\|) \leq \frac{\varepsilon}{2}$$

and

$$\int_t^\infty l(s)ds < \frac{\varepsilon}{2(M+1)}, \ \forall t \geq T(\varepsilon).$$

Hence,

$$\|y_n(t) - y_n(\infty)\| \leq \varepsilon, \text{ a.e. } t \geq \max(k_0, T(\varepsilon)).$$

Then $\{y_n : n \in \mathbb{N}\}$ is equiconvergent. Lemma 9.60 then implies that $S(y_\infty)$ is compact, and hence acyclic. This completes the proof of the theorem. □

Chapter 10

Impulsive Semilinear Differential Inclusions

10.1 Nondensely Defined Operators

In this chapter, we shall be concerned with the existence of integral solutions for first order impulsive semilinear functional differential inclusions in Banach spaces. Initially, we consider first order impulsive semilinear functional differential inclusions of the form,

$$y' - Ay \in F(t, y_t), \qquad \text{a.e.} \quad t \in J\backslash\{t_1, \dots, t_m\}, \tag{10.1}$$

$$\Delta y|_{t=t_k} = I_k(y(t_k^-)), \qquad k = 1, \dots, m, \tag{10.2}$$

$$y(t) = \phi(t) \qquad t \in (-\infty, 0], \tag{10.3}$$

where $J := [0, b]$, $F : J \times \mathcal{B} \to \mathcal{P}(E)$ is a multivalued map with compact values, $(E, |\cdot|)$ is a Banach space, $\mathcal{B}$ is a phase space to be defined, $\mathcal{P}(E)$ is the family of all nonempty subsets of E, $A : D(A) \subset E \to E$ is a nondensely defined closed linear operator on E that generates a semigroup, B is a bounded linear operator from E into E, and $\phi \in \mathcal{B}$. For every $t \in [0, \infty)$, the history function $y_t : (-\infty, 0] \to E$ is defined by

$$y_t(\theta) = y(t + \theta), \text{ for } \theta \in (-\infty, 0].$$

We assume that the histories $y_t : (-\infty, 0] \to E$, $y_t(\theta) = y(t + \theta)$, belong to the abstract phase space $\mathcal{B}$. Throughout this chapter, we assume that for every $y_t \in \mathcal{B}$, the multivalued function $t \to F(t, y_t)$ is measurable.

The existence of integral solutions on compact intervals for the problem (10.1)–(10.3), with periodic boundary conditions in a Banach space, was considered by Ezzinbi and Liu [195]. For more details on nondensely defined operators and the concept of integrated semigroups, we refer the reader to the monograph [13] and to the papers [25, 26, 119, 162, 387, 463].

In Section 10.3, we are concerned with an application of a recent Frigon nonlinear alternative for admissible contraction maps in Fréchet spaces [206] to the existence of integral solutions of first order controllability associated with impulsive semilinear functional and neutral functional differential inclusions in Fréchet spaces. In Subsection 10.3.1, we will consider first order controllability for the impulsive semilinear functional differential inclusion,

$$y'(t) - Ay(t) \in F(t, y_t) + (Bu)(t), \quad \text{a.e.} \quad t \in J := [0, \infty)\backslash\{t_1, t_2, \dots\}, \tag{10.4}$$

$$y(t_k^+) - y(t_k^-) = I_k(y(t_k^-)), \qquad k = 1, 2, \dots, \tag{10.5}$$

$$y(t) = \phi(t), \qquad t \in (-\infty, 0], \tag{10.6}$$

where $F : J \times \mathcal{B} \to (\mathcal{P}(\overline{D(A)})$ is a multivalued map with compact values, $(\mathcal{P}(\overline{D(A)})$ is the family of all nonempty subsets of $\overline{D(A)})$, $A : D(A) \subset E \to E$ is a nondensely defined closed linear operator on E, B is a bounded linear operator from $\overline{D(A)}$ into $\overline{D(A)}$, the control parameter $u(\cdot)$ belongs to $L^2(J, U)$ a space of admissible controls with U a Banach space, $\phi \in \mathcal{B}$, and $I_k \in C(\overline{D(A)}, \overline{D(A)})$, $k \in \{1, 2, \ldots\}$. In Subsection 10.3.2, we study the first order controllability of impulsive semilinear neutral functional differential inclusions of the form,

$$\begin{aligned} &\frac{d}{dt}[y(t) - g(t, y_t)] - Ay(t) \\ &\qquad \in F(t, y_t) + (Bu)(t), \quad \text{a.e.} \quad t \in [0, \infty)\backslash\{t_1, \ldots\}, && (10.7) \\ &y(t_k^+) - y(t_k^-) = I_k(y(t_k^-)), \qquad k = 1, \ldots, && (10.8) \\ &y(t) = \phi(t), \qquad t \in [-r, 0], && (10.9) \end{aligned}$$

where F, A, B, I_k and ϕ are as in problem (10.4)–(10.6), and $g : J \times \mathcal{B} \to \overline{D(A)}$.

In the case where $I_k \equiv 0$, $k = 1, \ldots,$ and A is a densely defined linear operator generating a semigroup, the controllability of differential inclusions with different conditions was studied by Benchohra and Ntouyas [75], and Li and Xue [344]. Recently, Guo et al. [259] initiated the study of controllability of impulsive evolution inclusions with nonlocal conditions, where they considered a class of first order evolution inclusions with a convex valued right hand side. For the nonconvexity of the right hand side, the controllability of first order impulsive functional differential inclusions with a fixed number of impulses was studied by Benchohra et al. [58]. Very recently, global and exact controllability for semilinear differential inclusions with nondensely defined operator was studied by Henderson and Ouahab [279]. For the exact controllability of impulsive functional differential inclusions with infinite delay and nondensely defined in an extrapolated space, the problem (10.4)–(10.6) was studied by Henderson and Ouahab [278].

10.2 Integral Solutions

In this section, we are concerned with the existence of integral solutions for problem (10.1)–(10.3). We shall first consider the space

$$\begin{aligned} PC = \big\{ y : [0, b] \to \overline{D(A)} : y_k \in C(J_k, \overline{D(A)}), k = 0, \ldots, m, \text{ and there} \\ \text{exist } y(t_k^-) \text{ and } y(t_k^+), k = 1, \ldots, m, \text{ with } y(t_k^-) = y(t_k) \big\}, \end{aligned}$$

which is a Banach space with the norm

$$\|y\|_{PC} = \max\{\|y_k\|_{J_k}, k = 0, \ldots, m\},$$

where y_k is the restriction of y to $J_k = (t_k, t_{k+1}]$, $k = 0, \ldots, m$. Set

$$\Omega = \{y : (-\infty, b] \to \overline{D(A)} : y \in PC \cap \mathcal{B}\}.$$

We will assume that $\mathcal{B}$ satisfies the following axioms:

(A) If $y : (-\infty, b] \to \overline{D(A)}$ and $y_0 \in \mathcal{B}$, then for every t in $[0, \infty)$ the following conditions hold:

(i) y_t is in $\mathcal{B}$;

(ii) $\|y_t\|_{\mathcal{B}} \leq K(t) \sup\{|y(s)| : 0 \leq s \leq t\} + M(t)\|y_0\|_{\mathcal{B}}$;

(iii) $|y(t)| \leq H\|y_t\|_{\mathcal{B}}$;

where $H \geq 0$ is a constant, $K : [0, \infty) \to [0, \infty)$ is continuous, $M : [0, b] \to [0, \infty)$ is locally bounded, and H, K, and M are independent of $y(\cdot)$.

(A-1) For the function $y(\cdot)$ in (A), y_t is a $\mathcal{B}$-valued continuous function on $[0, b]\backslash\{t_1, \ldots\}$.

(A-2) The space $\mathcal{B}$ is complete.

Set

$$K_1 = \sup\{\|K(t)\| : t \in [0, b]\}, \quad \widetilde{M}_1 = \sup\{\|M(t)\| : t \in [0, b]\}, \ n \in \mathbb{N}.$$

We define what we mean by an integral solution of problem (10.1)–(10.3).

Definition 10.1. A function $y \in \Omega$ is said to be an integral solution of (10.1)–(10.3) if there exists $f(t) \in F(t, y_t)$ a.e. on J such that y is the solution of the impulsive integral equation

$$y(t) = S'(t)\phi(0) + A\int_0^t y(s)ds + \int_0^t f(s)ds + \sum_{0<t_k<t} S'(t-t_k)I_k(y(t_k^-)),$$

$$\int_0^t y(s)ds \in D(A), \quad t \in [0, b] \quad \text{and} \quad y(t) = \phi(t), \quad t \in (-\infty, 0].$$

From the definition, it follows that $y(t) \in \overline{D(A)}$, $t \geq 0$. Moreover, y satisfies the variation of constants formula

$$\begin{aligned} y(t) &= S'(t)y_0 + \frac{d}{dt}\int_0^t S(t-s)f(s, y_s)ds \\ &+ \sum_{0<t_k<t} S'(t-t_k)I_k(y(t_k^-)), \quad t \geq 0. \end{aligned} \tag{10.10}$$

Let $B_\lambda = \lambda R(\lambda, A)$ where $R(\lambda, A) = (\lambda I - A)^{-1}$. Then for all $x \in \overline{D(A)}$, $B_\lambda x \to x$, as $\lambda \to \infty$. As a consequence, if y satisfies (10.10), then

$$\begin{aligned} y(t) = S'(t)y_0 &+ \lim_{\lambda\to\infty}\int_0^t S'(t-s)B_\lambda f(s, y_s)ds \\ &+ \sum_{0<t_k<t} S'(t-t_k)I_k(y(t_k^-)), \quad t \geq 0. \end{aligned}$$

Theorem 10.2. *Let $F : J \times \mathcal{B} \longrightarrow \mathcal{P}_{cp,cv}(\overline{D(A)})$. Assume that the Hille–Yosida condition and the following conditions hold.*

(J-1) *There exists $M \geq 1$ such that*

$$\|S'(t)\|_{B(E)} \leq M \text{ for each } t \in [0,b].$$

(J-2) *$(t,u) \to F(t,u)$ is measurable.*

(J-3) *For every $t \in [0,b]$, $u \to F(t,u)$ is u.s.c.*

(J-4) *There exist a continuous nondecreasing function $\psi : [0,\infty) \to (0,\infty)$ and $p \in L^1(J,\mathbb{R}_+)$ such that*

$$\|F(t,u)\|_{\mathcal{P}} \leq p(t)\psi(\|u\|_{\mathcal{B}}) \text{ for a.e. } t \in J \text{ and each } u \in \mathcal{B},$$

with

$$\int_0^b m(s)ds < \int_c^\infty \frac{du}{\psi(u)},$$

where $m(s) = \max(\omega, Mp(s))$.

(J-5) *there exists $\overline{p} \in L^1([0,b],\mathbb{R}_+)$ such that, for every bounded D in $\mathcal{B}$,*

$$\chi(F(t,D)) \leq \overline{p}(t) \sup\{\chi(D(\theta)) : \theta \in (-\infty,0]\},$$

where χ is the Hausdorff measure of noncompactness.

(J-6) *There exist $L_k > 0$, $k = 0,\ldots,m$, such that*

$$q_k = \sup_{[t_k,t_{k+1}]} 2M \int_{t_k}^{t_{k+1}} e^{-L_k s} p(s)ds < 1, k = 0,\ldots,m, t_{m+1} = b.$$

Then the IVP (10.1)–(10.3) has at least one integral solution.

Proof. To transform the problem (10.1)–(10.3) into a fixed point problem, consider the operator $N : \Omega \to \mathcal{P}(\Omega)$ defined by,

$$N(y) = \left\{ h \in \Omega |\, h(t) = \left\{ \begin{array}{ll} \phi(t), & t \in (-\infty,0], \\ S'(t)\phi(0) + \frac{d}{dt}\int_0^t S(t-s)v(s)ds & \\ \quad + \sum_{0<t_k<t} S'(t-t_k)I_k(y(t_k)), & t \in [0,\infty), \end{array} \right. \right\}$$

where

$$v \in S_{F,y} = \{u \in L^1(J,\overline{D(A)}) \mid u \in F(t,y_t)\, a.e\ t \in J\}.$$

Clearly, the fixed points of the operator N are integral solutions of the problem (10.1)–(10.3). Let $x(\cdot) : (-\infty,b] \to \overline{D(A)}$ be the function defined by

$$x(t) = \begin{cases} \phi(0), & \text{if } t \in (-\infty,0], \\ S'(t)\phi(0) & \text{if } t \in [0,\infty). \end{cases}$$

Then $x_0 = S'(t)\phi(0)$. For each $z \in PC$ with $z_0 = 0$, we denote by $\bar{z}$ the function defined by

$$\bar{z}(t) = \begin{cases} z(t), & \text{if } t \in [0,\infty), \\ 0, & \text{if } t \in (-\infty, 0]. \end{cases}$$

If $y(\cdot)$ satisfies the integral equation

$$y(t) = S'(t)\phi(0) + \int_0^t S'(t-s)v(s)ds + \sum_{0<t_k<t} S'(t-t_k)I_k(y(t_k^-)),$$

we can decompose $y(\cdot)$ as $y(t) = \bar{z}(t) + x(t), 0 \le t < \infty$. This implies $y_t = \bar{z}_t + x_t$, for every $0 \le t \le b$, and the function $z(\cdot)$ satisfies

$$z(t) = \int_0^t S'(t-s)v(s)ds + \sum_{0<t_k<t} S'(t-t_k)I_k(\bar{z}(t_k^-) + x(t_k^-)), \ v \in S_{F,\bar{z}+x}. \tag{10.11}$$

Let

$$\mathcal{B}_0 = \big\{z : (-\infty, b] \to \overline{D(A)} \mid z(t_k^-), z(t_k^+) \text{ exist with } z(t_i) = z(t_i^-), \\ z \in C(J_k, \overline{D(A)}),\ k = 1, \dots, m\big\}.$$

For any $z \in \mathcal{B}_0$, we have

$$\|z\|_0 = \|z_0\|_{\mathcal{B}} + \sup\{\|z(s)\| : 0 \le s \le b\} = \sup\{\|z(s)\| : 0 \le s \le b\}.$$

Thus $(\mathcal{B}_0, \|\cdot\|_k)$ is a Banach space. Set

$$C_0 = \{z \in \mathcal{B}_0 : z_0 = 0\}.$$

Let the operator $P : C_0 \to \mathcal{P}(C_0)$ be defined by $P(z) = \{h \in C_0\}$, where

$$h(t) = \begin{cases} 0, & t \in (-\infty, 0], \\ \int_0^t S'(t-s)v(s)ds \\ \quad + \sum_{0<t_k<t} S'(t-t_k)I_k(\bar{z}(t_k^-) + x(t_k^-)), & t \in [0,b], \end{cases}$$

and

$$v \in S_{F,\bar{z}+x} = \{v \in L^1(J, \overline{D(A)}) | v(t) \in F(t, \bar{z}_t + x_t),\ a.e.\ t \in J\}.$$

It is clear that the operator N having a fixed point is equivalent to P having one, so we turn to proving that P has a fixed point. From the definition of P, we can easily show that $P = L \circ S_F + G$, where $L : \widetilde{L}^1((-\infty, b], \overline{D(A)}) \to C_0$ is defined by

$$L(v)(t) = \begin{cases} 0, & t \in (-\infty, 0], \\ \int_0^t S'(t-s)v(s)ds, & t \in [0,b], \end{cases}$$

where

$$\widetilde{L}^1((-\infty,b],\overline{D(A)}) = \big\{v \in L^1((-\infty,b],\overline{D(A)}) : v|_{(-\infty,0]} = 0 \\ \text{and } v|_{[0,b]} \in L^1([0,b],\overline{D(A)})\big\},$$

$S_F : C_0 \to \mathcal{P}(L^1([0,b],\overline{D(A)})$ is defined by

$$S_F(z) = \{v \in L^1(J,\overline{D(A)}) \mid v(t) \in F(t,\bar{z}_t + x_t), \text{ a.e. } t \in J\},$$

and $G : C_0 \to C_0$ is defined by $G(z) = \{h \in C_0\}$, where

$$h(t) = \begin{cases} 0, & t \in (-\infty,0], \\ \sum_{0<t_k<t} S'(t-t_k)I_k(\bar{z}(t_k^-) + x(t_k^-)), & t \in [0,b]. \end{cases}$$

We shall show that P has at least one fixed point. The proof will be given in several steps.

Step 1. Let $P_0 : \mathcal{B} \cap C_0([0,t_1],\overline{D(A)}) \to \mathcal{P}_{cp,cv}(\mathcal{B} \cap C_0([0,t_1],\overline{D(A)}))$ be defined by

$$P_0(z) = \left\{h \in \mathcal{B} \cap C_0([0,t_1],\overline{D(A)}) : h(t) = \begin{cases} 0, & t \in (-\infty,0] \\ \int_0^t S'(t-s)v(s)ds, & t \in [0,t_1] \end{cases}\right\},$$

where

$$v \in S_F = \big\{v \in L^1([0,t_1],\overline{D(A)}) : v(t) \in F(t,\bar{z}_t + x_t) \text{ a.e. } t \in [0,t_1]\big\}.$$

Since, for each $z \in \mathcal{B} \cap C_0([0,t_1],\overline{D(A)})$, the nonlinearity F takes convex values, the selection set $S_F(z)$ is convex, and so P_0 has convex values. Set

$$\mathcal{K}_0 := \{z \in \mathcal{B} \cap C_0([0,t_1],\overline{D(A)}) : K_1|z(t)| \le a_0(t),\ t \in [0,t_1]\},$$

where

$$\|z\|_0 = \sup\big\{\|z(s)\| : 0 \le s \le t\big\}, \quad a_0(t) = \Gamma^{-1}\left(\int_0^t \widehat{M}(s)ds\right),$$

$$\Gamma(z) = \int_c^z \frac{du}{\psi(u+c)},$$

$$c = \widetilde{M}_1\|\phi\|_{\mathcal{B}} + K_1 M\|\phi(0)\| \text{ and } \widehat{M}(t) = K_1 M p(t),\ t \in [0,t_1].$$

It is clear that $\mathcal{K}_0$ is a closed bounded convex set in $\mathcal{B} \cap C_0([0,t_1],\overline{D(A)})$. Let $h \in P_0(z)$; then there exists $v \in S_{F,\bar{z}+x}$ such that

$$h(t) = \int_0^t S'(t-s)v(s)ds.$$

We have

$$|h(t)| \le \int_0^t |S'(t-s)\|v(s)|ds \le M \int_0^t p(s)\psi(\|\bar{z}_s + x_s\|_{\mathcal{B}})ds.$$

But

$$\begin{aligned}\|\bar{z}_s + x_s\|_{\mathcal{B}} &\le \|\bar{z}_s\|_{\mathcal{B}} + \|x_s\|_{\mathcal{B}} \\ &\le K(t)\sup\{\|z(s)\| : 0 \le s \le t\} + M(t)\|z_0\|_{\mathcal{B}} \\ &\quad + K(t)\sup\{\|x(s)\| : 0 \le s \le t\} + M(t)\|x_0\|_{\mathcal{B}} \\ &\le K_1 \sup\{\|z(s)\| : 0 \le s \le t\} + \widetilde{M}_1\|\phi\|_{\mathcal{B}} \\ &\quad + K_1 M\|\phi(0)\|,\end{aligned}$$

so

$$\|\bar{z}_s + x_s\|_{\mathcal{B}} \le K_1 \sup\left\{\|z(s)\| : 0 \le s \le t\right\} + \widetilde{M}_1\|\phi\|_{\mathcal{B}} + K_1 M\|\phi(0)\|. \quad (10.12)$$

Then we have

$$\|\bar{z}_s + x_s\|_{\mathcal{B}} \le a_0(t) + \widetilde{M}_1\|\phi\|_{\mathcal{B}} + K_1 M\|\phi(0)\|,$$

and so

$$|h(t)| \le M \int_0^t p(s)\psi(a_0(s) + c)ds.$$

Using the above inequality in the definition of w, we have

$$\begin{aligned}K_1\|h(t)\| + \widetilde{M}_1\|\phi\|_{\mathcal{B}} + K_1 M\|\phi(0)\| \le K_1 M &\int_0^t p(s)\psi(a_0(s) + c)ds \\ &+ \widetilde{M}_1\|\phi\|_{\mathcal{B}} + K_1 M\|\phi(0)\|.\end{aligned}$$

Thus,

$$K_1\|h(t)\| + \widetilde{M}_1\|\phi\|_{\mathcal{B}} + K_1\widetilde{M}_1\|\phi(0)\| \le c + \int_0^t \widehat{M}(s)\psi(a_0(s) + c)ds.$$

This implies that, for each $t \in [0, t_1]$,

$$K_1\|h(t)\| + \widetilde{M}_1\|\phi\|_{\mathcal{B}} + K_1\widetilde{M}_1\|\phi(0)\| \le c + \int_0^t a_0'(s)ds = a_0(t).$$

Hence, $P_0(\mathcal{K}_0) \subset \mathcal{K}_0$. We shall use the Lemma 2.98 to prove that

$$P_0 : \mathcal{K}_0 \to \mathcal{P}(\mathcal{K}_0),$$

has at least one fixed point. Since F is a multifunction with convex values, P_0 is a multivalued map with convex values.

Claim. *P_0 has a closed graph.*

Let $\{z_n : n \in \mathbb{N}\} \subseteq \mathcal{K}_0$ such that $z_n \longrightarrow z_*$, $h_n \in P_0(z_n)$ and $h_n \longrightarrow h_*$. We shall prove that $h_* \in P_0(z_*)$. Now $h_n \in P_0(z_n)$ means that there exists $v_n \in S_{F,\overline{z}_n+x}$ such that

$$h_n(t) = \int_0^t S'(t-s)v_n(s)ds \quad \text{for each } t \in [0,t_1].$$

We must prove that there exists $v_* \in S_{F,\overline{z}+x}$ such that, for each $t \in [0,t_1]$, we have

$$h_*(t) = \int_0^t S'(t-s)v_*(s)ds.$$

Since $\{z_n : n \in \mathbb{N}\} \subseteq \mathcal{K}_0$ and $\{v_n : n \in \mathbb{N}\} \subseteq F(t, \overline{z}_{t_n} + x_t)$,

$$|v_n(t)| \le p(t)\psi(a_0(t_1)),\ t \in [0,t_1].$$

Furthermore, the set $\{v_n(t) : n \in \mathbb{N}\}$ is relatively compact for a.e. $t \in [0,t_1]$ because, from (10.2.4),

$$\chi(\{v_n(t) : n \in \mathbb{N}\}) \le \chi(F(t, \overline{z}_{t_n} + x_t) \le \bar{p}(t)\chi(\bar{z}_{t_n} + x_t) = 0.$$

Then, $\{v_n : n \in \mathbb{N}\}$ is weakly compact in $L^1([0,t_1]; \overline{D(A)})$ and thus there exists $v_* \in L^1$ such that v_n converges weakly to v_*. From (J-1), we have

$$\|h_1 - h_2\|_0 \le M\|f_1 - f_2\|_{L^1},$$

where

$$h_i(t) = \int_0^t S'(t-s)f_i(s)ds,\ f_i \in L^1([0,t_1], \overline{D(A)}).$$

Thus, P_0 has a closed graph.

Next, we prove that P_0 is a β_0-condensing operator. For a bounded set $B \subset \mathcal{K}_0$, let $mod_C(B)$ be the modulus of quasiequicontinuity of the set of functions B given by

$$mod_C(B) = \lim_{\delta\to 0}\sup_{x\in B}\max_{|\tau_2-\tau_1|\le\delta}\|x(\tau_1) - x(\tau_2)\|,$$

and let γ_0 be the real MNC defined on bounded subset on $\mathcal{K}_0$ by

$$\gamma_0(B) = \sup_{t\in[0,t_1]} e^{-L_0 t}\chi(B(t)).$$

We define the following MNC β_0 on bounded subsets of $\mathcal{K}_0$:

$$\beta_0(B) = \max_{\Delta(\mathcal{K}_0)}(\gamma_0(B), \mathrm{mod}_C(B)),$$

where $\Delta(\mathcal{K}_0)$ is the collection of all the countable subsets of B. Then the MNC β is monotone, regular, and nonsingular (see Example 2.1.4 in [311]). This measure is also used in [131, 133, 174, 175, 223, 225] in the discussion of semilinear evolution differential inclusions.

To show that P_0 is β_0–condensing, let $B \subset \mathcal{K}_0$ be a bounded set in $\mathcal{K}_0$ such that

$$\beta_0(B) \leq \beta_0(P_0(B)). \tag{10.13}$$

We will show that B relatively compact. Let $\{v_n : n \in \mathbb{N}\} \subset P_0(B)$ be the countable set such that

$$\beta_0(P_0(\{f_n : n \in \mathbb{N}\})) = \beta_0(\gamma_0(\{f_n : n \in \mathbb{N}\}), mod_C(\{f_n : n \in \mathbb{N}\})).$$

Since $P = L \circ S_F$, we can represent each f_n as

$$f_n = L(v_n),\ v_n \in S_F(z_n),$$

where $\{z_n : n \in \mathbb{N}\} \subset B$. Therefore, by using (10.13) we have

$$\gamma_0(\{f_n : n \in \mathbb{N}\}) \geq \gamma_0(\{z_n : n \in \mathbb{N}\}).$$

For $t \in [0, t_1]$, we have

$$\begin{aligned}
\chi(\{v_n(t) : n \in \mathbb{N}\}) &\leq \chi(F(t, \{\overline{z}_{n_s}\}_{n=1}^{\infty} + x_s)) \\
&\leq \overline{p}(t) \sup_{-\infty \leq \theta \leq 0} \chi(\{\overline{z}_{n_t}(\theta)\}_{n=1}^{\infty} + x_t(\theta))) \\
&\leq \overline{p}(t) \sup_{0 \leq s \leq t} \chi(\{\overline{z}_n(s)\}_{n=1}^{\infty} + x(s))) \\
&\leq e^{-L_0 t}\overline{p}(t) \sup_{0 \leq s \leq t} e^{-L_0 s}\chi(\{\overline{z}_n(s)\}_{n=1}^{\infty} + x(s))) \\
&\leq e^{-L_0 t}\overline{p}(t)\gamma_0(\{f_n\}_{n=1}^{\infty})).
\end{aligned}$$

From Lemma 2.97,

$$e^{-L_0 t}\chi(\{L_0(v_n)(t)\}_{n=1}^{\infty}) \leq 2M\gamma_0(\{z_n\}_{n=1}^{\infty}) \int_0^t \overline{p}(s)ds.$$

Therefore,

$$\gamma_0(\{z_n\}_{n=1}^{\infty}) = \sup_{t \in [0,t_1]} e^{-L_0 t}\chi(\{z_n(t)\}_{n=1}^{\infty} \leq q_0\gamma_0(\{z_n\}_{n=1}^{\infty}). \tag{10.14}$$

Then,

$$\gamma_0(\{z_n\}_{n=1}^{\infty}) = 0, \text{ and } \mathrm{mod}_C(\{z_n\}_{n=1}^{\infty}) = 0 \text{ implies } \beta_0(\{z_n\}_{n=1}^{\infty}) = 0.$$

As a consequence of Lemma 2.98, we deduce that P_0 has at least one fixed point. Denote this point by z_0.

Step 2. Let the operator $P_1 : C^* \to \mathcal{P}(C^*)$ be defined by

$$h(t) = \begin{cases} 0, & t \in (-\infty, t_1], \\ \int_{t_1}^t S'(t-s)v(s)ds & \\ \quad +S'(t-t_1)I_1(\overline{z}_0(t_1^-) + x(t_1^-)), & t \in [t_1, t_2], \end{cases}$$

where

$$v \in S_{F,\bar{z}+x} = \{v \in L^1([t_1,t_2], \overline{D(A)}) \mid v(t) \in F(t, \bar{z}_t + x_t),\ a.e.\ t \in [t_1,t_2]\},$$

Let

$$C_1 = \{y \in C((t_1,t_2], \overline{D(A)}) : y(t_1^+) \text{ exists}\}.$$

Set $C^* = \mathcal{B} \cap C([0,t_1], \overline{D(A)}) \cap C_1$, $C^*_{t_1} = \{z \in C^* : z_{t_1} = 0\}$, and

$$\mathcal{K}_1 := \{z \in C^*_{t_1} : K_1|z(t)| \le a_1(t),\ t \in [t_1,t_2]\},$$

where

$$\|z\|_1 = \sup\{\|z(s)\| : t_1 \le s \le t\},\ a_1(t) = \Gamma^{-1}\left(\int_{t_1}^t \widehat{M}(s)ds\right),$$

and

$$\Gamma_1(z) = \int_{c_*}^z \frac{du}{\psi(u + c_*)},\ c_* = M\|I_1(\bar{z}_0 + x(t_1))\| + c.$$

We can easily prove that $P_1(\mathcal{K}_1) \subset \mathcal{K}_1$. Next, we prove that P_1 is a β-condensing operator. For a bounded set $B \subset \mathcal{K}_1$, let $mod_C(B)$ be the modulus of quasiequicontinuity of the set of functions B given by

$$mod_C(B) = \lim_{\delta \to 0} \sup_{x \in B} \max_{|\tau_2 - \tau_1| \le \delta} \|x(\tau_1) - x(\tau_2)\|,$$

and let γ_1 be the real MNC defined on bounded subsets on $\overline{U}$ by

$$\gamma_1(B) = \sup_{t \in [t_1,t_2]} e^{-L_1 t} \chi(B(t)).$$

We define an MNC β on bounded subsets of $\mathcal{K}_1$ by

$$\beta_1(B) = \max_{\Delta(\mathcal{K}_1)} (\gamma_1(B), \text{mod}_C(B))$$

where $\Delta(\mathcal{K}_1)$ is the collection of all countable subsets of B. Let $B \subset \mathcal{K}_1$ be a bounded set in $\mathcal{K}_1$ such that

$$\beta_1(B) \le \beta_1(P_1(B)). \tag{10.15}$$

We will show that B is relatively compact. Let $\{v_n : n \in \mathbb{N}\} \subset P_1(B)$ be the countable set such that

$$\beta_1(P_1(\{f_n : n \in \mathbb{N}\})) = \beta_1(\gamma_1(\{f_n : n \in \mathbb{N}\}), \text{mod}_C(\{f_n : n \in \mathbb{N}\})).$$

We have

$$f_n(t) = L(v_n(t)) + S'(t - t_1)I_1(\overline{z}_0(t_1) + x(t_1)).$$

For $t \in [t_1, t_2]$,

$$\begin{aligned}\chi(\{v_n(t) : n \in \mathbb{N}\}) &\leq \chi(F(t, \{\overline{z}_{n_s}\}_{n=1}^{\infty} + x_s) + S'(t - t_1)I_1(\overline{z}_0(t_1) + x(t_1)))\\ &\leq \chi(F(t, \{\overline{z}_{n_s}\}_{n=1}^{\infty} + x_s)) + M\chi(I_1(\overline{z}_0(t_1) + x(t_1)))\\ &\leq \overline{p}(t) \sup_{-\infty \leq \theta \leq 0} \chi(\{\overline{z}_{n_t}(\theta)\}_{n=1}^{\infty} + x_t(\theta)))\\ &\leq \overline{p}(t) \sup_{t_1 \leq s \leq t} \chi(\{\overline{z}_n(s)\}_{n=1}^{\infty} + x(s)))\\ &\leq e^{-L_1 t}\overline{p}(t) \sup_{t_1 \leq s \leq t} e^{-Ls}\chi(\{\overline{z}_n(s)\}_{n=1}^{\infty} + x(s)))\\ &\leq e^{-L_1 t}\overline{p}(t)\gamma_1(\{f_n\}_{n=1}^{\infty}).\end{aligned}$$

From Lemma 2.97,

$$e^{-L_1 t}\chi(\{L_1(v_n)(t)\}_{n=1}^{\infty}) \leq 2M\gamma_1(\{z_n\}_{n=1}^{\infty}) \int_{t_1}^{t} \overline{p}(s)ds.$$

Therefore,

$$\gamma_1(\{z_n\}_{n=1}^{\infty}) = \sup_{t \in [t_1, t_2]} e^{-L_1 t}\chi(\{z_n(t)\}_{n=1}^{\infty} \leq q_1\gamma_1(\{z_n\}_{n=1}^{\infty}).$$

Then,

$$\gamma_1(\{z_n\}_{n=1}^{\infty}) = 0, \text{ and } \mathrm{mod}_C(\{z_n\}_{n=1}^{\infty}) = 0 \text{ implies } \beta(\{z_n\}_{n=1}^{\infty}) = 0.$$

As a consequence of Lemma 2.98, we deduce that P_1 has a fixed point z in $\mathcal{K}_1$. Denote this solution by z_1. As in Step 1 we can prove that $\mathrm{Fix}P_1$ is compact.

Step 3. We continue this process and take into account that $z_m := z\big|_{[t_m,b]}$ is a fixed point of the operator P_m defined by

$$h(t) = \begin{cases} 0, & t \in (-\infty, t_m],\\ \int_{t_m}^{t} S'(t-s)v(s)ds & \\ \quad +S'(t-t_m)I_m(\bar{z}_{m-1}(t_1^-) + x(t_{m-1}^-)), & t \in (t_m, b],\end{cases}$$

and

$$v \in S_{F,\bar{z}+x} = \{v \in L^1([t_1, t_2], \overline{D(A)}) | v(t) \in F(t, \bar{z}_t + x_t),\ a.e.\ t \in J\}.$$

Let

$$C_m = \{z \in C((t_m, b], \overline{D(A)}) : z(t_m^+) \text{ exists}\}.$$

Set $C^* = \mathcal{B} \cap C([t_{m-1}, t_m], \overline{D(A)}) \cap C_m$, $C_{t_m}^* = \{z \in C^* : z_{t_m} = 0\}$, and

$$\mathcal{K}_m := \{z \in C_{t_m}^* : K_1|z(t)| \leq a_m(t),\ t \in [t_m, b]\},$$

where

$$\|z\|_m = \sup\{\|z(s)\| : t_m \le s \le t\},\ a_m(t) = \Gamma^{-1}\left(\int_{t_m}^{t} \widehat{M}(s)ds\right),$$

and

$$\Gamma(z) = \int_{c_m}^{z} \frac{du}{\psi(u + c_m)},\ c_m = M\,\|I_m(\bar{z}_{m-1} + x(t_m))\| + c.$$

Then, there exists a fixed point z_m of P_m, and FixP_m is a compact set in C_{t_m}. The fixed point z of the operator P is then defined by

$$z(t) = \begin{cases} z_0(t), & \text{if } t \in (-\infty, t_1], \\ z_1(t), & \text{if } t \in (t_1, t_2], \\ \dots \\ z_m(t), & \text{if } t \in (t_m, b]. \end{cases}$$

□

Theorem 10.3. *Assume that* $F : J \times \mathcal{B} \to \mathcal{P}_{cp,cv}(E)$ *and conditions* (J-1)–(J-6) *hold. Then* Fix$N$ *is a compact set and u.s.c., where* N *is defined in the proof of Theorem 10.2.*

Proof. Fix$N$ is compact. We prove only that Fix$P$ is compact where P is defined in Theorem 10.2. It clear that Fix$P \subset \bigcap_{k=0}^{k=m}$ FixP_k, so we prove that Fix$P$ is a closed set in C_0. We shall prove that for every $k = 1, \dots, m$, FixP_k is a compact set in C_{t_k}. First, we prove that FixP_0 is a compact set. From Lemma 2.99, we only need to show that FixP_0 is bounded. Let $z \in$ FixP_0. Then,

$$\|z\|_\infty \le \frac{a_0(t_1)}{K_1}.$$

This implies that FixP_0 is compact in C_{t_0}. By analogy, FixP_k is a compact set in C_{t_k}, $k = 1, \dots, m$.

Let $z_n \in$ Fix$P$ such that $z_n \to z$ as $n \to \infty$. We shall show that $z \in P(z)$. Now $z_n \in$ Fix$P$ means that there exists $v_n \in S_{F,\bar{z}_n + x}$ such that, for each $t \in J$,

$$z_n(t) = \int_0^t S'(t-s)v_n(s)ds + \sum_{0<t_k<t} S'(t-t_k)I_k(\bar{z}_n(t_k^-) + x(t_k)).$$

We need to show there exists $v_* \in S_{F,\bar{z}+x}$ such that, for each $t \in J$,

$$z(t) = \int_0^t S'(t-s)v_*(s)ds + \sum_{0<t_k<t} S'(t-t_k)I_k(\bar{z}(t_k^-) + x(t_k)).$$

Since FixP_k, $k = 0, \dots, m$, are compact, we can find subsequences $z_{n_k} \in$ FixP_k and $v_{n_k} \in L^1([t_{k-1}, t_k], \overline{D(A)})$ converging to v_k, such that z_{n_k} converges to z_k in C_{t_k}, and

$$z_k(t) = S'(t-t_k)I_k(\bar{z}_k(t_k) + x(t_k)) + \int_{t_{k-1}}^{t} S'(t-s)v_k(s)ds$$

for some $v_k \in L^1([t_{k-1}, t_k], \overline{D(A)})$. Set

$$v_*(t) = \begin{cases} v_0(t), & \text{if } t \in [0, t_1], \\ v_1(t), & \text{if } t \in (t_1, t_2], \\ \dots \\ v_m(t), & \text{if } t \in (t_m, b], \end{cases}$$

and

$$z(t) = \begin{cases} z_0(t), & \text{if } t \in [0, t_1], \\ z_1(t), & \text{if } t \in (t_1, t_2], \\ \dots \\ z_m(t), & \text{if } t \in (t_m, b]. \end{cases}$$

Since $I_k,\ k = 1, \dots, m$, are continuous functions, we have

$$\begin{aligned} \|z_n - z_*\|_\infty \le M \sum_{k=1}^{m} &\|I_k(\bar{z}_n(t_k^-) + x(t_k)) - I_k(\bar{z}(t_k^-) + x(t_k))\| \\ &+ M\|v_n - v_*\|_{L^1} \longrightarrow 0, \text{ as } n \to \infty, \end{aligned}$$

where

$$z_*(t) = \int_0^t S'(t-s)v_*(s)ds + \sum_{0<t_k<t} S'(t-t_k)I_k(\bar{z}(t_k^-) + x(t_k)).$$

Hence,

$$z(t) = \int_0^t S'(t-s)v_*(s)ds + \sum_{0<t_k<t} S'(t-t_k)I_k(\bar{z}(t_k^-) + x(t_k)).$$

Then the solution set of the problem (10.1)–(10.3) is compact.

Now we prove that $\phi \to \mathrm{Fix}N(\phi)$, is u.s.c. As in Theorem 9.31, (see also [178, Theorem 4.3]), we can prove that $\mathrm{Fix}N(\cdot)$ is quasicompact; we prove only that its graph is closed. Let $\phi_n \to \phi,\ y_n \to y$ and $y_n \in \mathrm{Fix}N(\phi_n)$. We show that $y \in \mathrm{Fix}N(\phi)$. As in Theorem 10.2, we can find

$$\begin{aligned} z_n(t) = \int_0^t S'(t-s)v_n(s)ds + \sum_{0<t_k<t} S'(t-t_k)I_k(\bar{z}_n(t_k^-) \\ + x_n(t_k)),\ v_n \in S_{z_n + x_n}. \end{aligned}$$

Consider the linear continuous operator $\Gamma : L^1([0,b], \overline{D(A)}) \longrightarrow C_0$ defined by

$$(\Gamma v)(t) = \int_0^t S'(t-s)v(s)ds.$$

We have

$$\|(h_n - \sum_{0<t_k<t} S'(t-t_k)I_k(\bar{z}_n(t_k^-) + x_n(t_k)))$$
$$- (h_* - \sum_{0<t_k<t} S'(t-t_k)I_k(\bar{z}(t_k^-) + x(t_k)))\|_\infty \longrightarrow 0 \text{ as } n \longrightarrow \infty.$$

From Lemma 2.65, it follows that $\Gamma \circ S_F$ is a closed graph operator. Also, from the definition of Γ, we have that

$$h_n(t) - \sum_{0<t_k<t} S'(t-t_k)I_k(\bar{z}_n(t_k^-) + x_n(t_k))) \in \Gamma(S^1_{F,\bar{z}_n+x_n}).$$

Since $\bar{z}_n + x_n \longrightarrow \bar{z} + x$, it follows from Lemma 2.65 that

$$h_*(t) = \int_0^t S'(t-s)v_*(s)ds + \sum_{0<t_k<t} S'(t-t_k)I_k(\bar{z}(t_k^-) + x(t_k)),\ t \in [0,b]$$

for some $v_* \in S^1_{F,\bar{z}+x}$. Thus, $y \in \mathrm{Fix}N(\phi)$. □

10.3 Exact Controllability

10.3.1 Controllability of Impulsive FDIs

In this section, we shall establish sufficient conditions for the controllability of the first order functional semilinear differential inclusions (10.4)–(10.6). We consider the space

$$\begin{aligned} PC &= \{y : (-\infty,\infty) \to \overline{D(A)} \mid y(t_k^-),\ y(t_k^+) \text{ exist with} \\ &\quad y(t_k) = y(t_k^-),\ y_k \in C(J_k, \overline{D(A)}),\ k = 1,\ldots\}, \end{aligned}$$

where y_k is the restriction of y to $J_k = (t_k, t_{k+1}]$, $k = 0, \ldots$. We will assume that $\mathcal{B}$ satisfies the following axioms:

($\bar{A}$) If $y : (-\infty,\infty) \to \overline{D(A)}$ and $y_0 \in \mathcal{B}$, then for every t in $[0,\infty)$ the following conditions hold:

(i) y_t is in $\mathcal{B}$;

(ii) $\|y_t\|_{\mathcal{B}} \leq K(t)\sup\{|y(s)| : 0 \leq s \leq t\} + M(t)\|y_0\|_{\mathcal{B}}$;

(iii) $|y(t)| \leq H\|y_t\|_{\mathcal{B}}$;

where $H \geq 0$ is a constant, $K : [0,\infty) \to [0,\infty)$ is continuous, $M : [0,\infty) \to [0,\infty)$ is locally bounded, and H, K, and M are independent of $y(\cdot)$.

($\bar{A}$-1) For the function $y(\cdot)$ in (A), y_t is a $\mathcal{B}$-valued continuous function on $[0,\infty)\backslash\{t_1,\ldots\}$.

(A-2) The space $\mathcal{B}$ is complete.

Let

$$K_n = \sup\{\|k(t)\| : t \in [0, t_n]\}, \quad \widetilde{M}_n = \sup\{\|M(t)\| : t \in [0, t_n]\}, \quad n \in \mathbb{N}.$$

Set

$$\mathcal{B}_* = \{y : (-\infty, \infty) \to \overline{D(A)} : y \in PC \cap \mathcal{B}\}.$$

Definition 10.4. A function $y \in \mathcal{B}_*$ is said to be an integral solution of (10.4)–(10.6) if there exists a function $v \in L^1(J, \overline{D(A)})$ such that $v(t) \in F(t, y_t)$ a.e. $t \in [0, \infty)$ and

$$y(t) = S'(t)\phi(0) + A\int_0^t y(s)ds + \int_0^t v(s)ds + \int_0^t Bu(s)ds + \sum_{0<t_k<t} S'(t-t_k)I_k(y(t_k^-)),$$

$\int_0^t y(s)ds \in \overline{D(A)}$ for $t \in [0, \infty)$, and $y(t) = \phi(t)$ for $t \in (-\infty, 0]$.

Before stating and proving our main result in this section, we define controllability on the interval J_n.

Definition 10.5. The system (10.4)–(10.6) is said to be infinitely controllable on the interval $(-\infty, \infty)$, if for every initial function $\phi \in \mathcal{B}$ and every $y_1 \in \overline{D(A)}$, and for each $n \in \mathbb{N}$, there exists a control $u \in L^2([0, t_n], U)$, such that the integral solution $y(t)$ of (10.4)–(10.6) satisfies $y(t_n) = y_1$.

Theorem 10.6. *In addition to* (J-1)–(J-6) *assume that the following conditions hold:*

(J-7) $\phi(0) \in \overline{D(A)}$ *and there exist constants* c_k, *such that* $|I_k(y)| \le c_k$, *for each* $y \in \overline{D(A)}$.

(J-8) $F : [0, \infty) \times \mathcal{B} \longrightarrow \mathcal{P}(\overline{D(A)})$ *is a nonempty compact-valued, multivalued map and for all* $R > 0$, *there exists* $l_R \in L^1_{loc}((-\infty, \infty), \mathbb{R}_+)$ *such that*

$$H(F(t,x), F(t,\overline{x})) \le l_R(t)\|x - \overline{x}\|_{\mathcal{B}} \text{ for all } x, \overline{u} \in \mathcal{D}$$

with

$$\|x\|_{\mathcal{B}}, \|\overline{x}\|_{\mathcal{B}} \le R,$$

and

$$H(0, F(t,0)) \le l_R(t) \text{ for a.e. } t \in J.$$

(J-9) *There exist a continuous nondecreasing function* $\psi : [0, \infty) \to (0, \infty)$ *and* $p \in L^1(J, \mathbb{R}_+)$ *such that*

$$\|F(t,u)\| \le p(t)\psi(\|u\|_{\mathcal{B}}) \text{ for a.e. } t \in J \text{ and each } u \in \mathcal{B},$$

where for any $c > 0$

$$\int_c^\infty \frac{du}{u + \psi(u)} = \infty.$$

(J-10) *For every* $n > 0$ *the linear operator* $W : L^2(J_n, U) \to E$ $(J_n = [0, t_n])$, *defined by*

$$Wu = \int_0^{t_n} S'(t_n - s)Bu(s)ds,$$

has an invertible operator $W^{-1} : E \to L^2(J_n, U)\backslash KerW$, *and there exist positive constants* $\overline{M}$, $\overline{M}_1$ *such that* $\|B\| \leq \overline{M}$ *and* $\|W^{-1}\| \leq \overline{M}_1$.

(J-11) *There exist constants* $b_k \geq 0$ *such that*

$$\|I_k(y) - I_k(x)\| \leq b_k \|x - \overline{x}\| \text{ for each } x, \overline{x} \in \overline{D(A)}.$$

(J-12) *There exist constants* $M > 0, \omega \in \mathbb{R}$ *such that*

$$\|S'(t)\| \leq Me^{-\omega t}, \text{ for each } t \geq 0.$$

If $\sum_{k=1}^{\infty} Mb_k < 1$, *then the IVP* (10.4)–(10.6) *has at least one integral solution.*

Remark 10.7. For the construction of W see [424].

Proof. Using hypothesis (J-10), for each $y(\cdot)$ and

$$v \in S_{F,y} = \{v \in L^1(J, \overline{D(A)}) : v(t) \in F(t, y_t) \ a.e \ t \in J\},$$

define the control

$$u_y(t) = W^{-1}\Big[y_1 - S'(t_n)\phi(0) - \lim_{\lambda \to +\infty} \int_0^{t_n} S'(t_n - s)B_\lambda v(s)ds - \sum_{0<t_k<t} S'(t_n - t_k)I_k(y(t_k^-))\Big](t).$$

To transform the problem (10.4)–(10.6) into a fixed point problem, consider the operator $N : \Omega \to \mathcal{P}(\Omega)$ defined by

$$N(y) = \left\{ h \in \Omega : h(t) = \begin{cases} \phi(t), & t \in (-\infty, 0], \\ S'(t)\phi(0) + \frac{d}{dt}\int_0^t S(t-s)v(s)ds \\ \quad + \frac{d}{dt}\int_0^t S(t-s)(Bu_y)(s)ds \\ \quad + \sum_{0<t_k<t} S'(t-t_k)I_k(y(t_k^-)), & t \in J, \end{cases} \right\}$$

where $v \in S_{F,y}$. Clearly, the fixed points of the operator N are solutions of the problem (10.4)–(10.6).

Let

$$\mathcal{B}_*^k = \big\{z : (-\infty, t_k] \to \overline{D(A)} \mid z(t_i^-), z(t_i^+) \text{ exist with } z(t_i) = z(t_i^-),$$
$$z_k \in C(J_i, \overline{D(A)}),\ i = 1, \dots, k-1, \text{ and } z_0 = 0\big\}.$$

For any $z \in \mathcal{B}_*^k$, we have

$$\|z\|_k = \|z_0\|_{\mathcal{B}} + \sup\big\{\|z(s)\| : 0 \le s \le t_k\big\} = \sup\big\{\|z(s)\| : 0 \le s \le t_k\big\}.$$

Thus, $(\mathcal{B}_*^k, \|\cdot\|_k)$ is a Banach space. Set

$$C_0 = \{z \in \mathcal{B}_* : z_0 = 0\}.$$

We define a family of seminorms on C_0, thus rendering C_0 a Fréchet space.

Let τ be sufficiently large. For each $n \in \mathbb{N}$, define in Ω the seminorms by

$$\|z\|_n = \sup\{e^{-|\omega|t - \tau L_n(t)}|y(t)| : 0 \le t \le t_n\},$$

where $L_n(t) = \int_0^t \widetilde{l}_n(s)ds$ and

$$\widetilde{l}_n(t) = \max\big\{t_n \overline{M} M^2 \overline{M}_1 e^{|\omega|t_n} l_n(t),\ \overline{M} M^2 \overline{M}_1 e^{|\omega|t_n},\ M l_n(t)\big\}.$$

Then C_0 is a Fréchet space with this family of seminorms $\|\cdot\|_n$. Let the operator $P : C_0 \to \mathcal{P}(C_0)$ be defined by $P(z) = \{h \in C_0\}$, where

$$h(t) = \begin{cases} 0, & t \in (-\infty, 0], \\ \int_0^t S'(t-s)v(s)ds + \int_0^t S'(t-s)Bu_{\bar{z}+x}(s)ds & \\ \quad + \sum_{0<t_k<t} S'(t-t_k)I_k(\bar{z}(t_k^-) + x(t_k^-)), & t \in [0, \infty), \end{cases}$$

and

$$v \in S_{F,\bar{z}+x} = \{z \in L^1(J, \overline{D(A)}) | v \in F(t, \bar{z}(t) + x(t),\ a.e.\, t \in J\}.$$

That the operator N having a fixed point is equivalent to P having one is clear, and so we next prove that P does have a fixed point. We shall use the Frigon nonlinear alternative (Lemma 2.90) to do this.

Let $z \in \gamma P(z)$, for some $\gamma \in [0,1]$ and for some $v \in S_{F,\bar{z}+x}$. Given $n \in \mathbb{N}$ and $t \le t_n$ such that

$$z(t) = \gamma\Big[\int_0^t S'(t-s)v(s)ds + \int_0^t S'(t-s)Bu_{\bar{z}_s+x_s}(s)ds$$
$$+ \sum_{0<t_k<t} S'(t-t_k)I_k(\bar{z}(t_k^-) + x(t_k^-))\Big],$$

we have

$$
\begin{aligned}
|z(t)| &\le Me^{|\omega|t} \int_0^t e^{-|\omega|s} p(s)\psi(\|\bar{z}_s\|_{\mathcal{B}} + \|x_s\|_{\mathcal{B}})ds \\
&\quad + Me^{|\omega|t} \int_0^t e^{-|\omega|s} \|(Bu_{\bar{z}_s+x_s}(s)\|ds + Me^{|\omega|t} \sum_{k=1}^{n} e^{-|\omega|t_k} c_k \\
&\le Me^{|\omega|t} \int_0^t e^{-|\omega|s} p(s)\psi(\|\bar{z}_s + x_s\|_{\mathcal{B}})ds \\
&\quad + M\overline{M}e^{|\omega|t}\overline{M}_1 t_n \Big(\|y_1\| + Me^{|\omega|t_n}|\phi(0)| + Me^{|\omega|t_n} \sum_{k=1}^{n} e^{-|\omega|t_k} c_k \\
&\quad + Me^{|\omega|t_n} \int_0^t e^{-|\omega|s} p(s)\psi(\|\bar{z}_s\| + \|x_s\|_{\mathcal{B}})ds \Big) + Me^{|\omega|t} \sum_{k=1}^{n} e^{-|\omega|t_k} c_k.
\end{aligned}
$$

But

$$
\begin{aligned}
\|\bar{z}_s + x_s\|_{\mathcal{B}} &\le \|\bar{z}_s\|_{\mathcal{B}} + \|x_s\|_{\mathcal{B}} \\
&\le K(t)\sup\{\|z(s)\| : 0 \le s \le t + M(t)\|z_0\|_{\mathcal{B}} \\
&\quad + K(t)\sup\{\|x(s)\| : 0 \le s \le t\} + M(t)\|x_0\|_{\mathcal{B}} \\
&\le K_n \sup\{\|z(s)\| : 0 \le s \le t\} + \widetilde{M}_n\|\phi\|_{\mathcal{B}} + K_n\widetilde{M}_n\|\phi(0)\|.
\end{aligned}
\tag{10.16}
$$

If we let $w(t)$ denote the right hand side of (10.16), then we have

$$
\|\bar{z}_s\|_{\mathcal{B}} + \|x_s\|_{\mathcal{B}} \le w(t).
$$

Therefore,

$$
\begin{aligned}
|z(t)| &\le Me^{|\omega|t} \int_0^t e^{-|\omega|s} p(s)\psi(w(s))ds \\
&\quad + M\overline{M}e^{|\omega|t}\overline{M}_1 t_n \Big(\|y_1\| + Me^{|\omega|t_n}|\phi(0)| + Me^{|\omega|t_n} \sum_{k=1}^{n} e^{-|\omega|t_k} c_k \\
&\quad + Me^{|\omega|t_n} \int_0^t e^{-|\omega|s} p(s)\psi(w(s))ds \Big) + Me^{|\omega|t} \sum_{k=1}^{n} e^{-|\omega|t_k} c_k.
\end{aligned}
$$

Using this in the definition of w, we have

$$
\begin{aligned}
w(t) &\le \widetilde{M}_n\|\phi\|_{\mathcal{B}} + K_n\widetilde{M}_n\|\phi(0)\| + K_n Me^{|\omega|t} \int_0^t e^{-|\omega|s} p(s)\psi(w(s))ds \\
&\quad + K_n M\overline{M}e^{|\omega|t}\overline{M}_1 t_n \Big(\|y_1\| + Me^{|\omega|t_n}|\phi(0)| + K_n Me^{|\omega|t_n} \sum_{k=1}^{n} e^{-|\omega|t_k} c_k \\
&\quad + K_n Me^{|\omega|t_n} \int_0^t e^{-|\omega|s} p(s)\psi(w(s))ds \Big) + K_n Me^{|\omega|t} \sum_{k=1}^{n} e^{-|\omega|t_k} c_k.
\end{aligned}
$$

Thus,

$$\begin{aligned} e^{-|\omega|t}w(t) \leq & \widetilde{M}_n\|\phi\|_{\mathcal{B}} + K_n\widetilde{M}_n\|\phi(0)\| \\ & + K_nM\sum_{k=1}^{n} e^{-\omega t_k}c_k + K_n\overline{M}M\overline{M}_1 t_n\|y_1\| \\ & + K_n\overline{M}M\overline{M}_1 t_n[Me^{|\omega|t_n}|\phi(0)| + K_nMe^{|\omega|t_n}\sum_{k=1}^{n} e^{-|\omega|t_k}c_k] \\ & + \int_0^t e^{-|\omega|s}m(t)\psi(w(s))ds, \end{aligned}$$

where

$$m(t) = K_n[M + \overline{M}M^2\overline{M}_1 t_n e^{|\omega|t_n}]p(t) + |\omega|.$$

Denote the right-hand side of the above inequality by $v(t)$. Then we have

$$w(t) \leq e^{\omega t}v(t) \text{ for all } \quad t \in [0, t_n],$$

and

$$\begin{aligned} v(0) = & \widetilde{M}_n\|\phi\|_{\mathcal{B}} + K_n\widetilde{M}_n\|\phi(0)\| + K_n\overline{M}M\overline{M}_1 t_n\|y_1\| \\ & + K_n\overline{M}M^2\overline{M}_1 t_n\phi(0)| + K_n[\overline{M}M^2\overline{M}_1 t_n + M]\sum_{k=1}^{n} e^{-|\omega|t_k}c_k, \end{aligned}$$

and

$$v'(t) = e^{-|\omega|t}m(t)\psi(\mu(t)), \quad t \in [0, t_n].$$

Using the increasing character of ψ gives

$$v'(t) \leq e^{-|\omega|t}m(t)\psi(e^{|\omega|t}v(t)) \text{ a.e. } \quad t \in [0, t_n].$$

So for each $t \in [0, t_n]$, we have

$$\begin{aligned} (e^{|\omega|t}v(t))' &= |\omega|e^{|\omega|t}v(t) + v'(t)e^{|\omega|t} \\ &\leq |\omega|e^{|\omega|t}v(t) + m(t)\psi(e^{|\omega|t}v(t)) \\ &\leq m(t)[e^{|\omega|t}v(t) + \psi(e^{|\omega|t}v(t))], \quad t \in [0, t_n]. \end{aligned}$$

Thus,

$$\int_{v(0)}^{e^{|\omega|t}v(t)} \frac{du}{u + \psi(u)} \leq \int_0^{t_n} m(s)ds < \infty.$$

Condition (J-9) implies there exists a constant d_n such that $v(t) \leq d_n$, $t \in [0, t_n]$, and hence $\sup\{\|z(t)\| : t \in [0, t_n]\} \leq d_n$. Set

$$U = \{y \in B_* : \|z\|_n < d_n + 1 \text{ for all } n \in \mathbb{N}\}.$$

Clearly, U is a open subset of C_0. We shall show that $P : U \to \mathcal{P}(C_0)$ is a contraction and an admissible operator.

First, we prove that P is a contraction; that is, there exists $0 \leq \gamma < 1$, such that

$$H(P(z), P(z_*)) \leq \gamma \|z - z_*\|_n, \quad \text{for each } z, z_* \in U.$$

Let z, $z_* \in U_0$ and $h \in P(z)$. From (J-9), it follows that

$$H(F(t, \bar{z}_t + x_t), F(t, \bar{z_*}_t + x_t)) \leq l(t)\|\bar{z}_t - \bar{z_*}_t\|_{\mathcal{B}}.$$

Hence, there exists $w \in F(t, \bar{z}_{*_t} + x_t)$ such that

$$|v(t) - w| \leq l_n(t)\|\bar{z}_t - \bar{z_*}_t\|_{\mathcal{B}}, \quad t \in J.$$

Consider $U_* : [0, t_n] \to \mathcal{P}(E)$, given by

$$U_*(t) = \{w \in \overline{D(A)} : |v(t) - w| \leq l_n(t)\|\bar{z}_t - \bar{z_*}_t\|_{\mathcal{B}}\}.$$

Since the multivalued operator $V_*(t) = U_*(t) \cap F(t, \bar{z_*}_t + x_t)$ is measurable (see Proposition III.4 in [136]), there exists a function $v_*(t)$, which is a measurable selection for V_*. So, $v_*(t) \in F(t, \bar{z}_t + x_t)$ and

$$|v(t) - \overline{v}(t)| \leq l(t)\|\bar{z}_t - \bar{z_*}_t\|_{\mathcal{B}}, \quad \text{for each } t \in [0, t_n].$$

Let us define, for each $t \in [0, t_n]$,

$$\begin{aligned} h_*(t) = & \int_0^t S'(t-s)v_*(s)ds + \int_0^t S'(t-s)Bu_{\bar{z}_*+x}(s)ds \\ & + \sum_{0<t_k<t} S'(t-t_k)I_k(\bar{z}_*(t_k^-) + x(t_k^-)). \end{aligned}$$

Then, we have

$$\begin{aligned} \|h(t) - \overline{h}(t)\| = & \left\| \frac{d}{dt}\int_0^t S(t-s)[(Bu_{\overline{z}+x})(s) - (Bu_{\overline{z}_*+x})(s)]ds \right. \\ & + \frac{d}{dt}\int_0^t S(t-s)[v(s) - \overline{v}(s)]ds \\ & \left. + \sum_{0<t_k<t_n} S'(t-t_k)[I_k(\overline{z}(t_k) + x(t_k)) - I_k((\overline{z}_*(t_k) + x_t))] \right\| \\ \leq & \, Me^{|\omega|t}\int_0^t e^{-|\omega|s}\|B\|\|u_{\overline{z}+x}(s) - u_{\overline{z}_*+x}(s)\|ds \\ & + Me^{|\omega|t}\int_0^t l_n(s)e^{-|\omega|s}\|z_s - z*_s\|_{\mathcal{B}}ds \\ & + Me^{\omega t}\sum_{k=1}^{n} e^{-\omega t_k}\|I_k(\overline{z}(t_k) + x(t_k)) - I_k((\overline{z}_*(t_k) + x(t_k)))\| \end{aligned}$$

$$
\begin{aligned}
&\leq t_n\overline{M}_1 M^2\overline{M} e^{|\omega|t}\int_0^t e^{-|\omega|s}e^{|\omega|t_n}\|v(s)-\overline{v}(s)\|ds \\
&\quad + \overline{M}_1 M^2\overline{M} e^{|\omega|t}\int_0^t \sum_{k=1}^{n} e^{|\omega|(t_n-t_k)} b_k\|z(t_k)-z_*(t_k)\|ds \\
&\quad + e^{|\omega|t}\int_0^t e^{-|\omega|s} l_n(s)\|z_s-z_{*_s}\|_{\mathcal{B}}ds \\
&\quad + Me^{|\omega|t}\sum_{k=1}^{n} e^{-|\omega|t_k} b_k\|z(t_k)-z_*(t_k)\| \\
&\leq K_n e^{|\omega|t}\int_0^t t_n\overline{M}_1 M^2\overline{M} e^{\omega t_n} l_n(s)e^{-|\omega|s}\sup_{s\in[0,t]}\|z(s)-z(s)\|ds \\
&\quad + e^{\omega t}\int_0^t \overline{M}_1 M^2\overline{M} e^{|\omega|t_n}\sum_{k=1}^{n} e^{|\omega|t_k} b_k\|z(s)-z_*(s)\|ds \\
&\quad + e^{|\omega|t}\int_0^t Ml_n(s)e^{-|\omega|s}\|z_s-z_{*_s}\|_{\mathcal{B}}ds \\
&\quad + e^{|\omega|t}\sum_{k=1}^{n} Mb_k e^{-|\omega|t_k}\|z(s)-z_*(s)\| \\
&\leq e^{|\omega|t}\int_0^t \widetilde{l}_n(s)e^{-|\omega|s}\|z_s-z_{*_s}\|_{\mathcal{B}}ds \\
&\quad + e^{\omega t}\int_0^t \widetilde{l}_n(s)\sum_{k=1}^{n} e^{|\omega|t_k} b_k\|z(t_k)-z_*(t_k)\|ds \\
&\quad + e^{|\omega|t}\int_0^t \widetilde{l}_n(s)e^{-|\omega|s}\|z_s-z_{*_s}\|_{\mathcal{B}}ds \\
&\quad + e^{|\omega|t}\sum_{k=1}^{n} Mb_k e^{-|\omega|t_k}\|z(t_k)-z_*(t_k)\| \\
&\leq e^{|\omega|t}\int_0^t \widetilde{l}_n(s)e^{\tau L_n(s)}e^{-|\omega|s}e^{-\tau L_n(s)}\sup_{s\in[0,t]}\|z(s)-z_*(s)\|ds \\
&\quad + e^{\omega t}\int_0^t \widetilde{l}_n(s)e^{\tau L_n(s)}\sum_{k=1}^{n} e^{-|\omega|t_k}e^{-\tau L_n(s)} b_k\|z(t_k)-z_*(t_k)\|ds \\
&\quad + e^{|\omega|t}\int_0^t \widetilde{l}_n(s)e^{\tau L_n(s)}e^{-|\omega|s}e^{-\tau L_n(s)}\sup_{s\in[0,t]}\|z(s)-z_*(s)\|ds \\
&\quad + e^{|\omega|t}\sum_{k=1}^{n} Mb_k e^{\tau L_n(t)}e^{-\tau L_n(t_k)}e^{-|\omega|t_k}\|z(t_k)-z_*(t_k)\|
\end{aligned}
$$

$$
\begin{aligned}
&\le e^{|\omega|t} \int_0^t (e^{\tau L_n(s)})' ds \|z - z_*\|_n \\
&\quad + e^{\omega t} \int_0^t (e^{\tau L_n(s)})' ds \sum_{k=1}^n b_k \|z - z_*\|_n \\
&\quad + e^{|\omega|t} \int_0^t (e^{\tau L_n(s)})' ds \|z - z_*\|_n \\
&\quad + e^{|\omega|t} \sum_{k=1}^n M b_k e^{\tau L_n(t)} \|z - z_*\|_n \\
&\le e^{|\omega|t + \tau L_n(t)} \frac{2}{\tau} \|z - z_*\|_n + e^{|\omega|t + L_n(t)} \sum_{k=1}^n \frac{b_k}{\tau} \|z - z_*\|_n \\
&\quad + e^{|\omega|t + \tau L_n(t)} \sum_{k=1}^n M b_k \|z - z_*\|_n.
\end{aligned}
$$

Thus,

$$
e^{-|\omega|t - \tau L_n(t)} \|h(t) - \overline{h}(t)\| \le \left(\frac{2}{\tau} + \sum_{k=1}^n \frac{b_k}{\tau} + \sum_{k=1}^n M b_k \right) \|z - z_*\|_n.
$$

Therefore,

$$
\|h - \overline{h}_n\|_n \le \left(\frac{2}{\tau} + \sum_{k=1}^n \frac{b_k}{\tau} + \sum_{k=1}^n M b_k \right) \|z - z_*\|_n.
$$

By an analogous relation, obtained by interchanging the roles of z and z_*, it follows that

$$
H_d(P(z), P(z_*)) \le \left(\frac{2}{\tau} + \sum_{k=1}^n \frac{b_k}{\tau} + \sum_{k=1}^n M b_k \right) \|z - \overline{z}_*\|_n,
$$

so P is a contraction.

Now, $P : \mathcal{B}_*^n \to \mathcal{P}_{cp}(\mathcal{B}_*^n)$ is given by

$$
P(y) = \left\{ h \in \mathcal{B}_*^n : h(t) = \begin{cases} 0, & t \in (-\infty, 0], \\ \int_0^t S'(t-s) v(s) ds & \\ \quad + \int_0^t S'(t-s)(B u_{\bar{z}+x}^n)(s) ds & \\ \quad + \sum_{0<t_k<t} S'(t-t_k) I_k(\bar{z}(t_k^-) & \\ \quad + x(t_k^-)), & t \in [0, t_n], \end{cases} \right\}
$$

where $v \in S_{F,\bar{z}+x}^n = \{u \in L^1([0,t_n], \overline{D(A)}) : u \in F(t, \bar{z}_t + x_t)$ a.e. $t \in [0, t_n]\}$. Let $x \in \overline{U}$ and $\varepsilon > 0$. If $x \notin P(x)$, then $d_n(x, P(x)) \neq 0$. Since $P(x)$ is compact, there exists $y \in N(x)$ such that $d_n(x, P(x)) = |x - y|_n$, and we have

$$|x - y|_n \le d_n(x, P(x)) + \varepsilon.$$

If $x \in P(x)$, then we may take $y = x$. Therefore P is an admissible operator contraction.

By Lemma 2.90, P has at least one fixed point, y that is an integral solution to (10.4)–(10.6). □

10.3.2 Controllability of Impulsive Neutral FDIs

In this subsection, we study the problem (10.7)–(10.9). We give first the definition of an integral solution of this problem.

Definition 10.8. A function $y \in \Omega$ is said to be an integral solution of (10.7)–(10.9) if there exists $v \in L^1(J, \overline{D(A)})$ such that $v(t) \in F(t, y_t)$ a.e. $t \in J$, and $y(t)$ satisfies the integral equation

$$\begin{aligned} y(t) = {} & \phi(0) - g(0, \phi(0) + g(t, y_t) + A\int_0^t y(s)ds + \int_0^t v(s)ds \\ & + \int_0^t (Bu)(s)ds + \sum_{\le t_k < t} S'(t - t_k) I_k(y(t_k^-)), \end{aligned}$$

$y(t) = \phi(t)$ for $t \in [-r, 0]$, and $\int_0^t y(s)ds \in D(A)$ for $t \in [0, \infty)$.

Definition 10.9. The system (10.7)–(10.9) is said to be controllable on the interval $(-\infty, \infty)$, if for every continuous initial function $\phi \in \mathcal{B}$, every $y_1 \in \overline{D(A)}$ and each $n \in \mathbb{N}$, there exists a control $u \in L^2([0, t_n], U)$, such that the integral solution $y(t)$ of (10.6)–(10.7) satisfies $y(t_n) = y_1$.

We also need the following notation.

Let $\overline{L}_n(t) = \int_0^t \widehat{l}_n(s)ds$,where

$$\widehat{l}_n(t) = \max\{t_n \overline{M} M^2 \overline{M}_1 e^{|\omega| t_n} l_n(t),\ \overline{M} M^2 \overline{M}_1 e^{|\omega| t_n}, \overline{M} M \overline{M}_1, M l_n(t)\}.$$

For each $n \in \mathbb{N}$ we define in B_* the seminorms by

$$\|z\|_n = \sup\{e^{-|\omega| t - \tau \overline{L}_n(t)} |z(t)| : 0 \le t \le t_n\}.$$

Then Ω is a Fréchet space with a family of seminorms $\{\| \cdot \|_n\}$.

Theorem 10.10. *In addition to the Hille–Yosida condition and* (J-7)–(J-12), *assume that:*

(J-13) *There exist constants* $0 \leq d_1 < 1$ *and* $d_2 \geq 0$ *such that*

$$|g(t,u)| \leq d_1 \|u\|_{\mathcal{B}} + d_2 \ \ t \in [0,\infty), \ u \in \mathcal{D};$$

(J-14) *For each* $R > 0$, *there exists a function* $c_R^* > 0$ *such that*

$$|g(t,x) - g(t,\overline{x})| \leq c_R^* \|x - \overline{x}\|_{\mathcal{B}}, \ \ t \in [0,\infty), \ x, \ \overline{x} \in \mathcal{D}$$

with

$$\|x\|_{\mathcal{B}}, \ \|\overline{x}\|_{\mathcal{B}} \leq R.$$

If for each $n > 0$, *we have* $c_n^* + M \sum_{k=1}^{\infty} b_k < 1$, *then the IVP (10.7)–(10.9) is controllable on* $(-\infty,\infty)$.

Proof. By using hypothesis (J-10) for each $y(\cdot)$ and

$$v \in S_{F,y} = \{v \in L^1(J, \overline{D(A)}) : v(t) \in F(t, y_t) \ a.e \ t \in J\},$$

define the control

$$\begin{aligned} u_y(t) = W^{-1}&[y_1 - S'(t_n)[\phi(0) - g(0,\phi(0))] - g(t_n, y_{t_n}) \\ &- \lim_{\lambda \to +\infty} \int_0^{t_n} S'(t_n - s) B_\lambda v(s) ds \\ &- \sum_{0<t_k<t} S'(t_n - t_k) I_k(y(t_k^-))](t). \end{aligned}$$

We transform the problem (10.7)–(10.9) into a fixed point problem by considering the operator $N_1 : \Omega \to \mathcal{P}(\Omega)$ defined by,

$$N_1(y) := \left\{ h \in B_* : h(t) = \left\{ \begin{array}{ll} \phi(t), & t \in [-r,0], \\ S'(t)[\phi(0) - g(0,\phi(0))] & \\ \quad + g(t,y_t) & \\ \quad + \frac{d}{dt}\int_0^t S(t-s)v(s)ds & \\ \quad + \frac{d}{dt}\int_0^t S(t-s)(Bu_y)(s)ds & \\ \quad + \sum_{0<t_k<t} S'(t-t_k) I_k(y(t_k^-)), & t \in [0,\infty), \\ & v \in S_{F,y}. \end{array} \right. \right\}$$

It is clear that the fixed points of N_1 are integral solutions to (10.7)–(10.9). Let $x(\cdot) : (-\infty,\infty) \to \overline{D(A)}$ be the function defined by

$$x(t) = \begin{cases} \phi(0), & \text{if } t \in (-\infty, 0], \\ S'(t)[\phi(0) - g(0,\phi(0))] + g(t,y_t) & \text{if } t \in [0,\infty). \end{cases}$$

Then $x_0 = S'(t)\phi(0)$. For each $z \in C_0$ with $z_0 = 0$, we denote by $\bar{z}$ the function defined by

$$\bar{z}(t) = \begin{cases} z(t), & \text{if } t \in [0,\infty), \\ 0, & \text{if } t \in (-\infty, 0]. \end{cases}$$

If $y(\cdot)$ satisfies the integral equation

$$y(t) = S'(t)\phi(0) + \int_0^t S'(t-s)v(s)ds + \int_0^t S'(t-s)(Bu_y)(s)ds + \sum_{0<t_k<t} S'(t-t_k)I_k(y(t_k^-)),$$

we can decompose $y(\cdot)$ as $y(t) = \bar{z}(t) + x(t), 0 \le t < \infty$, so that $y_t = \bar{z}_t + x_t$, for every $0 \le t < \infty$, and the function $z(\cdot)$ satisfies

$$\begin{aligned} z(t) &= \int_0^t S'(t-s)v(s)ds + \int_0^t S'(t-s)Bu_{\bar{z}+x}(s)ds. \\ &\quad + \sum_{0<t_k<t} S'(t-t_1)I_k(\bar{z}(t_k^-) + x(t_k^-)). \end{aligned} \tag{10.17}$$

Let the operator $P_1 : C_0 \to \mathcal{P}(C_0)$ be defined by $P_1(z) = \{h \in C_0\}$, where

$$h(t) = \begin{cases} 0, & t \in (-\infty, 0], \\ \int_0^t S'(t-s)v(s)ds & \\ + \int_0^t S'(t-s)Bu_{\bar{z}+x}(s)ds & \\ + \sum_{0<t_k<t} S'(t-t_k)I_k(\bar{z}(t_k^-) + x(t_k^-)), & t \in [0,\infty), \end{cases}$$

and

$$v \in S_{F,\bar{z}+x} = \{z \in L^1(J, \overline{D(A)}) | v \in F(t, \bar{z}_t + x_t),\ a.e.\ t \in J\}.$$

That N_1 having a fixed point is equivalent to P_1 having one is clear, so we now prove that P_1 has a fixed point. We shall use the Frigon nonlinear alternative to prove that this is so.

Let $z \in \gamma P_1(z)$ for some $\gamma \in [0,1]$ and for some $v \in S_{F,\bar{z}+x}$. Then, given $n \in \mathbb{N}$ and $t \le t_n$,

$$\begin{aligned} |z(t)| &\le Me^{|\omega|t} \int_0^t e^{-|\omega|s} p(s)\psi(\|\bar{z}_s\|_{\mathcal{B}} + \|x_s\|_{\mathcal{B}})ds \\ &\quad + Me^{|\omega|t} \int_0^t e^{-|\omega|s} \|(Bu_{\bar{z}+x}(s)\|ds + Me^{|\omega|t} \sum_{k=1}^n e^{-|\omega|t_k} c_k \end{aligned}$$

$$
\begin{aligned}
&\leq Me^{|\omega|t}\int_0^t e^{-|\omega|s}p(s)\psi(\|\bar{z}_s\|_{\mathcal{B}}+\|x_s\|_{\mathcal{B}})ds\\
&+ M\overline{M}e^{|\omega|t}\overline{M}_1 t_n\int_0^t d_1\|\bar{z}_{t_n}\|_{\mathcal{B}}ds\\
&+ M\overline{M}e^{|\omega|t}\overline{M}_1 t_n\bigg(\|y_1\|+Me^{|\omega|t_n}[\|\phi(0)\|\\
&+\|g(0,\phi(0))\|+d_2+Me^{|\omega|t_n}\sum_{k=1}^n e^{-|\omega|t_k}c_k\\
&+Me^{|\omega|t_n}\int_0^t e^{-|\omega|s}p(s)\psi(\|z_s\|_{\mathcal{B}}+\|x_s\|_{\mathcal{B}})ds\bigg)\\
&+Me^{|\omega|t}\sum_{k=1}^n e^{-|\omega|t_k}c_k.
\end{aligned}
$$

It clear that

$$\|\bar{z}_s\|_{\mathcal{B}}+\|x_s\|_{\mathcal{B}}\leq K_n\sup\left\{\|z(s)\|:0\leq s\leq t\right\}+\widetilde{M}_n\|\phi\|_{\mathcal{B}}+K_n\widetilde{M}_n\|\phi(0)\|. \tag{10.18}$$

If we call $w(t)$ the right hand side of (10.18), then we have

$$\|\bar{z}_s\|_{\mathcal{B}}+\|x_s\|_{\mathcal{B}}\leq w(t).$$

Therefore,

$$
\begin{aligned}
\|z(t)\|\leq{}& Me^{|\omega|t}\int_0^t e^{-|\omega|s}p(s)\psi(w(s))ds\\
&+ M\overline{M}e^{|\omega|t}\overline{M}_1 t_n\int_0^t d_1 w(t_n)ds\\
&+ M\overline{M}e^{|\omega|t}\overline{M}_1 t_n\bigg(\|y_1\|+Me^{|\omega|t_n}[\|\phi(0)\|\\
&+\|g(0,\phi(0))\|+d_2+Me^{|\omega|t_n}\sum_{k=1}^n e^{-|\omega|t_k}c_k\\
&+Me^{|\omega|t_n}\int_0^t e^{-|\omega|s}p(s)\psi(w(s))ds\bigg)+Me^{|\omega|t}\sum_{k=1}^n e^{-|\omega|t_k}c_k.
\end{aligned}
$$

We consider the function μ defined by

$$\mu(t)=\sup\left\{|w(s)|:0\leq s\leq t\right\},\quad 0\leq t\leq t_n.$$

Then we see that

$$\begin{aligned}\mu(t) \le{}& \widetilde{M}_n\|\phi\|_{\mathcal{B}} + K_n\widetilde{M}_n\|\phi(0)\| + K_nMe^{|\omega|t}\int_0^t e^{-|\omega|s}p(s)\psi(\mu(s))ds \\ &+ K_nM\overline{M}e^{|\omega|t}\overline{M}_1t_n\int_0^t d_1\mu(s)ds + K_nM\overline{M}e^{|\omega|t}\overline{M}_1t_n \\ &\times\left(\|y_1\| + Me^{|\omega|t_n}[\|\phi(0)\| + \|g(0,\phi(0))\| + d_2 + Me^{|\omega|t_n}\sum_{k=1}^n e^{-|\omega|t_k}c_k\right. \\ &\left.+ Me^{|\omega|t_n}\int_0^t e^{-|\omega|s}p(s)\psi(\mu(s))ds\right) + K_nMe^{|\omega|t}\sum_{k=1}^n e^{-|\omega|t_k}c_k.\end{aligned}$$

Thus,

$$\begin{aligned}\mu(t) \le{}& \widetilde{M}_n\|\phi\|_{\mathcal{B}} + K_n\widetilde{M}_n\|\phi(0)\| + K_nM\overline{M}e^{|\omega|t}\overline{M}_1t_n \\ &\times\left(\|y_1\| + Me^{|\omega|t_n}[\|\phi(0)\| + \|g(0,\phi(0))\| + d_2 + Me^{|\omega|t_n}\sum_{k=1}^n e^{-|\omega|t_k}c_k\right) \\ &+ K_nMe^{|\omega|t}\sum_{k=1}^n e^{-|\omega|t_k}c_k + \int_0^t m(s)[\mu(s) + \psi(\mu(s))]ds,\end{aligned}$$

where

$$m(t) = K_ne^{2t_n}[M^2\overline{M}M\overline{M}_1t_n + M]p(t) + K_nM\overline{M}e^{|\omega|t_n}\overline{M}_1t_nd_1.$$

Let us take the right-hand side of the above inequality to be $v(t)$. Then we have

$$\mu(t) \le v(t) \text{ for all } \quad t \in [0, t_n],$$

$$\begin{aligned}v(0) ={}& \widetilde{M}_n\|\phi\|_{\mathcal{B}} + K_n\widetilde{M}_n\|\phi(0)\| + K_nM\overline{M}e^{|\omega|t}\overline{M}_1t_n \\ &\times\left(\|y_1\| + Me^{|\omega|t_n}[\|\phi(0)\| + \|g(0,\phi(0))\| + d_2 + Me^{|\omega|t_n}\sum_{k=1}^n e^{-|\omega|t_k}c_k\right) \\ &+ K_nMe^{|\omega|t}\sum_{k=1}^n e^{-|\omega|t_k}c_k,\end{aligned}$$

and

$$v'(t) = m(t)[\mu(t) + \psi(\mu(t))], \quad t \in [0, t_n].$$

Using the increasing character of ψ we obtain

$$v'(t) \le m(t)\psi(v(t)) \text{ a.e. } \quad t \in [0, t_n].$$

Then, for each $t \in [0, t_n]$, we have

$$\int_{v(0)}^{v(t)} \frac{du}{u + \psi(u)} \le \int_0^{t_n} m(s)ds < \infty.$$

Hence, by condition (J-9), there exists a constant d_n such that $v(t) \leq \overline{d}_n$, $t \in [0, t_n]$, and hence $\|z\|_n \leq \overline{d}_n$. Set

$$U_1 = \{z \in C_0 : \|z\|_n < \overline{d}_n + 1 \text{ for all } n \in \mathbb{N}\}.$$

Clearly, U_1 is a open subset of C_0. As in Theorem 10.6, we can show that $N_1 : \overline{U}_1 \to \mathcal{P}(\Omega)$ is an admissible contraction operator. From the choice of U_1 there is no $y \in \partial U_1$ such that $z \in \lambda P_1(z)$ for some $\lambda \in (0,1)$. As a consequence of the nonlinear alternative of Frigon [206], we deduce that P_1 has at least one fixed point. We conclude that our problem has at least one fixed point that is an integral solution of (10.7)–(10.9). □

Example 10.11. As an application of our results we consider the impulsive partial functional differential inclusion,

$$\frac{\partial z(t,x)}{\partial t} - d\Delta z(t,x) \in Q(t, z(t-r,x)) + (Bu)(t), \quad t \in [0,\infty) \setminus \{1,2,\ldots\},\ x \in \Gamma, \tag{10.19}$$

$$b_k z(t_k^-, x) = z(t_k^+, x) - z(t_k^-, x,\ t_k = 1,2,\ldots,\ x \in \partial\Gamma, \tag{10.20}$$

$$z(t,x) = 0,\ t \in t \in [0,\infty)/\{1,2,\ldots\}, x \in \partial\Gamma, \tag{10.21}$$

$$z(t,x) = \phi(t,x),\ \ -r \leq t \leq 0,\ \ x \in \overline{\Gamma}, \tag{10.22}$$

where $d, r, b_k > 0$, Γ is a bounded open in $\mathbb{R}^n$ with a regular boundary $\partial\Gamma$, B is as in (10.4), $\Delta = \sum_{i=1}^{n} \frac{\partial^2}{\partial x_i^2}$, $\phi \in \mathcal{D}([-r,0] \times \overline{\Gamma}, \mathbb{R}^n) = \{\psi : [-r,0] \times \overline{\Gamma} \to \mathbb{R}^n, \psi$ is continuous everywhere except for a countable number of points $\bar{t}$ at which $\psi(\bar{t}^-, \cdot)$ and $\psi(\bar{t}^+, \cdot)$ exist, $\psi(\bar{t}^-, \cdot) = \psi(\bar{t}, \cdot)$ and $\psi(\cdot, x) \in L^1([-r,0], E)\}$, $(0 < r < \infty)$, $0 = t_0 < t_1 < \ldots < t_m < \ldots$, $z(t_k^+, x) = \lim_{(h,x)\to(0^+,x)} z(t_k + h, x)$ and $z(t_k^-, x) = \lim_{(h,x)\to(0^-,x)} z(t_k - h, x)$ represent the right and left hand limits of $z(t,x)$ at (t,x), $I_k(z) = b_k z$ $k \in \{1,2,\ldots\}$, $Q : [0,\infty) \times \mathbb{R}^n \to \mathcal{P}(\mathbb{R}^n)$ is a multivalued map with compact values, and there exists a constant $k_q > 0$ such that

$$H(Q(t,x), Q(t,y)) \leq k_q \|x - y\| \ \text{ for all } x, y \in [0,\pi],\ \ t \in [0,\infty),$$

and

$$H(0, Q(t,0)) \leq k_q.$$

Consider $E = C(\overline{\Gamma}, \mathbb{R}^n)$ to be the Banach space of continuous functions on $\overline{\Gamma}$ with values in $\mathbb{R}^n$. Define the linear operator A in E by $Az = d\Delta z$ where

$$D(A) = \{z \in E : \Delta z \in E \text{ and } z|_{\partial\Gamma} = 0\}.$$

Now we have

$$\overline{D(A)} = \{z \in E : z|_{\partial\Gamma} = 0\} \neq E.$$

It is well-known from [162] that Δ satisfies the following properties:

(i) $(0, \infty) \subset \rho(\Delta)$,

(ii) $\|(\lambda I - \Delta)^{-1}\| \leq \frac{1}{\lambda}$, for some $\lambda > 0$.

It follows that Δ satisfies the Hille–Yosida condition and hence it generates an integrated semigroup $(S(t))_t$, $t \geq 0$ and that $|S'(t)| \leq e^{-\mu t}$, for $t \geq 0$ and some constant $\mu > 0$. Let

$$F(t, w_t)(x) = Q(t, w(t - x)),\ 0 \leq x \leq \pi.$$

Then problem (10.19)–(10.22) takes the abstract form (10.4)–(10.6). We can easily see that all hypotheses of Theorem 10.6 are satisfied so the problem (10.19)–(10.22) has at least one integral solution on $(-\infty, \infty)$.

10.4 Controllability in Extrapolation Spaces

Our main result in this section is contained in the following theorem.

Theorem 10.12. *Assume that the following conditions hold.*

(J-15) *There exist a function $p \in L^1(J, \mathbb{R}_+)$ and a continuous nondecreasing function $\psi : [0, \infty) \to [0, \infty)$ such that*

$$\|F(t, x)\| \leq p(t)\psi(\|x\|_{\mathcal{B}}) \text{ for a.e. } t \in J \text{ and each } x \in \mathcal{B}$$

whre for every $c > 0$

$$\int_c^{\infty} \frac{ds}{s + \psi(s)} = \infty.$$

(J-16) *There exists $M > 0$ such that*

$$\|T_1(t)\|_{B(E)} \leq M \text{ for each } t > 0.$$

(J-17) *For all $R > 0$ there exists $l_R \in L^1_{loc}([0, \infty), \mathbb{R}_+)$ such that, for all $x, \overline{x} \in \mathcal{B}$ with $\|x\|, \|\overline{x}\| \leq R$,*

$$H(F(t, x), F(t, \overline{x})) \leq l_R(t)\|x - \overline{x}\|_{\mathcal{B}},$$

and

$$H(0, F(t, 0)) \leq l_R(t) \text{ for a.e. } t \in J.$$

(J-18) *There exist constants $c_k \geq 0$, $k = 1, \ldots,$ such that*

$$|I_k(y) - I_k(x)| \leq c_k|x - \overline{x}| \text{ for each } x, \overline{x} \in E.$$

(J-19) *For every $n > 0$, the linear operator $W : L^2(J_n, U) \to E$ ($J_n = [0, t_n]$), defined by*

$$Wu = \int_0^{t_n} T_1(t_n - s)Bu(s)ds,$$

has an invertible operator $W^{-1} : E \to L^2(J_n, U)\backslash KerW$, and there exist positive constants $\overline{M}$, $\overline{M}_1$ such that $\|B\| \leq \overline{M}$ and $\|W^{-1}\| \leq \overline{M}_1$.

(J-20) *The function $F : J \times \mathcal{B} \to \mathcal{P}_{cp}(E)$ is a Carathéodory map and for each $t \in [0, \infty)$, $t \to F(t, y_t)$ has measurable selection.*

If $M \sum_{k=1}^{\infty} c_k < 1$, then the IVP (10.4)–(10.6) has a global solution on $(-\infty, \infty)$.

Proof. Using hypothesis (J-19), for each $y(\cdot)$ and

$$v \in S_{F,y} = \{v \in L^1(J, E) : v(t) \in F(t, y_t) \; a.e \; t \in J\},$$

define the control

$$u_y(t) =$$
$$W^{-1}\left[y_1 - T_1(t_n)\phi(0) - \int_0^{t_n} T_1(t_n - s)v(s)ds - \sum_{0<t_k<s} T_1(s - t_k)I_k(y(t_k^-))\right](t).$$

To transform the problem (10.4)–(10.6) into a fixed point problem, define the operator $N : \mathcal{B}_* \to \mathcal{P}(\mathcal{B}_*)$ defined by

$$N(y) = \left\{ h \in \mathcal{B}_* : h(t) = \left\{ \begin{array}{ll} \phi(t), & t \in [-r, 0], \\ T_1(t)\phi(0) + \int_0^t T_1(t - s)v(s)ds & \\ \quad + \int_0^t T_1(t - s)(Bu_y)(s)ds & \\ \quad + \sum_{0<t_k<t} T_1(t - t_k)I_k(y(t_k^-)), & t \in J, \end{array} \right. \right\}$$

where $v \in S_{F,y}$. The fixed points of the operator N are solutions of the problem (10.4)–(10.6). Let $x(\cdot) : (-\infty, \infty) \to \mathbb{R}^n$ be the function defined by

$$x(t) = \begin{cases} \phi(0), & \text{if } t \in (-\infty, 0], \\ T_1(t)\phi(0), & \text{if } t \in [0, \infty). \end{cases}$$

Then $x_0 = T_1(t)\phi(0)$. For each $z \in C([0, \infty), \mathbb{R}^n)$ with $z(0) = 0$, we set

$$\bar{z}(t) = \begin{cases} z(t), & \text{if } t \in [0, \infty), \\ 0, & \text{if } t \in (-\infty, 0]. \end{cases}$$

If $y(\cdot)$ satisfies the integral equation

$$y(t) = T_1(t)\phi(0) + \int_0^t T_1(t - s)v(s)ds + \int_0^t T_1(t - s)(Bu_y)(s)ds$$
$$+ \sum_{0<t_k<t} T_1(t - t_k)I_k(y(t_k^-)),$$

we can decompose $y(\cdot)$ as $y(t) = \bar{z}(t) + x(t), 0 \le t < \infty$. This implies $y_t = \bar{z}_t + x_t$, for every $0 \le t < \infty$, and the function $z(\cdot)$ satisfies

$$z(t) = \int_0^t T_1(t-s) f(s, \bar{z}_s + x_s) ds + \int_0^t T_1(t-s) B u_{\bar{z}+x}(s) ds \\ + \sum_{0<t_k<t} T_1(t-t_1) I_k(\bar{z}(t_k^-) + x(t_k^-)). \tag{10.23}$$

Let

$$\mathcal{B}_*^k = \{z : (-\infty, t_k] \to E \mid z(t_i^-), z(t_i^+) \text{ exist with } z(t_i) = z(t_i^-), \\ z_k \in C(J_i, E),\ i = 1, \dots, k-1, \text{ and } z_0 = 0\}.$$

For any $z \in \mathcal{B}_*^k$, we have

$$\|z\|_k = \|z_0\|_{\mathcal{B}} + \sup\{\|z(s)\| : 0 \le s \le t_k\} = \sup\{\|z(s)\| : 0 \le s \le t_k\}.$$

Thus, $(\mathcal{B}_*^k, \|\cdot\|_k)$ is a Banach space. Set

$$C_0 = \{z \in \mathcal{B}_* : z_0 = 0\}.$$

We define a family of seminorms on C_0, thus rendering C_0 a Fréchet space.

Let θ be sufficiently large. For each $n \in \mathbb{N}$, define on C_0 the seminorms

$$\|z\|_n = \sup\{e^{-\theta L_n(t)} |z(t)| : [0, t_n]\},$$

where $L_n(t) = \int_0^t \hat{l}_n(s) ds$ and

$$\hat{l}_n(t) = [M^2 \overline{M} t_n \overline{M}_1 K_n + M K_n] l(t) + M^2 \overline{M} t_n \overline{M}_1 \sum_{0<t_k<t} c_k.$$

Then C_0 is a Fréchet space with this family of seminorms $\|\cdot\|_n$. Let the operator $P : C_0 \to \mathcal{P}(C_0)$ be defined by $Pz = \{h \in C_0\}$, where

$$h(t) = \begin{cases} 0, & t \in (-\infty, 0], \\ \int_0^t T_1(t-s) v(s) ds + \int_0^t T_1(t-s) B u_{\bar{z}+x}(s) ds & \\ \quad + \sum_{0<t_k<t} T_1(t-t_k) I_k(\bar{z}(t_k^-) + x(t_k^-)), & t \in [0, \infty), \end{cases}$$

and

$$v \in S_{F, \bar{z}+x} = \{z \in L^1(J, E) | v \in F(t, \bar{z}(t) + x(t), \ a.e.\ t \in J\}.$$

The operator N having a fixed point is equivalent to P having a fixed point, and so now we prove that P does have a fixed point. We will use the Frigon nonlinear alternative to do this.

Let $z \in \gamma P(z)$ for some $\gamma \in [0,1]$ and for some $v \in S_{F,\bar{z}+x}$. Given $n \in \mathbb{N}$ and $t \leq t_n$ such that

$$z(t) = \gamma\Big[\int_0^t T_1(t-s)v(s)ds + \int_0^t T_1(t-s)Bu_{\bar{z}_s+x_s}(s)ds + \sum_{0<t_k<t} T_1(t-t_k)I_k(\bar{z}(t_k^-)+x(t_k^-))\Big],$$

we have

$$\begin{aligned}
|z(t)| &\leq \int_0^t |T_1(t-s)||v(s)|ds + \int_0^t |T_1(t-s)||Bu_{\bar{z}+x}(s)|ds \\
&\quad + \sum_{0<t_k<t} |T_1(t-t_k)||I_k(\bar{z}(t_k)+x(t_k^-))| \\
&\leq M\int_0^t p(s)\psi(\|\bar{z}_s+x_s\|_{\mathcal{B}})ds + M\int_0^t \|B\|\|u(s)\|ds \\
&\quad + M\sum_{k=1}^n |I_k(\bar{z}(t_k^-)+x(t_k))| \\
&\leq M\int_0^t p(s)\psi(\|\bar{z}_s+x_s\|_{\mathcal{B}})ds \\
&\quad + M\overline{M}\int_0^t \|W^{-1}\Big[y_1 - T_1(t_n)\phi(0) - \int_0^{t_n} T_1(t_n-\tau)v(\tau)d\tau \\
&\quad - \sum_{0<t_k<\tau} T_1(\tau-t_k)I_k(\bar{z}(t_k)+x(t_k))\Big](s)\|ds + M\sum_{k=1}^n |I_k(\bar{z}(t_k)+x(t_k))| \\
&\leq M\int_0^t p(s)\psi(\|\bar{z}_s+x_s\|_{\mathcal{B}})ds + M\overline{M}M_1t_n[|y_1| + M|\phi(0)|] \\
&\quad \times M^2\overline{M}M_1\int_0^t\int_0^{t_n} p(\tau)\psi(\|\bar{z}_\tau+x_\tau\|_{\mathcal{B}})d\tau ds \\
&\quad + M^2\overline{M}M_1\int_0^t \sum_{0<t_k<s} |I_k(\bar{z}(t_k)+x(t_k))|ds + M\sum_{k=1}^n |I_k(\bar{z}(t_k)+x(t_k))| \\
&\leq M\int_0^t p(s)\psi(\|\bar{z}_s+x_s\|_{\mathcal{B}})ds + M\overline{M}M_1t_n[|y_1| + M|\phi(0)|] \\
&\quad \times M^2\overline{M}M_1t_n\int_0^t p(s)\psi(\|\bar{z}_s+x_s\|_{\mathcal{B}})ds \\
&\quad + M^2\overline{M}M_1\int_0^t \sum_{0<t_k<s} |I_k(\bar{z}(t_k)+x(t_k))|ds + M\sum_{k=1}^n |I_k(\bar{z}(t_k)+x(t_k))|
\end{aligned}$$

$$\leq M\int_0^t p(s)\psi(\|\bar{z}_s+x_s\|_{\mathcal{B}})ds+M\overline{M}M_1t_n[|y_1|+M|\phi(0)|]$$
$$\times M^2\overline{M}M_1t_n\int_0^t p(s)\psi(\|\bar{z}_s+x_s\|_{\mathcal{B}})ds$$
$$+M^2\overline{M}M_1\int_0^t\sum_{0<t_k<s}c_k\|\bar{z}(t_k)+x(t_k))|ds+M^2\overline{M}M_1t_n\sum_{k=1}^n c_k\|I_k(0)\|$$
$$+M\sum_{k=1}^n c_k\|\bar{z}(t_k)+x(t_k)\|+M\sum_{k=1}^n c_k\|I_k(0)\|.$$

But

$$\begin{aligned}\|\bar{z}_s+x_s\|_{\mathcal{B}} &\leq \|\bar{z}_s\|_{\mathcal{B}}+\|x_s\|_{\mathcal{B}}\\ &\leq K(t)\sup\{\|z(s)\|:0\leq s\leq t\}+M(t)\|z_0\|_{\mathcal{B}}\\ &\quad+K(t)\sup\{\|x(s)\|:0\leq s\leq t\}+M(t)\|x_0\|_{\mathcal{B}}\\ &\leq K_n\sup\{\|z(s)\|:0\leq s\leq t\}+\widetilde{M}_n\|\phi\|_{\mathcal{B}}+K_n\widetilde{M}_n\|\phi(0)\|.\end{aligned} \tag{10.24}$$

If we let $w(t)$ denote the right hand side of (10.24), then we have

$$\|\bar{z}_s+x_s\|_{\mathcal{B}}\leq w(t).$$

Therefore,

$$|z(t)|\leq M\int_0^t p(s)\psi(w(s))ds+M\overline{M}M_1t_n[|y_1|+M|\phi(0)|]$$
$$+M^2\overline{M}M_1t_n\int_0^t p(s)\psi(w(s))ds$$
$$+M^2\overline{M}M_1\int_0^t\sum_{0<t_k<s}c_k\|\bar{z}(t_k)+x(t_k))|ds+M^2\overline{M}M_1t_n\sum_{k=1}^n c_k\|I_k(0)\|$$
$$+M\sum_{k=1}^n c_k\|\bar{z}(t_k)+x(t_k)\|+M\sum_{k=1}^n c_k\|I_k(0)\|.$$

Using the above inequality in the definition of w, we have that

$$w(t)\leq K_nM\int_0^t p(s)\psi(w(s))ds+K_nM\overline{M}M_1t_n[|y_1|+M|\phi(0)|]$$
$$+K_nM^2\overline{M}M_1t_n\int_0^t p(s)\psi(w(s))ds+\widetilde{M}_n\|\phi\|_{\mathcal{B}}+K_n\widetilde{M}_n\|\phi(0)\|$$

$$+ M^2\overline{M}M_1 \int_0^t \sum_{0<t_k<s} c_k w(s)ds + K_n M^2\overline{M}M_1 t_n \sum_{k=1}^{n} c_k \|I_k(0)\|$$
$$+ K_n M \sum_{k=1}^{n} c_k w(t) + K_n M \sum_{k=1}^{n} c_k \|I_k(0)\|.$$

Then,

$$w(t) \le \frac{1}{1 - M\sum_{k=1}^{n} c_k}\Big[K_n M \int_0^t p(s)\psi(w(s))ds + K_n M\overline{M}M_1 t_n[|y_1| + M|\phi(0)|]$$
$$\times K_n M^2\overline{M}M_1 t_n \int_0^t p(s)\psi(w(s))ds + \widetilde{M}_n\|\phi\|_{\mathcal{B}} + K_n\widetilde{M}_n\|\phi(0)\|$$
$$+ M^2\overline{M}M_1 \int_0^t \sum_{0<t_k<s} c_k w(s)ds + K_n M^2\overline{M}M_1 t_n \sum_{k=1}^{n} c_k \|I_k(0)\|$$
$$+ K_n M \sum_{k=1}^{n} c_k \|I_k(0)\|\Big].$$

Thus, we have

$$w(t) \le c + \int_0^t \widehat{M}(s)[w(s) + \psi(w(s))]ds,$$

where

$$c = \frac{K_n}{1 - M\sum_{k=1}^{n} c_k}\Big[M\overline{M}M_1 t_n[|y_1| + M|\phi(0)|] + \widetilde{M}_n\|\phi\|_{\mathcal{B}} + \widetilde{M}_n\|\phi(0)\|$$
$$+ M^2\overline{M}M_1 t_n \sum_{k=1}^{n} c_k\|I_k(0)\| + M\sum_{k=1}^{n} c_k\|I_k(0)\|\Big],$$

and

$$\widehat{M}(t) = \frac{K_n}{1 - M\sum_{k=1}^{n} c_k}\Big[(K_n M + M^2\overline{M}M_1 t_n)p(t) + M^2\overline{M}M_1 \sum_{0<t_k<t} c_k\Big].$$

Denoting by $\beta(t)$ the right hand side of the last inequality, we have

$$w(t) \le \beta(t), \quad t \in [0, t_n],\ \beta(0) = c,$$

and

$$\begin{aligned}\beta'(t) &= \widehat{M}(t)[w(t) + \psi(w(t))]\\ &\le \widehat{M}(t)[\beta(t) + \psi(\beta(t))], \quad t \in [0, t_n].\end{aligned}$$

This implies that for each $t \in [0, t_n]$,

$$\int_{\beta(0)}^{\beta(t)} \frac{ds}{s + \psi(s)} \leq \int_0^{t_n} \widehat{M}(s)ds.$$

Thus, from (J-15), there exists a constant M^* such that $\beta(t) \leq M^*$, $t \in [0, t_n]$. From the definition of w, we conclude that

$$\sup\{\|z(t)\| : t \in [0, t_n]\} \leq M^*.$$

Set

$$U_0 = \{z \in C_0 : \|z\|_n < M^* + 1\}.$$

Clearly, U_0 is a open subset of C_0. We shall show that $P : U_0 \to \mathcal{P}(C_0)$ is a contraction and an admissible operator.

First, we prove that P is a contraction; that is, there exists $\gamma < 1$, such that

$$H(P(z), P(z_*)) \leq \gamma \|z - z_*\|_n, \quad \text{for each } y, \overline{y} \in U_0.$$

Let $z, z_* \in U_0$ and $h \in P(z)$. Then there exists $v(t) \in F(t, y_t)$ such that for each $t \in [0, t_n]$,

$$\begin{aligned} h(t) = & \int_0^t T_1(t-s)v(s)ds + \int_0^t T_1(t-s)Bu_{\bar{z}+x}(s)ds \\ & + \sum_{0<t_k<t} T_1(t-t_k)I_k(\bar{z}(t_k^-) + x(t_k^-)). \end{aligned}$$

From (J-17), it follows that

$$H(F(t, \bar{z}_t + x_t), F(t, \bar{z_*}_t + x_t) \leq l(t)\|\bar{z}_t - \bar{z_*}_t\|_{\mathcal{B}}.$$

Hence, there is $w \in F(t, \bar{z_*}_t + x_t)$ such that

$$|v(t) - w| \leq l(t)\|\bar{z}_t - \bar{z_*}_t\|_{\mathcal{B}}, \quad t \in J.$$

Consider $U_* : [0, t_n] \to \mathcal{P}(E)$, given by

$$U_*(t) = \{w \in E : |v(t) - w| \leq l(t)\|\bar{z}_t - \bar{z_*}_t\|_{\mathcal{B}}\}.$$

Since the multivalued operator $V_*(t) = U_*(t) \cap F(t, \bar{z_*}_t + x_t)$ is measurable (see Proposition III.4 in [136]), there exists a function $v_*(t)$, which is a measurable selection for V_*. So, $v_*(t) \in F(t, \overline{z}_t + x_t)$ and

$$|v(t) - \overline{v}(t)| \leq l(t)\|\bar{z}_t - \bar{z_*}_t\|_{\mathcal{B}}, \quad \text{for each } t \in [0, t_n].$$

Let us define, for each $t \in [0, t_n]$,

$$\begin{aligned} h_*(t) = & \int_0^t T_1(t-s)v_*(s)ds + \int_0^t T_1(t-s)Bu_{\bar{z}_*+x}(s)ds \\ & + \sum_{0<t_k<t} T_1(t-t_k)I_k(\bar{z}_*(t_k^-) + x(t_k^-)). \end{aligned}$$

Then, we have

$$
\begin{aligned}
|h(t) - h_*(t)| &= \Big| \int_0^t T_1(t-s)[(Bu_{\bar z + x})(s) - (Bu_{z_*+x})(s)]ds \\
&\quad + \int_0^t T_1(t-s)[v(s) - v_*(s)]ds \\
&\quad + \sum_{0<t_k<t} T_1(t-t_k)[I_k(z(t_k^-) + x(t_k)) - I_k(z_*(t_k^-) + x(t_k))] \Big| \\
&\le M\overline{M} \int_0^t |u_{\bar z + x}(s) - u_{z_*+x}(s)|ds \\
&\quad + M \int_0^t |v(s) - v_*(s)|ds + M \sum_{0<t_k<t} c_k |z(t_k^-) - z_*(t_k^-)| \\
&\le M\overline{M} \int_0^t |u_{\bar z + x}(s) - u_{z_*+x}(s)|ds \\
&\quad + M \int_0^t l_n(t)\|\bar z_s - z_{*s}\|_{\mathcal{B}} ds + M \sum_{0<t_k<t} c_k |z(t_k^-) - z_*(t_k^-)| \\
&\le M^2\overline{M} t_n \overline{M}_1 \int_0^t l_n(s)\|\bar z_s - z_{*s}\|_{\mathcal{B}} ds \\
&\quad + M^2\overline{M} t_n \overline{M}_1 \int_0^t \sum_{0<t_k<s} |\bar z(t_k) - \bar z_*(t_k)|ds \\
&\quad + M \int_0^t l_n(s)\|\bar z_s - z_{*s}\|_{\mathcal{B}} ds + M \sum_{0<t_k<t} c_k |z(t_k^-) - z_*(t_k^-)| \\
&\le M^2\overline{M} t_n \overline{M}_1 \int_0^t l_n(s) K_n \sup_{s\in[0,t]} \|z(s) - z_*(s)\| ds \\
&\quad + M^2\overline{M} t_n \overline{M}_1 \int_0^t \sum_{0<t_k<s} c_k |z(t_k) - z_*(t_k)|ds \\
&\quad + M \int_0^t l_n(s) K_n \sup_{s\in[0,t]} \|z(s) - z_*(s)\| ds \\
&\quad + M \sum_{0<t_k<t} c_k |z(t_k^-) - z_*(t_k^-)| \\
&\le \int_0^t \hat l_n(s) e^{\theta L_n(s)} e^{-\theta L_n(s)} \sup_{s\in[0,t]} \|z(s) - z_*(s)\| ds \\
&\quad + \int_0^t \hat l_n(s) e^{\theta L_n(s)} e^{-\theta L_n(s)} \sup_{s\in[0,t]} \|z(s) - z_*(s)\| ds \\
&\quad + \int_0^t \hat l_n(s) e^{\theta L_n(s)} e^{-\theta L_n(s)} \sup_{s\in[0,t]} \|z(s) - z_*(s)\| ds
\end{aligned}
$$

$$
\begin{aligned}
&+ M \sum_{0<t_k<t} c_k e^{\theta L_n(t)} e^{-\theta L_n(t)} \sup_{s\in[0,t]} \|z(s) - z_*(s)\| ds \\
&\le \int_0^t (e^{\theta L_n(s)})' ds \|z - z_*\|_n + \int_0^t (e^{\theta L_n(s)})' ds \|z - z_*\|_n \\
&\quad + \int_0^t (e^{\theta L_n(s)})' ds \|z - z_*\|_n + M \sum_{0<t_k<t} c_k e^{\theta L_n(t)} \|z - z_*\|_n.
\end{aligned}
$$

Therefore,

$$
\|h - \overline{h}\|_n \le \left(\frac{3}{\theta} + M \sum_{k=1}^{n} c_k\right) \|z - \overline{z}_*\|_n.
$$

By an analogous relation, obtained by interchanging the roles of z and z_*, it follows that

$$
H(P(z), P(z_*)) \le \left(\frac{3}{\theta} + M \sum_{k=1}^{n} c_k\right) \|z - \overline{z}_*\|_n.
$$

So, P is a contraction. Now, $P : \mathcal{B}_*^n \to \mathcal{P}_{cp}(\mathcal{B}_*^n)$ is given by

$$
P(y) = \left\{ h \in \mathcal{B}_*^n : h(t) = \begin{cases} 0, & t \in (-\infty, 0], \\ \int_0^t T_1(t-s) v(s) ds & \\ \quad + \int_0^t T_1(t-s)(Bu^n_{\bar{z}+x})(s) ds & \\ \quad + \sum_{0<t_k<t} T_1(t-t_k) I_k(\bar{z}(t_k^-) & \\ \quad + x(t_k^-)), & t \in [0, t_n], \end{cases} \right\}
$$

where $v \in S^n_{F,\bar{z}+x} = \{u \in L^1([0,t_n], E) : u \in F(t, \bar{z}_t + x_t) \ a.e. \ t \in [0,t_n]\}$. Let $x \in \overline{U}_0$ and $\varepsilon > 0$. If $x \notin P(x)$, then $d_n(x, P(x)) \neq 0$. Since $P(x)$ is compact, there exists $y \in P(x)$ such that $d_n(x, P(x)) = |x - y|_n$ and we have

$$
|x - y|_n \le d_n(x, P(x)) + \varepsilon.
$$

If $x \in P(x)$, then we may take $y = x$. Therefore P is an admissible operator contraction, and by Lemma 2.90, P has at least one fixed point, y, which is a global solution to (10.4)–(10.6). □

10.5 Second Order Impulsive Semilinear FDIs

In this section, we are concerned with Filippov's theorem and the relaxation theorem for second order impulsive semilinear functional differential inclusions in Banach spaces. More precisely, we consider the problem,

$$
\begin{aligned}
&y'' - Ay \in F(t, y_t), && \text{a.e. } t \in J = [0,b] \backslash \{t_1, \dots, t_m\}, && (10.25) \\
&\Delta y|_{t=t_k} = I_k(y(t_k^-)), && k = 1, \dots, m, && (10.26) \\
&\Delta y'|_{t=t_k} = \overline{I}_k(y(t_k^-)), && k = 1, \dots, m, && (10.27) \\
&y(t) = \phi(t), && t \in [-r, 0], \ y'(0) = \alpha, && (10.28)
\end{aligned}
$$

where E a real separable Banach space with norm $|\cdot|$, $F : J \times \mathcal{D} \to \mathcal{P}(E)$ is a multifunction, $0 < r < \infty$, $\mathcal{D}$ is defined in chapter 8, $\phi \in \mathcal{D}$, $0 = t_0 < t_1 < \cdots < t_m < t_{m+1} = b$, $\alpha \in E$, A is the infinitesimal generator of a family of cosine operators $\{C(t) : t \geq 0\}$, $I_k, \overline{I}_k \in C(E, E)$, $k = 1, \ldots, m$, $\Delta y|_{t=t_k} = y(t_k^+) - y(t_k^-)$ $y(t_k^+) = \lim_{h\to 0^+} y(t_k + h)$ and $y(t_k^-) = \lim_{h\to 0^+} y(t_k - h)$ represent the right and left hand limits of $y(t)$ at $t = t_k$, respectively.

The study of the dynamical buckling of a hinged extendible beam that is either stretched or compressed by an axial force in a Hilbert space can be modeled by the hyperbolic equation

$$\frac{\partial^2 u}{\partial t^2} + \frac{\partial^4 u}{\partial x^4} - \left(\alpha + \beta \int_0^L \left|\frac{\partial u}{\partial t}(\xi, t)\right|^2 d\xi\right) \frac{\partial^2 u}{\partial x^2} + g\left(\frac{\partial u}{\partial t}\right) = 0, \qquad (E)$$

where $\alpha, \beta, L > 0$, $u(t, x)$ is the deflection of the point x of the beam at the time t, g is a nondecreasing numerical function, and L is the length of the beam. Equation (E) has its analogue in $\mathbb{R}^n$ and can be included in a general mathematical model

$$u'' + A^2 u + M(\|A^{1/2} u\|_H^2) A u + g(u') = 0, \qquad (E_1)$$

where A is a linear operator in a Hilbert space H, and M and g are real functions. Equation (E) was studied by Patcheu [409] and the equation (E_1) by Matos and Pereira [369]. These equations are special cases of the equations (10.25)–(10.28).

Frankowska [201] investigated the case where the impulses are absent (i.e., $\overline{I}_k = I_k = 0, k = 1, \ldots, m$) and A is an infinitesimal generator of a C_0-semigroup.

The goal of this section is to give impulsive versions of Filippov's theorem and the Filippov–Wazewski theorem, where A is the infinitesimal generator of a cosine semigroup on a separable Banach space E.

10.5.1 Mild Solutions

In order to define mild solutions for the problem (10.25)–(10.28), we consider the space

$$PC = \big\{y : [0, b] \to E : y_k \in C(J_k, E), k = 0, \ldots, m, \text{ and there exist } y(t_k^-) \text{ and } y(t_k^+), k = 1, \ldots, m, \text{ with } y(t_k^-) = y(t_k)\big\},$$

which is a Banach space with the norm

$$\|y\|_{PC} = \max\{\|y_k\|_{J_k}, k = 0, \ldots, m\},$$

where y_k is the restriction of y to $J_k = (t_k, t_{k+1}]$, $k = 0, \ldots, m$. If we set

$$\Omega = \big\{y : [-r, b] \to E : y \in PC \cap \mathcal{D}\big\},$$

then Ω is a Banach space with the norm

$$\|y\|_{\Omega} = \max\{|y(t)| \ : \ t \in [-r,b]\}.$$

We now define what we mean by a mild solution of problem (10.25)–(10.28).

Definition 10.13. A function $y \in \Omega$ is said to be a mild solution of (10.25)–(10.28) if there exists $v \in L^1(J,E)$ such that $v(t) \in F(t,y_t)$ a.e. on J, $y(t) = \phi(t)$, $y'(0) = \alpha$, and

$$\begin{aligned} y(t) = C(t)\phi(0) + S(t)\eta + \int_0^t S(t-s)v(s)ds \\ + \sum_{0<t_k<t} [C(t-t_k)I_k(y(t_k^-)) + S(t-t_k)\overline{I}_k(y(t_k^-))]. \end{aligned}$$

We also wish to consider the situation where $\psi \in \mathcal{D}$, $g \in L^1(J,E)$, and $x \in PC \cap \mathcal{D}$ is a mild solution of the problem

$$\begin{aligned} &x''(t) - Ax(t) = g(t), \quad \text{a.e.} \quad t \in J = [0,b]\backslash\{t_1,\dots,t_m\}, && (10.29) \\ &\Delta x|_{t=t_k} = I_k(x(t_k^-)), \quad k = 1,\dots,m, && (10.30) \\ &\Delta x'|_{t=t_k} = \overline{I}_k(x(t_k^-)), \quad k = 1,\dots,m, && (10.31) \\ &x(t) = \psi(t) \quad t \in [-r,0], \ x'(0) = \alpha. && (10.32) \end{aligned}$$

10.5.2 Filippov's Theorem

Now, we present a Filippov's result for the problem (10.25)–(10.28).

Theorem 10.14. *Assume the following conditions hold:*

(J-20) *The function $F : J \times \mathcal{D} \to \mathcal{P}_{cl}(E)$ satisfies:*
 (a) *for all $u \in \mathcal{D}$, $t \mapsto F(t,u)$ is measurable;*
 (b) *the map $t \to \gamma(t) = d(g(t), F(t,x_t))$ is integrable.*

(J-21) *There exists a function $p \in L^1(J,\mathbb{R}_+)$ such that*

$$H(F(t,u),F(t,z)) \le p(t)\|u-z\|_\infty \quad \textit{for all} \ x, \ \overline{x} \in \mathcal{D}.$$

Then, for every $\psi \in \mathcal{D}$, and for $\epsilon > 0$, there exists $y \in \Omega$ such that

$$\begin{aligned} \|x_t - y_t\|_\infty \le M \sum_{0<t_k<t} |x(t_k) - y(t_k)| + M \sum_{0<t_k<t} |I_k(x(t_k)) - I_k(y(t_k))| \\ + \sum_{0<t_k<t} \frac{Me^{\omega t_k}}{\omega}|x'(t_k) - y'(t_k)| + \sum_{0<t_k<t} \frac{Me^{\omega t_k}}{\omega}|\overline{I}_k(x(t_k)) - \overline{I}_k(y(t_k))| \\ + \sum_{0<t_k<t} \eta_k(t), \quad \textit{for all } t \in [0,b], \end{aligned}$$

where x is a mild solution of the problem (10.29)–(10.32) *and*

$$|g(t) - y''(t) - Ay(t)| \leq p(t)\eta_k(t) + \gamma(t) + \epsilon,\ t \in (t_k, t_{k+1}],\ k = 0, \ldots, m,$$

where

$$\eta_0(t) = Me^{\omega t_1}\delta \exp\Big(Me^{\omega t_1}m(s)\Big)ds + \frac{Me^{\omega t_1}}{\omega}\int_0^t (\gamma(s) + \epsilon)$$
$$\times \exp Me^{\omega t_1}\Big(m(t) - m(s)\Big)ds,\ t \in [-r, t_1],\ \|\phi_\psi\|_\infty \leq \delta,$$

and

$$\eta_k(t) = \frac{Me^{\omega t_{k+1}}}{\omega}\int_{t_k}^t \gamma(s) \exp Me^{\omega t}\Big(m(t) - m(s)\Big)ds,\ t \in (t_k, t_{k+1}].$$

Proof. The proof will be given in several steps.

Step 1. Consider the problem (10.25)–(10.28) on $[-r, t_1]$,

$$y''(t) - Ay(t) \in F(t, y_t), \quad t \in J = [0, t_1], \tag{10.33}$$
$$y(t) = \phi(t), \quad t \in [-r, 0],\ y'(0) = \alpha. \tag{10.34}$$

Let $f_0 = g$ and

$$y^0(t) = \begin{cases} \phi(t), & t \in [-r, 0], \\ C(t)\phi(0) + S(t)\alpha + \int_0^t S(t-s) f_0(s)ds, & t \in [0, t_1]. \end{cases}$$

Let $\epsilon > 0$ and $U_1 : [0, t_1] \to \mathcal{P}(E)$ given by $U_1(t) = F(t, y_t^0) \cap B(x'(t), \gamma(t) + \epsilon)$. Since the multivalued operator $U_1(t)$ is measurable (see Proposition III.4 in [136]), there exists a function $t \to f_1(t)$, which is a measurable selection for U_1. We consider

$$y^1(t) = \begin{cases} \psi(t), & t \in [-r, 0], \\ C(t)\phi(0) + S(t)\alpha + \int_0^t S(t-s) f_1(s)ds, & t \in [0, t_1]. \end{cases}$$

Then, for every $t \in [0, t_1]$, we have

$$|y^1(t) - y^0(t)| \leq Me^{\omega t}|\phi(0) - \psi(0)| + M\int_0^t \int_0^{t-s} e^{\omega u}|f_0(s) - f_1(s)|\,duds$$
$$\leq Me^{\omega t}|\phi(0) - \psi(0)| + \frac{M}{\omega}\int_0^t e^{\omega(t-s)}|f_0(s) - f_1(s)|ds,$$

so

$$|y^1(t) - y^0(t)| \leq Me^{\omega b}\|\phi - \psi\|_\infty + \frac{Me^{\omega b}}{\omega}\int_0^t |f_0(s) - f_1(s)|ds.$$

Thus,

$$\begin{aligned}\sup\{|y^1(t)-y^0(t)| : t\in[-r,t_1]\} &\leq Me^{\omega b}\|\psi-\phi\|_\infty+\frac{Me^{\omega b}}{\omega}\int_0^t[\gamma(s)+\epsilon]ds\\ &\leq Me^{\omega b}\delta+\frac{Me^{\omega b}}{\omega}\int_0^t[\gamma(s)+\epsilon]ds.\end{aligned}$$

Define the set valued map $U_2(t)=F(t,y_t^1)\cap B(f_1(t),p(t)\|y_t^1-y_t^0\|_{\mathcal{D}})$. Since the multivalued operator U_2 is measurable (see Proposition III.4 in [136]), there exists a function $f_2(t)\in U_2(t)$. We consider

$$y^2(t)=\begin{cases}\psi(t), & t\in[-r,0],\\ C(t)\psi(0)+S(t)\alpha+\int_0^t S(t-s)f_2(s)ds, & t\in[0,t_1].\end{cases}$$

Hence,

$$\begin{aligned}|y^2(t)-y^1(t)| &\leq \int_0^t |S(t-s)||f_2(s)-f_1(s)|\,ds\\ &\leq \int_0^t Me^{\omega b}p(s)\left[Me^{\omega b}\delta+\frac{Me^{\omega b}}{\omega}\int_0^s[\gamma(u)+\epsilon]du\right]ds\\ &\leq \delta e^{2\omega b}\int_0^t p(s)ds+\frac{M^2e^{2\omega b}}{\omega}\int_0^t p(s)\int_0^u[\gamma(u)+\epsilon]dsdu\\ &\leq M^2\delta e^{2\omega b}\int_0^t p(s)e^{\int_s^t p(\tau)d\tau}ds\\ &\quad+\frac{M^2e^{2\omega b}}{\omega}\int_0^t p(s)e^{\int_s^t p(\tau)d\tau}\int_0^s[\gamma(u)+\epsilon]dsdu\\ &= -M^2\delta e^{2\omega b}\int_0^t (e^{\int_s^t p(\tau)d\tau})'ds\\ &\quad-\frac{M^2e^{2\omega b}}{\omega}\int_0^t (e^{\int_s^t p(\tau)d\tau})'\int_0^s[\gamma(u)+\epsilon]dsdu\\ &\leq M^2\delta e^{2\omega b}e^{\int_0^t p(s)ds}-\frac{M^2e^{2\omega b}}{\omega}\\ &\quad\times\left[\int_0^t[\gamma(s)+\epsilon]ds-\int_0^t[\gamma(s)+\epsilon]e^{\int_s^t p(\tau)d\tau}ds\right]\\ &= M^2\delta e^{2\omega b}e^{\int_0^t p(s)ds}+\frac{M^2e^{2\omega b}}{\omega}\int_0^t[\gamma(s)+\epsilon][e^{\int_s^t p(\tau)d\tau}-1]ds\\ &\leq M^2\delta e^{2\omega b}m(t)+\frac{M^2e^{2\omega b}}{\omega}\int_0^t(\gamma(s)+\epsilon)[m(t)-m(s)]ds.\end{aligned}$$

We consider $U_3(t)=F(t,y_t^2)\cap B(f_2(t),p(t)\|y_t^2-y_t^1\|_\infty)$. U_3 is measurable (see Proposition III.4 in [136]), and so there exists a function $f_3(t)\in U_3(t)$. Then we consider

$$y^3(t) = \begin{cases} \psi(t), & t \in [-r, 0], \\ C(t)\psi(0) + S(t)\alpha + \int_0^t S(t-s) f_3(s) ds, & t \in [0, t_1]. \end{cases}$$

We have

$$\begin{aligned}
|y^3(t) - y^2(t)| &\le M e^{\omega b} \int_0^t |f_2(s) - f_3(s)|\, ds \\
&\le M e^{\omega b} \int_0^t p(s) \|y_s^2 - y_s^1\|_\infty\, ds \\
&\le M^3 e^{3\omega b} \int_0^t p(s)[\delta m(s) + \frac{1}{\omega} \int_0^s (\gamma(u) + \epsilon)[m(s) - m(u)] du] ds \\
&\le M^3 e^{3b\omega} g \int_0^t p(s) e^{2\int_0^s p(\tau) d\tau} ds + \frac{M^3 e^{3\omega b}}{\omega} \int_0^t p(s) e^{2\int_0^s p(\tau) d\tau} \\
&\quad \times \int_0^s (\gamma(u) + \epsilon)[m(s) - m(u)] du ds \\
&\le \frac{M^3 e^{3\omega b} g}{2} \int_0^t (e^{2\int_0^s p(\tau) d\tau})' ds + \frac{M^3 e^{3\omega b}}{\omega} \int_0^t (e^{\int_0^s p(\tau) d\tau})' \\
&\quad \times \int_0^s (\gamma(u) + \epsilon)[m(s) - m(u)] du ds \\
&\le \frac{M^3 \delta e^{3\omega b}}{2} m^2(t) + \frac{M^3 e^{3\omega b}}{\omega} m(t) \int_0^t (\gamma(s) + \epsilon)[m(t) - m(s)] ds \\
&\quad - \frac{M^3 e^{3b\omega}}{\omega} \int_0^t p(s) m^2(s) \int_0^s [\gamma(u) + \epsilon] ds du \\
&= \frac{M^3 \delta e^{3\omega b}}{2} m^2(t) + \frac{M^3 e^{3\omega b}}{\omega} m(t) \int_0^t (\gamma(s) + \epsilon)[m(t) - m(s)] ds \\
&\quad - \frac{M^3 e^{3b\omega}}{2\omega} \int_0^t (m^2(s))' \int_0^s [\gamma(u) + \epsilon] ds du.
\end{aligned}$$

Thus,

$$\begin{aligned}
|y^3(t) - y^2(t)| \le{} & \frac{M^3 \delta e^{3b\omega} g}{2} m^2(t) \\
& + \frac{M^3 e^{3\omega b}}{2\omega} \int_0^t (\gamma(s) + \epsilon)[m(t) - m(s)]^2 ds.
\end{aligned}$$

For $n = 1, 2, 3$, we have

$$\begin{aligned}
|y^n(t) - y^{n-1}(t)| \le{} & \frac{M^n e^{n\omega b}}{(n-1)!} \int_0^t (\gamma(s) + \epsilon)[m(s) - m(u)]^{n-1} ds \\
& + \frac{M^n e^{n\omega b} g}{(n-1)!} m^{n-1}(t).
\end{aligned} \tag{10.35}$$

We suppose that (10.35) holds for some n, and now, we check (10.35) for $n+1$. Let $U_{n+1}(t) = F(t, y_n(t)) \cap B(f_n, p(t)\|y_t^n - y_t^{n-1}\|_\infty)$. Now U_n is measurable (see Proposition III.4 in [136]), and so there exists a function $f_n(t) \in U_n(t)$. Thus, we consider

$$y^{n+1}(t) = \begin{cases} \psi(t), & t \in [-r, 0], \\ C(t)\psi(0) + S(t)\alpha + \int_0^t S(t-s) f_n(s) ds, & t \in [0, t_1]. \end{cases}$$

Therefore,

$$\begin{aligned}
|y^{n+1}(t) - y^n(t)| &\le Me^{\omega b} \int_0^t |f_n(s) - f_{n-1}(s)|\, ds \\
&\le \frac{M^{n+1} e^{(n+1)\omega b}}{(n-1)!} \int_0^t p(s)\|y_s^n - y_s^{n-1}\|_\infty\, ds \\
&\le \frac{M^{n+1} e^{(n+1)\omega b}}{(n-1)!} \int_0^t p(s)[\delta m^{n-1}(s) \\
&\quad + \frac{1}{\omega} \int_0^s (\gamma(u) + \epsilon)[m(s) - m(u)]^{n-1}] duds \\
&\le \frac{M^{n+1} e^{(n+1)b\omega} \delta}{n!} \int_0^t p(s) e^{n \int_0^s p(\tau) d\tau} ds + \frac{M^{n+1} e^{(n+1)\omega b}}{n!} \\
&\quad \times \int_0^t np(s) \int_0^s (\gamma(u) + \epsilon)[m(s) - m(u)]^{n+1} dsdu \\
&\le \frac{M^{n+1} e^{(n+1)b\omega} \delta}{n!} m^n(t) + \frac{M^{n+1} e^{(n+1)\omega b}}{n!\omega} \\
&\quad \times \int_0^t (\gamma(s) + \epsilon)[m(s) - m(u)]^n ds.
\end{aligned}$$

We have seen that (10.35) holds for $n+1$. Therefore, by induction, (10.35) holds for all $n \in \mathbb{N}$. Since γ is an integrable function and $p \in L^1$, then, for all $n \in \mathbb{N}$, we have

$$\begin{aligned}
|y^{n+1}(t) - y^n(t)| &\le \frac{M^{n+1} e^{(n+1)\omega t_1} \delta m^n(t_1)}{n!} \\
&\quad + \frac{M^{n+1} e^{(n+1)\omega t_1}}{n!\omega} \int_0^{t_1} (\gamma(s) + \epsilon)[m(t_1) - m(s)]^n ds.
\end{aligned}$$

We conclude that $\{y^n\}$ is a Cauchy sequence in Ω_1, converging uniformly to a function $y \in \Omega_1$, where

$$\Omega_1 = C([0, t_1], E) \cap \mathcal{D}.$$

From the definition of U_n, $n \in \mathbb{N}$,

$$|f_{n+1}(t) - f_n(t)| \le p(t)\|y_t^n - y_t^{n-1}\|_\infty, \text{ for } n \in \mathbb{N}, \text{ a.e, } t \in [0, t_1].$$

Hence, for almost every $t \in [0,t_1]$, the sequence $\{f_n(t)\}$ is Cauchy in E. Then $\{f_n(t)\}$ converges almost everywhere to a measurable function $f(\cdot)$ in E. Moreover, since $f_0 = g$ and by the last inequality, we obtain

$$\begin{aligned}|f_n(t) - f_0(t) + f_0(t)| &\leq |f_n(t) - f_0(t)| + |f_0(t)| \\ &\leq |f_n(t) - f_{n-1}(t)| + |f_{n-1}(t) - f_{n-2}(t)| \\ &\quad + \cdots + |f_2(t) - f_1(t)| + |f_1(t) - f_0(t)| + |f_0(t)| \\ &\leq \sum_{k=1}^{n} p(t)|y^k(t) - y^{k-1}(t)| + |f_0(t)| \\ &\leq p(t)M \sum_{k=1}^{\infty} |y^k(t) - y^{k-1}(t)| + |f_0(t)| \\ &\leq p(t)A + \gamma(t),\end{aligned}$$

where

$$A = Me^{\omega t_1}\delta \exp\big(m(t_1)e^{\omega t_1}\big) + \frac{Me^{\omega t_1}}{\omega}\int_0^{t_1} (\gamma(s) + \epsilon) \exp Me^{\omega t_1}\big(m(t_1) - m(s)\big)ds.$$

Then, for all $n \in \mathbb{N}$,

$$|f_n(t)| \leq \beta p(t) + \gamma(t). \tag{10.36}$$

From (10.36), we see that f_n converges to f in $L^1([0,t_1], E)$. Consequently

$$y(t) = C(t)\psi(0) + S(t)\eta + \int_0^t S(t-s)f(s)ds, \ t \in [0,t_1],$$

is a solution of the problem (10.25)–(10.26) with the condition $y(t) = \psi(t)$, $t \in [-r,0]$. Then $y \in S_{[0,t_1]}(\psi)$.

Next, we prove that

$$|x(t) - y(t)| \leq \eta_0(t), \text{ for } t \in [0,t_1].$$

We have

$$\begin{aligned}|x(t) - y(t)| &= \left|C(t)\phi(0) + \int_0^t S(t-s)g(s)ds - C(t)\psi(0) - \int_0^t S(t-s)f(s)ds\right| \\ &\leq Me^{\omega t_1}|\phi(0) - \psi(0)| + Me^{\omega t_1}\int_0^t |f(s) - f_0(s)|ds \\ &\leq Me^{\omega t_1}\|\phi - \psi\|_\infty + Me^{\omega t_1}\int_0^t |f(s) - f_n(s)|ds \\ &\quad + Me^{\omega t_1}\int_0^t |f_n(s) - f_0(s)|ds\end{aligned}$$

$$
\begin{aligned}
&\leq Me^{\omega t_1}\|\phi-\psi\|_\infty + Me^{\omega t_1}\int_0^t |f(s)-f_n(s)|ds \\
&\quad + Me^{\omega t_1}\int_0^t Me^{\omega t_1}p(s)\delta\exp\big(Me^{\omega t_1}m(s)\big)ds \\
&\quad + \frac{Me^{\omega t_1}}{\omega}\int_0^t Me^{\omega t_1}p(s)\int_0^s(\gamma(\tau)+\epsilon)d\tau \\
&\quad \times \exp Me^{\omega t_1}\big(m(t)-m(s)\big)ds \\
&\leq Me^{\omega t_1}\int_0^t |f(s)-f_n(s)|ds \\
&\quad + Me^{\omega t_1}\delta\exp\big(Me^{\omega t_1}m(s)\big)ds \\
&\quad + \frac{Me^{\omega t_1}}{\omega}\int_0^t(\gamma(s)+\epsilon)\exp Me^{\omega t_1}\big(m(t)-m(s)\big)ds.
\end{aligned}
$$

Thus,

$$
\begin{aligned}
|x(t)-y(t)| \leq{}& Me^{\omega t_1}\int_0^{t_1}|f(s)-f_n(s)|ds + Me^{\omega t_1}\delta\exp\big(Me^{\omega t_1}m(s)\big)ds \\
&+ \frac{Me^{\omega t_1}}{\omega}\int_0^t(\gamma(s)+\epsilon)\exp Me^{\omega t_1}\big(m(t)-m(s)\big)ds.
\end{aligned}
$$

As $n\to\infty$, we conclude that

$$|x(t)-y(t)|\leq \eta_0(t).$$

Finally, we prove that

$$|g(t)-y''(t)-Ay(t)| \leq p(t)\eta_0(t)+\gamma(t)+\epsilon, \quad \text{a.e.} \quad t\in[0,t_1].$$

We have

$$
\begin{aligned}
|g(t)-y''(t)-Ay(t)| &= |f_0(t)-f(t)| \\
&\leq |f_0(t)-f_n(t)+f_n(t)-f(t)| \\
&\leq |f_0(s)-f_n(s)|+|f_n(t)-f(t)| \\
&\leq \sum_{k=1}^{n} p(t)|y_k(t)-y_{k-1}(t)|+|f_n(t)-f(t)| \\
&\leq p(t)\sum_{k=1}^{\infty}|y_k(t)-y_{k-1}(t)|+|f_n(t)-f(t)| \\
&\leq |f_n(t)-f(t)|+p(t)\sum_{n=0}^{\infty}\frac{M^{n+1}e^{(n+1)\omega t_1}m^n(t)}{n!} \\
&\quad + p(t)\sum_{n=0}^{\infty}\frac{1}{n!\omega}\int_0^t M^{n+1}e^{(n+1)\omega t}(\gamma(s)+\epsilon) \\
&\quad \times[m(t)-m(s)]^n ds+|f_1(t)-f_0(t)| \\
&\leq p(t)\eta_0(t)+\gamma(t)+\epsilon+|f_1(t)-f_0(t)|.
\end{aligned}
$$

Thus,

$$|g(t) - y''(t) - Ay(t)| \le p(t)\eta_0(t) + \gamma(t)\; t \in [0, t_1].$$

We denote this solution of by y_1.

Step 2. Now consider the problem (10.25)–(10.28) on $[t_1 - r, t_2]$,

$$y''(t) - Ay(t) \in F(t, y_t), \quad \text{a.e.} \quad t \in (t_1, t_2], \tag{10.37}$$

$$y'(t_1^+) = y_1'(t_1) + \overline{I}_1(y_1(t_1)), \tag{10.38}$$

$$y(t_1^+) = y_1(t_1) + I_1(y_1(t_1)), \; y(t) = y_1(t) \quad t \in [t_1 - r, t_1]. \tag{10.39}$$

Let $f_0 = g$ and

$$y^0(t) = \begin{cases} y_1(t), & t \in [t_1 - r, t_1], \\ C(t - t_1)[y_1(t_1) - I_1(y_1(t_1))] & \\ \quad + S(t - t_1)[y_1'(t_1) + \overline{I}_1(y_1(t_1))] & \\ \quad + \int_{t_1}^{t} S(t - s) f_0(s) ds, & t \in (t_1, t_2]. \end{cases}$$

Let $U_1 : [t_1, t_2] \to \mathcal{P}(E)$ given by $U_1(t) = F(t, y_t^0) \cap B(g(t), \gamma(t) + \epsilon)$. Since the multivalued operator $U_1(t)$ is measurable (see Proposition III.4 in [136]), there exists a function $t \to f_1(t)$, which is a measurable selection for U_1. Consider

$$y^1(t) = \begin{cases} y_1(t), & t \in [t_1 - r, t_1], \\ C(t - t_1)[y_1(t_1) + I_1(y(t_1))] & \\ \quad + S(t - t_1)[y_1'(t_1) + \overline{I}_1(y(t_1))], & \\ \quad + \int_{t_1}^{t} S(t - s) f_1(s) ds, & t \in (t_1, t_2]. \end{cases}$$

Then, for every $t \in (t_1, t_2]$, we have

$$\begin{aligned} |y^1(t) - y^0(t)| &\le M \int_{t_1}^{t} \int_{t_1}^{t-s} e^{\omega u} |f_0(s) - f_1(s)|\, du ds \\ &\le \frac{M}{\omega} \int_{t_1}^{t} e^{\omega(t-s)} |f_0(s) - f_1(s)| ds \\ &\le \frac{M e^{\omega b}}{\omega} \int_{t_1}^{t} |f_0(s) - f_1(s)| ds. \end{aligned}$$

Thus,

$$\begin{aligned} \sup\{|y^1(t) - y^0(t)| : t \in [t_1 - r, t_2]\} &\le \frac{M e^{\omega b}}{\omega} \int_{t_1}^{t} (\gamma(s) + \epsilon) ds \\ &\le \frac{M e^{\omega b}}{\omega} \int_{0}^{t} (\gamma(s) + \epsilon) ds. \end{aligned}$$

Define the set valued map $U_2(t) = F(t, y_t^1) \cap B(f_1(t), p(t)\|y_t^1 - y_t^0\|_\infty)$. Since the multivalued operator U_2 is measurable, there exists a function $f_2(t) \in U_2(t)$. Then we consider

$$y^2(t) = \begin{cases} y_1(t), & t \in [t_1 - r, t_1], \\ C(t-t_1)[y_1(t_1) + I_1(y(t_1))] & \\ \quad + S(t-t_1)[y_1'(t_1) + \overline{I}_1(y(t_1))] & \\ \quad + \int_{t_1}^t S(t-s) f_2(s)ds, & t \in [0, t_1]. \end{cases}$$

Hence,

$$\begin{aligned} |y^2(t) - y^1(t)| &\leq \int_{t_1}^t |S(t-s)||f_2(s) - f_1(s)|\, ds \\ &\leq \int_{t_1}^t Me^{\omega b} p(s) \Big[\frac{Me^{\omega b}}{\omega} \int_0^s (\gamma(u) + \epsilon) du\Big] ds \\ &\leq \frac{M^2 e^{2\omega b}}{\omega} \int_{t_1}^t p(s) \int_{t_1}^u (\gamma(u) + \epsilon) ds du \\ &\leq \frac{M^2 e^{2\omega b}}{\omega} \int_{t_1}^t p(s) e^{\int_s^t p(\tau)d\tau} \int_0^s (\gamma(u) + \epsilon) ds du \\ &= -\frac{M^2 e^{2\omega b}}{\omega} \int_{t_1}^t (e^{\int_s^t p(\tau)d\tau})' \int_0^s (\gamma(u) + \epsilon) ds du \\ &\leq -\frac{M^2 e^{2\omega b}}{\omega} \Big[\int_{t_1}^t (\gamma(s) + \epsilon) ds. - \int_{t_1}^t (\gamma(s) + \epsilon) e^{\int_s^t p(\tau)d\tau} ds\Big] \\ &= \frac{M^2 e^{2\omega b}}{\omega} \int_{t_1}^t (\gamma(s) + \epsilon)[e^{\int_s^t p(\tau)d\tau} - 1] ds \\ &\leq \frac{M^2 e^{2\omega b}}{\omega} \int_{t_1}^t (\gamma(s) + \epsilon)[m(t) - m(s)] ds. \end{aligned}$$

We consider $U_3(t) = F(t, y_t^2) \cap B(f_2(t), p(t)\|y_t^2 - y_t^1\|_\infty)$. Since U_3 is measurable, there exists a function $f_3(t) \in U_3(t)$. Then we consider

$$y^3(t) = \begin{cases} y_1(t), & t \in [t_1 - r, t_1], \\ C(t-t_1)[y_1(t_1) + I_1(y(t_1))] & \\ \quad + S(t-t_1)[y_1'(t_1) + \overline{I}_1(y(t_1))] & \\ \quad + \int_{t_1}^t S(t-s) f_3(s)ds, & t \in (t_1, t_2]. \end{cases}$$

We have

$$
\begin{aligned}
|y^3(t)-y^2(t)| &\le Me^{\omega b}\int_{t_1}^{t} |f_2(s)-f_3(s)|\,ds\\
&\le Me^{\omega b}\int_{t_1}^{t} p(s)\|y_s^2-y_s^1\|_\infty\,ds\\
&\le M^3e^{3\omega b}\int_{t_1}^{t} p(s)[\frac{1}{\omega}\int_{t_1}^{s}(\gamma(u)+\epsilon)[m(s)-m(u)]du]ds\\
&\le \frac{M^3e^{3\omega b}}{\omega}\int_{t_1}^{t} p(s)e^{2\int_{t_1}^{s}p(\tau)d\tau}\int_{t_1}^{s}(\gamma(u)+\epsilon)\times\\
&\quad [m(s)-m(u)]duds\\
&\le \frac{M^3e^{3\omega b}}{\omega}\int_{t_1}^{t}(e^{\int_{t_1}^{s}p(\tau)d\tau})'\int_{t_1}^{s}(\gamma(u)+\epsilon)\times\\
&\quad [m(s)-m(u)]duds\\
&\le \frac{M^3e^{3\omega b}}{\omega}m(t)\int_{t_1}^{t}(\gamma(s)+\epsilon)[m(t)-m(s)]ds\\
&\quad -\frac{M^3e^{3b\omega}}{\omega}\int_{t_1}^{t}p(s)m^2(s)\int_{t_1}^{s}\gamma(u)dsdu\\
&= \frac{M^3e^{3\omega b}}{\omega}m(t)\int_{t_1}^{t}(\gamma(s)+\epsilon)[m(t)-m(s)]ds\\
&\quad -\frac{M^3e^{3b\omega}}{2\omega}\int_{t_1}^{t}(m^2(s))'\int_{t_1}^{s}(\gamma(u)+\epsilon)dsdu.
\end{aligned}
$$

Then, we obtain

$$|y^3(t)-y^2(t)| \le \frac{M^3e^{3\omega b}}{2\omega}\int_{t_1}^{t}(\gamma(s)+\epsilon)[m(t)-m(s)]^2ds.$$

For $n = 1, 2, 3$, we have

$$|y^n(t)-y^{n-1}(t)| \le \frac{M^ne^{n\omega b}}{(n-1)!}\int_{0}^{t}(\gamma(s)+\epsilon)[m(s)-m(u)]^{n-1}ds, \qquad (10.40)$$

so we suppose that (10.40) holds for some n, and we now check (10.40) for $n+1$. Let $U_{n+1}(t) = F(t, y_n(t)) \cap B(f_n, p(t)\|y_t^n - y_t^{n-1}\|_\infty)$. Since U_n is measurable, there exists a function $f_n(t) \in U_n(t)$. Thus, we consider

$$
y^{n+1}(t) = \begin{cases}
y_1(t), & t\in[t_1-r,t_1],\\
C(t-t_1)[y_1(t_1)+I_1(y_1(t_1))] & \\
\quad + S(t-t_1)[y_1'(t_1)+\overline{I}_1(y_1(t_1))] & \\
\quad + \int_0^t S(t-s)f_n(s)ds, & t\in(t_1,t_2].
\end{cases}
$$

Set

$$\Omega_2 = \{y | y \in \mathcal{D} \cap C([0, t_1] \cap C((t_1, t_2] \text{ and } y(t_1^+), \text{ exists}\}.$$

As in Step 1, we can prove that $\{y^n\}$ converges to some $y \in \Omega_2$ such that y is a solution of the problem (10.37)–(10.39), and

$$|x(t) - y(t)| \leq \eta_0(t_1) + M|I_1(x(t_1)) - I_1(y(t_1))| + \frac{Me^{\omega t_1}}{\omega}|\overline{I}_1(x(t_1)) - \overline{I}_1(y(t_1))|$$
$$+ \frac{Me^{\omega t_2}}{\omega}\int_{t_1}^{t} (\gamma(s) + \epsilon) \exp Me^{\omega t_2}\big(m(t) - m(s)\big)ds, \; t \in (t_1, t_2],$$

and

$$|g(t) - y''(t) - Ay(t)| \leq p(t)\eta_1(t) + \gamma(t) + \epsilon, \; t \in (t_1, t_2].$$

Rename $y(t)$ as $y_2(t)$.

Step 3. We continue this process taking into account that $y_{m+1} := y\big|_{[t_m - r, t_m] \cup (t_m, b]}$ is a solution to the problem

$$y''(t) - Ay(t) \in F(t, y_t), \qquad \text{a.e. } t \in (t_m, b], \tag{10.41}$$
$$y(t_m^+) = y_{m-1}(t_m) + I_m(y_{m-1}(t_m)), \tag{10.42}$$
$$y(t_m^+) = y_{m-1}(t_m) + I_m(y_{m-1}(t_m)), \; y(t) = y_{m-1}(t), \quad t \in [t_m - r, t_m]. \tag{10.43}$$

The solution y of the problem (10.25)–(10.28) is then defined by

$$y(t) = \begin{cases} y_1(t), & \text{if } t \in [-r, t_1], \\ y_2(t), & \text{if } t \in (t_1, t_2], \\ \dots \\ y_{m+1}(t), & \text{if } t \in (t_m, b]. \end{cases}$$

By Steps 1 to 3 we have

$$\begin{aligned} |x(t) - y(t)| &\leq \sum_{k=1}^{m} |x(t) - y_k(t)| \\ &\leq M \sum_{0<t_k<t} |x(t_k) - y(t_k)| + M \sum_{0<t_k<t} |I_k(x(t_k)) - I_k(y(t_k))|] \\ &\quad + \sum_{0<t_k<t} \frac{Me^{\omega t_k}}{\omega}|x'(t_k) - y'(t_k)| \\ &\quad + \sum_{0<t_k<t} \frac{Me^{\omega t_k}}{\omega}|\overline{I}_k(x(t_k)) - \overline{I}_k(y(t_k))|] + \sum_{k=0}^{m} \eta_k(t), \end{aligned}$$

and

$$\begin{aligned}|g(t) - y''(t) - Ay(t)| &\leq \sum_{k=1}^{m} |g(t) - y_k''(t) - Ay(t)| \\ &\leq p(t) \sum_{0<t_k<t} \eta_k(t) + \sum_{0<t_k<t} \gamma_k(t),\end{aligned}$$

where $\gamma_k = \gamma_{|J_k}$. □

10.5.3 Filippov–Wazewski's Theorem

In this subsection, we compare trajectories of (10.25)–(10.28) to the relaxation impulsive differential inclusion

$$y''(t) - Ay(t) \in \overline{co}F(t, y_t), \quad \text{a.e. } t \in J =: [0, b]\backslash\{t_1, \ldots, t_m\}, \tag{10.44}$$
$$y'(t_k^+) - y'(t_k) = \bar{I}_k(y(t_k^-)), \quad k = 1, \ldots, m, \tag{10.45}$$
$$y(t_k^+) - y(t_k) = I_k(y(t_k^-)), \quad k = 1, \ldots, m, \tag{10.46}$$
$$y(t) = \psi(t), \quad t \in [-r, 0],\ y'(0) = \alpha. \tag{10.47}$$

We will need the following auxiliary result in order to prove our main relaxation theorem.

Lemma 10.15 ([439]). *Let $(E, |\cdot|)$ be a separable space and $G : [0, b] \to \mathcal{P}(E)$ be a measurable multifunction. Then $s \to S(t-s)G(s)$ is measurable. Moreover, if $f(s) \in S(t-s)G(s)$ a.e. in $[0, b]$, then there exists a measurable selection $g(s) \in G(s)$ such that $f(s) = S(t-s)g(s)$ a.e. in $[0, b]$.*

Theorem 10.16. *Let $(E, |\cdot|)$ be a reflexive Banach space. Assume that conditions* (J-20), (J-21), *and the following condition hold:*

(J-22) *there exist constants $c_k, \bar{c}_k \geq 0$ such that, for each $u, z \in E$,*

$$|I_k(u) - I_k(z)| \leq c_k|u - z| \text{ and } |\bar{I}_k(u) - \bar{I}_k(z)| \leq \bar{c}_k|u - z|.$$

Then, for all $\delta > 0$, there exists a solution y to (10.25)–(10.28) *on* $[0, b]$ *satisfying*

$$x(t) = y(t),\ t \in [-r, 0] \text{ and } \|x - y\|_{PC} \leq \delta.$$

This implies that $S^{co}_{[0,b]}(\phi) = \overline{S_{[0,b]}(\phi)}$, where

$$S^{co}_{[0,b]} = \{y \mid y \text{ is a solution to (10.44)–(10.47) on } [-r, b], y(t) = \phi(t),\ t \in [-r, 0]\}.$$

Proof. Let y be a solution of the problem (10.44)–(10.47). Fix $\epsilon > 0$. Now g is a measurable selection of $t \to \overline{co}F(t, y_t)$. Hence, from Theorem 5.4 and Lemma 10.15, there exists f, a measurable selection of $t \to F(t, y_t)$, such that

$$\sup_{t\in[0,b]}\left|\int_0^t S(t-s)f(s)ds - \int_0^t S(t-s)g(s)ds\right| \le Me^{\omega b}\delta.$$

Assume that δ is a small positive number. Set

$$x(t) = \begin{cases} \phi(t), & t \in [-r, 0], \\ C(t)\psi(0) + S(t)\alpha + \int_0^t S(t-s)f(s)ds & \\ \quad + \sum_{0<t_k<t}[S(t-t_k)I_k(y(t_k)) + C(t-t_k)\overline{I}_k(y(t_k))], & t \in [0, b]. \end{cases}$$

Then, x is a mild solution of the problem,

$$y''(t) - Ay(t) = f(t), \quad \text{a.e.} \quad t \in J = [0, b]\backslash\{t_1, \dots, t_m\}, \tag{10.48}$$

$$\Delta y_{t=t_k} = I_k(y(t_k^-)), \qquad k = 1, \dots, m, \tag{10.49}$$

$$\Delta y'_{t=t_k} = \overline{I}_k(y(t_k^-)), \qquad k = 1, \dots, m, \tag{10.50}$$

$$y(t) = \psi(t) \qquad t \in [-r, 0], \ y'(0) = \alpha. \tag{10.51}$$

Hence,

$$d(f(t), F(t, x_t)) \le p(t)\|x_t - y_t\|_\infty \le p(t)\delta, \quad \text{a.e.} \quad t \in [0, b].$$

From Theorem 10.14, there exists a solution, y_*, of the problem (10.25)–(10.27) such that, for all $t \in [0, t_1]$, we have

$$|y_*(t) - y(t)| \le \eta_0(t), \ t \in [0, t_1].$$

For $t \in (t_1, t_2]$, we obtain

$$|y_*(t) - y(t)| \le M[1 + c_1 + Mt_1 + e^{\omega(t_2-t_1)}\bar{c}_1)]\eta_0(t_1) + M\eta_1(t).$$

For $t \in (t_2, t_3]$, we have

$$\begin{aligned} |y_*(t) - y(t)| \le\, & M^2[1 + c_1 + Mt_1 + e^{\omega(t_2-t_1)}\bar{c}_1)]\eta_0(t_1) \\ & + M^2\eta_1(t) + Mc_1 + Me^{\omega(t_3-t_2)}\eta_2(t_2) \\ & (t_2 - t_1)M^2[1 + c_1 + Mt_1 + e^{\omega(t_2-t_1)}\bar{c}_1)]\eta_0(t_1) \\ & + M^2\eta_1(t) + M\eta_3(t), \end{aligned}$$

where

$$\begin{aligned} \eta_0(t) \le\, & Me^{\omega t_1}\delta \exp\Big(Me^{\omega t_1}m(s)\Big)ds \\ & + \frac{\delta Me^{\omega t_1}}{\omega}\int_0^t p(s)\exp Me^{\omega t_1}(m(t) - m(s))ds, \ t \in [0, t_1], \end{aligned}$$

$$\eta_k(t) = \frac{Me^{\omega t_{k+1}}}{\omega}\int_{t_k}^{t} p(s)\exp Me^{\omega t}\big(m(t)-m(s)\big)ds$$
$$\le \delta\frac{Me^{\omega t_{k+1}}}{\omega}\int_{t_k}^{t} p(s)\exp Me^{\omega t}\big(m(t)-m(s)\big)ds, \quad k = 1,\dots,m.$$

Thus, there exist $L > 0$ such that, for all $t \in [-r, b]$, we have

$$|x(t)-y(t)| \le L\sum_{k=0}^{m}\eta_k(t_k) \le L\delta\sum_{k=0}^{m} R_k,$$

where

$$R_0 = Me^{\omega t_1}\exp\big(Me^{\omega t_1}m(s)\big)ds + \frac{Me^{\omega t_1}}{\omega}\int_0^t p(s)\exp Me^{\omega t_1}\big(m(t)-m(s)\big)ds,$$

and for $k = 1,\dots,m$,

$$R_k = \frac{Me^{\omega t_{k+1}}}{\omega}\int_{t_k}^{t_{k+1}} p(s)\exp Me^{\omega t_{k+1}}\big(m(t_{k+1})-m(s)\big)ds.$$

Set

$$\delta = \frac{\epsilon}{L\sum\limits_{k=0}^{m} R_k}.$$

We obtain

$$\|y_* - y\|_\Omega \le \epsilon.$$

Thus, $S^{co}_{[0,b]}(\phi) = \overline{S_{[0,b]}(\phi)}$. □

Chapter 11

Selected Topics

There are many directions in which the methods of the previous chapters can be applied and extended in modeling a variety of situations. In this chapter, we have chosen to show how our methods can be adapted to problems addressed by impulsive stochastic differential equations in Hilbert spaces, by a nonsmooth analysis concept of sweeping processes, by Volterra integral inclusions in Banach spaces, by a Filippov's theorem on a bounded interval, and by a few others.

11.1 Stochastic Differential Equations

This section deals with the existence of solutions, for first order stochastic semilinear functional differential equations and inclusions, in Hilbert spaces. More precisely, in Section 11.1.2, we consider first order stochastic semilinear differential equations of the form,

$$\begin{aligned} d[y(t) - Ay(t)] &= f(t, y(t))dt + g(t, y(t))dW(t), \\ &\qquad t \in J := [0, b] \setminus \{t_1, \ldots, t_m\}, \end{aligned} \tag{11.1}$$

$$y(t_k^-) - y(t_k) = I_k(y(t_k)), \; k = 1, \ldots, m, \tag{11.2}$$

$$y(0) = y_0, \tag{11.3}$$

where $f, g : J \times H \to H$ are given functions, $A : H \to H$ generates a C_0-semigroup $\{T(t) : t \geq 0\}$ on H, $0 = t_0 < \cdots < t_m < b$, $I_k \in C(H, H)$, $W(t)$ is a Hilbert space Q-Wiener process, and H is a real separable Hilbert space with inner product $\langle \cdot, \cdot \rangle$ and norm $|\cdot|$. Random differential and integral equations play an important role in characterizing many social, physical, biological and engineering problems; see for instance, the monographs of Da Prato and Zabczyk [163] and Sobzyk [447]. For example, a stochastic model for drug distribution in a biological system was described by Tsokos and Padgett [470] to a closed system with a simplified heat, one organ or capillary bed, and re-circulation of blood with a constant rate of flow, where the heart is considered as a mixing chamber of constant volume. For the basic theory concerning stochastic differential equations, see the monographs of Tsokos and Padgett [470], Bharucha [88], Da Prato and Zabczyk [163], Krylov [329], and the papers of Liu [348] and Taniguchi [461].

Recently, where the impulse is absent (i.e., $I_k \equiv 0$), Balasubramaniam and Ntouyas [42] studied semilinear stochastic evolution delay equations with nonlocal conditions, where A is a densely defined linear operator.

For additional background on stochastic processes and stochastic differential equations in Hilbert spaces, Brownian motion and white noise, and Wiener processes, we refer the reader to the books by Hida [289] and Hida et al. [290].

Fundamental to our development for solutions of (11.1)–(11.3) is the Itô integral.

11.1.1 Itô Integral

Let K be another real separable Hilbert space and $W(t)$, $t \geq 0$, be a K-valued Wiener process with mean zero and covariance operator $\mathcal{C}$ with $tr\mathcal{C} < \infty$ ($tr\mathcal{C}$ denotes the trace of the operator $\mathcal{C}$) defined by

$$E\langle W(t), g\rangle\langle W(s), h\rangle = (t \wedge s)\langle \mathcal{C}g, h\rangle, \quad \text{for every } g, h \in K,$$

where $\langle \cdot, \cdot \rangle$ denotes the inner product and E stands for integration with respect to the probability measure P. Let $L(K, H)$ denote the space of bounded linear operators from K into H. For $g_1, g_2 \in L(K, H)$ define $\langle\langle g_1, g_2\rangle\rangle = tr(g_1 Q g_2^*)$, where g_2^* is the adjoint of the operator g_2 and Q is the nuclear operator associated with the Brownian motion. Here $Q \in L_n^+(K)$, the space of positive nuclear operators in K. Let $L(K_Q, H)$ denote the completion of $L(K, H)$ with respect to the topology induced by the norm $\|\cdot\|_2$ where $\|g\|_2^2 = \langle\langle g, g\rangle\rangle$. Let $(\Omega, \mathcal{F}, \mathcal{F}_t, P, H)$ be a complete probability space furnished with a complete family of right continuous increasing σ-algebras $\{\mathcal{F}_t : t \in [0, T]\}$ satisfying $\mathcal{F}_t \subset \mathcal{F}$.

Let $L^2(\Omega, \mathcal{F}, \mathcal{F}_t, PH)$ be a space of all square random variables with values in H, that are measurable with respect to $\{\mathcal{F}_t : t \in [0, b]\}$. Let $\widehat{M}_2([0, b], H)$ denote the classes of H-valued stochastic processes $\{\xi(t) : t \in [0, b]\}$ which are $\mathcal{F}_t$-adapted and have finite second moments, that is,

$$\|\xi\|_{\widehat{M}_2} = \sup_{t \in [0,b]} (E|\xi(t)|^2)^{\frac{1}{2}} < \infty.$$

It is easy to verify that $\widehat{M}_2$, furnished with the norm topology as defined above, is a Banach space. If $\{W_t : t \in [0, T]\}$ denotes Brownian motion on the finite interval $[0, T]$ and if $\{f(\omega; t) : t \in [0, T]\}$ denotes a well-behaved stochastic process whose specific qualfications will be given later, then the Itô integral is a random variable that is commonly denoted by

$$I(f)(\omega) = \int_0^T f(\omega, t) dW_t.$$

This notation may be misleading since it perhaps suggests that the integral on the right may be interpreted in a way that is analogous to the classical Riemann–Stieltjes integral.

Definition 11.1. Let $f_t = f_t(\omega)$ be a function defined on $\Omega \times [0, T]$. We write $f \in H_0$ if there exist nonrandom points $0 = t_0 \leq t_1 \leq \ldots \leq t_n \leq T$ such that f_{t_i} are

$\mathcal{F}_{t_i}$–measurable, $E|f_{t_i}|^2 < \infty$, and $f_t = f_{t_i}$ for $t \in [t_i, t_{i+1})$, $i = 0, 1, \dots, n$. If H_0 is the set of step-functions, then the stochastic integral of $f \in H_0$ with respect to $\{W(t) : t \in [0, T]\}$ is naturally defined by the equality

$$\int_0^T f_t dW(t) = \sum_{i=0}^{n} f_{t_i}(W_{t_{i+1}} - W_{t_i}).$$

White noise is usually regarded as the informal time derivative $W'(t)$ of Brownian motion or Wiener process $W(t)$. In the Itô theory of stochastic integration, an integral with respect to $W'(t)$ is rewritten as an integral with respect to $dW(t)$, that is,

$$\int_a^b \psi(t)dW(t) = \int_a^b \psi(t)W'(t)dt.$$

The Itô integral $\int_a^b \psi(t)dW(t)$ is defined for any process $\psi(t)$ which satisfies the conditions:

(1) ψ is nonanticipating,

(2) almost all sample paths of ψ belong to $L^2([a, b])$.

Moreover, $\displaystyle\int_a^b \psi(t)dW(t) \in L^2(\Omega)$ if and only if $\psi \in L^2([a, b] \times \Omega)$. In fact, the following equality holds

$$E\left|\int_a^b \psi(t)dW(t)\right|^2 = E\int_a^b |\psi(t)|^2 dt.$$

We denote by $G_R(H)$ the space of nonanticipating functions satisfying (2).

Lemma 11.2. ([471]) *For an arbitrary $\psi \in G_R(H)$ there exists a sequence of step-functions $(\psi_n)_{n\in\mathbb{N}} \in G_R(H)$ such that $\psi_n \to \psi$ and $\int_a^b \psi_n(t)dW_t$ converges in $L^2(\Omega; \mathcal{B}; P)$ as $n \to \infty$.*

As indicated above, additional details on Brownian motion and white noise can be found in Hida [289] and Hida et al. [290].

In what follows, $B(H)$ denotes the Banach space of bounded linear operators from H into H with norm

$$\|N\|_{B(H)} = \sup\{|N(y)| : |y| = 1\}.$$

11.1.2 Definition of a Mild Solution

In order to define a mild solution of (11.1)–(11.3), we consider the space

$$PC = \big\{y : [0, b] \to H : y_k \in \widehat{M}_2(J_k, H),\ k = 0, \dots, m, \text{and there exist}$$
$$y(t_k^-) \text{ and } y(t_k^+),\ k = 1, \dots, m,\ \text{with } y(t_k^-) = y(t_k)\big\},$$

which is a Banach space with the norm

$$\|y\|_{PC} = \max\{\|y_k\| : k = 0, \dots, m\},$$

where y_k is the restriction of y to $J_k = (t_k, t_{k+1}]$, $k = 0, \dots, m$, and

$$\|y_k\| = \|y_k\|_{\widehat{M}_2} = \sup\{\sqrt{E|y(t)|^2} : t \in [t_k, t_{k+1}]\}.$$

Lemma 11.3. *If $y \in PC$ is an mild solution of the problem*

$$d[y(t) - Ay(t) - f(t)] = g(t)dW(t), \; t \in J := [0, b], \; t \neq t_k, \; k = 1, \dots, m, \tag{11.4}$$

$$y(t_k^-) - y(t_k) = I_k(y(t_k)), \; k = 1, \dots, m, \tag{11.5}$$

$$y(0) = y_0, \tag{11.6}$$

then it is given by

$$y(t) = T(t)y_0 + \int_0^t T(t-s)f(s)ds + \int_0^t T(t-s)g(s)dW(s)$$
$$+ \sum_{0<t_k<t} T(t-t_k)I_k(y(t_k^-)), \; \textit{for } t \in J.$$

Proof. Let y be a mild solution of the problem (11.4)–(11.6). Define $L(s) = T(t-s)y(s)$. Then we have

$$\begin{aligned} L'(s) &= -T'(t-s)y(s) + T(t-s)y'(s) \\ &= -AT(t-s)y(s) + T(t-s)y'(s) \\ &= T(t-s)[y'(s) - Ay(s)] \\ &= T(t-s)[f(s) + g(s)\tfrac{dW(s)}{ds}]. \end{aligned} \tag{11.7}$$

Consider $t_k < t$, $k = 1, \dots, m$. Then, integrating the previous equation, we have

$$\int_0^t L'(s)ds = \int_0^t T(t-s)\left(f(s) + g(s)\frac{dW(s)}{ds}\right)ds.$$

For $k = 1$,

$$L(t) - L(0) = \int_0^t T(t-s)f(s)ds + \int_0^t T(t-s)g(s)dW(s).$$

Hence,

$$y(t) = T(t)y(0) + \int_0^t T(t-s)f(s)ds + \int_0^t T(t-s)g(s)dW(s).$$

Now for $k = 2, \dots, m$, we have that

$$\int_0^{t_1} L'(s)ds + \int_{t_1}^{t_2} L'(s)ds + \cdots + \int_{t_k}^{t} L'(s)ds = \int_0^t T(t-s)f(s)ds + \int_0^t T(t-s)g(s)dW(s),$$

which is equivalent to

$$L(t_1^-) - L(0) + L(t_2^-) - L(t_1^+) + \cdots + L(t_k^+) - L(t) = \int_0^t T(t-s)f(s)ds + \int_0^t T(t-s)g(s)dW(s),$$

or

$$\begin{aligned} y(t) &= L(0) + \sum_{0<t_k<t} [L(t_k^+) - L(t_k^-)] \\ &\quad + \int_0^t T(t-s)f(s)ds + \int_0^t T(t-s)g(s)dW(s) \\ &= T(t)y_0 + \sum_{0<t_k<t} T(t-t_k)I(y(t_k^-)) \\ &\quad + \int_0^t T(t-s)f(s)ds + \int_0^t T(t-s)g(s)dW(s), \end{aligned}$$

which proves the lemma. □

We can now define what we mean by a mild solution of the problem (11.1)–(11.3).

Definition 11.4. For any $y_0 \in H$ with $E|y_0|^2 < \infty$, an element $y \in PC$ is said to be a mild solution of (11.1)–(11.3), if y satisfies the impulsive integral equation

$$\begin{aligned} y(t) &= T(t)y_0 + \int_0^t T(t-s)f(s,y(s))ds \\ &\quad + \int_0^t T(t-s)g(s,y(s))dW(s) + \sum_{0<t_k<t} T(t-t_k)I_k(y(t_k)). \end{aligned}$$

11.1.3 Existence and Uniqueness

In this subsection, we establish our first result for the problem (11.1)–(11.3).

Theorem 11.5. *Suppose the following conditions are satisfied:*

(S-1) *There exist a function $l \in L^2(J, \mathbb{R}^+)$ and a positive constant $K \geq 0$ such that*

$$|f(t,x) - f(t,\overline{x}) \leq l(t)\|x - \overline{x}\| \ a.e.\ t \in J, \text{ for all } \overline{x} \in H,$$

and

$$|g(t,x) - g(t,\overline{x})| \leq K\|x - \overline{x}\| \text{ for all } x, \overline{x} \in H.$$

(S-2) *There exists $M \geq 1$ such that*

$$\|T(t)\|_{B(H)} \leq M \text{ for each } t \geq 0.$$

Then the initial value problem (11.1)–(11.3) *has a unique mild solution on* $[0,b]$. *Moreover the unique solution y belongs to PC.*

Proof. We give the proof in several steps.

Step 1. Consider the problem (11.1)–(11.3) on $[0, t_1]$,

$$d[y(t) - Ay(t) - f(t, y(t))]dt = g(t, y(t))dW(t), \quad t \in J = [0, t_1], \tag{11.8}$$

$$y(0) = y_0. \tag{11.9}$$

Let

$$y^0(t) = T(t)y_0, \ t \in [0, t_1].$$

Next we define y^1 by

$$\begin{aligned} y^1(t) = T(t)y_0 &+ \int_0^t T(t-s)f(s, y^0(s))ds \\ &+ \int_0^t T(t-s)g(s, y^0(s))dW(s), \quad t \in [0, t_1]. \end{aligned}$$

Then for each $t \in [0, t_1]$, we have

$$\begin{aligned} |y^0(t) - y^1(t)|^2 &= \left|\int_0^t T(t-s)f(s, y^0(s))ds \right.\\ &\quad \left. + \int_0^t T(t-s)g(s, y^0(s))dW(s)\right|^2 \\ &\leq 2M^2\left|\int_0^t f(s, y^0(s))ds\right|^2 + 2M^2\left|\int_0^t g(s, y^0(s))dW(s)\right|^2 \end{aligned}$$

$$
\begin{aligned}
&\le 4M^2\left|\int_0^t [l(s)\|y^0(s)\|ds\right|^2 + 4M^2\left|\int_0^t l(s)|f(s,0)|]ds\right| \\
&\quad + 4M^2\left|\int_0^t K\|y^0(s)\|dW(s)\right|^2 + 4M^2\left|\int_0^t K|g(s,0)|dW(s)\right|^2 \\
&\le 4M^2\|l\|_{L^2}\int_0^t \|y^0(s)\|^2 ds + 4M^2\|l\|_{L^2}\int_0^t |f(s,0)|^2 ds \\
&\quad + 4M^2K^2\left|\int_0^t \|y^0(s)\|dW(s)\right|^2 + 4M^2K^2\left|^2\int_0^t g(s,0)dW(s)\right|^2.
\end{aligned}
$$

Thus,

$$
\begin{aligned}
E|y^1(t)-y^0(t)|^2 &\le 4M^2\|l\|_{L^2}E\int_0^t \|y^0(s)\|^2 ds \\
&\quad + 4M^2\|l\|_{L^2}E\int_0^t |f(s,0)|^2 ds + 4M^2K^2E \\
&\quad \times\left|\int_0^t \|y^0(s)\|dW(s)\right|^2 + 4M^2K^2E\left|\int_0^t |g(s,0)|dW(s)\right|^2 \\
&\le 4M^2\|l\|_{L^2}\int_0^t E\|y^0(s)\|^2 ds + 4M^2\|l\|_{L^2}\int_0^t E|f(s,0)|^2 ds \\
&\quad + 4M^2K^2\int_0^t E\|y^0(s)\|^2 ds + 4M^2K^2\int_0^t E|g(s,0)|^2 ds \\
&\le 4M^4t\|l\|_{L^2}E(\|y_0\|^2) + 4M^2t\|l\|_{L^2}\sup_{s\in[0,t_1]} E(|f(s,0)|^2) \\
&\quad + 4M^4tK^2E(\|y_0\|^2) + 4M^2tK^2\sup_{s\in[0,t_1]} E(|g(s,0)|^2).
\end{aligned}
$$

Hence,

$$
E|y^1(t)-y^0(t)|^2 \le tC, \text{ for all } t\in[0,t_1],
$$

where

$$
\begin{aligned}
C &= 4M^4\|l\|_{L^2}E(\|y_0\|^2) + 4M^2\|l\|_{L^2}\sup_{s\in[0,t_1]} E(|f(s,0)|^2) \\
&\quad + 4M^4K^2E(\|y_0\|^2) + 4M^2K^2\sup_{s\in[0,t_1]} E(|g(s,0)|^2).
\end{aligned}
$$

Set

$$
\begin{aligned}
y^2(t) &= T(t)y_0 + \int_0^t T(t-s)f(s,y^1(s))ds \\
&\quad + \int_0^t T(t-s)g(s,y^1(s))dW(s), \quad t\in[0,t_1].
\end{aligned}
$$

Then, we obtain that for each $t \in [0, t_1]$,

$$\begin{aligned}
|y^1(t) - y^2(t)|^2 &= \Big|\int_0^t T(t-s)[f(s, y^1(s)) - f(s, y^0(s))]ds \\
&\quad + \int_0^t T(t-s)[g(s, y^1(s)) - g(s, y^0(s))]dW(s)\Big|^2 \\
&\le 2M^2\Big|\int_0^t |f(s, y^1(s)) - f(s, y^0(s))|ds\Big|^2 \\
&\quad +2M^2\Big|\int_0^t |g(s, y^1(s)) - g(s, y^0(s))|dW(s)\Big|^2 \\
&\le 2M^2\Big|\int_0^t l(s)\|y^1(s) - y^0(s)\|ds\Big|^2 \\
&\quad +2M^2K^2\Big|\int_0^t \|y^1(s) - y^0(s)\|dW(s)\Big|^2 \\
&\le 2M^2\|l\|_{L^2}\int_0^t |y^1(s) - y^0(s)|^2 ds \\
&\quad +2M^2K^2\Big|\int_0^t |y^1(s) - y^0(s)|dW(s)\Big|^2.
\end{aligned}$$

Since

$$E|y^1(s) - y^0(s)|^2 \le sC, \text{ for all } s \in [0, t_1],$$

we have

$$\begin{aligned}
E|y^2(t) - y^1(t)|^2 &\le 2M^2\|l\|_{L^2} E\int_0^t |y^1(s) - y^0(s)|^2 ds \\
&\quad +2M^2K^2 E\Big|\int_0^t |y^1(s) - y^0(s)|dW(s)\Big|^2 \\
&\le 2M^2\|l\|_{L^2} C\int_0^t s ds + 2M^2K^2C\int_0^t s ds.
\end{aligned}$$

Hence,

$$E|y^2(t) - y^1(t)|^2 \le \frac{[2M^2\|l\|_{L^2} + 2M^2K^2]Ct^2}{2}, \text{ for all } t \in [0, t_1].$$

Now consider

$$\begin{aligned}
y^3(t) = T(t)y_0 &+ \int_0^t T(t-s)f(s, y^2(s))ds \\
&+ \int_0^t T(t-s)g(s, y^2(s))dW(s), \quad t \in [0, t_1].
\end{aligned}$$

Then, we have for each $t \in [0, t_1]$,

$$\begin{aligned}
|y^3(t) - y^2(t)|^2 &= \Big| \int_0^t T(t-s)[f(s, y^2(s)) - f(s, y^1(s))]ds \\
&\quad + \int_0^t T(t-s)[g(s, y^2(s)) - g(s, y^1(s))]dW(s)\Big|^2 \\
&\le 2M^2 \Big| \int_0^t |f(s, y^2(s)) - f(s, y^1(s))|ds\Big|^2 \\
&\quad + 2M^2 \Big| \int_0^t |g(s, y^2(s)) - g(s, y^1(s))|dW(s)\Big|^2 \\
&\le 2M^2 \Big| \int_0^t l(s)\|y^1(s) - y^0(s)\|ds\Big|^2 \\
&\quad + 2M^2K^2 \Big| \int_0^t \|y^2(s) - y^1(s)\|dW(s)\Big|^2.
\end{aligned}$$

Thus,

$$E|y^3(t) - y^2(t)|^2 \le [2M^2\|l\|_{L^2} + 2M^2K^2]^2 C \int_0^t \frac{s^2}{2} ds.$$

This implies

$$E|y^3(t) - y^2(t)|^2 \le \frac{[2M^2\|l\|_{L^2} + 2M^2K^2]^2 C t^3}{3!}, \text{ for all } t \in [0, t_1].$$

By induction, if follows that, for $n = 0, 1, \ldots$, we have

$$E|y^{n+1}(t) - y^n(t)|^2 \le \frac{[2M^2\|l\|_{L^2} + 2M^2K^2]^n C t^{n+1}}{(n+1)!}, \text{ for all } t \in [0, t_1]. \tag{11.10}$$

The above inequality implies that, for all $t \in [0, t_1]$,

$$\sqrt{E|y^{n+1}(t) - y^n(t)|^2} \le \sqrt{\frac{[2M^2\|l\|_{L^2} + 2M^2K^2]^n C t^{n+1}}{(n+1)!}}.$$

Then, for all $n \in \mathbb{N} \cup \{0\}$ and $t \in [0, t_1]$, we obtain that

$$\sup_{0 \le t \le t_1} \{\sqrt{E|y^{n+1}(t) - y^n(t)|^2}\} \le \sqrt{\frac{[2M^2\|l\|_{L^2} + 2M^2K^2]^n C t_1^{n+1}}{(n+1)!}}.$$

Therefore,

$$\|y^{n+1}(t) - y^n\|_1 \le \sqrt{\frac{[2M^2\|l\|_{L^2} + 2M^2K^2]^n C t_1^{n+1}}{(n+1)!}},$$

where

$$\|y\|_1 := \sup\{\sqrt{E|y(t)|^2} :$$

Since

$$\sum_{n=0}^{\infty}\sqrt{\frac{[2M^2\|l\|_{L^2}+2M^2K^2]^n C t_1^{n+1}}{(n+1)!}}$$

is a convergent series, we deduce that $\{y^n\}$ is an Cauchy sequence in $PC([0,t_1])$, which is a Banach space with the norm $\|\cdot\|_1$. Then there exists $y \in PC([0,t_1])$, such that

$$y^n \to y \text{ as } n \to \infty.$$

To prove that y actually is a mild solution of the problem (11.8)–(11.9) in the sense of Definition 11.4, we let

$$\begin{aligned} y_*(t) = T(t)y_0 + \int_0^t T(t-s)f(s,y(s))ds \\ + \int_0^t T(t-s)g(s,y(s))dW(s), \quad t \in [0,t_1]. \end{aligned}$$

and shall show that $y^n \to y_*$ in $PC([0,t_1])$. Let $t \in [0,t_1]$. Then,

$$\begin{aligned} |y^n(t)-y_*(t)|^2 &= \bigg|\int_0^t T(t-s)[f(s,y^n(s))-f(s,y(s))]ds \\ &\quad + \int_0^t T(t-s)[g(s,y^n(s))-g(s,y(s))]dW(s)\bigg|^2 \\ &\le 2M^2\bigg|\int_0^t |f(s,y^n(s))-f(s,y(s))|ds\bigg|^2 \\ &\quad +2M^2\bigg|\int_0^t |g(s,y^n(s))-g(s,y(s))|dW(s)\bigg|^2 \\ &\le 2M^2\bigg|\int_0^t l(s)\|y^n(s)-y(s)\|ds\bigg|^2 \\ &\quad +2M^2K^2\bigg|\int_0^t \|y^n(s)-y(s)\|dW(s)\bigg|^2. \\ &\le 2M^2\|l\|_{L^2}\int_0^t \|y^n(s)-y(s)\|^2 ds \\ &\quad +2M^2K^2\bigg|\int_0^t \|y^n(s)-y(s)\|dW(s)\bigg|^2. \end{aligned}$$

Thus,

$$\sup_{t\in[0,t_1]}\sqrt{E|y^n(t)-y_*(t)|^2} \le \sqrt{2t_1M^2[\|l\|_{L^2}+K^2]}\sup_{t\in[0,t_1]}\sqrt{E|y^n(t)-y(t)|^2}.$$

Since y^n converges to y, then we have

$$\|y^n-y_*\|_1 \le \sqrt{2t_1M^2[\|l\|_{L^2}+K^2]}\|y^n-y\|_1 \to 0 \quad \text{as} \quad n\to\infty.$$

Therefore, y is a mild solution of the problem (11.8)–(11.9). Denote this mild solution by y_1.

For uniqueness consideration, let y and $\bar{y}$ be two solutions of the problem, that is,

$$\begin{aligned} y(t) = T(t)y_0 &+ \int_0^t T(t-s)f(s,y(s))ds \\ &+ \int_0^t T(t-s)g(s,y(s))dW(s), \quad t \in [0,t_1], \end{aligned}$$

and

$$\begin{aligned} \bar{y}(t) = T(t)y_0 &+ \int_0^t T(t-s)f(s,\bar{y}(s))ds \\ &+ \int_0^t T(t-s)g(s,\bar{y}(s))dW(s), \quad t \in [0,t_1]. \end{aligned}$$

For $t \in [0,t_1]$,

$$\sup_{s\in[0,t]} E|y(s)-\bar{y}(s)|^2 \le 2M^2[\|l\|_{L^2} + K^2] \int_0^t \sup_{t\in[0,s]} E|y(s)-\bar{y}(s)|^2 ds.$$

By Gronwall's inequality,

$$\sup_{s\in[0,t]} E|y(s)-\bar{y}(s)|^2 = 0 \text{ for each } t \in [0,t_1].$$

Thus,

$$\|y-\bar{y}\|_1 = 0 \text{ so } y(t) = \bar{y}(t) \text{ for each } t \in [0,t_1],$$

and solutions of (11.8)–(11.9) are unique.

Step 2. Now consider the problem (11.8)–(11.9) determined by

$$d[y(t) - Ay(t) - f(t,y(t))]dt = g(t,y(t))dW(t), \ a.e.\ t \in (t_1,t_2], \tag{11.11}$$

$$y(t_1^+) = y_1(t_1) + I_1(y_1(t_1)). \tag{11.12}$$

First, we define the space

$$PC_1 = \{y \in PC((t_1,t_2],H)) : y(t_1^+) \text{ exists}\},$$

which is a Banach space with the norm

$$\|y\|_2 := \sup\{\sqrt{E|y(t)|^2} : t \in [t_1,t_2]\}, \ y \in PC([t_1,t_2],H).$$

Let

$$y^0(t) = T(t-t_1)[y_1(t_1) + I_1(y_1(t_1))], \quad t \in (t_1,t_2].$$

Next we define y^1 by

$$y^1(t) = T(t-t_1)[y_1(t_1) + I_1(y_1(t_1))] + \int_{t_1}^{t} T(t-s)f(s, y^0(s))ds$$
$$+ \int_{t_1}^{t} T(t-s)g(s, y^0(s))dW(s), \quad t \in (t_1, t_2].$$

Then for each $t \in [t_1, t_2]$, we have

$$\begin{aligned}
|y^1(t) - y^0(t)|^2 &= \Big|\int_{t_1}^{t} T(t-s)f(s, y^0(s))ds \\
&\quad + \int_{t_1}^{t} T(t-s)g(s, y^0(s))dW(s)\Big|^2 \\
&\le 2M^2\Big|\int_{t_1}^{t} f(s, y^0(s))ds\Big|^2 + 2M^2\Big|\int_{t_1}^{t} g(s, y^0(s))dW(s)\Big|^2 \\
&\le 4M^2\Big|\int_{t_1}^{t} [l(s)\|y^0(s)\|ds\Big|^2 + 4M^2\Big|\int_{t_1}^{t} l(s)|f(s,0)|]ds\Big| \\
&\quad + 4M^2\Big|\int_{t_1}^{t} K\|y^0(s)\|dW(s)\Big|^2 \\
&\quad + 4M^2\Big|\int_{t_1}^{t} K|g(s,0)|dW(s)\Big|^2 \\
&\le 4M^2\|l\|_{L^2}\int_{t_1}^{t} \|y^0(s)\|^2 ds + 4M^2\|l\|_{L^2}\int_{t_1}^{t} |f(s,0)|^2 ds \\
&\quad + 4M^2K^2\Big|\int_{t_1}^{t} \|y^0(s)\|dW(s)\Big|^2 \\
&\quad + 4M^2K^2\Big|^2\int_{t_1}^{t} g(s,0)dW(s)\Big|^2.
\end{aligned}$$

Thus,

$$\begin{aligned}
E|y^1(t) - y^0(t)|^2 &\le 4M^2\|l\|_{L^2}\int_{t_1}^{t} E(\|y^0(s)\|^2)ds \\
&\quad + 4M^2\|l\|_{L^2}\int_{t_1}^{t} E(|f(s,0)|^2)ds + 4M^2K^2\int_{t_1}^{t} E(\|y^0(s)\|^2)ds \\
&\quad + 4M^2K^2\int_{t_1}^{t} E(|g(s,0)|^2)ds
\end{aligned}$$

$$\begin{aligned}
&\leq 4M^2\|l\|_{L^2}(t-t_1)E(\|y^0\|^2) \\
&\quad + 4M^2\|l\|_{L^2}(t-t_1)\sup_{s\in[t_1,t_2]}E(|f(s,0)|^2) \\
&\quad + 4M^2(t-t_1)K^2E(\|y^0\|^2) \\
&\quad + 4M^2(t-t_1)K^2\sup_{s\in[t_1,t_2]}E(|g(s,0)|^2).
\end{aligned}$$

Hence,

$$E|y^1(t)-y^0(t)|^2 \leq (t-t_1)C_1, \text{ for all } t\in[t_1,t_2],$$

where

$$\begin{aligned}
C_1 = {}& 4M^2\|l\|_{L^2}E(\|y^0\|^2) + 4M^2\|l\|_{L^2}\sup_{s\in[t_1,t_2]}E(|f(s,0)|^2) \\
& + 4M^2K^2E(\|y^0\|^2) + 4M^2K^2\sup_{s\in[t_1,t_2]}E(|g(s,0)|^2).
\end{aligned}$$

Set

$$\begin{aligned}
y^2(t) = {}& T(t-t_1)[y_1(t_1)+I_1(y_1(t_1))] \\
& + \int_{t_1}^t T(t-s)f(s,y^1(s))ds \\
& + \int_{t_1}^t T(t-s)g(s,y^1(s))dW(s), \quad t\in[t_1,t_2].
\end{aligned}$$

Then, we obtain that, for $t\in[t_1,t_2]$,

$$\begin{aligned}
|y^1(t)-y^2(t)|^2 = {}& \Big|\int_{t_1}^t T(t-s)[f(s,y^1(s))-f(s,y^0(s))]ds \\
& + \int_{t_1}^t T(t-s)[g(s,y^1(s))-g(s,y^0(s))]dW(s)\Big|^2 \\
\leq {}& 2M^2\Big|\int_{t_1}^t |f(s,y^1(s))-f(s,y^0(s))|ds\Big|^2 \\
& + 2M^2\Big|\int_{t_1}^t |g(s,y^1(s))-g(s,y^0(s))|dW(s)\Big|^2 \\
\leq {}& 2M^2\Big|\int_{t_1}^t l(s)\|y^1(s)-y^0(s)\|ds\Big|^2 \\
& + 2M^2K^2\Big|\int_{t_1}^t \|y^1(s)-y^0(s)\|dW(s)\Big|^2. \\
\leq {}& 2M^2\|l\|_{L^2}\int_{t_1}^t |y^1(s)-y^0(s)|^2ds \\
& + 2M^2K^2\Big|\int_{t_1}^t |y^1(s)-y^0(s)|dW(s)\Big|^2.
\end{aligned}$$

Since

$$E|y^1(s) - y^0(s)|^2 \le (s - t_1)C_1, \text{ for all } s \in [t_1, t_2],$$

we have

$$\begin{aligned}E|y^2(t) - y^1(t)|^2 \le & \, 2M^2\|l\|_{L^2} E \int_{t_1}^{t} |y^1(s) - y^0(s)|^2 ds \\ & + 2M^2K^2 E\left|\int_{t_1}^{t} |y^1(s) - y^0(s)| dW(s)\right|^2 \\ \le & \, 2M^2\|l\|_{L^2} C_1 \int_{t_1}^{t} (s - t_1) ds \\ & + 2M^2K^2 C_1 \int_{t_1}^{t} (s - t_1) ds.\end{aligned}$$

Therefore, for all $t \in [t_1, t_2]$,

$$E|y^2(t) - y^1(t)|^2 \le \frac{[2M^2\|l\|_{L^2} + 2M^2K^2]C_1(t - t_1)^2}{2}.$$

Proceeding inductively, for $n \ge 1$ define

$$\begin{aligned}y^{n+1}(t) = & \, T(t - t_1)[y_1(t_1) + I_1(y_1(t_1))] \\ & + \int_{t_1}^{t} T(t - s) f(s, y^n(s)) ds \\ & + \int_{t_1}^{t} T(t - s) g(s, y^n(s)) dW(s), \quad t \in [t_1, t_2].\end{aligned}$$

Analogous to Step 1, we can show that

$$\|y^{n+1} - y^n\|_2 \le \sqrt{\frac{[2M^2\|l\|_{L^2} + 2M^2K^2]^n C_1(t_2 - t_1)^{n+1}}{(n + 1)!}},$$

and $\{y^n\}$ is a Cauchy sequence converging to y in $PC([t_1, t_2], H)$. By the same technique used in Step 1, we conclude that y is a unique mild solution of the problem (11.11)–(11.12) on $(t_1, t_2]$. Denote this solution by y_2.

Step 3. We continue this process and take into account that $y_{m+1} := y\big|_{(t_m, b]}$ is a solution to the problem

$$d[y(t) - Ay(t) - f(t, y(t))]dt = g(t, y(t))dW(t), \quad \text{a.e. } t \in (t_m, b], \quad (11.13)$$

$$y(t_m^+) = y_{m-1}(t_m) + I_m(y_{m-1}(t_m)). \quad (11.14)$$

The solution y of the problem (11.1)–(11.3) is then defined by

$$y(t) = \begin{cases} y_1(t), & \text{if } t \in [0, t_1], \\ y_2(t), & \text{if } t \in (t_1, t_2], \\ \dots \\ y_{m+1}(t), & \text{if } t \in (t_m, b]. \end{cases}$$

As in Theorem 3.7, we can show that, if there are two mild solutions y and x of the problem (11.1)–(11.3), then $y(t_k^+) = x(t_k^+)$, for all $k \in \{1, \ldots, m\}$. Hence, y is a unique solution of the problem (11.1)–(11.3).

Now we prove that $y \in PC$. Let

$$y^n(t) = \begin{cases} y_1^n(t), & \text{if } t \in [0, t_1], \\ y_2^n(t), & \text{if } t \in (t_1, t_2], \\ \ldots & \\ y_{m+1}^n(t), & \text{if } t \in (t_m, b], \end{cases}$$

where y_k^n, $k = 1, \ldots, m$, are as in the above problems and y_k^n converges to y_k in $PC([t_k, t_{k+1}], H)$, $k = 0, \ldots, m$. From Steps 1 to 3, we can easily show that $\{y^n\}$ is a Cauchy sequence in PC.

Let $t \in [0, t_1]$, then

$$|y(t)|^2 \leq 2|y^n(t)|^2 + 2|y^n(t) - y(t)|^2.$$

Thus,

$$E|y(t)|^2 \leq 2E|y^n(t)|^2 + 2\|y^n - y\|_1^2.$$

Also,

$$\begin{aligned} |y^n(t)|^2 &= \Big|T(t)y_0 + \int_0^t T(t-s)[f(s, y^{n-1}(s)) - f(s,0) + f(s,0)]ds \\ &\quad + \int_0^t T(t-s)[g(s, y^{n-1}(s)) - g(s,0) + g(s,0)]dW(s)\Big|^2 \\ &\leq 2M^2\|\phi\| + 8M^2\Big|\int_0^t l(s)\|y^{n-1}(s)\|ds\Big|^2 \\ &\quad + 8M^2\Big|\int_0^t l(s)|f(s,0)|ds\Big|^2 \\ &\quad + 8M^2\Big|\int_0^t K\|y^{n-1}(s)\|dW(s)\Big|^2 + 8M^2\Big|\int_0^t K|g(s,0)|dW(s)\Big|^2. \end{aligned}$$

And then,

$$\begin{aligned} E(|y^n(t)|^2) \leq 2M^2E(\|y_0\|^2) &+ 8M^2E\Big|\int_0^t l(s)\|y^{n-1}(s)\|ds\Big|^2 \\ &+ 8M^2\Big|\int_0^t l(s)|f(s,0)|ds\Big|^2 \\ &+ 8M^2E\Big|\int_0^t K\|y^{n-1}(s)\|dW(s)\Big|^2 \\ &+ 8M^2E\Big|\int_0^t K|g(s,0)|dW(s)\Big|^2 \end{aligned}$$

$$
\begin{aligned}
&\leq 2M^2E(\|y_0\|^2) + 8M^2\|l\|_{L^2}\int_0^t E(\|y^{n-1}(s)\|^2)ds \\
&+ 8M^2\|l\|_{L^2}\int_0^t E(|f(s,0)|^2)ds \\
&+ 8M^2K^2\int_0^t E(\|y^{n-1}(s)\|^2)ds + 8M^2K^2\int_0^t E(|g(s,0)|^2)ds.
\end{aligned}
$$

Gronwall's inequality implies

$$
E(|y(t)|^2) \leq d_1 e^{t_1 d_2} \quad \text{so} \quad \sup_{t\in[0,t_1]} E(|y(t)|^2) \leq d_1 e^{t_1 d_2},
$$

where

$$
\begin{aligned}
d_1 = 2M^2E(\|y_0\|^2) + 8M^2\|l\|_{L^2} \sup_{t\in[0,t_1]} E(|f(t,0)|^2)& \\
+ 8M^2K^2 \sup_{\in[0,t_1]} E(|g(t,0)|^2)&,
\end{aligned}
$$

and

$$
d_2 = 8M^2[\|l\|_{L^2} + K^2].
$$

By analogy, we easily show that, for every $t \in (t_k, t_{k+1}]$, $k = 1, \ldots, m$, there exists $M_k > 0$ such that

$$
\sup\left\{E|y(t)|^2 : t \in [t_k - r, t_{k+1}]\right\} \leq M_k, \; k = 1, \ldots, m.
$$

Hence, $E\|y\|^2 < \infty$ which implies $y \in PC$. □

11.1.4 Global Existence and Uniqueness

In this subsection, we are concerned with the existence and uniqueness of solutions of impulsive stochastic differential equations on a half-line. More precisely, we consider the class of first order stochastic differential equations with impulsive effects,

$$
\begin{aligned}
d[y(t) - Ay(t) - f(t,y(t))]dt = g(t,y(t))dW(t),& \\
t \in J := [0,\infty)\backslash\{t_1,\ldots,t_k,\ldots\},& \qquad (11.15)
\end{aligned}
$$

$$
y(t_k^-) - y(t_k) = I_k(y(t_k)), \; k = 1, \ldots, \qquad (11.16)
$$

$$
y(0) = y_0, \qquad (11.17)
$$

where $f, g : J \times H \to H$ are given functions, $0 = t_0 < t_1 < \cdots < t_m < \cdots$, $\lim_{k\to\infty} t_k = \infty$, $y(t_k^+) = \lim_{h\to 0^+} y(t_k + h)$ and $y(t_k^-) = \lim_{h\to 0^-} y(t_k - h)$ represent the right and left hand limits of $y(t)$ at $t = t_k$.

Theorem 11.6. *Suppose the following hypotheses hold:*

(S-3) *There exists a function* $l \in L^2(J, \mathbb{R}^+)$ *such that*

$$|f(t,x) - f(t,\overline{x})| \leq l(t)E|x - \overline{x}|^2,$$

and

$$|g(t,x) - g(t,\overline{x})| \leq l(t)E|x - \overline{x}|^2,$$

for a.e. $t \in J$ *and all* $x, \overline{x} \in H$ *with* $E|x|^2 \leq R$ *and* $E|\overline{x}|^2 \leq R$;

(S-4) *There exist* $p \in L^2(J, \mathbb{R}_+)$ *and a continuous nondecreasing function* $\psi : [0,\infty) \longrightarrow (0,\infty)$ *with*

$$\int_c^\infty \frac{dx}{\psi(x)} = \infty$$

for every positive constant c, *such that*

$$\|f(t,u)\| \leq p(t)\psi(E|u|^2)$$

and

$$|g(t,u)| \leq p(t)\psi(E|u|^2) \text{ for a.e. } t \in J \text{ and each } u \in H;$$

(S-5) *There exist constants* $\overline{c}_k \geq 0$, $k = 1, 2, \ldots$, *such that*

$$|I_k(x) - I_k(\overline{x})| \leq \overline{c}_k|x - \overline{x}| \text{ for all } x, \overline{x} \in H;$$

(S-6) *There exist* $M > 0$ *and* $\omega \in \mathbb{R}$ *such that*

$$\|T(t)\|_{B(H)} \leq M \exp(\omega t), \quad t \geq 0.$$

If $\sum_{k=1}^{\infty} c_k^2 < \infty$, *then the IVP* (11.15)–(11.17) *has a unique solution on* $[0,\infty)$.

Proof. We begin by defining a family of seminorms on PC, thus rendering PC into a Fréchet space. Let τ be sufficiently large. Then for each $n \in \mathbb{N}$, we define seminorms in Ω by

$$\|y\|_n = \sup\{e^{-\tau 2L_n(t) - 2\omega} E|y(t)|^2 : 0 \leq t \leq t_n\},$$

where $L_n(t) = \displaystyle\int_0^t \tau l_n^2(s)ds$. Thus, $PC = \bigcap_{n \geq 1} PC_n$, where

$$PC_n = PC([0,t_n], H).$$

Then PC is a Fréchet space with the family of seminorms $\{\|\cdot\|_n\}_{n \in \mathbb{N}}$.

To transform the problem (11.15)–(11.17) into a fixed point problem, consider the operator $G : PC \to PC$ defined by:

$$G(y)(t) = T(t)y_0 + \int_0^t T(t-s)f(s, y(s))ds + \int_0^t T(t-s)g(s, y(s))dW(s) + \sum_{0<t_k<t} T(t-t_k)I_k((y(t_k)), \quad t \in [0, \infty).$$

Clearly, the fixed points of the operator G are solutions of the problem (11.15)–(11.17). We next exhibit *a priori* bounds on solutions.

Let y be a possible solution of the problem (11.15)–(11.17). Then for each $t \in [0, t_1]$, we have

$$y(t) = T(t)y_0 + \int_0^t T(t-s)f(s, y(s))ds + \int_0^t T(t-s)g(s, y(s))dW(s),$$

and

$$\begin{aligned} E(|y(t)|^2) &= E\Big|T(t)y_0 + \int_0^t T(t-s)f(s, y(s))ds \\ &\quad + \int_0^t T(t-s)g(s, y(s))dW(s)\Big|^2 \\ &\le 2E|y_0|^2 + 4E\Big|\int_0^t T(t-s)f(s, y(s))ds\Big|^2 \\ &\quad + 4E\Big|\int_0^t T(t-s)g(s, y(s))dW(s)\Big|^2 \\ &\le 2M^2e^{2\omega t}E|y_0|^2 + 4M^2e^{2\omega t}\int_0^t p(s)\psi(E|y(s)|^2)ds \\ &\quad + 4M^2e^{2\omega t}\int_0^t p(s)\psi(E|y(s)|^2)ds. \end{aligned}$$

Thus,

$$E(|y(t)|^2) \le 2M^2e^{2\omega t}E|y_0|^2 + 8M^2\int_0^t p(s)\psi(E|y(s)|^2)ds.$$

Let us take the right-hand side of the above inequality as $v(t)$. Then we have

$$E(|y(t)|^2) \le v(t) \text{ for all } \ t \in [0, t_1],$$

$$c = v(0) = 4M^2E|y_0|^2,$$

and

$$v'(t) = 8M^2e^{2\omega t}p(t)\psi(E(|y(t)|^2)), \quad \text{a.e.} \quad t \in [0, t_1].$$

Using the increasing character of ψ, we get

$$v'(t) \leq 8M^2e^{2\omega t}p(t)\psi(v(t)) \quad \text{a.e.} \quad t \in [0, t_1].$$

Thus,

$$\int_{v(0)}^{v(t)} \frac{du}{\psi(u)} = \int_0^t \frac{v'(t)dt}{\psi(v(t))} \leq 8M^2e^{2\omega t}\int_0^t p(s)ds$$
$$\leq 8M^2e^{2\omega t_1}\int_0^{t_1} p(s)ds < \infty$$

by (S-4). Therefore, there exists $R_0 > 0$ such that $v(t) \leq R_0$, $t \in [0, t_1]$. Hence,

$$\|v\|_1 = \sup\{e^{-\tau 2L_n(t)-2\omega}E|y(t)|^2 : 0 \leq t \leq t_1\}$$
$$\leq \sup\{E|y(t)|^2 : 0 \leq t \leq t_1\} \leq R_0,$$

so $\|E(|y(t)|^2)\|_1 \leq \|v\|_1 \leq R_0$.

Now $y|_{[t_1,t_2]}$ is a solution to

$$d[y(t) - Ay(t) - f(t, y(t))]dt = g(t, y(t))dW(t), \text{ a.e, } t \in [t_1, t_2],$$

and

$$y(t_1^+) - y(t_1^-) = I_1(y(t_1^-)).$$

Note that

$$|y(t_1^+)| \leq c_1R_0 + c_1|I_1(0)| + R_0 := N_1.$$

Next,

$$E(|y(t)|^2) = E\Big|T(t-t_1)y(t_1) + \int_{t_1}^t T(t-s)f(s, y(s))ds$$
$$+ \int_{t_1}^t T(t-s)g(s, y(s))dW(s)\Big|^2$$
$$\leq 2M^2N_1^2e^{2\omega t} + 4M^2E\Big|\int_{t_1}^t T(t-s)f(s, y(s))ds\Big|^2$$
$$+ 4M^2E\Big|\int_{t_1}^t T(t-s)g(s, y(s))dW(s)\Big|^2$$
$$\leq 2M^2N_1^2e^{2\omega t} + 4M^2e^{2\omega t}\int_{t_1}^t p(s)\psi(E|y(s)|^2)ds$$
$$+4M^2e^{2\omega t}\int_{t_1}^t p(s)\psi(E|y(s)|^2)ds.$$

By the previous inequality, we have for $t \in [t_1, t_2]$,

$$E|y(t)|^2 \leq 4M^2N_1^2e^{2\omega t} + 8M^2e^{2\omega t_2}\int_{t_1}^t p(s)\psi(E|y(s)|^2)]ds.$$

Letting the right-hand side of the above inequality be $w(t)$, we have

$$c = v(t_1) = 4M^2 N_1^2 e^{2\omega t}, \quad E|y(t)|^2 \leq w(t), \quad t \in [t_1, t_2],$$

and

$$w'(t) = 8M^2 e^{2\omega t} p(t)\psi(E|y(t)|^2), \quad a.e.\, t \in [t_1, t_2].$$

Since ψ is nondecreasing, we obtain

$$w'(t) \leq p(t)\psi(w(t)) \quad a.e. \ t \in [t_1, t_2].$$

As before, this implies there exists a constant K_1 such that $w(t) \leq K_1$, $t \in [t_1, t_2]$, and hence $E|y(t)|^2 \leq R_1 := \max\{R_0, K_1\}$, $t \in [t_1, t_2]$, since for every $t \in [t_1, t_2]$, $\sup_{t\in[t_1,t_2]} E|y(t)|^2 \leq v(t)$.

Continuing this process and taking into account that $y|_{[t_{n-1},t_n]}$ is a solution to the problem

$$\begin{aligned} &d[y(t) - Ay(t) - f(t, y(t))]dt = g(t, y(t))dW(t), \quad \text{a.e.} \quad t \in [t_{n-1}, t_n],\\ &y(t_{n-1}^+) - y(t_{n-1}) = I_{n-1}(y(t_{n-1})), \end{aligned}$$

we obtain that there exists a constant R_n such that

$$\sup\{\sqrt{E|y(t)|^2} : t \in [t_{n-1}, t_n]\} \leq R_n.$$

Then for each $n \in \mathbb{N}$, there exist $\mathcal{R}_n > 0$ such that

$$\sup\{\sqrt{E|y(t)|^2} : t \in [0, t_n]\} \leq \mathcal{R}_n.$$

Set

$$Y = \{z \in PC : \|y\|_n \leq \mathcal{R}_n + 1 \text{ for all } n = 1, 2, \ldots\}.$$

Clearly, Y is a closed subset of PC.

Now we shall show that $G : Y \to PC$ is a contraction operator. Indeed, consider $y, \overline{y} \in PC_n$. For each $t \in [0, t_n]$ and $n \in \mathbb{N}$,

$$\begin{aligned} |Gy(t) - G\overline{y}(t)|^2 &= \Big| \int_0^t T(t-s)[f(s, y(s)) - f(s, \overline{y}(s))]ds \\ &\quad + \int_0^t T(t-s)[g(s, y(s)) - g(s, \overline{y}(s))]dW(s) \\ &\quad + \sum_{0<t_k<t} T(t-t_k)[I_k(y(t_k)) - I_k(\overline{y}(t_k))] \Big|^2 \\ &\leq 2M^2 \Big| \int_0^t e^{|\omega| s} l(s) \|y(s) - \overline{y}(s)\| ds \Big|^2 \\ &\quad + 4M^2 \Big| \int_0^t e^{|\omega| s} l(s) \|y(s) - \overline{y}(s)\| dW(s) \Big|^2 \\ &\quad + 4 \Big| \sum_{0<t_k<t} M e^{|\omega|(t-t_k)} c_k \|y(t_k) - \overline{y}(t_k)\| \Big|^2 \end{aligned}$$

$$
\begin{aligned}
&\leq 2M^2\Big|\int_0^t l(s)e^{L_n(s)}e^{-L_n(s)+|\omega|s}\|y(s)-\overline{y}(s)\|ds\Big|^2 \\
&+4M^2\Big|\int_0^t e^{L_n(s)}e^{-L_n(s)+|\omega|s}l(s)\|y(s)-\overline{y}(s)\|dW(s)\Big|^2 \\
&+4\Big|\sum_{0<t_k<t} Me^{|\omega|(t-t_k)}c_k\|y(t_k)-\overline{y}(t_k)\|\Big|^2.
\end{aligned}
$$

Then,

$$
\begin{aligned}
E|Gy(t)-G\overline{y}(t)|^2 \leq{}& 2M^2\int_0^t \frac{\tau}{\tau}l^2(s)e^{2L_n(s)}ds\|y-\overline{y}\|_n^2 \\
&+4M^2E\int_0^t \frac{\tau}{\tau}l^2(s)e^{2L_n(s)}e^{-2L_n(s)+2|\omega|s}\|y(s)-\overline{y}(s)\|^2ds \\
&+4E\sum_{0<t_k<t} 2^{n-1}Me^{2|\omega|(t-t_k)}c_k^2\|y(t_k)-\overline{y}(t_k)\|^2.
\end{aligned}
$$

Assume that $\tau \gg 1$ is such that $e^{4|\omega|t_n} < \tau$, and

$$
\tau \leq e^{2L_n(t)-2L_n(t_k)} \Rightarrow 1 \leq \frac{e^{2L_n(t)-2L_n(t_k)}}{\tau}.
$$

Then,

$$
\begin{aligned}
E|Gy(t)-G\overline{y}(t)|^2 \leq{}& \frac{M^2}{\tau}\int_0^t (e^{2L_n(s)})'ds\|y-\overline{y}\|_n^2 \\
&+\frac{2M^2}{\tau}\int_0^t (l^2(s)e^{2L_n(s)})'ds\|y-\overline{y}\|_n^2 \\
&+\frac{2^nM^2}{\tau}\sum_{0<t_k<t} e^{2L_n(t)+2\omega t}c_k^2\|y-\overline{y}\|_n^2 \\
\leq{}& \frac{M^2}{\tau}e^{2L_n(t)}\|y-\overline{y}\|_n^2+\frac{2M^2}{\tau}e^{2L_n(t)}\|y-\overline{y}\|_n^2 \\
&+\frac{2^nM^2}{\tau}\sum_{k=1}^n e^{2L_n(t)+2|\omega|t}e^{-4|\omega|t_k}c_k^2\|y-\overline{y}\|_n^2.
\end{aligned}
$$

Thus,

$$
\begin{aligned}
&e^{-2L_n(t)+2\omega t}E|Gy(t)-G\overline{y}(t)|^2 \leq \\
&\qquad\left(\frac{M^2}{\tau}+\frac{2M^2}{\tau}+\frac{2^nM^2}{\tau}\sum_{k=1}^n e^{4|\omega|(t_n-t_k)}c_k^2\right)\|y-\overline{y}\|_n^2.
\end{aligned}
$$

Hence,

$$
\|Gy-G\overline{y}\|_n \leq M\sqrt{\frac{3}{\tau}+\frac{2^n}{\tau}\sum_{k=1}^n e^{4|\omega|(t_n-t_k)}c_k^2}\|y-\overline{y}\|_n,
$$

showing that, for τ sufficiently large, the operator G is a contraction.

From the choice of Y there is no $z \in \partial Y$ such that $y = \lambda G(y)$, for some $\lambda \in (0, 1)$. As a consequence of Theorem 2.4, G has a unique fixed point that is a solution to (11.15)–(11.17). □

11.2 Impulsive Sweeping Processes

The so-called sweeping process ("processus de Rafle"in French) is a particular differential inclusion of the general form

$$-y'(t) \in N_{K(t)}(y(t)), \quad \text{a.e.} \quad t \in [0, \infty), \tag{11.18}$$

$$y(0) \in K(0), \tag{11.19}$$

where $K(t)$ is a convex time dependant set, and $N_{K(t)}(y(t))$ is the normal cone to $K(t)$ at $y(t)$. The sweeping process was introduced in 1971 by Moreou and extensively studied by him and other authors [134, 378–381]. Castaing [135] studied the stochastic version in the same setting. These models prove to be quite useful in elastoplasticity, nonsmooth mechanics, convex optimization, mathematical economics, queuing theory, etc. There are also many different studies in mechanics for which sweeping processes are good models (see for instance [377]); for other applications, see [182, 185, 367, 382, 383] and the references therein.

Several results have been established in the finite and infinite dimensional context. We refer, for example, to Bounkhel [101], Benabdellah [53], Colombo and Goncharov [154, 155], Thibault [462], Valadier [472], and the references therein.

In this section, we consider an impulsive sweeping process. More precisely, we consider the existence and uniqueness of solutions of the impulsive sweeping process (ISP),

$$-y'(t) \in N^P(y(t), K(t)), \quad \text{a.e.} \quad t \in J := [0, b], \tag{11.20}$$

$$y(t_k^+) - y(t_k) = I_k(y(t_k)), \quad k = 1, \ldots, m, \tag{11.21}$$

$$y(t) \in K(t), \quad t \in [0, b], \ y(0) = y_0, \tag{11.22}$$

where K is as in problem (11.18)–(11.19), $N^P(x, K)$ is the normal cone to K at x, $0 = t_0 < t_1 < \cdots < t_m < t_{m+1} = b$, $I_k \in C(\mathbb{R}^n, \mathbb{R}^n)$ $(k = 1, \ldots, m)$, and $y(t_k^+) = \lim_{h \to 0+} y(t_k + h)$ and $y(t_k^-) = \lim_{h \to 0+} y(t_k - h)$ represent the right and left hand limits of $y(t)$ at $t = t_k$, respectively.

11.2.1 Preliminaries in Nonsmooth Analysis

For $y \in \mathbb{R}^n$, the projection of y into $K \subset \mathbb{R}^n$ is the set

$$proj(y, K) = \{x \in K : d(x, K) = \|x - y\|\}.$$

This set is nonempty if, for example, K is weakly closed. Let K be a closed subset of Euclidean space $\mathbb{R}^n$, and let $x \in K$. We say that a vector $v \in \mathbb{R}^n$ is a *proximal normal* to K at x if $v = y - x$ for some $y \in \mathbb{R}^n$ with $x \in proj(y, K)$. It can be shown that $\xi \in N^P(y, K)$ if and only if there exists $M > 0$ such that the following proximal normal inequality holds

$$\langle \xi, x - y \rangle \leq M\|x - y\|^2, \text{ for all } x \in K.$$

(In general, M will depend on x). On the other hand,

$$N^P(y, K) = \bigcup_{n=1}^{\infty} \left\{ v \in \mathbb{R}^n : d\left(y + \frac{v}{n}\right) = \frac{\|v\|}{n} \right\}.$$

This cone is convex, but in general, is not closed (see [151]). We define a Bouligand cone by

$$\begin{aligned} T_K(x) &= \left\{ v \in \mathbb{R}^n : \liminf_{h \to 0^+} \frac{d(x + hv, K)}{h} \right\} \\ &= \bigcap_{\epsilon > 0} \bigcap_{\delta > 0} \bigcup_{0 < h < \delta} \left(\frac{K - x}{h} + \epsilon \overline{B}(0, 1) \right). \end{aligned}$$

11.2.2 Uniqueness Result

In order to define a solution of (11.20)–(11.22), we consider the space

$$\begin{aligned} PC = \{ y : [0, b] \longrightarrow \mathbb{R}^n :\ & y(t_k^-) \text{ and } y(t_k^+) \text{ exist, with } y(t_k^-) = y(t_k), \\ & k = 1, \ldots, m, \text{ and } y \in C((t_k, t_{k+1}], \mathbb{R}^n),\ k = 0, \ldots, m \}, \end{aligned}$$

which is a Banach space with the norm

$$\|y\|_{PC} = \max\{\|y_k\|_{J_k} : k = 0, \ldots, m\},$$

where y_k is the restriction of y to $J_k = [t_k, t_{k+1}]$, $k = 0, \ldots, m$.

Definition 11.7. A function $y \in PC \cap AC([0, b] \setminus \{t_1, \ldots, t_m\}, \mathbb{R}^n)$ is said to be a solution of (11.20)–(11.22) if y satisfies the equation $y'(t) \in N^P(y(t), K(t))$ a.e. $[0, b]$, $t \neq t_k$, $k = 1, \ldots, m$, and the conditions $y(t_k^+) - y(t_k^-) = I_k(y(t_k^-))$, $k = 1, \ldots, m$, and $y(t) \in K(t)$ for all $t \in [0, b]$.

In what follows $H(X, Y)$ is the usual Hausdorf measure of noncompactness.

Theorem 11.8. *Assume that:*

(S-7) *$K(t)$ is convex for every $t \in [0, b]$, and there exists $\gamma > 0$ such that*

$$H(K(t), K(s)) \leq \gamma |t - s|, \textit{ for all } t, s \in [0, b];$$

(S-8) *For every $t \in K(t)$, we have*

$$K(t) + \sum_{0<k<L} I_k(K(t)) \subset K(t), \text{ for each } L \in \{1,\dots,m\}$$

and there exist c_k, $d_k > 0$ such that

$$|I_k(x)| \le c_k|x| + d_k \text{ for all } x \in \mathbb{R}^n.$$

Then the problem (11.20)–(11.22) *has a unique solution.*

Proof. Consider for every $n \in \mathbb{N}$, the following partition of $[0,b]$:

$$t_{n,i} := \tfrac{ib}{2^n},\ 0 \le i \le 2^n \text{ and } I_{n,i} = (t_{n,i}, t_{n,i+1}],\ \text{if } 0 \le i \le 2^n - 1.$$

With no loss of generality, assume that $t_{n,i} \ne t_k$, $k = 1,\dots,m$. For every $n \in \mathbb{N}\cup\{0\}$, we define by induction:

$$y_{n,0} := y_0,\ y_{n,1} = \operatorname{proj}\Bigg(y_{n,0} + \sum_{t_{n,0}\le t_k<t_{n,1}} I_k(y_{n,0}), K(t_{n,1})\Bigg)$$

$$y_{n,2} = \operatorname{proj}\Bigg(y_{n,1} + \sum_{t_{n,1}\le t_k<t_{n,2}} I_k(y_{n,1}), K(t_{n,2})\Bigg)$$

$$\vdots$$

$$y_{n,i+1} = \operatorname{proj}\Bigg(y_{n,i} + \sum_{t_{n,i}\le t_k<t_{n,i+1}} I_k(y_{n,i}), K(t_{n,i+1})\Bigg).$$

It is clear that $y_{n,i} \in K(t_{n,i})$. Then, for every $0 \le i < 2^n$,

$$|y_{n,i+1} - y_{n,i}| = d(y_{n,i}, K(t_{i+1})) \le H(K(t_{n,i}), K(t_{n,i+1})) \le \frac{b}{2^n}.$$

Put

$$y_n(t) = y_{n,i} + \sum_{t_{n,i}\le t_k<t} I_k(y_n(t_k)) + \frac{t - t_{n,i}}{\epsilon_n}(y_{n,i+1} - y_{n,i}),$$

$$y_n(t_k) = y_{n,i}, \quad \epsilon_n = \frac{b}{2^n}.$$

From the definition of a proximal normal cone, we have

$$\begin{cases} \dfrac{1}{\epsilon_n}(y_{n,i+1} - y_{n,i}) \in -N(y_{n,i+1}, K(t_{n,i+1})), \\ y_n(t_k^+) = y_n(t_k) + I_k(y_n(t_k)),\ k = 1,\dots,m. \end{cases} \tag{11.23}$$

We can easily prove that $\{y_n\}_{n\in\mathbb{N}}$ is equicontinuous and bounded. Then by the Arzelá–Ascoli theorem, we conclude that there is a subsequence of $\{y_n\}$, denoted by $\{y_n\}$ for simplicity, that converges in PC to a function y that clearly satisfies $y(0) = y_0$ and

$$y(t) = y_0 + \sum_{0 \le t_k < t} I_k(y(t_k)) + \int_0^t y'(s)ds.$$

Let $t \in [0, b]$. For each $n \in \mathbb{N}, t_{n,i} \in I_{n,i}$ for some i, and

$$\begin{aligned} d &\left(y_n(t) - \sum_{t_{n,i} \le t_k < t} I_k(y_{n,i}(t_k)), K(t) \right) \\ &\le \left| y_n(t) - y_n(t_{n,i}) - \sum_{t_{n,i} \le t_k < t} I_k(y_{n,i}(t_k)) \right| + d(y_n(t_{n,i}), K(t)) \\ &\le \left| \frac{t - t_{n,i}}{\epsilon_n} \left(y_{n,i+1} - y_{n,i} \right) \right| + H(K_n(t_{n,i}), K(t)) \\ &\le \frac{b}{2^{n-1}}. \end{aligned} \tag{11.24}$$

Letting $n \to \infty$ in (11.24) and applying (S-8) shows that

$$y(t) \in K(t) \text{ for all } t \in [0, b]. \tag{11.25}$$

It remains to prove that (11.20)–(11.21) holds a.e. in $[0, b]$. For $t \in [0, b)$ and $n \ge 1$, define

$$\sigma_n(t) = t_{n,i} \text{ and } \theta_n(t) = t_{n,i+1} \text{ if } t \in [t_{n,i}, t_{n,i+1}).$$

Then, by (11.23), we have

$$\begin{cases} y_n'(t) \in -N(y_n(\theta_n(t)), K(\theta_n(t)) & \text{a.e. } \ t \in [0, b], \\ y_n(t_k^+) = y_n(t_k) + I_k(y_n(t_k)), & k = 1, \dots, m. \end{cases} \tag{11.26}$$

Since $|\theta_n(t) - t| \le \frac{b}{2^n}$ and $|\sigma_n(t) - t| \le \frac{b}{2^n}$, then

$$\theta_n(t) \to t, \ \sigma_n(t) \to t \text{ uniformly on } [0, b) \backslash \{t_1, \dots, t_m\}. \tag{11.27}$$

It follows that

$$\begin{aligned} \lim_{n \to \infty} \|y_n(\theta_n(t)) - y(t)\| &= \lim_{n \to \infty} \|y_n(\theta_n(t)) - y_n(t)\| \\ &\le H(K(\theta_n(t)), K(t)) \\ &\le \gamma \lim_{n \to \infty} |\theta_n(t) - t| = 0. \end{aligned}$$

By the continuity of I_k, we can prove that $y(t_k^+) = y(t_k) + I_k(y(t_k)), \ k = 1, \dots, m$.

Finally, we prove the uniqueness of solutions of the problem (11.20)–(11.22). Assume that x and y are two solution of (11.20)–(11.22). Then,

$$-x'(t) \in N(x(t), K(t)), \ \ -y'(t) \in N(y(t), K(t)) \quad \text{a.e.} \quad t \in [0, b],$$

and

$$x(t_k^+) = x(t_k) + I_k(x(t_k)),\ y(t_k^+) = y(t_k) + I_k(y(t_k)), \qquad k = 1, \dots, m.$$

Since $K(t)$ is a convex set, $T_K(z) = \bigcup_{h>0} \overline{\frac{K(t)-z}{h}}$ and $T_K(z) \subset \{v \in \mathbb{R}^n : \langle v, \xi \rangle \leq 0$ for all $\xi \in N^P(z, K(t))\}$. Hence,

$$\langle x'(t) - y'(t), x(t) - y(t) \rangle \leq 0 \text{ for every } t \in J_k = (t_k, t_{k+1}],\ k = 1, \dots, m,$$

and all $t \in J_0 = [0, t_1]$. Then, integrating on $[0, t] \subset [0, t_1]$, we obtain

$$\frac{d}{2dt}\|x(t) - y(t)\|^2 \leq 0 \text{ implies } \|x(t) - y(t)\| = 0.$$

Thus, $x(t) = y(t)$ on $[0, t_1]$, and by the continuity of I_1, we have $x(t_1^+) = y(t_1^+)$. Similarly, we can prove that, for every $t \in J_k$, $k = 1, \dots, m$, we have $y(t) = x(t)$ and $y(t_k^+) = x(t_k^+)$. □

11.3 Integral Inclusions of Volterra Type in Banach Spaces

In the past few years, several papers have been devoted to the study of integral equations on real compact intervals under different conditions on the kernel (see for instance [115, 153, 248], and references therein). However, very few results are available for integral inclusions on compact intervals; see [23, 74, 401]. The topological structure of solution sets of integral inclusions of Volterra type is studied in [315].

In the next several sections, we study the existence of solutions, the compactness of the solutions set, a Filippov's theorem, and a relaxation result for linear and semilinear integral inclusions of Volterra type,

$$y(t) \in \int_0^t a(t-s)[Ay(s) + F(s, y(s))]ds, \ a.e.\ t \in J := [0, b], \tag{11.28}$$

where $a \in L^1([0, b], \mathbb{R})$ and $A : D(A) \subset E \to E$ is the generator of an integral resolvent family defined on a complex Banach space E, and $F : [0, b] \times E \to \mathcal{P}(E)$ is a multivalued map.

In 1980, Da Prato and Iannelli [166] introduced the concept of resolvent families, which can be regarded as an extension of C_0-semigroups in the study of a class of integrodifferential equations. It is well known that the abstract Volterra equation

$$y(t) = f(t) + \int_0^t a(t-s)Ay(s)ds,\ t \geq 0, \tag{11.29}$$

where $f : \mathbb{R}_+ \to E$ is a continuous function, is well-posed if and only if it admits a resolvent family, that is, there is a strongly continuous family $S(t)$, $t > 0$, of bounded linear operators defined in E, which commutes with A and satisfies the resolvent equation

$$S(t)x = x + \int_0^t a(t-s)AS(s)xds, \quad t \geq 0, x \in D(A).$$

The study of diverse properties of resolvent families such as the regularity, positivity, periodicity, approximation, uniform continuity, compactness, and others are studied by several authors under different conditions on the kernel and the operator A (see [24, 45, 148, 158, 159, 286, 357–363, 400, 419]). For the kernel a we have in mind the important kernel given by

$$a(t) = f_{1-\alpha}(t)e^{-kt},\ t > 0,\ k \geq 0,\ \alpha \in (0,1),$$

where

$$f_\beta(t) = \begin{cases} \dfrac{t^{\beta-1}}{\Gamma(\beta)}, & \text{if } t \geq 0 \\ 0, & \text{if } t \leq 0. \end{cases}$$

This is the Riemann–Liouville kernel. In this case, (11.28) and (11.29) can be represented in the form of fractional differential equations and inclusions or abstract fractional differential equations and inclusions; see for instance [6, 337, 420].

Our goal in the next two sections is to complement and extend some recent results to the case of infinite-dimensional spaces; moreover the nonlinearity on the right-hand side may be either convex or nonconvex.

11.3.1 Resolvent Family

The Laplace transformation of a function $f \in L^1_{loc}(\mathbb{R}_+, E)$ is defined by

$$\mathcal{L}(f)(\lambda) :=: \widehat{a}(\lambda) = \int_0^\infty e^{-\lambda t} f(t)dt, \quad Re(\lambda) > \omega,$$

if the integral is absolutely convergent for $Re(\lambda) > \omega$. In order to define a mild solution of the problem (11.28), we recall the following definition. Here, $B(E)$ is the set of bounded linear operators on E.

Definition 11.9. Let A be a closed and linear operator with domain $D(A)$ defined on a Banach space E. We call A the generator of an integral resolvent if there exist $\omega > 0$ and a strongly continuous function $S : \mathbb{R}^+ \to B(E)$ such that

$$\left(\frac{1}{\widehat{a}(\lambda)}I - A\right)^{-1} x = \int_0^\infty e^{-\lambda t} S(t)xdt, \quad Re\lambda > \omega,\ x \in E.$$

In this case, $S(t)$ is called the integral resolvent family generated by A.

The following result is a direct consequence of Proposition 3.1 and Lemma 2.2 in [357].

Proposition 11.10. *Let $\{S(t)\}_{t\geq 0} \subset B(E)$ be an integral resolvent family with generator A. Then the following conditions are satisfied:*

(a) *$S(t)$ is strongly continuous for $t \geq 0$ and $S(0) = I$.*

(b) *$S(t)D(A) \subset D(A)$ and $AS(t)x = S(t)Ax$ for all $x \in D(A)$, $t \geq 0$.*

(c) *For every $x \in D(A)$ and $t \geq 0$,*

$$S(t)x = a(t)x + \int_0^t a(t-s)AS(s)x\,ds.$$

(d) *Let $t \geq 0$ and $x \in D(A)$. Then $\int_0^t a(t-s)S(s)x\,ds \in D(A)$ and*

$$S(t)x = a(t)x + A\int_0^t a(t-s)S(s)x\,ds.$$

In particular, $S(0) = a(0)$.

Remark 11.11. The uniqueness of the resolvent is well-known (see Prüss [419]).

If an operator A with domain $D(A)$ is the infinitesimal generator of an integral resolvent family $S(t)$ and $a(t)$ is a continuous, positive and nondecreasing function satisfying

$$\lim_{t\to 0+} \frac{\|S(t)\|_{B(E)}}{a(t)} < \infty,$$

then for all $x \in D(A)$, we have

$$Ax = \lim_{t\to 0+} \frac{S(t)x - a(t)x}{(a*a)(t)}$$

(see [363, Theorem 2.1]). For example, the case $a(t) \equiv 1$ corresponds to the generator of a C_0-semigroup, and $a(t) = t$ actually corresponds to the generator of a sine family; see [31]. A characterization of generators of integral resolvent families, analogous to the Hille–Yosida Theorem for C_0-semigroups, can be directly deduced from [363, Theorem 3.4].

Definition 11.12. A resolvent family of bounded linear operators, $\{S(t)\}_{t>0}$, is called uniformly continuous if

$$\lim_{t\to s} \|S(t) - S(s)\|_{B(E)} = 0.$$

Definition 11.13. The solution operator $S(t)$ is called exponentially bounded if there are constants $M > 0$ and $\omega \geq 0$ such that

$$\|S(t)\|_{B(E)} \leq Me^{\omega t}, \quad t \geq 0.$$

11.3.2 Existence results

Mild Solutions

In order to define mild solutions for problem (11.28), we first establish an auxiliary lemma:

Lemma 11.14. *Let $a \in L^1(J,\mathbb{R})$. Assume that A generates an integral resolvent family $\{S(t)\}_{t\geq 0}$ on E, which, in addition, is integrable and $\overline{D(A)} = E$. Let $f : J \to E$ be a continuous function (or $f \in L^1(J,E)$). Then the unique bounded solution of the problem*

$$y(t) = \int_0^t a(t-s)Ay(s)ds + \int_0^t a(t-s)f(s)ds, \quad t \in [0,b], \qquad (11.30)$$

is given by

$$y(t) = \int_0^t S(t-s)f(s)ds, \ t \in [0,b]. \qquad (11.31)$$

Proof. Let y be a solution of the integral equation (11.31). Then

$$y(t) = \int_0^t S(t-s)f(s)ds.$$

Using the fact that S is a solution operator and Fubini's theorem, we obtain

$$\begin{aligned}
\int_0^t a(t-s)Ay(s)ds &= \int_0^t a(t-s)A\int_0^s S(s-r)f(r)drds \\
&= \int_0^t \int_r^t a(t-s)AS(s-r)dsf(r)dr \\
&= \int_0^t \int_0^{t-r} a(t-s-r)AS(s)dsf(r)dr \\
&= \int_0^t [S(t-r)-a(t-r)]f(r)dr \\
&= \int_0^t S(t-s)f(s)ds - \int_0^t a(t-s)f(s)ds.
\end{aligned}$$

Thus,

$$y(t) = \int_0^t a(t-s)Ay(s)ds + \int_0^t a(t-s)f(s)ds, \ t \in [0,b].$$

□

This lemma leads us to the definition of a mild solution of the problem (11.28).

Definition 11.15. A function $y \in C([0,b],E)$ is said to be a mild solution of problem (11.28) if there exists $f \in L^1(J,E)$ such that $f(t) \in F(t,y(t))$ a.e. on J such that

$$y(t) = \int_0^t S(t-s)f(s)ds, \ t \in [0,b].$$

Throughout the sequel, we assume that $S(\cdot)$ is exponentially bounded. Our first main existence result is the following.

Theorem 11.16. *Assume $F\colon J \times E \to \mathcal{P}_{cp,cv}(E)$ is a Carathéodory map satisfying:*

(S-9) *The operator solution $\{S(t)\}_{t\in J}$ is compact for $t > 0$;*

(S-10) *There exist a function $p \in L^1(J,\mathbb{R}_+)$ and a continuous nondecreasing function $\psi : [0,\infty) \to [0,\infty)$ such that*

$$\|F(t,x)\|_{\mathcal{P}} \leq p(t)\psi(\|x\|) \text{ for a.e. } t \in J \text{ and each } x \in E$$

with

$$\int_0^\infty \frac{du}{\psi(u)} = \infty;$$

Then the problem (11.28) *has at least one solution. If, in addition, E is a reflexive space, then the solution set is compact in* $C([0,b],E)$.

Proof. The proof consists of two parts, with each part involving several Steps.

Part 1. Existence of solutions.

It is clear that all solutions of problem (11.28) are fixed points of the multivalued operator $N : C([0,b],E) \to \mathcal{P}(C([0,b],E))$ defined by

$$N(y) := \left\{h \in C([0,b],E) : h(t) = \int_0^t S(t-s)f(s)ds, \ \text{for } t \in J\right\}$$

where

$$f \in S_{F,y} = \left\{f \in L^1([0,b],E) : f(t) \in F(t,y(t)), \ \text{for a.e. } t \in [0,b]\right\}.$$

Notice that the set $S_{F,y}$ is nonempty. Since, for each $y \in C([0,b],E)$, the nonlinearity F takes convex values, the selection set $S_{F,y}$ is convex and therefore N has convex values.

Step 1. N is completely continuous.

(a) N sends bounded sets into bounded sets in $C([0,b],E)$. Let $q > 0$, $B_q := \{y \in C([0,b],E) : \|y\|_\infty \leq q\}$ be a bounded set in $C([0,b],E)$, and $y \in B_q$. Then for each $h \in N(y)$, there exists $f \in S_{F,y}$ such that

$$h(t) = \int_0^t S(t-s)f(s)ds, \ \text{for } t \in J.$$

Thus, for each $t \in J$,

$$\|h\|_\infty \leq e^{\omega b}\psi(q)\int_0^b p(t)dt.$$

(b) N maps bounded sets into equicontinuous sets of $C(J, E)$. Let $\tau_1, \tau_2 \in J, 0 < \tau_1 < \tau_2$, and B_q be a bounded set of $C(J, E)$ as in Step 1. Let $y \in B_q$; then for each $t \in J$,

$$\begin{aligned}\|h(\tau_2) - h(\tau_1)\| \leq \psi(q)&\int_{\tau_1}^{\tau_2} \|S(\tau_2 - s)\|_{B(E)}p(s)ds \\ &+ \psi(q)\int_0^{\tau_1} \|S(\tau_2 - s) - S(\tau_1 - s)\|_{B(E)}p(s)ds.\end{aligned}$$

The right-hand side tends to zero as $\tau_2 - \tau_1 \to 0$ since $S(t)$ is uniformly continuous.

(c) As a consequence of (a) and (b) above, together with the Arzéla–Ascoli theorem, it suffices to show that N maps B_q into a precompact set in E. Let $0 < t \leq b$ and let $0 < \varepsilon < t$. For $y \in B_q$, define

$$h_\varepsilon(t) = \int_0^{t-\varepsilon} S(t - s - \varepsilon)f(s)ds.$$

Then,

$$\begin{aligned}|h(t) - h_\varepsilon(t)| \leq \psi(q)&\int_0^{t-\varepsilon} \|S(t - s) - S(t - s - \epsilon)\|_{B(E)}p(s)ds \\ &+ \psi(q)\int_{t-\varepsilon}^{t} \|S(t - s)\|_{B(E)}p(s)ds,\end{aligned}$$

which tends to 0 as $\varepsilon \to 0$. Therefore, there are precompact sets arbitrarily close to the set $H(t) = \{h(t) : h \in N(y)\}$. This set is then precompact in E.

Step 2. N has a closed graph.

Let $h_n \longrightarrow h_*$, $h_n \in N(y_n)$ and $y_n \longrightarrow y_*$. We shall prove that $h_* \in N(y_*)$. Now $h_n \in N(y_n)$ means there exists $f_n \in S_{F,y_n}$ such that for each $t \in J$,

$$h_n(t) = \int_0^t S(t - s)f_n(s)ds.$$

First, we have

$$\|h_n - h_*\|_\infty \longrightarrow 0, \text{ as } n \to \infty.$$

Consider the continuous linear operator $\Gamma : L^1(J, E) \longrightarrow C(J, E)$ defined by

$$(\Gamma v)(t) = \int_0^t S(t - s)v(s)ds.$$

From the definition of Γ, we know that

$$h_n(t) \in \Gamma(S_{F,y_n}).$$

Since $y_n \longrightarrow y_*$ and $\Gamma \circ S_F$ is a closed graph operator by Lemma 2.65, there exists $f_* \in S_{F,y_*}$ such that

$$h_*(t) = \int_0^t S(t-s) f_*(s) ds.$$

Hence, $h_* \in N(y_*)$, proving our claim. Also note that Lemma 2.46 implies N is u.s.c.

Step 3. A priori bounds on solutions.

Let $y \in C(J, E)$ be such that $y \in N(y)$. Then there exists $f \in S_{F,y}$ such that

$$y(t) = \int_0^t S(t-s) f(s) ds, \quad \text{for } t \in J.$$

Then,

$$\begin{aligned} \|y(t)\| &\leq \int_0^t \|S(t-s)\|_{B(E)} \|f(s)\| ds \\ &\leq \int_0^t \|S(t-s)\|_{B(E)} p(s) \psi(\|y(s)\|) ds \\ &\leq M e^{\omega b} \int_0^t p(s) \psi(\|y(s)\|) ds. \end{aligned}$$

Set

$$w(t) = M e^{\omega b} \int_0^t p(s) \psi(\|y(s)\|) ds.$$

Then $w(0) = 0$ and for a.e. $t \in [0, b]$, we have

$$w'(t) = M e^{\omega b} p(t) \psi(\|y(t)\|) \leq M e^{\omega b} p(t) \psi(w(t)) ds.$$

Thus,

$$\int_0^t \frac{w(s)}{\psi(w(s))} ds \leq M e^{\omega b} \int_0^t p(s) ds.$$

Using a change of variable we get

$$\int_0^{w(t)} \frac{du}{\psi(u)} \leq M e^{\omega b} \int_0^b p(s) ds.$$

From (S-10), there exists $\widetilde{M} > 0$ such that

$$\|y(t)\| \leq w(t) \leq \widetilde{M} \text{ for each } t \in [0, b].$$

Let

$$U := \{y \in C([0, b], E) : \|y\|_\infty < \widetilde{M} + 1\},$$

and consider the operator $N : \overline{U} \to \mathcal{P}_{cv,cp}(C([0,b],E))$. From the choice of U, there is no $y \in \partial U$ such that $y \in \gamma N(y)$ for some $\gamma \in (0,1)$. As a consequence of the Leray–Schauder nonlinear alternative (Lemma 2.87), N has a fixed point y in U that is a mild solution of problem (11.28).

Part 2. Compactness of the solution set.
Let

$$S_F = \{y \in C([0,b],E) : \ y \text{ is a solution of problem } (11.28)\}.$$

From Part 1, $S_F \neq \emptyset$ and there exists $\widetilde{M}$ such that for every $y \in S_F$, $\|y\|_\infty \leq \widetilde{M}$. Since N is completely continuous, then $N(S_F)$ is relatively compact in $C([0,b],E)$. Let $y \in S_F$; then $y \in N(y)$ and $S_F \subset \overline{N(S_F)}$. It remains to prove that S_F is a closed set in $C(J,E)$. Let $y_n \in S_F$ be such that y_n converges to y. For every $n \in \mathbb{N}$, there exists $v_n(t) \in F(t, y_n(t))$ a.e. $t \in J$ such that

$$y_n(t) \ = \ \int_0^t S(t-s)v_n(s)ds. \tag{11.32}$$

Condition (S-9) implies that $v_n(t) \in p(t)B(0,1)$, so $(v_n)_{n\in\mathbb{N}}$ is integrably bounded. Note that this still remains true when (S-10) holds since S_F is a bounded set. Since E is reflexive, $(v_n)_{n\in\mathbb{N}}$ is semicompact. By Lemma A.39, there exists a subsequence, still denoted as $(v_n)_{n\in\mathbb{N}}$, which converges weakly to some limit $v \in L^1(J,E)$. Moreover, the mapping $\Gamma : L^1(J,E) \to C(J,E)$ defined by

$$\Gamma(g)(t) = \int_0^t S(t-s)g(s)ds$$

is a continuous linear operator. It is continuous even if these spaces are endowed with their weak topologies. Therefore, for a.e. $t \in J$, the sequence $y_n(t)$ converges to $y(t)$, and it follows that

$$y(t) = \int_0^t S(t-s)v(s)ds.$$

We need to show that $v \in F(t, y(t))$, for a.e. $t \in J$. Lemma 2.66 yields the existence of $\alpha_i^n \geq 0$, $i = n, \ldots, k(n)$ such that $\sum_{i=1}^{k(n)} \alpha_i^n = 1$ and the sequence of convex combinations $g_n(\cdot) = \sum_{i=1}^{k(n)} \alpha_i^n v_i(\cdot)$ converges strongly to v in L^1. Since F takes convex values, using Lemma 2.64, we obtain that

$$\begin{aligned} v(t) &\in \bigcap_{n\geq 1} \overline{\{g_n(t)\}}, \quad \text{a.e. } t \in J \\ &\subset \bigcap_{n\geq 1} \overline{co}\{v_k(t), \ k \geq n\} \\ &\subset \bigcap_{n\geq 1} \overline{co}\{\bigcup_{k\geq n} F(t, y_k(t))\} \\ &= \overline{co}(\limsup_{k\to\infty} F(t, y_k(t))). \end{aligned} \tag{11.33}$$

Since F is u.s.c. with compact and (11.33), we have $v(t) \in \overline{co}\, F(t, y(t))$. Since $F(\cdot,\cdot)$ has closed, convex values, we deduce that $v(t) \in F(t, y(t))$, for a.e. $t \in J$, as claimed. Hence, $y \in S_F$, so S_F is closed, and hence compact in $C([0,b], E)$. □

11.3.3 The Convex Case: an MNC Approach

Throughout this part, we assume that there exists $M > 0$ such that

$$\|S(t)\|_{B(E)} \le M \quad \text{for every } t \in [0,b]. \tag{11.34}$$

and

$$F : [0,b] \times E \to \mathcal{P}_{cp,cv}(E) \quad \text{is a multivaled map.}$$

In what follows, we let χ denote the Hausdorf measure of noncompactness.

Lemma 11.17. *In addition to condition* (S-10), *assume that*

(S-11) *There exists* $\overline{p} \in L^1([0,b], \mathbb{R}_+)$ *such that for every bounded* D *in* E,

$$\chi(F(t,D)) \le \overline{p}(t)\chi(D).$$

Then the operator N *is closed and* $N(y) \in \mathcal{P}_{cp,cv}(C([0,b]))$ *for every* $y \in C([0,b], E)$, *where* N *is as defined in the proof of Theorem 11.16.*

Proof.
Step 1. N is closed.
Let $h_n \longrightarrow h_*$, $h_n \in N(y_n)$ and $y_n \longrightarrow y_*$. We shall prove that $h_* \in N(y_*)$. Now $h_n \in N(y_n)$ implies there exists $f_n \in S_{F,y_n}$ such that for a.e. $t \in J$

$$h_n(t) = \int_0^t S(t-s) f_n(s)ds. \tag{11.35}$$

Since $\{f_n(t) : n \in \mathbb{N}\} \subseteq F(t, y_n(t))$, (S-9) implies that $(f_n)_{n\in\mathbb{N}}$ is integrably bounded. In addition, the set $\{f_n(t) : n \in \mathbb{N}\}$ is relatively compact for a.e. $t \in J$, because (S-12) and the convergence of $\{y_n\}_{n\in\mathbb{N}}$ imply that

$$\chi(\{f_n(t) : n \in \mathbb{N}\}) \le \chi(F(t, y_n(t)) \le \bar{p}(t)\chi(\{y_n(t) : n \in \mathbb{N}\}) = 0.$$

Hence, the sequence $\{f_n : n \in \mathbb{N}\}$ is semicompact so it is weakly convergent in $L^1([0,b]; E)$ to some limit f_* by Lemma A.39. Arguing as in the proof of Theorem 11.16, Part 2, and passing to the limit in (11.35), we obtain that $f_* \in S_{F,y_*}$, and for each $t \in J$,

$$h_*(t) = \int_0^t S(t-s) f_*(s)ds. \tag{11.36}$$

As a consequence, $h_* \in N(y_*)$, as claimed.

Step 2. N has compact, convex values.
The convexity of $N(y)$ follows immediately by the convexity of the values of F. To prove the compactness of the values of F, let $N(y) \in \mathcal{P}(E)$ for some $y \in C([0,b],E)$ and $h_n \in N(y)$. Then there exists $f_n \in S_{F,y}$ satisfying (11.35). Arguing again as in Step 1, we prove that $\{f_n\}$ is semicompact and converges weakly to some limit $f_* \in F(t, y(t))$, a.e. $t \in [0,b]$. Hence, passing to the limit in (11.35), h_n tends to some limit h_* in the closed set $N(y)$ with h_* satisfying (11.36). Therefore the set $N(y)$ is sequentially compact and hence compact. □

Lemma 11.18. *Under the conditions* (S-10), *and* (S-11), *the operator N is u.s.c.*

Proof. Using Lemmas 2.46 and 11.17, we only prove that N is quasicompact. Let K be a compact set in $C([0,b],E)$ and $h_n \in N(y_n)$ such that $y_n \in K$. Then there exists $f_n \in S_{F,y_n}$ such that

$$h_n(t) = \int_0^t S(t-s) f_n(s)ds.$$

Since K is compact, we may pass to a subsequence, if necessary, to get that $\{y_n\}$ converges to some limit y_* in $C([0,b],E)$. Arguing as in Step 1 of the proof of Theorem 11.16, we can prove the existence of a subsequence $\{f_n\}$ which converges weakly to some limit f_* and so h_n converges to h_*, where

$$h_*(t) = \int_0^t S(t-s) f_*(s)ds.$$

As a consequence, N is u.s.c. □

We are now in position to prove our second existence result in the convex case.

Theorem 11.19. *Assume that F satisfies hypotheses* (S-10), *and* (S-12). *If*

$$q := 2M \int_0^b \overline{p}(s)ds < 1,$$

then the set of solutions for problem (11.28) *is nonempty and compact.*

Proof. It is clear that all solutions of problem (11.28) are fixed points of the multivalued operator N defined in Theorem 11.16. By Lemmas 11.17 and 11.18, $N(\cdot) \in \mathcal{P}_{cv,cp}(C([0,b],E))$ and it is u.s.c. Next, we prove that N is a β-condensing operator for a suitable MNC β. Given a bounded subset $D \subset C([0,b],E)$, let the modulus of quasiequicontinuity of the set of functions D be denoted by

$$mod_C(D) = \lim_{\delta \to 0} \sup_{x \in D} \max_{\|\tau_2 - \tau_1\| \le \delta} \|x(\tau_1) - x(\tau_2)\|.$$

It is well-known (see e.g. Example 2.1.2 in [311]) that $mod_C(D)$ defines an MNC in $C([0,b],E)$ that satisfies all of the properties in Definition 2.92, except regularity. Given the Hausdorff MNC χ, let γ be the real MNC defined on bounded subsets on $C([0,b],E)$ by

$$\gamma(D) = \sup_{t\in[0,b]} \chi(D(t)).$$

Finally, define the MNC on bounded subsets of $C([0,b],E)$ by

$$\beta(D) = \max_{D\in\Delta(C([0,b],E))} (\gamma(D), mod_C(D)),$$

where $\Delta(C([0,b],E))$ is the collection of all countable subsets of B. Then the MNC β is monotone, regular and nonsingular (see Example 2.1.4 in [311]).

To show that N is β–condensing, let $B \subset$ be a bounded set in $C(J,E)$ such that

$$\beta(B) \leq \beta(N(B)). \tag{11.37}$$

We will show that B is relatively compact. Let $\{y_n : n \in \mathbb{N}\} \subset B$ and let $N = \Gamma \circ S_F$ where $S_F : C([0,b],E) \to L^1(J,E)$ is defined by

$$S_F(y) = S_{F,y} = \{v \in L^1(J,E) : v(t) \in F(t, y(t)) \text{ a.e } t \in [0,b]\},$$

and $\Gamma : L^1(J,E) \to C([0,b],E)$ is defined by

$$\Gamma(f)(t) = \int_0^t S(t-s)f(s)ds, \quad t \in [0,b].$$

Then,

$$\begin{aligned} \|\Gamma f_1(t) - \Gamma f_2(t)\| &\leq \int_0^t \|S(t-s)\| \cdot \|f_1(s) - f_2(s)| \, ds \\ &\leq M \int_0^t \|f_1(s) - f_2(s)\| \, ds. \end{aligned}$$

Moreover, each element h_n in $N(y_n)$ can be represented as

$$h_n = \Gamma(f_n), \text{ with } f_n \in S_F(y_n). \tag{11.38}$$

Also, (11.37) yields

$$\beta(\{h_n : n \in \mathbb{N}\}) \geq \beta(\{y_n : n \in \mathbb{N}\}). \tag{11.39}$$

From Assumption (S-12), for a.e. $t \in [0,b]$,

$$\begin{aligned} \chi(\{f_n(t) : n \in \mathbb{N}\}) &\leq \chi(F(t, \{y_n(t)\}_{n=1}^{\infty}) \\ &\leq \overline{p}(t)\chi(\{y_n(t)\}_{n=1}^{\infty}) \\ &\leq \overline{p}(t) \sup_{0\leq s\leq t} \chi(\{y_n(s)\}_{n=1}^{\infty}) \\ &\leq \overline{p}(t)\gamma(\{y_n\}_{n=1}^{\infty}). \end{aligned} \tag{11.40}$$

Lemmas 2.96 and 2.97 imply that

$$\chi(\{\Gamma(f_n)(t)\}_{n=1}^{\infty}) \leq \gamma(\{y_n\}_{n=1}^{\infty})2M \int_0^t \overline{p}(s)ds.$$

Therefore,

$$\gamma(\{y_n\}_{n=1}^{\infty}) \leq \gamma(\{h_n\}_{n=1}^{\infty}) = \sup_{t\in[0,b]} \chi(\{h_n(t)\}_{n=1}^{\infty} \leq q\gamma(\{y_n\}_{n=1}^{\infty}). \tag{11.41}$$

Since $0 < q < 1$, we infer that

$$\gamma(\{y_n\}_{n=1}^{\infty}) = 0. \tag{11.42}$$

Now $\gamma(y_n) = 0$ implies that $\chi(\{y_n(t)\}) = 0$, for a.e. $t \in [0,b]$. In turn, (11.40) implies that

$$\chi(\{f_n(t)\}) = 0, \quad \text{for a.e. } t \in [0,b].$$

Hence (11.38) implies that $\chi(\{h_n\}_{n=1}^{\infty}) = 0$. To show that $\text{mod}_{\text{C}}(\text{B}) = 0$, i.e., the set $\{h_n\}$ is equicontinuous, we proceed as in Step 1(b) of the proof of Theorem 11.16. It follows that $\beta(\{h_n\}_{n=1}^{\infty}) = 0$ which implies, by (11.39), that $\beta(\{y_n\}_{n=1}^{\infty}) = 0$. We have proved that B is relatively compact. Hence, $N : \overline{U} \to \mathcal{P}_{cp,cv}(C([0,b],E))$ is u.s.c and $\beta-$condensing, where U is as in the proof of Theorem 11.16. From the choice of U, there is no $y \in \partial U$ such that $y \in \lambda N(y)$ for some $\lambda \in (0,1)$. As a consequence of the nonlinear alternative of Leray–Schauder type for condensing maps (Lemma 2.98), we have that N has a fixed point y in U that is a solution to problem (11.28). Finally, since $FixN$ is bounded, by Lemma 2.99, $FixN$ is also compact. □

11.3.4 The Nonconvex Case

In this section, we present a third existence result for problem (11.28) where this time the multivalued nonlinearity is not necessarily convex. In the proof, we will make use of the nonlinear alternative of Leray–Schauder type combined with a selection theorem due to Bressan and Colombo for lower semicontinuous multivalued maps with decomposable values.

Theorem 11.20. *Suppose that conditions (S-9) and* (S-10) *hold. In addition, assume that*

(S-12) $F : [0,b] \times E \longrightarrow \mathcal{P}_{cp}(E)$ *is a multivalued map such that*
(a) *the mapping* $(t,y) \mapsto F(t,y)$ *is* $\mathcal{L} \otimes \mathcal{B}$ *measurable;*
(b) *the mapping* $y \mapsto F(t,y)$ *is lower semicontinuous for a.e.* $t \in [0,b]$.

Then problem (11.28) *has at least one solution.*

Proof. Conditions (S-10) and (S-12) imply, by Lemma 2.84, that F is of lower semi-continuous type. From Lemma 2.85, there is a continuous selection $f : C(J,E) \to L^1([0,b],E)$ such that $f(y) \in \mathcal{F}(y)$ for all $y \in C(J,E)$. Consider the problem

$$y(t) = \int_0^t a(t-s)[Ay(s) + f(y)(s)]ds, \quad t \in [0,b], \tag{11.43}$$

and the operator $G : C(J,E) \to C(J,E)$ defined by

$$G(y)(t) = \int_0^t S(t-s)f(y)(s))ds, \text{ for } t \in [0,b].$$

As in Theorem 11.16, we can prove that the single-valued operator G is compact and there exists $M_* > 0$ such that for all possible solutions y, we have $\|y\|_\infty < M_*$. Now, we only check that G is continuous. Let $\{y_n\}$ be a sequence such that $y_n \longrightarrow y$ in $C(J,E)$, as $n \to +\infty$. Then,

$$\|G(y_n(t)) - G(y(t))\| \leq M \int_0^b \|f(y_n(s)) - f(y(s))\|ds.$$

Since the function f is continuous, we have

$$\|G(y_n) - G(y)\|_\infty \leq M\|f(y_n) - f(y)\|_{L^1} \to 0, \quad \text{as } n \to \infty.$$

Let

$$U = \{y \in C(J,E) | \ \|y\|_\infty < M_*\}.$$

From the choice of U, there is no $y \in \partial U$ such that $y = \lambda Ny$ for in $\lambda \in (0,1)$. As a consequence of the nonlinear alternative of Leray–Schauder type (Lemma 2.86), G has a fixed point $y \in U$ that is a solution of problem (11.43), and hence a solution to the problem (11.28). □

In this part, we present a second existence result for problem (11.28) in the case where the right-hand side is nonconvex valued.

Theorem 11.21. *Assume the following conditions are satisfied:*

(S-13) $F : J \times E \longrightarrow \mathcal{P}_{cp}(E)$; $t \longmapsto F(t,x)$ *is measurable for each* $x \in E$;

(S-14) *There exists a function* $l \in L^1(J,\mathbb{R}_+)$ *such that*

$$H(F(t,x),F(t,y)) \leq l(t)\|x-y\|, \text{ a.e. } t \in J \text{ and all } x,y \in E,$$

with

$$F(t,0) \subset l(t)\overline{B}(0,1), \text{ for a.e. } t \in J.$$

Then problem (11.28) *has at least one solution.*

Proof. In order to transform the problem (11.28) into a fixed point problem, let the multivalued operator $N : C(J,E) \to \mathcal{P}(C(J,E))$ be defined as in the proof of Theorem 11.16. We shall show that N satisfies the assumptions of Lemma 2.88.

(a) $N(y) \in \mathcal{P}_{cl}(C(J,E))$ for each $y \in C(J,E)$. Indeed, let $\{h_n : n \in \mathbb{N}\} \subset N(y)$ be a sequence converging to h. Then there exists a sequence $g_n \in S_{F,y}$ such that

$$h_n(t) = \int_0^t S(t-s)g_n(s)ds, \quad t \in [0,b].$$

Since $F(\cdot,\cdot)$ has compact values, let $w(\cdot) \in F(\cdot,0)$ be such that $\|g(t) - w(t)\| = d(g(t), F(t,0))$. From (S-14) and (S-15), we infer that for a.e. $t \in [0,b]$,

$$\begin{aligned} \|g_n(t)\| &\le \|g_n(t) - w(t)\| + \|w(t)\| \\ &\le l(t)\|y\|_\infty + l(t) := \widehat{M}(t) \end{aligned}$$

for all $n \in \mathbb{N}$. Then the Lebesgue dominated convergence theorem implies that, as $n \to \infty$,

$$\|g_n - g\|_{L^1} \to 0 \text{ and thus } h_n(t) \to h(t)$$

with

$$h(t) = \int_0^t S(t-s)g(s)ds, \ t \in [0,b],$$

proving that $h \in N(y)$.

(b) There exists $\gamma < 1$, such that

$$H(N(y), N(\overline{y})) \le \gamma\|y - \overline{y}\|_\infty, \quad \text{for all} \quad y, \overline{y} \in C(J,E).$$

For this, let $y, \overline{y} \in C(J,E)$ and $h \in N(y)$. Then there exists $g(t) \in F(t, y(t))$ (g is a measurable selection) such that for each $t \in J$

$$h(t) = \int_0^t S(t-s)g(s)ds.$$

Condition (S-15) guarantees that

$$H(F(t,y(t)), F(t,\overline{y}(t))) \le l(t)\|y(t) - \overline{y}(t)\|, \text{ a.e. } t \in J.$$

Hence, there is $w \in F(t,\overline{y}(t))$ such that

$$\|g(t) - w\| \le l(t)\|y(t) - \overline{y}(t)\|, \quad t \in J.$$

Consider the mapping $U : J \to \mathcal{P}(E)$ given by

$$U(t) = \{w \in E : \|g(t) - w\| \le l(t)\|y(t) - \overline{y}(t)\|\}, \ t \in J,$$

that is $U(t) = \overline{\mathcal{B}}(g(t), l(t)\|y(t) - \overline{y}(t)\|)$. Since g, l, y, and $\bar{y}$ are measurable, Theorem III.4.1 in [136] tells us that the closed ball U is measurable. Finally, the set $V(t) = U(t) \cap F(t, \overline{y}(t))$ is nonempty since it contains w. Therefore, the intersection multivalued operator V is measurable with nonempty, closed values (see [35, 136, 230]). By Lemma 2.73, there exists a function $\overline{g}(t)$ that is a measurable selection for V. Thus, $\overline{g}(t) \in F(t, \overline{y}(t))$ and

$$\|g(t) - \overline{g}(t)\| \leq l(t)\|y(t) - \overline{y}(t)\|, \quad \text{for a.e. } t \in J.$$

Let us define, for a.e. $t \in J$,

$$\overline{h}(t) = \int_0^t S(t-s)\overline{g}(s)ds.$$

Then,

$$\begin{aligned} \|h(t) - \overline{h}(t)\| &\leq \int_0^b l(s)\|y(s) - \overline{y}(s)\|ds \\ &\leq \int_0^b l(s)e^{\tau L(s)}e^{-\tau L(s)}\|y(s) - \overline{y}(s)\|ds \\ &\leq \frac{1}{\tau}e^{\tau L(t)}\|y - \overline{y}\|_*, \end{aligned}$$

where $\|\cdot\|_*$ is the Bielecki-type norm on $C([0,b], E)$ defined by

$$\|\beta\|_* = \sup\{e^{-\tau L(t)}\|\beta(t)\| : t \in [0,b]\}, \quad \text{and} \quad L(t) = Me^{\omega b}\int_0^t l(s)ds.$$

Thus,

$$\|h - \overline{h}\|_* \leq \frac{1}{\tau}\|y - \overline{y}\|_*.$$

By an analogous relation, obtained by interchanging the roles of y and $\overline{y}$, we finally arrive at

$$H(N(y), N(\overline{y})) \leq \frac{1}{\tau}\|y - \overline{y}\|_*,$$

where $\tau \gg 1$ sufficient large. Therefore, N is a contraction and so by Lemma 2.88, N has a fixed point y, which is a mild solution to (11.28). □

Arguing as in Theorem 11.16, we also have the following result, the proof of which is omitted.

Theorem 11.22. *Let $(E, \|\cdot\|)$ be a reflexive Banach space. Suppose that all conditions of Theorem 11.21 are satisfied and $F : J \times E \to \mathcal{P}_{cp}(E)$. Then the solution set of problem (11.28) is nonempty and compact. Moreover the solution set is AR.*

11.4 Filippov's Theorem

11.4.1 Filippov's Theorem on a Bounded Interval

Let $x \in C([0,b],E)$ be a mild solution of the integral equation

$$x(t) = \int_0^t a(t-s)[Ax(s) + g(s)]ds, \text{ a.e. } t \in J := [0,b]. \tag{11.44}$$

Theorem 11.23. *Assume the following conditions hold.*

(S-16) *The function* $F : J \times E \to \mathcal{P}_{cl}(E)$ *satisfies:*
 (a) *for all* $y \in E$, *the map* $t \mapsto F(t,y)$ *is measurable;*
 (b) *the map* $\gamma : t \mapsto d(g(t), F(t,x(t)))$ *is integrable.*

(S-17) *There exists a function* $p \in L^1(J,\mathbb{R}^+)$ *such that*

$$H(F(t,z_1), F(t,z_2)) \leq p(t)\|z_1 - z_2\|, \text{ for all } z_1, z_2 \in E.$$

Then, for every $\epsilon > 0$, *problem* (11.28) *has at least one solution* y_ϵ *satisfying*

$$\|y_\epsilon(t) - x(t)\| \leq M \int_0^t [\gamma(u) + \epsilon] \exp\left(2Me^{P(t)-P(s)}\right) ds$$

for a.e. $t \in [0,b]$, *where* $P(t) = \int_0^t p(s)ds$.

Proof. We construct a sequence of functions $\{y_n\}_{n\in\mathbb{N}}$ that will be shown to converge to some solution of problem (11.28) on the interval $[0,b]$, namely to

$$y(t) \in \int_0^t a(t-s)[Ay(s) + F(s,y(s))]ds, \quad t \in [0,b]. \tag{11.45}$$

Let $f_0 = g$ on $[0,b]$ and $y_0(t) = x(t)$, $t \in [0,b]$, i.e.,

$$y_0(t) = \int_0^t S(t-s) f_0(s)ds, \quad t \in [0,b],$$

Then define the multivalued map U_1: $[0,b] \to \mathcal{P}(E)$ by

$$U_1(t) = F(t, y_0(t)) \cap B(g(t), \gamma(t) + \epsilon),$$

which we claim is nonempty. Since g and γ are measurable, Theorem III.4.1 in [136] tells us that the ball $B(g(t), \gamma(t)+\epsilon)$ is measurable. Moreover $F(t, y_0(t))$ is measurable (see [35]) and U_1 is nonempty. In particular, since $v = \epsilon > 0$ is a measurable function, from Lemma 2.76, there exists a function u which is a measurable selection of $F(t, y_0(t))$ and such that

$$|u(t) - g(t)| \leq d(g(t), F(t, y_0(t))) + \epsilon = \gamma(t) + \epsilon.$$

Then $u \in U_1(t)$, proving our claim. We deduce that the intersection multivalued operator $U_1(t)$ is measurable (see [35, 136, 230]). By Lemma 2.77 (Kuratowski–Ryll–Nardzewski selection theorem), there exists a function $t \to f_1(t)$ which is a measurable selection for U_1. Consider

$$y_1(t) = \int_0^t S(t-s) f_1(s) ds, \quad t \in [0, b].$$

For each $t \in [0, b]$, we have

$$\|y_1(t) - y_0(t)\| \le M \int_0^t \|f_0(s) - f_1(s)\| ds. \tag{11.46}$$

Hence,

$$\|y_1(t) - y_0(t)\| \le M \int_0^t \gamma(s) ds + M t \epsilon.$$

Using the fact that $F(t, y_1(t))$ is measurable, the ball $B(f_1(t), p(t)\|y_1(t) - y_0(t)\|)$ is also measurable by Theorem III.4.1 in [136]. From (S-17) we have

$$H(F(t, y_1(t)), F(t, y_0(t)) \le p(t)\|y_1(t) - y_0(t)\|.$$

Thus, there exists $w \in F(t, y_1(t))$ such that

$$\|f_1(t) - w\| \le p(t)\|y_1(t) - y_0(t)\|.$$

Now, we consider the multivalued map $U_2(t) = F(t, y_1(t)) \cap B(f_1(t), p(t)\|y_1(t) - y_0(t)\|)$ which is nonempty. Therefore, the intersection multivalued operator U_2 is measurable with nonempty, closed values (see [35, 136, 230]). By Lemma 2.73, there exists a function $f_2(t)$, which is a measurable selection for U_2. Thus, $f_2(t) \in F(t, y_1(t))$ and

$$\|f_1(t) - f_2(t)\| \le p(t)\|y_1(t) - y_0(t)\|. \tag{11.47}$$

Define

$$y_2(t) = \int_0^t S(t-s) f_2(s) ds, \quad t \in [0, b].$$

Using (11.46) and (11.47), a simple integration by parts yields the following estimates, valid for every $t \in [0, b]$,

$$\begin{aligned}
\|y_2(t) - y_1(t)\| &\le \int_0^t \|S(t-s)\|_{B(E)} \|f_2(s) - f_1(s)\| ds \\
&\le \int_0^t M p(s) \left(M \int_0^s \gamma(u) du + M \epsilon s \right) ds \\
&\le M^2 \int_0^t [\gamma(s) + \epsilon] e^{P(t) - P(s)} ds.
\end{aligned}$$

Let $U_3(t) = F(t, y_2(t)) \cap B(f_2(t), p(t)\|y_2(t) - y_1(t)\|)$. Arguing as for U_2, we can prove that U_3 is a measurable multivalued map with nonempty values; so there exists a measurable selection $f_3(t) \in U_3(t)$. This allows us to define

$$y_3(t) = \int_0^t S(t-s) f_3(s) ds, \quad t \in [0, b].$$

For $t \in [0, b]$, we have

$$\begin{aligned}\|y_3(t) - y_2(t)\| &\le M \int_0^t \|f_2(s) - f_3(s)\|\, ds \\ &\le M \int_0^t p(s)\|y_2(s) - y_1(s)\|\, ds.\end{aligned}$$

Then,

$$\|y_2(s) - y_3(s)\| \le M \left(\int_0^t M^2 \int_0^s (\gamma(r) + \epsilon) e^{P(s)-P(r)} ds dr \right).$$

Since P is nondecreasing, integrating by parts we obtain, the estimate

$$\begin{aligned}\|y_3(t) - y_2(t)\| &\le \frac{M^3}{2} \int_0^t 2p(s) \left(\int_0^s [\gamma(u) + \epsilon] e^{P(s)-P(u)} du \right) ds \\ &\le \frac{M^3}{2} \left(\int_0^t 2p(s) ds \int_0^s [\gamma(u) + \epsilon] e^{2(P(s)-P(u))} du \right) \\ &\le \frac{M^3}{2} \left(\int_0^t \left(e^{2P(s)}\right)' ds \int_0^s [\gamma(u) + \epsilon] e^{-2P(u))} du \right) \\ &\le \frac{M^3}{2} \left(e^{2P(t)} \int_0^t [\gamma(s) + \epsilon] e^{-2P(s)} ds - \int_0^t [\gamma(s) + \epsilon] ds \right) \\ &\le \frac{M^3}{2} \left(\int_0^t [\gamma(s) + \epsilon] e^{2(P(t)-P(s))} ds \right).\end{aligned}$$

Let $U_4(t) = F(t, y_3(t)) \cap B(f_3(t), p(t)\|y_3(t) - y_2(t)\|)$. Then, arguing again as for U_1, U_2, U_3, we show that U_4 is a measurable multivalued map with nonempty values and that there exists a measurable selection $f_4(t) \in U_4(t)$. Define

$$y_4(t) = \int_0^t S(t-s) f_4(s) ds, \quad t \in [0, b].$$

For $t \in [0, b]$, we have

$$\begin{aligned}\|y_4(t) - y_3(t)\| &\le M \int_0^t \|f_4(s) - f_3(s)\|\, ds \\ &\le M \int_0^t p(s)\|y_3(s) - y_2(s)\| + M\epsilon\, ds\end{aligned}$$

$$
\begin{aligned}
&\le \frac{M^4}{2}\int_0^t p(s)\left(\int_0^s [\gamma(s)+\epsilon]e^{2(P(s)-P(u))}du\right)ds \\
&\le \frac{M^4 e^{4\omega b}}{6}\int_0^t 3p(s)e^{3P(s)}ds \int_0^s [\gamma(s)+\epsilon]e^{-3P(u)}du \\
&\le \frac{M^4}{6}\left(\int_0^t [\gamma(s)+\epsilon]e^{3(P(t)-P(s))}ds\right).
\end{aligned}
$$

Repeating the process for $n = 1, 2, 3, \ldots$, we arrive at the bound

$$
\|y_n(t) - y_{n-1}(t)\| \le \frac{M^n}{(n-1)!}\int_0^t [\gamma(s)+\epsilon]e^{(n-1)(P(t)-P(s))}ds, \quad t \in [0,b]. \tag{11.48}
$$

By induction, suppose that (11.48) holds for some n and check (11.48) for $n+1$. Let $U_{n+1}(t) = F(t, y_n(t)) \cap B(f_n, p(t)\|y_n(t)-y_{n-1}(t)\|+\epsilon)$. Since U_{n+1} is a nonempty measurable set, there exists a measurable selection $f_{n+1}(t) \in U_{n+1}(t)$, which allows us to define for $n \in \mathbb{N}$

$$
y_{n+1}(t) = \int_0^t S(t-s) f_{n+1}(s)ds, \quad t \in [0,b]. \tag{11.49}
$$

Therefore, for a.e. $t \in [0,b]$, we have

$$
\begin{aligned}
\|y_{n+1}(t) - y_n(t)\| &\le M\int_0^t \|f_{n+1}(s) - f_n(s)\|\, ds \\
&\le \frac{M^{n+1}}{(n-1)!}\int_0^t p(s)ds\left(\int_0^s [\gamma(u)+\epsilon]e^{(n-1)(P(s)-P(u))}du\right) \\
&\le \frac{M^{n+1}}{n!}\int_0^t np(s)e^{nP(s)}ds \int_0^s [\gamma(u)+\epsilon]e^{-nP(u)}du.
\end{aligned}
$$

Again, an integration by parts leads to

$$
\|y_{n+1}(t) - y_n(t)\| \le \frac{M^{(n+1)}}{n!}\int_0^t [\gamma(s)+\epsilon]e^{n(P(t)-P(s))}ds.
$$

Consequently, (11.48) holds true for all $n \in \mathbb{N}$. We infer that $\{y_n\}$ is a Cauchy sequence in $C([0,b], E)$, converging uniformly to a limit function $y \in C([0,b], E)$. Moreover, from the definition of $\{U_n\}$, we have

$$
\|f_{n+1}(t) - f_n(t)\| \le p(t)\|y_n(t) - y_{n-1}(t)\|, \quad \text{for a.e } t \in [0,b].
$$

Hence, for almost every $t \in [0,b]$, $\{f_n(t)\}$ is also a Cauchy sequence in E and so converges almost everywhere to some measurable function $f(\cdot)$ in E. In addition, since $f_0 = g$, we have for a.e. $t \in [0,b]$

$$\begin{aligned}\|f_n(t)\| &\leq \sum_{k=1}^{n} \|f_k(t) - f_{k-1}(t)\| + \|f_0(t)\| \\ &\leq \sum_{k=1}^{n} p(t)\|y_{k-1}(t) - y_{k-2}(t)\| + \gamma(t) + \|g(t)\| + \epsilon \\ &\leq p(t)\sum_{k=1}^{\infty} \|y_k(t) - y_{k-1}(t)\| + \gamma(t) + \|g(t)\| + \epsilon.\end{aligned}$$

Hence,

$$\|f_n(t)\| \leq MH(t)p(t) + \gamma(t) + \|g(t)\| + \epsilon,$$

where

$$H(t) := M\int_0^t [\gamma(s) + \epsilon] \exp\left(Me^{P(t)-P(s)}\right)ds. \tag{11.50}$$

From the Lebesgue dominated convergence theorem, we deduce that $\{f_n\}$ converges to f in $L^1([0,b],E)$. Passing to the limit in (11.49), we find that the function

$$y(t) = \int_0^t S(t-s)f(s)ds, \quad t \in [0,b],$$

is a solution to problem (11.28) on $[0,b]$. Moreover, for a.e. $t \in [0,b]$, we have

$$\begin{aligned}\|x(t) - y(t)\| &= \left\|\int_0^t S(t-s)g(s)ds - \int_0^t S(t-s)f(s)ds\right\| \\ &\leq M\int_0^t \|f(s) - f_0(s)\|ds \\ &\leq \int_0^t \|f(s) - f_n(s)\|ds + M\int_0^t \|f_n(s) - f_0(s)\|ds \\ &\leq M\int_0^t \|f(s) - f_n(s)\|ds + M\int_0^t p(s)H(s)ds \\ &\leq M\int_0^t \|f(s) - f_n(s)\|ds + M\int_0^t p(s)[M\int_0^s [\gamma(u) \\ &\quad + \epsilon]\exp\left(Me^{P(s)-P(u)}\right)dsdu \\ &\leq M\int_0^t \|f(s) - f_n(s)\|ds + M\int_0^t (e^{P(s)})' \\ &\quad \times \left[M\int_0^s e^{-P(u)}[\gamma(u) + \epsilon]\exp\left(Me^{P(s)-P(u)}\right)dsdu\right].\end{aligned}$$

Passing to the limit as $n \to \infty$, we get

$$\|x(t) - y(t)\| \leq \eta(t), \quad \text{a.e. } t \in [0,b], \tag{11.51}$$

with

$$\eta(t) := M \int_0^t [\gamma(u) + \epsilon] \exp\left(2Me^{P(t)-P(s)}\right) ds. \qquad \square$$

11.5 The Relaxed Problem

To be precise, in this section we compare the trajectories of the problem,

$$y(t) \in \int_0^t a(t-s)[Ay(s) + F(s, y(s))]ds, \text{ a.e. } t \in J := [0, b], \tag{11.52}$$

to those of the convexified Voltera integral inclusion problem,

$$y(t) \in \int_0^t a(t-s)[Ay(s) + \overline{co}\, F(s, y(s))]ds, \text{ a.e. } t \in J, \tag{11.53}$$

Remark 11.24. Throughout we take E to be a reflexive Banach space.

Our main contribution in this section is the following theorem.

Theorem 11.25. *In addition to* (S-17), *assume that*

(S-18) *The function* $F : J \times E \to \mathcal{P}_{cl}(E)$ *satisfies*
(a) *for all* $y \in E$, *the map* $t \mapsto F(t, y)$ *is measurable;*
(b) *the map* $t \mapsto F(t, 0)$ *is integrably bounded (i.e., there exists* $k \in L^1([0, b], \mathbb{R}_+)$ *such that* $F(t, 0) \subset k(t)B(0, 1)$*).*

Then problem (11.53) *has at least one solution. In addition, for all* $\varepsilon > 0$ *and every solution* x *of problem* (11.53), *there is a solution* y *of* (11.52) *defined on* $[0, b]$ *satisfying*

$$\|x - y\|_\infty \leq \varepsilon.$$

In particular $S^{co} = \overline{S}$, *where*

$$S^{co} = \{y \colon y \text{ is a solution to } (11.53)\}.$$

Proof.

Part 1. $S^{co} \neq \emptyset$.
For this, we first transform problem (11.53) into a fixed point problem and then make use of Lemma 2.88. It is clear that all solutions of problem (11.53) are fixed points of the multivalued operator $N : C([0, b], E) \to \mathcal{P}(C[0, b], E)$ defined by

$$N(y) := \left\{h \in C([0, b], E) \mid h(t) = \int_0^t S(t-s) f(s) ds, t \in [0, b]\right\}$$

where

$$g \in S_{\overline{co}F,y} = \left\{g \in L^1([0, b], E) : g(t) \in \overline{co}\, F(t, y(t)) \text{ for a.e. } t \in [0, b]\right\}.$$

To show that N satisfies the assumptions of Lemma 2.88, the proof will be given in two steps.

Step 1. $N(y) \in \mathcal{P}_{cl}(C[0,b],E))$ for each $y \in C[0,b],E)$.
Indeed, let $\{y_n\} \in N(y)$ be such that $y_n \to \tilde{y}$ in $C([0,b],E)$, as $n \to \infty$. Then $\tilde{y} \in C([0,b],E)$ and there exists a sequence $g_n \in S_{\overline{co}\,F,y}$ such that

$$y_n(t) = \int_0^t S(t-s)g_n(s)ds, \quad t \in [0,b].$$

Then $\{g_n\}$ is integrably bounded. Since $\overline{co}F(\cdot,\cdot)$ has closed values and is integrably bounded, from Corollary 2.79, for every $n \in \mathbb{N}$, we have the existence of some $w_n(\cdot) \in \overline{co}F(\cdot,0)$ such that

$$\|g_n(t) - w_n(t)\| \le d(g_n(t), \overline{co}F(t,0)).$$

Since $\{y_n\}_{n\in\mathbb{N}}$ is a convergent sequence, there exists $M_* > 0$ such that

$$\|y_n\|_\infty \le M_*, \quad \text{for all } n \in \mathbb{N}.$$

Then,

$$\begin{aligned}\|g_n(t)\| &\le \|g_n(t) - w_n(t)\| + \|w_n(t)\| \\ &\le p(t)\|y_n\|_\infty + k(t) \\ &\le M_* p(t) + k(t) := M(t), \ \forall\, n \in \mathbb{N},\end{aligned}$$

that is,

$$g_n(t) \in M(t)B(0,1), \qquad \text{a.e.} \quad t \in [0,b].$$

Since $B(0,1)$ is weakly compact in the reflexive Banach space E, there exists a subsequence, still denoted $\{g_n\}$, which converges weakly to g by the Dunford–Pettis theorem. By Mazur's Lemma (Lemma 2.66), there exists a second subsequence which converges strongly to g in E, hence almost everywhere. The Lebesgue dominated convergence theorem then implies that, as $n \to \infty$,

$$\|g_n - g\|_{L^1} \to 0 \ \text{ and thus } \ y_n(t) \to \tilde{y}(t),$$

with

$$\tilde{y}(t) = \int_0^t S(t-s)g(s)ds, \ t \in [0,b],$$

proving that $\tilde{y} \in N(y)$.

Step 2. There exists $\gamma < 1$ such that $H(N(y), N(\overline{y})) \le \gamma\|y - \overline{y}\|_{BC}$ for each $y, \overline{y} \in C([0,b],E)$, where the norm $\|y - \overline{y}\|_{BC}$ will be chosen conveniently.

Indeed, let $y,\ \overline{y} \in C([0,b],E)$ and $h_1 \in N(y)$. Then there exists $g_1(t) \in \overline{co}\,F(t,y(t))$ such that for each $t \in [0,b]$,

$$h_1(t) = \int_0^t S(t-s)g_1(s)ds.$$

Since, for each $t \in [0,b]$,

$$H(\overline{co}\, F(t,y_t), \overline{co}\, F(t,\overline{y}_t)) \leq p(t)\|y(t)-\overline{y}(t)\|,$$

there exists some $w(t) \in \overline{co}\, F(t,\overline{y}(t))$ such that

$$\|g_1(t)-w(t)\| \leq p(t)\|y(t)-\overline{y}(t)\|, \quad t \in [0,b].$$

Consider the multimap $U_1 : [0,b] \to \mathcal{P}_{cl}(E)$ defined by

$$U_1(t) = \{w \in E : \|g_1(t)-w\| \leq p(t)\|y(t)-\overline{y}(t)\|\}.$$

As in the proof of Theorem 11.23, we can show that the multivalued operator $V_1(t) = U_1(t) \cap \overline{co}\, F(t,\overline{y}(t))$ is measurable and takes nonempty values. Then there exists a function $g_2(t)$, which is a measurable selection for V_1. Thus, $g_2(t) \in \overline{co}\, F(t,\overline{y}(t))$ and

$$\|g_1(t)-g_2(t)\| \leq p(t)\|y(t)-\overline{y}(t)\|, \quad \text{for a.e. } t \in [0,b].$$

For each $t \in [0,b]$, let

$$h_2(t) = \int_0^t S(t-s)g_2(s)ds.$$

Therefore, for each $t \in [0,b]$, we have

$$\begin{aligned} \|h_1(t)-h_2(t)\| &\leq M\int_0^t \|g_1(s)-g_2(s)\|\, ds \\ &\leq \int_0^t Mp(s)\|y(s)-\bar{y}(s)\|ds \\ &\leq \int_0^t p(s)e^{\tau\int_0^s p(u)du}\Big(\sup_{0\leq z\leq b} e^{-\tau\int_0^z p(u)du}\|y(z)-\bar{y}(z)\|\Big)ds \\ &\leq \frac{1}{\tau}\int_0^t (e^{\tau\int_0^s p(u)du})'\|y-\bar{y}\|_* ds. \end{aligned}$$

Hence,

$$\|h_1-h_2\|_* \leq \frac{1}{\tau}\|y-\bar{y}\|_*,$$

where

$$\|y\|_* = \sup\{e^{-\tau\int_0^t p(s)ds}\|y(t)\| : t \in [0,b],\ \tau \gg 1\}.$$

By an analogous relation, obtained by interchanging the roles of y and $\bar{y}$, we find that

$$H(N(y),N(\bar{y})) \leq \frac{1}{\tau}\|y-\bar{y}\|_*.$$

Therefore, N is a contraction, and by Lemma 2.88, N has a fixed point y that is a solution to problem (11.53).

Part 2. Let x be a solution of problem (11.53). Then, there exists $g \in S_{\overline{co}\,F,x}$ such that

$$x(t) = \int_0^t S(t-s)g(s)ds, \quad t \in [0,b],$$

that is, x is a solution of the problem

$$x(t) = \int_0^t a(t-s)[Ax(s) + \bar{g}(s)]ds, \quad \text{a.e. } t \in [0,b]. \tag{11.54}$$

Let $\epsilon > 0$ and $\delta > 0$ be given by the relation $\delta = b\epsilon/(2\|p\|_{L^1})$. From Theorem 5.4, there exists a measurable selection f_* of $t \mapsto F(t,x(t))$ such that

$$\sup_{t\in[0,b]} \left\| \int_0^t S(t-s)g(s)ds - \int_0^t S(t-s)f_*(s)ds \right\| \leq M\delta.$$

Let

$$z(t) = \int_0^t S(t-s)f_*(s)ds, \quad t \in [0,b].$$

Hence,

$$\|x(t) - z(t)\| \leq M\delta.$$

With assumption (S-18), we infer from Corollary 2.79 that there exists $u(t) \in \overline{co}F(t,z(t))$ such that

$$\|g(t) - u(t)\| \leq d(g(t), \overline{co}F(t,z(t))).$$

Then,

$$\begin{aligned}
\gamma(t) &:= d(g(t), F(t,x(t))) \\
&\leq d(g(t),u) + H(\overline{co}F(t,z(t)), F(t,x(t))) \\
&\leq H(\overline{co}\,F(t,x(t)), \overline{co}\,F(t,z(t)) + H(\overline{co}F(t,z(t)), \overline{co}F(t,x(t))) \\
&\leq 2p(t)\|x(t) - z(t)\| \leq 2\delta p(t).
\end{aligned}$$

Since, under conditions (S-18)(a) and (S-17), γ is measurable (see [35]), by the above inequality, we deduce that $\gamma \in L^1(J,E)$. From Theorem 11.23, problem (11.52) has a solution y that satisfies

$$\|y(t) - x(t)\| \leq \eta(t), \; t \in [0,b],$$

where

$$\begin{aligned}
\eta(t) &= M \int_0^t [\gamma(s) + \epsilon] \exp\left(2Me^{P(t)-P(s)}\right)ds \\
&\leq M[2\delta\|p\|_{L^1} + \epsilon b] \exp\left(2Me^{P(b)}\right).
\end{aligned}$$

Using the definition of δ, we obtain the upper bound

$$\|y - x\|_\infty \le 2b\epsilon \exp\left(2Me^{P(b)}\right).$$

Since ε is arbitrary, $\|y - x\|_\infty \le \varepsilon$, showing the density relation $S^{co} = \overline{S}$. □

Appendix

In this appendix, we gather together some auxiliary notions from homology and functional analysis used throughout this book.

A.1 Čech Homology Functor with Compact Carriers

The construction of the Čech homology theory is outlined. Interested readers are invited to consult the fundamental reference [191] or [181, 229, 306, 365].

Let (X, X_0) be an arbitrary pair in $\mathcal{E}$. We shall denote by $\mathcal{M} = \{(A_\alpha, A_{0\alpha})\}$ the directed set of all compact pairs such that $(A_\alpha, A_{0\alpha}) \subset (X, X_0)$ for each α, with the natural quasiorder relation defined by the inclusion $\leq$ defined by the condition

$$(A_\alpha, A_{0\alpha}) \leq (A_\beta, A_{0\beta}) \quad \text{if and only if} \quad (A_\alpha, A_{0\alpha}) \subset (A_\beta, A_{0\beta}).$$

If $(A_\alpha, A_{0\alpha}) \leq (A_\beta, A_{0\beta})$, then we shall denote by $i_{\alpha\beta}\colon (A_\alpha, A_{0\alpha}) \to (A_\beta, A_{0\beta})$ the inclusion map. For each pair $(A_\alpha, A_{0\alpha})$ consider the graded vector space $H_*(A_\alpha, A_{0\alpha})$, together with the linear map $i_{\alpha\beta *}$ given for $(A_\alpha, A_{0\alpha}) \leq (A_\beta, A_{0\beta})$. Then the family $\{H_*(A_\alpha, A_{0\alpha}), i_{\alpha\beta *}\}$ is a direct system in the category $\mathcal{A}$ over $\mathcal{M}$. We define a graded vector space

$$H(X, X_0) = \varinjlim\{H_*(A_\alpha, A_{0\alpha}), i_{\alpha\beta *}\}.$$

It is easy to see that $H(X, X_0) = \{H_q(X, X_0)\}$, where

$$H_q(X, X_0) = \varinjlim\{H_q(A_\alpha, A_{0\alpha}), i_{\alpha\beta *}\}, \quad \text{for each } q.$$

Let $f\colon (X, X_0) \to (Y, Y_0)$ be a map. Consider the directed sets $\mathcal{M} = \{(A_\alpha, A_{0\alpha})\}$ and $\mathcal{N} = \{(B_\gamma, B_{0\gamma})\}$ for (X, X_0) and (Y, Y_0), respectively. We define $F\colon \mathcal{M} \to \mathcal{N}$ by the formula

$$F((A_\alpha, A_{0\alpha})) = (f(A_\alpha), f(A_{0\alpha})), \quad \text{for each } (A_\alpha, A_{0\alpha}) \in \mathcal{M}.$$

We observe that if $(A_\alpha, A_{0\alpha}) \leq (A_\beta, A_{0\beta})$ then

$$F((A_\alpha, A_{0\alpha})) \leq F((A_\beta, A_{0\beta})).$$

For each α, by $f_\alpha\colon (A_\alpha, A_{0\alpha}) \to (f(A_\alpha), f(A_{0\alpha}))$ we denote a map given by $f_\alpha(x) = f(x)$ for each $x \in A$. Then the map F and the family $\{f_{\alpha *}\}$ is a map of directed systems $\{H_*(A_\alpha, A_{0\alpha}), i_{\alpha\beta *}\}$ and $\{H_*(B_\gamma, B_{0\gamma}), i_{\delta\gamma *}\}$. We define the induced linear map $H(f)$ for f, by putting

$$H(f) = f_* = \varinjlim\{f_{*\alpha}\}.$$

Then we have $f_{*q} = \lim_{\rightarrow}\{f_{\alpha * q}\}$, for every q.

From the functoriality of $\lim_{\rightarrow}$, we deduce that $H: \mathcal{E} \to \mathcal{A}$ is a covariant functor. The functor H is said to be the Čech homology functor with compact carriers.

We note that if (X, X_0) is a compact pair, then the family consisting of the single pair (X, X_0) is a cofinal subset of $\mathcal{M} = \{(A_\alpha, A_{0\alpha})\}$ for (X, X_0), and hence we obtain $H_*(X, X_0) = H(X, X_0)$. Similarly, if $f: (X, X_0) \to (Y, Y_0)$ is a map of compact pairs, then $H_*(f) = H(f)$.

The following properties of H clearly follow from the Eilenberg–Steenrod axioms for H_* and some simple properties of $\varinjlim$.

Proposition A.1. *If $f, g: (X, X_0) \to (Y, Y_0)$ are homotopic maps, then the induced linear maps are equal, that is, $f_* = g_*$.*

Proposition A.2. *Let (X, X_0) be a pair in $\mathcal{E}$ and let $i: X_0 \to X$, $j: X \to (X, X_0)$ be inclusions. Then, for each q, there exists a linear map*

$$\partial_q: H_q(X, X_0) \to H_{q-1}(X_0 0,)$$

such that the sequence

$$\ldots H_q(X) \xrightarrow{i_{*q}} H_q(X) \xrightarrow{j_{*q}} H_q(X, X_0) \xrightarrow{\partial_q} H_{q-1}(X_0) \ldots$$

is exact.

The linear map ∂_q has the additional property of being natural in the following sense:

Let X be a compact metric space. The notion of cohomological acyclicity is equivalent with homological acyclicity. Denote by $\check{H}(X)$ the reduced homology vector space of X (see [446]). Also we can use the following definition of acyclic set.

Definition A.3. A set A is *acyclic* provided

$$\check{H}^q(A) = \begin{cases} 0, & \text{for } q > 0, \\ \mathbb{Q}, & \text{for } q = 0. \end{cases}$$

where $\{\check{H}^q\}$ denotes the Čech cohomology functor with coefficients in $\mathbb{Q}$.

From the dimension axiom, singletons are prototypes of acyclic sets. Definition A.3 then says that X is acyclic if its homology is exactly the same as the homology of a one point space $\{p\}$.

A.2 The Bochner Integral

In this section, we present some properties of the Bochner integral of vector-valued functions. We shall consider only those properties needed in this book, and we shall assume that the reader is familiar with the basic facts about measure and integration of scalar-valued functions. Recall that a measure space (Ω, Σ, μ) is called σ-finite if there exists $\{\Omega_n \mid n \in \mathbb{N}\} \subset \Sigma$ such that $\mu(\Omega_n) < \infty$ and $\Omega = \cup_{n\in\mathbb{N}}\Omega_n$; it is called finite if $\mu(\Omega) < \infty$. The measure space (Ω, Σ, μ) is complete if every null μ-measure set is measurable.

Let E be a complex Banach space and let J be an interval (bounded or unbounded) in $\mathbb{R}$.

Definition A.4. A function f : $J \to E$ is called simple if there is a finite sequence $\Omega_n \subset J$, $n = 1, \dots, m$, of Lebesgue measurable sets such that

$$\Omega_p \cap \Omega_l = \emptyset, \text{ for } p \neq l,$$

and

$$J = \bigcup_{l=1}^{l=m} \Omega_l,$$

where

$$f(t) = \sum_{l=1}^{l=m} a_l \chi_{\Omega_l}(t), \ a_l \in E$$

and Lebesgue measurable sets $\Omega_l \subset J$ with finite Lebesgue measure $\mu(\Omega_l)$, i.e. f is constant on the measurable set Ω_l, for each $l \in \{1, \dots, m\}$.

Definition A.5. A function $f : J \to E$ is called measurable if there is a sequence of simple functions $\{f_n\}_{n\in\mathbb{N}}$ such that

$$\lim_{n\to\infty} \|g_n(t) - f(t)\| = 0$$

for almost all $t \in J$.

Proposition A.6. *If $f : J \to E$ is measurable, then the real function $\|f\| : I \to \mathbb{R}^+$ is measurable.*

Proof. Since f is measurable, then there exists a sequence of simple functions $\{f_n\}_{n\in\mathbb{N}}$ such that

$$\lim_{n\to\infty} \|f_n(t) - f(t)\| = 0, \text{ a.e. } t \in J.$$

Then $\|f_n\|$ are simple real functions for all $n \in \mathbb{N}$ and

$$|\|f_n(t)\| - \|f(t)\|| \leq \|f_n(t) - f(t)\|, \text{ a.e. } t \in J.$$

We conclude that $\lim_{n\to\infty} \|f_n(t)\| = \|f(t)\|$ a.e. in J and therefore $\|f(.)\|$ is measurable. □

Remark A.7. It has to be mentioned that (in the case $E = \mathbb{R}$), a function $f : J \to \mathbb{R}$ is measurable if and only if for every finite $a \in \mathbb{R}$, the set $\{t \in J \mid f(t) > a\}, \{t \in J \mid f(t) \geq a\}$ (or equivalently $\{t \in J \mid f(t) < a\}, \{t \in J \mid f(t) \leq a\}$) is measurable.

Definition A.8. A function $f : J \to E$ is called almost separably valued if there is a null set Ω_0 in J such that $f(J\backslash\Omega_0) := \{f(t) \mid t \in J\backslash\Omega_0\}$ is separable (equivalently, $f(\Omega\backslash\Omega_0)$ is contained in a separable closed subspace of E).

Corollary A.9 ([222]). *Let $f : J \to E$ be a function. Then the following statements hold:*

(i) *the function f is measurable if and only if it is the uniform limit almost everywhere of a sequence of measurable, countably valued functions.*

(ii) *If E is separable, then f is measurable if and only if it is weakly measurable.*

(iii) *If f is continuous, then it is measurable.*

(iv) *If $f_n : J \to E$ $(n \in \mathbb{N})$ are measurable functions and $f_n \to f$ pointwise a.e., then f is measurable.*

For a simple function $f : J \to E$ given by $f(t) = \sum_{l=1}^{l=m} a_l \chi_{\Omega_l}(t)$, define

$$\int_J f(t)dt = \sum_{l=1}^{m} a_l \mu(\Omega_l),$$

where $\mu(\Omega_l)$ is the Lebesgue measure of J. It is routine to verify that the definition is independent of the representation $f = \sum_{l=1}^{m} a_l \mu(\Omega_l)$, and the integral so defined is linear.

Definition A.10. A function $f : J \to E$ is Bochner integrable if there exists a sequence of simple functions $f_n : J \to E$ such that the following two conditions are satisfied:

$$\lim_{n\to\infty} f_n(t) = f(t) \text{ almost everywhere}$$

and

$$\lim_{n\to\infty} \int_J \|f_n(t) - f(t)\| dt = 0.$$

If f is Bochner integrable, then for all $x^* \in E^*$, we have

$$\left\langle \int_J f(t)dt, x^* \right\rangle = \int_J \langle f(t)dt, x^* \rangle dt.$$

Theorem A.11. *A function $f : J \to E$ is Bochner integrable if and only if f is measurable and $\|f\|$ is integrable. If f is Bochner integrable, then*

$$\left\| \int_J f(t)dt \right\| \leq \int_J \|f(t)\|dt.$$

Now we recall the following analogue of the dominated convergence theorem:

Theorem A.12 (Dominated Convergence [222]). *Let $f_n : J \to E$ $(n \in \mathbb{N})$ be Bochner integrable functions. Assume that $f(t) = \lim_{n\to\infty} f(t)$ exists a.e. and there exists an integrable function $g : J \to \mathbb{R}^+$ such that*

$$\|f_n(t)\| \leq g(t), \text{ for a.e. } t \text{ and all } n \in \mathbb{N}.$$

Then f is Bochner integrable and

$$\int_J f(t)dt = \lim_{n\to\infty} \int_J f_n(t)dt.$$

Furthermore

$$\int_J \|f_n(t) - f(t)\|dt = 0, \text{ as } n \to \infty.$$

Theorem A.13 (Fubini's Theorem [27]). *Let $J = J_1 \times J_2$ be a rectangle in $\mathbb{R}^2$, let $f : J \to E$ be measurable, and suppose that*

$$\int_{J_1} \int_{J_2} \|f(s,t)\|dtds < \infty.$$

Then f is Bochner integrable and the repeated integrals

$$\int_{J_1} \int_{J_2} f(s,t)dtds, \quad \int_{J_1} \int_{J_2} f(s,t)dsdt$$

exist and are equal.

A.3 Absolutely Continuous Functions

Definition A.14. Let E be a Banach space. A function $f : [0,b] \longrightarrow E$ is called *absolutely continuous* (we write $f \in$ AC) if for each $\varepsilon > 0$, there exists $\delta(\varepsilon) > 0$ such that for all pairwise disjoint intervals of $[0,b]$, we have

$$\sum_{n=1}^{\infty}(b_n - a_n) < \delta \Longrightarrow \sum_{n=1}^{\infty} \|f(b_n) - f(a_n)\|_E < \varepsilon.$$

Remark A.15. If $f \in$ AC, then f is of bounded variation, i.e. for any partition $0 = x_0 < x_1 < \ldots < x_n = b$ of $[0, b]$, we have

$$V(f) = \sup_{x_0,\ldots,x_n} \sum_{k=1}^{m} \|f(x_k) - f(x_{k-1})\|_E < \infty.$$

Definition A.16. Let $f : [a, b] \to E$ be a Bochner integrable function, we say that $F : [a, b] \to E$ is an antiderivative or primitive of f if

$$F(t) = F(a) + \int_a^t f(s)ds, \quad t \in [a, b].$$

A point $t \in [a, b]$ is said to be a Lebesgue point of $f \in L^1([a, b], E)$ if $\lim_{h\to 0} \frac{1}{h} \int_t^{t+h} \|f(s) - f(t)\|ds = 0$. It easy to see that every point of continuity is a Lebesgue point, and the following proposition shows that almost all points are Lebesgue points.

Proposition A.17 ([27]). *Let $f : [a, b] \to E$ be Bochner integrable function and $F(t) = \int_a^t f(s)ds$, $t \in [a, b]$. Then*

(a) *F is differentiable a.e. and $F' = f$ a.e.*

(b) $\lim_{h\to 0} \frac{1}{h} \int_t^{t+h} \|f(s) - f(t)\|ds = 0$ *a.e.*

(c) *F is absolutely continuous.*

(d) $V(F) = \int_a^b \|f(t)\|dt$.

Theorem A.18 ([222, Theorem 2.2.17]). *If $f \in$ AC and E is reflexive, then f is differentiable almost everywhere and we have the formula*

$$f(t) = f(0) + \int_0^t f'(s)ds.$$

Now, we define spaces of higher-order absolutely continuous functions and compare with Sobolev spaces.

Definition A.19.

$$AC^{1,p}([0, b], E) = \{f \in \text{AC such that } f \text{ is differentiable a.e. and } f' \in L^p([0, b], E)\}.$$

$$W^{1,p}(]0, b[, E) = \Big\{f \in L^p(]0, b[, E) \mid \exists g \in L^p(]0, b[, E) \text{ such that } \int_0^b f\varphi' = -\int_0^b g\varphi, \forall \varphi \in C_0^\infty(]0, b[, E)\Big\}.$$

Then, we have (see, e.g., [222, Theorem 2.2.24]):

Theorem A.20.

$$W^{1,p}([0,b],E) = AC^{1,p}([0,b],E).$$

Remark A.21. If E is reflexive, then $f \in AC^{1,p}([0,b],E)$ if and only if there exits $g \in L^p(]0,b[,E)$ such that

$$f(t) = f(0) + \int_0^t g(s)ds.$$

A.4 Compactness Criteria in $C([a,b],E)$, $C_b([0,\infty),E)$, and $PC([a,b],E)$

Let E be a real Banach space and let $[a,b]$ be an interval.

Definition A.22. A family $\mathcal{A}$ in $C([a,b],E)$ is equicontinuous at t in $[a,b]$ if for each $\varepsilon > 0$ there exists $\delta(\varepsilon,t) > 0$ such that, for each $s \in [a,b]$ with $|t-s| < \delta(\varepsilon,t)$, we have

$$\|f(t) - f(s)\| < \varepsilon,$$

uniformly with respect to $f \in \mathcal{A}$.

The family $\mathcal{A}$ is equicontinuous on $[a,b]$ if it is equicontinuous at each point $t \in [a,b]$, in the sense mentioned above.

The family $\mathcal{A}$ is uniformly equicontinuous on $[a,b]$ if it is equicontinuous on $[a,b]$, and $\delta(\varepsilon,t)$ can be chosen independent of $t \in [a,b]$.

Remark A.23. It is straightforward that a family $\mathcal{A}$ in $C([a,b],E)$ is equicontinuous on $[a,b]$ if and only if it is uniformly equicontinuous on $[a,b]$.

Theorem A.24 (Arzela–Ascoli [491]). *A bounded subset $\mathcal{A}$ in $C([a,b],E)$ is relatively compact if and only if*

(A.24.1) *$\mathcal{A}$ is equicontinuous on $[a,b]$ and there exists a dense subset D in $[a,b]$ such that, for each $t \in D$,*

$$\mathcal{A}(t) = \{f(t) \mid f \in \mathcal{A}\}$$

is relatively compact in E.

Corollary A.25 ([477]). *If $\mathcal{A} \subset C([a,b],E)$ is relatively compact, then the set*

$$\mathcal{A}([a,b]) = \{f(t) \mid f \in \mathcal{A}, t \in [a,b]\}$$

is relatively compact in E.

Corollary A.26. *Let C be nonempty and closed in E, $g : [a,b] \times C \to E$ a continuous function,*

$$\mathcal{C} = \{u \in C([a,b]) \mid u(t) \in C,\ t \in [a,b]\},$$

and let $G : \mathcal{C} \to C([a,b], E)$ the superposition operator associated to the function g, i.e. $G(x)(t) = g(t, x(t))$ for each $x \in \mathcal{C}$ and $t \in [a,b]$. Then G is continuous from $\mathcal{C}$ in $C([a,b], E)$, both the domain and range being endowed with the norm topology $\|\cdot\|_\infty$.

Let $J_k = (t_k, t_{k+1}]$, $k = 0, \ldots, m$, $t_0 = a < t_1 < \cdots, t_m < t_{m+1} := b$, and let y_k be the restriction of a function y to J_k. Define

$$PC = \{y : [a,b] \to E \mid y_k \in C(J_k, E),\ k = 0, \ldots, m, \text{ such that } y(t_k^-) \text{ and } y(t_k^+) \text{ exist and satisfy, } y(t_k^-) = y(t_k) \text{ for } k = 1, \ldots, m\}.$$

Then endowed with the norm

$$\|y\|_{PC} = \max\{\|y_k\|_\infty, \quad k = 0, \ldots, m\},$$

PC is a Banach space, where $y_k = y_{|J_k}$.

Theorem A.27. ([399]) *Let $\mathcal{A}$ be a bounded set in PC. Assume that*

(a) *$\mathcal{A}$ is equicontinuous on $[a,b]$ (i.e $\mathcal{A}$ is equicontinuous on $C(J_k, E)$, $k = 1, \ldots, m$),*

(b) *there exists a dense subset D in $[a,b]$ such that, for each $t \in D$, the set $\mathcal{A}(t) = \{f(t) \mid f \in \mathcal{A}, t \in [a,b]\}$ is relatively compact in E.*

Then $\mathcal{A}$ is relatively compact in PC.

The following compactness criterion for subsets of C_b is a consequence of the well-known Arzéla–Ascoli theorem (see Avramesu [37], Corduneanu [153], Przeradzki [421], Staikos [449])

Theorem A.28. *Let $B \subset C_b([0,\infty), \mathbb{R}^n)$ be a subset assume the following conditions are satisfied:*

(i) *for every $t \in \mathbb{R}_+$, the set $\{x(t) \mid x \in B\}$ is relatively compact,*

(ii) *for every $\alpha > 0$, the set B is equi-continuous on the interval $[0,\alpha]$,*

(iii) *for every $\varepsilon > 0$ there exist $T = T(\varepsilon)$ and $\delta = \delta(\varepsilon) > 0$ such that if $x, y \in B$ with $\|x(T) - y(T)\| \le \delta$, then $\|x(t) - y(t)\| \le \varepsilon$ for all $t \in [T,\infty)$.*

Then the set $\overline{B}$ is compact in $C_b := C_b([0,\infty), \mathbb{R}^n)$.

As a consequence, we have

Corollary A.29. *Let $M \subset C_b$ be the space of functions which have limits at positive infinity. Then M is relatively compact in C_b if the following conditions hold:*

(a) *M is uniformly bounded in* C_b.

(b) *The functions belonging to M are almost equi-continuous on* $\mathbb{R}_+$, *i.e. equi-continuous on every compact interval of* $\mathbb{R}_+$.

(c) *The functions from M are equi-convergent at* ∞ *that is, given* $\varepsilon > 0$, *there corresponds* $T = T(\varepsilon) > 0$ *such that* $\|x(t) - x(\infty)\| < \varepsilon$ *for any* $t \geq T(\varepsilon)$ *and* $x \in M$.

A.5 Weak-compactness in L^1

Let $L^1(\Omega, \mu, E)$ and (Ω, Σ, μ) be a finite measure space (i.e. $\mu(\Omega) < \infty$).

Definition A.30. Let E be a Banach space. A subset A in $L^1(\Omega, \mu, E)$ is called uniformly integrable, if for each $\varepsilon > 0$ there exists $\delta(\varepsilon) > 0$ such that, for each measurable subset $C \in \Sigma$ with measure $\mu(C) < \delta(\varepsilon)$, we have

$$\int_C |f(s)| d\mu(s) \leq \varepsilon.$$

Remark A.31. Let $A \subset L^1(\Omega, \mu, E)$.

(i) If (Ω, Σ, μ) is of totally bounded type i.e., for each $\varepsilon > 0$, there exist a finite covering $\{\Omega_k : k = 1, \ldots, n(\varepsilon)\} \subset \Sigma$ of Ω with $\mu(\Omega_k) \leq \varepsilon$ for $k = 1, \ldots, n(\varepsilon)$ and A is uniformly integrable, then it is norm bounded in $L^1(\Omega, \sigma, E)$,

(ii) if $\mu(\Omega) < \infty$ and A is bounded in $L^p(\Omega, \Sigma, \mu)$ for some $p > 1$, then it is uniformly integrable;

Definition A.32. A subset $K \subset L^p([0, b], E)$ $(p \geq 1)$ is said to be *p-equi-integrable* if it is uniformly integrable and

$$\lim_{h \to 0} \int_0^{b-h} \|f(t+h) - f(t)\|_E^p dt = 0, \text{ uniformly for all } f \in K.$$

We have the Kolmogorov criterion of compactness in $L^p([0, b], E)$ (see [111, 187, 222]):

Theorem A.33. *A subset* $K \subset L^p([0, b], E)$ $(p \geq 1)$ *is relatively compact if and only if it is p-equi-integrable and for all* $0 < s < t < b$, *the set,* $\{\int_s^t f(\tau) d\tau \mid f \in K\}$ *is relatively compact in* E.

Definition A.34. A sequence $\{v_n\}_{n \in \mathbb{N}} \subset L^1([0, b], E)$ is said to be *integrably bounded* if there exists $q \in L^1([0, b], \mathbb{R}^+)$ such that

$$|v_n(t)|_E \leq q(t), \quad \text{for a.e. } t \in [0, b] \text{ and every } n \in \mathbb{N};$$

Remark A.35. Every integrably bounded sequence is uniformly integrable. This follows from the fact that, for a finite measure space (Ω, Σ, μ), $K \subset L^1(\Omega, \mu, E)$ is uniformly integrable (see [187]) if and only if K is (uniformly) bounded and for each $\varepsilon > 0$, there exists $\delta > 0$ such that

$$\sup_{f \in K} \int_A |f(w)| d\mu \le \varepsilon$$

for all $A \in \Sigma$ with $\mu(A) \le \delta$.

Now, we present two weak compactness criteria that follow from the well known Dunford–Pettis theorem (see [187, 492]).

Lemma A.36 ([477]). *Let (Ω, Σ, μ) be a σ-finite measure space, let $\{\Omega_k \mid k \in \mathbb{N}\}$ be a subfamily of Σ such that*

$$\begin{cases} \mu(\Omega_{k-1}) < \infty, & \text{for } k \in \mathbb{N}, \\ \Omega_{k-1} \subset \Omega_k, & \text{for } k \in \mathbb{N}, \\ \bigcup_{k=0}^{\infty} \Omega_k = \Omega, \end{cases}$$

and let E be a Banach space. Let $A \in L^1(\Omega, \mu, E)$ be bounded and uniformly integrable in $L^1(\Omega_k, \mu, E)$, for $k \in \{0\} \cup \mathbb{N}$ and

$$\lim_{k \to \infty} \int_{\Omega \setminus \Omega_k} |f(s)| d\mu(s) = 0,$$

uniformly for $f \in A$. If for each $\gamma > 0$ and each $k \in \mathbb{N}$, there exist a weakly compact subset $C_{\gamma,k} \subset E$ and measurable subset $\Omega_{\gamma,k}$ with $\mu(\Omega \setminus \Omega_{\gamma,k}) \le \gamma$ and $f(\Omega_{\gamma,k}) \subset C_{\gamma,k}$ for all $f \in A$, then A is weakly compact in $L^1(\Omega, \mu, E)$.

Corollary A.37. *Let (Ω, Σ, μ) be a σ–finite measure space, E reflexive, and $K \subset L^1(\Omega, E)$ be a bounded subset. Then K is relatively weakly compact if and only if K is uniformly integrable.*

Definition A.38. Let E be a Banach space. A sequence $\{v_n\}_{n \in \mathbb{N}} \subset L^1([0, b], E)$ is said to be semicompact if:

(a) it is integrably bounded;

(b) the image sequence $\{v_n(t)\}_{n \in \mathbb{N}}$ is relatively compact in E for a.e. $t \in J$.

Finally, the following results follow from the Dunford–Pettis theorem.

Lemma A.39 ([311, Proposition 4.2.1] or [422, Proposition 3.6] in case dim E $< \infty$)). *Every semicompact sequence $L^1([0, b], E)$ is weakly compact in $L^1([0, b], E)$.*

Lemma A.40 ([407, Corollary 6.4.11]). *Let $A \subset L^1(\Omega, E)$ be a bounded decomposable set with Ω finite-measurable and E reflexive. Then A is weakly relatively compact in $L^1(\Omega, E)$.*

A.6 Proper Maps and Vector Fields

Let X, Y be two metric spaces and $f : X \longrightarrow Y$ a continuous map.

Definition A.41. We say that f is *proper* if $f^{-1}(K)$ is compact for every compact subset $K \subset Y$.

Notice that for finite-dimensional spaces, f proper means that $f^{-1}(B)$ is bounded for every bounded subset B.

Proposition A.42. *If $f : X \longrightarrow Y$ is proper, then it is closed.*

Proposition A.43. *Let $C \subset X$ be a nonempty, bounded, closed subset of a Banach space X and $f = I - K : C \longrightarrow X$ be a vector field associated with a compact mapping K. Then f is proper.*

Proof. Let $K \subset E$ be a compact set of E; we show that $G^{-1}(K)$ is compact. Let $\{x_n\}_{n\in\mathbb{N}} \subset G^{-1}(K)$ be a sequence; thus for every $n \in \mathbb{N}$, there exist $y_n \in K$ such that $x_n - F(x_n) = y_n$. Since F is a completely continuous map and K is compact, there exist subsequences of $\{x_n\}_{n\in\mathbb{N}}$, $\{y_n\}_{n\in\mathbb{N}}$ converge to $(x, y) \in C \times K$ respectively. Hence

$$x - F(x) = y \Rightarrow x \in G^{-1}(K).$$

□

A.7 Fundamental Theorems in Functional Analysis

For this section, we recommend [109, 111, 186, 187, 373, 385, 492]. We start with the Eberlein-Šmulian Theorem.

Theorem A.44 ([187, Theorem 8.12.1 and Theorem 8.12.7]). *Let K be a weakly closed subset of a Banach space X. Then the following are equivalent:*

(i) *K is weakly compact.*

(ii) *K is weakly sequentially compact.*

The following result is known as Eberlein–Kakutani theorem:

Theorem A.45. *A normed space is reflexive if and only if every bounded sequence admits a convergent subsequence.*

The following results are due to Mazur, 1933.

Theorem A.46 (Mazur–Smŭlian Theorem). *The closure and weak closure of a convex subset of a normed space are the same.*

As a consequence, a convex subset of a normed space is closed if and only if it is weakly closed.

Theorem A.47 (Mazur's Compactness Theorem [186, Theorem 6, p. 416]). *The closed convex hull of a (weakly) compact subset of a Banach space is itself (weakly) compact.*

In a reflexive Banach space, the unit ball is weakly sequentially compact.

Bibliography

[1] N. Abada, R. Agarwal, M. Benchohra, and H. Hammouche, Impulsive semilinear neutral functional differential inclusions with multivalued jumps, *Appl. Math.* **56** (2011) 227–250.

[2] N. Abada, M. Benchohra, and H. Hammouche, Existence and controllability results for nondensely defined impulsive semilinear functional differential inclusions *J. Differential Equations* **246** (2009), 3834–3863.

[3] N. Abada, M. Benchohra, and H. Hammouche, Nonlinear impulsive partial functional differential inclusions with state-dependent delay and multivalued jumps, *Nonlinear Anal. Hybrid Syst.* **4** (2010), 791–803.

[4] N. Abada, M. Benchohra, H. Hammouche and A. Ouahab, Controllability of impulsive semilinear functional differential inclusions with finite delay in Frećhet spaces, *Discuss. Math. Differ. Incl. Control Optim.* **27** (2007), 329–347

[5] M. Abbas, Existence for fractional order impulsive integrodifferential inclusions with nonlocal initial conditions, *Int. J. Math. Anal. (Ruse)* **6** (2012), 1813–1828.

[6] R. P. Agarwal, M. Benchohra, J. J. Nieto, and A. Ouahab, *Fractional Differential Equations and Inclusions*, Springer-Verlag, in press.

[7] R. P. Agarwal and D. O'Regan, *Infinite Interval Problems for Differential, Difference and Integral Equations,* Kluwer Acad. Pub. Dordrecht, 2001.

[8] R. P. Agarwal and D. O'Regan, The solutions set of integral inclusions on the half line, *Analysis* (2000), 1–7.

[9] R. P. Agarwal, and D. O'Regan, Multiple solutions for second order impulsive differential equations, *Appl. Math. Comput.* **114** (2000), 51–59.

[10] Z. Agur, L. Cojocaru, G. Mazaur, R. M. Anderson, and Y. L. Danon, Pulse mass measles vaccination across age cohorts, *Proc. Nat. Acad. Sci. USA.* **90** (1993), 11698–11702.

[11] N. Ahmed, Optimal relaxed controls for systems governed by impulsive differential inclusions, *Nonlinear Funct. Anal. Appl.* **10** (2005), 427–460.

[12] E. Ait Dads, M. Benchohra and S. Hamani, Impulsive fractional differential inclusions involving the Caputo fractional derivative, *Fract. Calc. Appl. Anal.* **12** (2009), 15–38.

[13] N. U. Ahmed, *Semigroup Theory with Applications to Systems and Control*, Pitman Res. Notes in Math. Ser. **246**, Longman Scientific and Technical and John Wiley, London, New York, 1991.

[14] R. R. Akhmerov, M. I. Kamenskii,A. S. Potapov, A. E. Rodkina, and B. N. Sadovskii, *Measures of Noncompactness and Condensing Operators*, Bikhäuser, Boston-Basel-Berlin, 1992.

[15] J. Andres, On the multivalued Poincaré operator, *Nonlinear Anal.* **10** (1997), 171–182.

[16] J. Andres, G. Gabor, and L. Górniewicz, Boundary value problems on infinite intervals, *Trans. Amer. Math. Soc.* **351** (1999), 4861–4903.

[17] J. Andres, G. Gabor, and L. Górniewicz, Topological structure of solution sets to multivalued asymptotic problems, *Z. Anal. Anwendungen* **18** (1999), 1–20.

[18] J. Andres, G. Gabor, and L. Górniewicz, Acyclicity of solutions sets to functional inclusions, *Nonlinear Anal.*, **49** (2002) 671–688.

[19] J. Andres and L. Górniewicz, *Topological Fixed Point Principles for Boundary Value Problems,* Kluwer Academic Publishers, Dordrecht, 2003.

[20] J. Andres and L. Górniewicz, *On the Banach contraction principle for multivalued mappings*, In: Approximation, Optimization and Mathematics Economis(Marc Lassonle,ed), Physica-Verlg, Springer, Berlin, 2001, 1–22.

[21] A. Anguraj and M. M. Arjunan, Existence of solutions for second order impulsive functional integrodifferential inclusions, *Far East J. Dyn. Syst.* **9** (2007), 265–278.

[22] N. Anichini and P. Zecca, Multivalued differential equations in a Banach space: an application to control theory, *J. Optim. Th. Appl.* **21** (1977), 477–738.

[23] J. Appell, E. De Pascale, H. T. Nguyên, and P. P. Zabreiko, Nonlinear integral inclusions of Hammerstein type, *Topol. Meth. Nonlinear Anal.* **5** (1995), 111–124.

[24] D. Araya and C. Lizama, Almost automorphic mild solutions to fractional differential equations, *Nonlinear Anal.* **69** (2008), 3692–3705.

[25] W. Arendt, Vector valued Laplace transforms and Cauchy problems, *Israel J. Math.* **59** (1987), 327–352.

[26] W. Arendt, Resolvent positive operators and integrated semigroup, *Proc. London Math. Soc.* **3**, (1987), 321–349.

[27] W. Arendt, C. J. K. Batty, M. Hieber and F. Neubrander, *Vector-valued Laplace transforms and Cauchy problems.* Second edition. Monographs in Mathematics, 96. Birkhäuser/Springer Basel AG, Basel, (2011).

[28] M. M. Arjunan and S. Selvi, Existence results for impulsive mixed Volterra-Fredholm integrodifferential inclusions with nonlocal conditions, *Int. J. Math. Sci. Appl.* **1** (2011), 19 pp.

[29] N. Aronszajn, Le correspondant topologique de l'unicité dans la théorie des équations différentielles, *Ann. Math.* **43** (1942), 730–738.

[30] O. Arino, R. Benkhalti, and K. Ezzinbi, Existence results for initial value problems for neutral functional differential equations, *J. Differential Equations* **114** (1994) 527–537.

[31] W. Arendt, C. Batty, M., Hieber, and F. Neubrander, *Vector-Valued Laplace Transforms and Cauchy Problems*, Monographs in Mathematics, **96** Birkhäser, Basel, 2001.

[32] J. P. Aubin, *Viability Theory*, Birkhäuser, Boston, Basel, Berlin, 1991.

[33] J. P. Aubin, *Neural Networks and Qualitative Physics: A Viability Approch*, Cambridge University Press, 1996.

[34] J. P. Aubin and A. Cellina, *Differential Inclusions,* Springer-Verlag, Berlin-Heidelberg, New York, 1984.

[35] J. P. Aubin and H. Frankowska, *Set-Valued Analysis,* Birkhäuser, Boston, 1990.

[36] J. P. Aubin, J. Lygeros, M. Quincampoix, S. Sastry, and N. Seube, Impulse Differential Inclusions: A Viability Approach to Hybrid Systems, *IEEE Transactions on Automatic Control* **47** (2002), 2–20.

[37] C. Avramescu, Sur l'existence des solutions convergentes des systémes d'équations différentielles non linéaire, *Ann. Mat. Pura. Appl.*, **81** (1969), 147–168.

[38] N. B. Azbelev, V. P. Maximov, and L. F. Rakhmatulina, *Introduction to the Functional Differential Dquations*, (Russian), Nauka, Moscouw, 1991.

[39] D. D. Bainov and P. S. Simeonov, *Systems with Impulse Effect*, Ellis Horwood Ltd., Chichister, 1989.

[40] D. D. Bainov and P. S. Simeonov, *Impulsive differential equations: periodic solutions and applications,* Pitman Monographs and Surveys in Pure and Applied Mathematics, **66**, Longman Scientific & Technical and John Wiley & Sons, Inc., New York, 1993.

[41] I. Bajo and E. Liz, Periodic boundary value problem for first order differential equations with impulses at variable times, *J. Math. Anal. Appl.* **204** (1996), 65–73.

[42] P. Balasubramaniam and S. K. Ntouyas, Global existence for semilinear stochastic delay evolution equations with nonlocal conditions, *Soochow J. Math.* **27** (2001) 331–342.

[43] G. Ballinger and X. Z. Liu, Existence and uniqueness results for impulsive delay differential equations, *Dynam. Contin. Discrete Impuls. Systems* **5** (1999), 579–591.

[44] G. Ballinger and X. Liu, Permanence of population growth models with impulsive effects, *Math. Comput. Modelling* **26** (1997), 59–72.

[45] E. Bazhlekova, *Fractional Evolution Equations in Banach Spaces*, Ph.D. Thesis, Eindhoven University of Technology, 2001.

[46] M. E. Ballotti, Aronszajn's theorem for a parabolic partial differential equations, *Nonlinear Anal.* **9** (1985), 1183–1187.

[47] J. W. Bebernes and L. K. Jackson, Infinite interval boundary value problems for $y'' = f(x, y)$, *Duke Math. J.* **34** (1997), 39–48.

[48] J. Bebernes and K. Schmit, Invariant sets and the Hukuhara-Kneser property for systems of parabolic partial, differential equations, *Rocky Mountain J. Math.* **7** (1967), 557–567.

[49] A. Belarbi and M. Benchohra, Existence theory for perturbed impulsive hyperbolic differential inclusions with variable times, *J. Math. Anal. Appl.* **327** (2007), 1116–1129.

[50] A. Belarbi, M. Benchohra and A. Ouahab, On a second order boundary value problem for impulsive dynamic inclusions on time scales, *Dynam. Systems Appl.* **14** (2005), 353–364.

[51] A. Belarbi, M. Benchohra and A. Ouahab, Nonconvex valued impulsive functional differential inclusions with variable times, *Neliniini Koliv.* **10** (2007), 443–463; *translation in Nonlinear Oscil.* **10** (2007), 447–468.

[52] A. Belarbi, M. Benchohra, and A. Ouahab, Existence results for impulsive dynamic inclusions on time scales, *Electron. J. Qual. Theory Differ. Equ.* (2005), 22 pp.

[53] H. Benabdellah, Existence of solutions to the nonconvex sweeping process, *J. Differential Equations* **164** (2000) 286–295.

[54] M. Benchohra, F. Berhoun and J. Henderson, Nonlinear impulsive differential inclusions with integral boundary conditions, *Commun. Math. Anal.* **5** (2008), 60–75.

[55] M. Benchohra, A. Boucherif, and A. Ouahab, On nonresonance impulsive functional differential inclusions with nonconvex valued right hand side, *J. Math. Anal. Appl.* **282** (2003), 85–94.

[56] M. Benchohra, S. Djebali, and S. Hamani, Boundary-value problems for differential inclusions with Riemann-Liouville fractional derivative, *Nonlinear Oscillations* **14** (2011), 6–20.

[57] M. Benchohra and L. Górniewicz, Existence results for nondensely defined impulsive semilinear functional differential inclusions with infinite delay, *JP J. Fixed Point Theory Appl.* **2** (2007), 11–51.

[58] M. Benchohra, L. Górniewicz, S. K. Ntouyas, and A. Ouahab, Existence results for impulsive hyperbolic differential inclusions, *Appl. Anal.* **82** (11) (2003), 1085–1097.

[59] M. Benchohra, L. Górniewicz, S. K. Ntouyas, and A. Ouahab, Impulsive hyperbolic differential inclusions with variable times, *Topol. Methods Nonlinear Anal.* **22** (2003), 319–329.

[60] M. Benchohra, L. Górniewicz, S. K. Ntouyas, and A. Ouahab, Controllability results for impulsive functional differential inclusions, *Rep. Math. Phys.* **54** (2004), 211–288.

[61] M. Benchohra, J. Graef, S. K. Ntouyas, and A. Ouahab, Upper and lower solutions method for impulsive differential inclusions with nonlinear boundary conditions and variable times, *Dyn. Contin. Discrete Impuls. Syst.*, Ser. A, **12** (2005), 383–396.

[62] M. Benchohra, J. Graef, and A. Ouahab, Nonresonance impulsive functional differential inclusions with variable times, *Comput. Math. Appl.* **47** (2004), 1725–1737.

[63] M. Benchohra, J. Graef, and A. Ouahab, Oscillatory and nonoscillatory solutions of multivalued differential inclusions, *Comput. Math. Appl.* **49** (2005), 1347–1354.

[64] M. Benchohra and S. Hamani, The method of upper and lower solutions and impulsive fractional differential inclusions, *Nonlinear Anal. Hybrid Syst.* **3** (2009), 433–440.

[65] M. Benchohra, S. Hamani, and J. Henderson, Oscillatory and nonoscillatory solutions for first order impulsive dynamic inclusions on time scales, *Arch. Math. (Brno)* **43** (2007), 237–250.

[66] M. Benchohra and B. Hedia, Impulsive functional differential inclusions with state-dependent delay and variable times, *Commun. Appl. Anal.* **16** (2012), 47–62.

[67] M. Benchohra, J. Henderson and S. K. Ntouyas, An existence result for first order impulsive functional differential equations in Banach spaces, *Comput. Math. Appl.* **42** (2001), 1303–1310.

[68] M. Benchohra, J. Henderson and S. Ntouyas, *Impulsive Differential Equations and Inclusions*, Contemporary Mathematics and Its Applications, Hindawi Publishing, New York, 2006.

[69] M. Benchohra, J. Henderson, and S. Ntouyas, On nonresonance second order impulsive functional differential inclusions with nonlinear boundary conditions, *Can. Appl. Math. Q.* **14** (2006), 21–32.

[70] M. Benchohra, J. Henderson and S. K. Ntouyas, Impulsive neutral functional differential equations in Banach spaces, *Appl. Anal.* **80** (2001), 353–365.

[71] M. Benchohra, J. Henderson, S. K. Ntouyas, and A. Ouahab, Impulsive functional differential equations with variable times and infinite delay, *Inter. J. Appl. Math. Sci.* **2** (2005), 130–148.

[72] M. Benchohra, J. Henderson, S. K. Ntouyas, and A. Ouahab, Impulsive functional differential equations with variable times, *Comput. Math. Appl.* **47** (2004), 1659–1665.

[73] M. Benchohra, J. Henderson, S. K. Ntouyas, and A. Ouahab, Existence results for impulsive functional and neutral functional differential inclusions with lower semicontinuous right hand side, *Electron. J. Math. and Phys. Sci.* (2002), **1**, 72–91.

[74] M. Benchohra, J. J. Nieto, and A. Ouahab, Existence results for functional integral inclusions of Volterra type, *Dyn. Syst. Appl.* **14**, (2005), 57–70.

[75] M. Benchohra and S. K. Ntouyas, Hyperbolic functional differential inclusions with nonlocal conditions, *Funct. Approx. Comment. Math.* **29** (2001), 29–39.

[76] M. Benchohra and S. K. Ntouyas, The method of lower and upper solutions to the Darboux problem for partial differential inclusions, *Miskolc Math. Notes* **4** (2003), 81–88.

[77] M. Benchohra, J. J. Nieto and A. Ouahab, Impulsive differential inclusions involving evolution operators in separable Banach spaces, *Ukr. Math. J.* **64**, (2012), 991–1018.

[78] M; Benchohra, J. J. Nieto and A. Ouahab, Topological structure of the solutions set of impulsive, semilinear differential inclusions with nonconvex right-hand side, *Afr. Diaspora J. Math.*, to appear.

[79] M. Benchohra, S. K. Ntouyas, and L. Górniewicz, Controllability of some nonlinear systems in Banach spaces (The fixed point theory approch), Plock Univ Press, (2003).

[80] M. Benchohra and A. Ouahab, Oscillatory and nonoscillatory solutions for first order impulsive differential inclusions, *Comment. Math. Univ. Carolin.* **46** (2005), 541–553.

[81] M. Benchohra and A. Ouahab, Impulsive neutral functional differential equations with variable times, *Nonlinear Anal.* **55** (2003), 679–693.

[82] M. Benchohra and A. Ouahab, Some uniqueness results for impulsive semilinear neutral functional differential equations, *Georgian Math. J.* **9** (2002) 423–430.

[83] I. Benedetti, V. Obukhovskii, and P. Zecca, Controllability for impulsive semilinear functional differential inclusions with a non-compact evolution operator, *Discuss. Math. Differ. Incl. Control Optim.* **31** (2011), 39–69.

[84] I. Benedetti and P. Rubbioni, Existence of solutions on compact and non-compact intervals for semilinear impulsive differential inclusions with delay, *Topol. Methods Nonlinear Anal.* **32** (2008), 227–245.

[85] A. Bensoussan and J. L. Lions, Novelle formulation de problmes de contrôle impulsionnel et applications, *C. R. Acad. Sci. Paris Sér. A–B* **276** (1973), A1189–A1192.

[86] A. Bensoussan and J. L. Lions, *Contrôle Impulsionnel et Inquations Quasi-variationnelles*, Dunod, Paris, 1982.

[87] A. Bensoussan and J. L. Lions, *Impulse Contrôle and Quasi-varitionnelles Inequalites*, Gauthier-Villars, 1984.

[88] A. T. Bharucha-Reid, *Random Integral Equations*, Academic Press, New York, 1972.

[89] R. Bielawski, L. Górniewicz and S. Plaskacz, Topological approach to differential inclusions on closed sets of $\mathbb{R}^n$, *Dynamics Reported* **1** (1992), 225–250.

[90] A. Bielecki, Une remarque sur la méthode de Banach-Cacciopoli-Tikhonov dans la théorie des équations différentialles ordinaires, *Bull. Acad. Polon. Sci.* **4** (1956), 261–264.

[91] F. S. De Blasi, Existence and stability of solutions for autonomous multivalued differential equations in a Banach space, *Rend. Accad. Naz. Lincei,* Serie VII, **60** (1976), 767–774.

[92] F. S. De Blasi, On a property of the unit sphere in a Banach space, *Bull. Soc. Math. R. S. Roumaine* **21** (1977), 259–262.

[93] F. S. De Blasi, Characterizations of certain classes of semicontinuous multifunctions by continuous approximations, *J. Math. Anal. Appl.* **106** (1985), 1–8.

[94] F. S. De Blasi and J. Myjak, One the solutions sets for differential inclusions, *Bull. Polon. Acad. Sci.* **33** (1985), 17–23.

[95] F. S. De Blasi and J. Myjak, One the structure of the set of solutions of Darboux problem for hyperbolic equations, *Proc. Edinburgh Math. Soc.*, Ser. 2 **29** (1986), 7–14.

[96] F. S. De Blasi and G. Pianigiani, On the solutions sets of boundary value problems for nonconvex differential inclusions, *J. Diff. Equs.* **128** (1996), 541–555.

[97] F. S. De Blasi and G. Pianigiani, Solutions sets of boundary value problems for nonconvex differential inclusions, *Nonlinear Anal.* **1** (1993), 303–313.

[98] F. S. De Blasi, G. Pianigiani, and V. Staicu, Topological properties of nonconvex differential inclusions of evolution type, *Nonlinear Anal.* TMA, **24** (1995), 711–720.

[99] A. W. Bogatyrev, Fixed points and properties of solutions of differential inclusions, *Izv. Akad. Nauk SSSR Ser. Mat.* **47** (1983), 895–909 (in Russian).

[100] A. Boucherif and J. Nieto, Periodic boundary value problems for impulsive first order differential inclusions, *Commun. Appl. Anal.* **9** (2005), 409–416.

[101] M. Bounkhel,*Regularity concepts in nonsmooth analysis. Theory and applications*,Springer Optimization and Its Applications **59**, Berlin: Springer 2012.

[102] A. Bressan, Directionally continuous selections and differential inclusions, *Funkcial. Ekvac.* **31** (1988), 459–470.

[103] A. Bressan A. Cellina, and A. Fryszkowski, A class of absolute retracts in spaces of integrable functions, *Proc. Amer. Math. Soc.* **112** (1991), 413–418.

[104] A. Bressan and G. Colombo, Extensions and selections of maps with decomposable values, *Studia Math.* **90** (1988), 69–86.

[105] A. Bressan and G. Colombo, Generalized Baire category and differential inclusions in Banach spaces, *J. Differential Equations* **76** (1987), 135–158.

[106] A. Bressan and G. Colombo, Boundary value problems for lower semicontinuous differential inclusions, *Funkcial. Ekvac.* **36** (1993), 359–373.

[107] A. Bressan, A. Cellina, and G. Colombo, Upper semicontinuous differential inclusions without convexity, *Proc. Amer. Math. Soc.* **106** (1989), 771–775.

[108] P. C. Bressloff and S. Coombes, A dynamic theory of spike train transitions in networks of integral and fire oscillators, *SIAM J. Applied Analysis* **98**, 1998.

[109] R. Brette *Integrale and Fire Model in Neurobiology: an example of hybrid system,* Proceedings of the HSCC Conference, 2001.

[110] H. Brézis, *Analyse Fonctionnelle. Théorie et Applications,* Masson, Paris, 1983.

[111] H. Brézis, *Functional Analysis, Sobolev Spaces, and Partial Differential Equations,* Universitext, Springer, New York, 2010.

[112] R. F. Brown, M. Furi, L. Górniewicz, and B. Jiang, *Handbook of Topological Fixed Point Theory.* Springer, Dordrecht, 2005.

[113] F. E. Browder and G. P. Gupta, Topological degree and nonlinear mappings of analytic type in Banach spaces, *J. Math. Anal. Appl.* **26** (1969), 730–738.

[114] K. Borsuk, *Theory of Retracts,* Polish Scientific Publishers, Warszawa, 1967.

[115] A. I. Bulgakov and I. N. Lyapin, Some properties of the set of solutions of a Volterra-Hammerstein integral inclusions, *Diff. Urav.* **14** (1978),1043–1048.

[116] A. I. Bulgakov and I. N. Lyapin, Certain properties of the set of solutions of the Volterra-Hammerstein integral inclusions, *Diff. Urav.*, **14** (1978), 1465–1472.

[117] T. A. Burton, *Stability and Periodic Solutions of Ordinary and Functional-Differential Equations*, Mathematics in Science and Engineering, 178, Academic Press, Inc., Orlando, FL, 1985.

[118] T. A. Burton, *Volterra Integral and Differential Equations*, Second edition. Mathematics Science and Engineering, 202, Elsevier, Amsterdam, 2005.

[119] S. Busenberg and B. Wu, Convergence theorems for integrated semigroups, *Differential Integral Equations* **5** (1992), 509–520.

[120] L. Byszewski, Theorems about the existence and uniqueness of solutions of semilinear evolution nonlocal Cauchy problem, *J. Math. Anal. Appl.* **162** (1991), 494–505.

[121] L. Byszewski, Existence and uniqueness of mild and classical solutions of semilinear functional differential evolution nonlocal Cauchy problem, *Selected Problems in Mathematics, Cracow Univ. of Tech. Monographs, Anniversary Issue* **6** (1995), 25–33.

[122] L. Byszewski and H. Akca, On a mild solution of a semilinear functional differential evolution nonlocal problem, *J. Appl. Math. Stoc. Anal.* **10** (1997), 265–271.

[123] L. Byszewski and V. Lakshmikantam, Monotone iterative technique for non-local hyperbolic differential problem, *J. Math. Phys. Sci.* **26** (1992), 345–359.

[124] L. Byszewski and S. N. Papageorgiou, An application of a noncompactness technique to an investigation of the existence of solutions to a nonlocal multivalued Darboux Problem, *J. Appl. Math. Stoch. Anal.* **12** (1999), 179–180.

[125] A. Cabada and E. Liz, Discontinuous impulsive differential equations with nonlinear boundary conditions, *Nonlinear Anal.* **28** (1997), 1491–1499.

[126] A. Cabada and E. Liz, Boundary value problem for higher order ordinary differential equations with impulses, *Nonlinear Anal.* **32** (1998), 775–86.

[127] A. Cabada, J. J. Niteo, D. Franco, and S. I. Trofimchuk, Ageneralization of the montone method for second order periodic boundary value problem with impulses at a fixed points, *Dynam. Contin. Discrete Impuls. Systems* **7** (2000), 145–158.

[128] J. Cao and H. Chen, Some results on impulsive boundary value problem for fractional differential inclusions, *Electron. J. Qual. Theory Differ. Equ.* **2011** (2011), 24 pp.

[129] Y. Cao and X. Fu, Nondensely defined impulsive neutral functional differential inclusions with nonlocal conditions, *Ann. Differential Equations* **25** (2009), 127–139.

[130] T. Cardinali and S. Panfili, Global mild solutions for semilinear differential inclusions and applications to impulsive problems, *Pure Math. Appl.* **19** (2008), 1–19.

[131] T. Cardinali and P. Rubbioni, Mild solutions for impulsive semilinear evolution differential inclusions, *J. Appl. Funct. Anal.* **1** (2006), 303–325.

[132] T. Cardinali and P. Rubbioni, Impulsive semilinear differential inclusions: topological structure of the solutions set and solutions on noncompact domains, *Nonlin. Anal.* **69** (2008), 73–84.

[133] T. Cardinali and P. Rubbioni, Impulsive mild solutions for semilinear differential inclusions with nonlocal conditions in Banach spaces, *Nonlinear Anal.* **75** (2012), 871–879.

[134] C. Castaing, Sur un nouvelle classe d'équation d'évolution dans les espaces de Hilbert, exposé no 10. *Sém. Anal. Convexe*, University of Montpellier, page 24 pages, 1983.

[135] C. Castaing, Version aleátoire du probléme de rafle par un convexe, *Sém. Anal. Convexe*, Montpellier, Exposé 1 (1974), 11 pp.

[136] C. Castaing and M. Valadier, Convex Analysis and Measurable Multifunctions, *Lecture Notes in Mathematics, Springer-Verlag, Berlin-Heidelberg-New York,* **580** (1977).

[137] A. Cernea, On the existence of solutions for fractional differential inclusions with anti-periodic boundary condtions, *J. Appl. Math. Comp.* **38** (2012), 133–143.

[138] D. Chalishajar, Controllability of second order impulsive neutral functional differential inclusions with infinite delay, *J. Optim. Theory Appl.* **154** (2012), 672–684.

[139] D. N. Chalishajar, H. D. Chalishajar, and F. S. Acharya, Controllability of second order neutral impulsive differential inclusions with nonlocal conditions, *Dyn. Contin. Discrete Impuls. Syst. Ser. A Math. Anal.* **19** (2012), 107–134.

[140] D. Chalishajar and F. Acharya, Controllability of second order semi-linear neutral impulsive differential inclusions on unbounded domain with infinite delay in Banach spaces, *Bull. Korean Math. Soc.* **48** (2011), 813–838.

[141] Y. Chang, Controllability of impulsive functional differential inclusions with infinite delay in Banach spaces, *J. Appl. Math. Comput.* **25** (2007), 137–154.

[142] Y. Chang, A. Anguraj and M. Arjunan, Existence results for non-densely defined neutral impulsive differential inclusions with nonlocal conditions, *J. Appl. Math. Comput.* **28** (2008), 79–91.

[143] Y. Chang, A. Anguraj, and K. Karthikeyan, Existence for impulsive neutral integrodifferential inclusions with nonlocal initial conditions via fractional operators, *Nonlinear Anal.* **71** (2009), 4377–4386.

[144] Y. Chang, A. Anguraj, and A. Mallika, Controllability of impulsive neutral functional differential inclusions with infinite delay in Banach spaces, *Chaos Solitons Fractals* **39** (2009), 1864–1876.

[145] Y. Chang and J. Nieto, Existence of solutions for impulsive neutral integro-differential inclusions with nonlocal initial conditions via fractional operators, *Numer. Funct. Anal. Optim.* **30** (2009), 227–244.

[146] Y. K. Chang and W. T. Li, Existence results for second order impulsive functional differential inclusions, *J. Math. Anal. Appl.* **301** (2005), 477–490.

[147] Y. Chang and L. Qi, Li Existence results for second-order impulsive functional differential inclusions, *J. Appl. Math. Stoch. Anal.* (2006), 12 pp.

[148] C. Chen and M. Li, On fractional resolvent operator functions, *Semigroup Forum*, **80** (2010), 121–142.

[149] D. Cheng and J. Yan, Global Existence and asymptotic behavior of solution of second order nonlinear impulsive differential equations, *Int. J. Math. Math. Sci.* **25** (2001), 175–182.

[150] M. Cichoń and I. Kubiaczyk, Some remarks on the structure of the solutions set for differential inclusions in Banach spaces, *J. Math. Anal. Appl.* **233** (1971), 1018–1020.

[151] F. H. Clarke, Yu. S. Ledyaev, R. J. Sten and P. R. Wolenski, *Nonsmooth Analysis and Control Theory*, Springer, New York, 1998.

[152] H. Covitz and S. B. Nadler Jr., Multivalued contraction mappings in generalized metric spaces, *Israel J. Math.* **8** (1970), 5–11.

[153] C. Corduneanu, *Integral Equations and Stability of Feedback Systems*, Academic Press, New York, 1973.

[154] G. Colombo and V. Goncharov, The sweeping processes without convexity, *Set-Valued Anal.* **7** (1999), 357–374.

[155] G. Colombo and M. D. P. Monterio Marques, Sweeping by a continuous prox-regular set, *J. Differential Equations* **187** (2003), 46–62.

[156] G. Conti, W. Kryszewski, and P. Zecca, One the solvability of systems of noncompact inclusions, *Ann. Mat. Pura. Appl.* **160** (1991), 371–408.

[157] C. Corduneanu and V. Lakhmikantham, Equations with unbounded delay, *Nonlinear Anal.* **4** (1980), 831–877.

[158] Ph. Clément and J. Prüss, Global existence for a semilinear parabolic Volterra equation, *Math. Z.* **209** (1992), 17–26.

[159] C. Cuevas and C. Lizama, Almost automorphic solutions to a class of semilinear fractional differential equations, *Appl. Math. Lett.* **21** (2008), 1315–1319.

[160] K. Czanowski, Structure of the set of solutions of an initial-boundary value problem for a parabolic partial differential equations in an unbounded domain, *Nonlinear Anal.* **27** (1996), 723–729.

[161] K. Czanowski, On the structure of fixed point sets of 'k-set contractions' in B_0 spaces, *Demonstratio Math.* **30** (1997), 233–244.

[162] G. Da Prato and E. Sinestrari, Differential operators with non-dense domains, *Ann. Scuola Norm. Sup. Pisa Sci.* **14** (1987), 285–344.

[163] G. Da Prato and J. Zabczyk, *Stochastic Equations in Infinite Dimensions*, Cambridge University Press, Cambridge, 1992.

[164] J. L. Davy, Properties of the solution set of a generalized differential equations, *Bull. Austral. Math. Soc.* **6** (1972), 379–398.

[165] R. H. Day, Multiple-phase economic dynamics, in: *Nonlinear and Convex Analysis in Economic Theory (Tokyo, 1993), T. Maruyama & W. Takahashi, Eds.*, pp. 25–45, Lecture Notes in Econom. and Math. Systems, Vol. **419**, Springer, Berlin, 1995.

[166] Da Prato and G. M. Iannelli, Linear integrodifferential equations in Banach space, *Rend. Sem. Mat. Univ. Padova* **62**, (1980), 207–219.

[167] K. Deimling, *Multivalued Differential Equations*, Walter De Gruyter, Berlin-New York, 1992.

[168] K. Deimling and M. R. M. Rao, On solution sets of multivalued differential equations, *Appl. Anal.* **28** (1988), 129–135.

[169] B. C. Dhage, First order impulsive differential inclusions involving discontinuous multifunctions, *Dynam. Systems Appl.* **16** (2007), 285–298.

[170] O. Dordan, *Analyse Qualitative*, Masson, 1992.

[171] F. S. De Belasi and J. Myjak, On the solutions sets for differential inclusions, *Bull. Polish Acad. Sci. Math.* **12** (1985) 17–23.

[172] B. C. Dhage, On periodic boundary value problems of first order discontinuous impulsive differential inclusions, *Fixed Point Theory* **8** (2007), 237–252.

[173] B. Dhage and S. Ntouyas, Existence results for impulsive neutral functional differential inclusions, *Topol. Methods Nonlinear Anal.* **25** (2005), 349–361.

[174] S. Djebali, L. Górniewicz, and A. Ouahab, Filippov-Wazewski theorems and structure of solution sets for first order impulsive semilinear functional differential inclusions, *Topol. Methods Nonlinear Anal.* **32** (2008), 261–312.

[175] S. Djebali, L. Górniewicz, and A. Ouahab, First order periodic impulsive semilinear differential inclusions existence and structure of solution sets, *Math. Comput. Modeling* **52** (2010), 683–714.

[176] S. Djebali, L. Górniewicz, and A. Ouahab, Topological structure of solution sets for impulsive differential inclusions in Fréchet spaces, *Nonlinear Anal.* **74** (2011), 2141–2169.

[177] S. Djebali, L. Górniewicz, and A. Ouahab, Existence and structure of solution sets for impulsive differential inclusions: a survey, *Lecture Notes in Nonlinear Analysis 13, Juliusz Schauder Center for Nonlinear Studies, Torun* (2012), 34–02

[178] S. Djebali, L. Górniewicz, and A. Ouahab, *Solution sets for differential equations and inclusions*, De Gruyter Series in Nonlinear Analysis and Applications 18, De Gruyter, Berlin, 2013.

[179] T. Donchev, Fixed time impulsive differential inclusions, *Surv. Math. Appl.* **2** (2007), 1–9.

[180] T. Donchev, Impulsive differential inclusions with constraints, *Electron. J. Differential Equations* (2006), 12 pp.

[181] A. Dold, *Lectures on Algebraic Topology,* Springer-Verlag Berlin, 1972.

[182] P. Drábek, P. Krejčĭ, and P. Takač, *Nonlinear Differential Equations*, Research Notes in Mathematics, Vol. 404, Chapman & Hall, London, 1999.

[183] J. Dubous and P. Morales, Structure de l'ensemble des solutions du probleme de Cauchy sous les conditions de Carathédory, *Ann. Sci. Math. Quebec* **7** (1983), 5–27.

[184] J. Dugundji and A. Granas, *Fixed Point Theory,* Springer-Verlag, New York 2003.

[185] P. Dupuis and H. Ishii, On Lipschitz continuity of the solution mapping to the Skorokhod problems with applications, *Stochastic Stochastic Rep.* **35** (1991), 31–62.

[186] N. Dunford and J. T. Schwartz, *Linear Operators. Part I. General Theory,* Wiley-Interscience, New York, 1967.

[187] R. Edwards, *Functional Analysis. Theory and Applications,* Holt-Rinehart-Winston, New York, 1965.

[188] L. È. Èl'sgol'ts, *Introduction to the Theory of Differential Equations with Deviating Arguments*, Translated from the Russian by Robert J. McLaughlim, Holden-Dely, Inc., San Francisco, Calif.-London-Amsterdam, 1966.

[189] L. È. Èl'sgol'ts and S. B. Norkin, *Introduction to the Theory and Application of Differential Equations with Deviating Arguments,* Translated from the Russian by John L. Casti. Mathematics in Science and Engineering, Vol 105. Academic Press [A Subsidiary of Harcourt Brace Jovanovich, Publishers], New York-London, 1973.

[190] K-J. Engel and R. Nagel, *One-Parameter Semigroups for Linear Evolution Equations.* Graduate Texts in Mathematics, 194, Springer-Verlag, New York, 2000.

[191] S. Eilenberg and N. E. Steenrod, *Foundations of algebraic topology*, Princeton University Press, Princeton, New Jersey, 1952

[192] L. Erbe, H. I. Freedman, X. Z. Liu, and J. H. Wu, Comparison principales for impulsive parabolic equations with applications to models of singles species growth, *J. Austral. Math. Soc. Ser. B* **32** (1991), 382–400.

[193] L.,H. Erbe, Q. Kong, and B.,G. Zhang, *Oscillation Theory for Functional Differential Equations*, Pure and Applied Mathematics, Marcel Dekker, New York, 1994.

[194] L. Erbe and W. Krawcewicz, Existence of solutions to boundary value problems for impulsive second order differential inclusions, *Rocky Mountain J. Math.* **22** (1992), 519–539.

[195] K. Ezzinbi and J. Liu, Periodic solutions of non-densely defined evolution equations, *J. Appl. Math. Stochastic Anal.* **15** (2002), 113–123.

[196] H. O. Fattorini, *Second Order Linear Differential equations in Banach spaces*, North-Holland, Mathematical Studies Vol. 108, North-Holland, Amsterdam, 1985.

[197] A. F. Filippov, *Differential Equations with Discontinuous Right-hand Sides,* Kluwer Academic Publishers, Dordrecht, 1988.

[198] T. Filippova, Set-valued solutions to impulsive differential inclusions, *Math. Comput. Model. Dyn. Syst.* **11** (2005), 149–158.

[199] D. Franco, E. Liz, J. J. Nieto, and Y. V. Rogovchenko, A contribution to the study of functional differential equations with impulses, *Math. Nachr.* **218** (2000), 49–60.

[200] D. Franco, J. J. Nieto, and D. O'Regan, Existence of nonegative solutions for resonant periodic boundary value problems with impulses, *Nonlinear Studies* **9** (2002), 1–10.

[201] H. Frankowska, A priori estimates for operational differential inclusions, *J. Differential Equations* **84** (1990), 100–128.

[202] H. Frankowska, Local controllability and infinitesimal generators of semi-groups of set- valued maps, *SIAM J. Control Optim.* **25** (1987), 412–431.

[203] H. Frankowska, S. Plaskacz and T. Rzeżuchowski, T. Measurable viability theorems and the Hamilton-Jacobi-Bellman equation, *J. Differential Equations* **116** (1995), 265–305.

[204] H. Frankowska and S. Plaskacz, A measurable upper semicontinuous viability theorem for tubes. *Nonlinear Anal.* **26** (1996), 565–582.

[205] M. Frigon, Théorémes d'existence de solutions d'inclusions différentielles. [Existence theorems for solutions of differential inclusions] Topological Methods in Differential Equations and Inclusions (Montreal, PQ, 1994), 51–87, *NATO Adv. Sci. Inst. Ser. C Math. Phys. Sci.*, **472**, Kluwer Acad. Publ., Dordrecht, 1995.

[206] M. Frigon, Fixed point results for multivalued contractions on gauge spaces, set valued mappings with applications in nonlinear analysis, 175–181, Ser. Math. Anal. Appl., 4, Taylor & Francis, London, 2002.

[207] M. Frigon and A. Granas, Théorémes d'existence pour des inclusions différentielles sans convexité, *C. R. Acad. Sci. Paris Sér. I Math.* **310** (1990), 819–822.

[208] M. Frigon and A. Granas, Résultats de type Leray-Schauder pour des contractions sur des espaces de Fréchet, *Ann. Sci. Math. Québec* **22** (1998), 161–168.

[209] M. Frigon and D. O'Regan, Existence results for first order impulsive differential equations, *J. Math. Anal. Appl.* **193** (1995), 96–113.

[210] M. Frigon and D. O'Regan, Boundary value problems for second order impulsive differential equations using set-valued maps, *Appl. Anal.* **58** (1995), 325–333.

[211] M. Frigon and D. O'Regan, Impulsive differential equations with variable times, *Nonlinear Anal.* **26** (1996), 1913–1922.

[212] M. Frigon and D. O'Regan, First order impulsive initial and periodic problems with variable moments, *J. Math. Anal. Appl.* **233** (1999), 730–739.

[213] M. Frigon and D. O'Regan, Second order Sturm-Liouville BVP's with impulses at variable moments, *Dynam. Contin. Discrete Impuls. Systems* **8** (2), (2001), 149–159.

[214] A. Fryszkowski and L. Górniewicz, Mixed semicontinuous mappings and their applications to differential inclusions, *Set-Valued Anal.* **8** (2000), 203–217.

[215] A. Fryszkowski and T. Rzeżuchowski, Pointwise estimates for retractions on the solution set to Lipschitz differential inclusions, *Proc. Am. Math. Soc.* **139** (2011), 597–608.

[216] X. Fu, Approximate controllability for neutral impulsive differential inclusions with nonlocal conditions, *J. Dyn. Control Syst.* **17** (2011), 359–386.

[217] X. Fu and Y. Cao, Existence for neutral impulsive differential inclusions with nonlocal conditions, *Nonlinear Anal.* **68** (2008), 3707–3718.

[218] G. Gabor, Some results on existence and structure of solution sets to differential inclusions on the halfline, *Boll. Unione Mat. Ital.* **5B** (8) (2002), 431–446.

[219] G. Gabor and A. Grudzka, Structure of the solution set to impulsive functional differential inclusions on the half-line, *NoDEA Nonlinear Differential Equations Appl.* **19** (2012), 609–627.

[220] E. Gatsori, L. Górniewicz and S. Ntouyas, Controllability results for nondensely defined evolution impulsive differential inclusions with nonlocal conditions, *Panamer. Math. J.* **15** (2005), 1–27.

[221] S. Gao, L. Chen, J. J. Nieto, and A. Torres, Analysis of a delayed epidemic model with pulse vaccination and saturation incidence, *Vaccine*, **24** (2006), 6037–6045.

[222] L. Gasiński and N. S. Papageorgiou, *Handbook of Nonlinear Analysis, Taylor & Francis Group, LLC.* **8**, 2005.

[223] C. Gori, V. Obukhovskii, M. Ragni, and P. Rubbioni, Existence and continuous dependence results for semilinear functional differential inclusions with infinite delay, *Nonlin. Anal.* **51** (2002), 765–782.

[224] L. Górniewicz, S. K. Ntouyas, and D. O'Regan, Existence results for impulsive functional differential inclusions, *Dyn. Contin. Discrete Impuls, Syst. Ser. A Math. Anal.* **18** (2011), 115–134.

[225] C. Gori, V. Obukhovskii, M. Ragni, and P. Rubbioni, On some properties of semilinear functional differential inclusions in abstract spaces, *J. Concr. Appl. Math.* **4** (2006), 183–214.

[226] I. I. Gihman and A. V. Skorohod, *The Theory of Stochastic Processes*, Springer-Verlag, Berlin-Heidelberg-New York, 1974.

[227] J. A. Goldstein, *Semigroups of Linear Operators and Applications*, Oxford Univ. Press, New York, 1985.

[228] K. Gopalsamy, *Stability and Oscillations in Delay Differential Equations of Population Dynamics*. Mathematics and its Applications, 74, Kluwer Academic Publishers Group, Dordrecht, 1992.

[229] L. Górniewicz, Homological methods in fixed point theory of multivalued maps, *Dissertations Math.* **129** (1976), 1–71.

[230] L. Górniewicz, *Topological Fixed Point Theory of Multivalued Mappings*, Mathematics and its Applications, 495, Kluwer Academic Publishers, Dordrecht, 1999.

[231] L. Górniewicz, *Topological Approach to Differential Inclusions*, Topological Methods in Differential and Inclusions (A. Granas and M. Frigon,eds.) NATO ASI Series C, Vol. 472, Kluwer Academic Publishers, Dordrecht, 1995.

[232] L. Górniewicz, On the solution sets of differential inclusions, *J. Math. Anal. Appl.* **113** (1986), 235–244.

[233] L. Górniewicz, Topological structure of solutions sets: current results, *Arch. Math.*, (2000), 343–382.

[234] L. Górniewicz, P. Nistri, and V. Obukhovskii, Differential inclusions on approximate retracts of Hilbert spaces, *Int. J. Nonlinear Diff. Equs.* **3** (1995), 13–26.

[235] L. Górniewicz, S. Ntouyas, and D. O'Regan, Existence results for first and second order semilinear impulsive differential inclusions, *Topol. Methods Nonlinear Anal.* **26** (2005), 135–162.

[236] J. R. Graef, J. Henderson, and A. Ouahab, Differential inclusions with nonlocal conditions: Existence results and topological properties of solution sets, *Topol. Meth. Nonl. Anal.* **37** (2011), 117–145.

[237] J. R. Graef and A. Ouahab, Global existence results for impulsive functional dynamic inclusions on time scales in Fréchet spaces, *Comm. Appl. Nonlinear Anal.* **13** (2006), 59–81.

[238] J. R. Graef and A. Ouahab, Some existence and uniqueness results for impulsive functional differential equations with variable times in Fréchet spaces, *Dynam. Contin. Discrete Impuls. Systems* **14** (2007), 27–45.

[239] J. R. Graef and A. Ouahab, Nonresonance impulsive functional dynamic boundary value inclusions on time scales, *Nonlinear Stud.* **15** (2008), 339–354.

[240] J. R. Graef and A. Ouahab, Global existence and uniqueness results for impulsive functional differential equations impulsive Differential equations with variable times and multiple delays, *Dyn. Contin. Discrete Impuls. Syst. Ser. A Math. Anal.* **16** (2009), 27–40.

[241] J. R. Graef and A. Ouahab, Some existence and uniqueness results for first-order boundary value problems for impulsive functional differential equations with infinite delay in Fréchet spaces, *Int. J. Math. Math. Sci.* **16** (2006), 0161–1712.

[242] J. R. Graef and A. Ouahab, *A class of second order impulsive functional differential equations*, submitted.

[243] J. R. Graef and A. Ouahab, Second order impulsive functional differential inclusions with infinite delay in Banach spaces, preprint.

[244] J. R. Graef and A. Ouahab, First order impulsive differential inclusions with periodic condition, *Electr. J. Qual. Theory Differ. Equ.* **2008** (2008), No. 31, pp. 1–40.

[245] J. R. Graef and A. Ouahab, Structure of solutions sets and a continuous version of Filippov's theorem for first order impulsive differential inclusions with periodic conditions, *Electr. J. Qual. Theory Differ. Equ.* **2009** (2009), No. 24, pp. 1–23.

[246] J. R. Graef and A. Ouahab, Existence results for functional semilinear differential inclusions in Fréchet spaces, *Math. Comput. Modelling* **48** (2008), 1708–1718.

[247] J. R. Graef and A. Ouahab, Nonresonance impulsive functional dynamic boundary value inclusions on time scales, *Nonlinear Studies* **15** (2008), 339–354.

[248] G. Gripenberg, S. O. Londen, and O. Staffans, *Volterra Integral and Functional Equations*, Cambridge Univ. Press, New York, 1990.

[249] O. A. Gross, The boundary value problem on an infinite interval, *J. Math. Anal. Appl.* **7** (1963), 100–109.

[250] L. Guedda, Some remarks in the study of impulsive differential equations and inclusions with delay, *Fixed Point Theory* **12** (2011), 349–354.

[251] D. Guo, Second order integro-differential equations of Votera type on unbounded domains in a Banach space, *Nonlinear Anal.* **41** (2000), 465–476.

[252] G. Guo, Multiple positive solutions for first order nonlinear integro-differential equations in a Banach space, *Nonlinear Anal.* **53** (2003), 183–195.

[253] D. Guo, Boundary value problem for impulsive integro-differential equation on unbounded domains in a Banach space, *Appl. Math. Comput.* **99** (1999), 1–15.

[254] G. Guo, A class of second-order implusive integro-differential equations on unbounded domain in a Banach space, *Appl. Math. Comput.* **125** (2002), 59–77.

[255] G. Guo, Multiple positive solutions for first order nonlinear implusive integro-differential equations in a Banach space, *Appl. Math. Comput.* **143** (2003), 233–249.

[256] G. Guhring, F. Rabiger, and W. Ruess, Linearized stability for semilinear non-autonomous evolution equations to retarded differential equations, *Differ. Integral Equations* **13** (2000), 503–527.

[257] D. Guo and X. Z. Liu, Impulsive integro-differential equations on unbounded domain in Banach space, *Nonlinear Stud.* **3** (1996), 49–57.

[258] D. Guo and X. Liu, Multiple positive solutions of boundary value problems for impulsive differential equations, *Nonlinear Anal.* **25** (1995), 327–337.

[259] M. Guo, X. Xue, and R. Li, Controllability of impulsive evolutions inclusions with nonlocal conditions, *J. Optim. Theory Appl.* **120** (2004), 355–374.

[260] H. G. Guseĭnov, A. I. Subbotin, and V. N. Ushakov, Derivatives for multivalued mappings with applications to game-theoretical problems of control, *Problems Control Inform. Theory* **14** (1985), 155–167.

[261] H. G. Guseĭnov, and V. N. Ushakov, Strongly and weakly invariant sets with respect to a differential inclusion, *Differentsial'nye Uravneniya* **26** (1990), 1888–1894.

[262] G. Haddad, Monotone trajectories of differential inclusions and functional-differential inclusions with memory, *Israel J. Math.* **39** (1981), 83–100.

[263] G. Haddad, Monotone viable trajectories for functional-differential inclusions, *J. Differential Equations* **42** (1981), 1–24.

[264] G. Haddad and J. M. Lasry, Periodic solutions of functional differential inclusions and fixed points of σ-selectionable correspondences, *J. Math. Anal. Appl.* **96** (1983), 295–312.

[265] J. Hale and J. Kato, Phase space for retarded equations with infinite delay, *Funkcial. Ekvac* **21** (1978), 11–41.

[266] A. Halanay and D. Wexler,*Teoria calitativa a systeme cu impulduri*, Editura Republicii Socialiste Romania, Bucharest, 1968.

[267] J. Hale, *Ordinary Differential Equations*, Pure and Applied Mathematics, John Wiley and Sons, New York-London-Sydney, 1969.

[268] J. K. Hale and S. M. Verduyn Lunel, *Introduction to Functional Differential Equations*, Applied Mathematical Sciences. 99, Springer-Verlag, New York, 1993.

[269] N. Halidias and N. Papageorgiou, Existence and relaxation results for nonlinear second order multivalued boundary value problems in $\mathbb{R}^n$, *J. Differential Equations* **147** (1998), 123–154.

[270] P. Hartman and A. Wintner, On the non-increasing solutions of $y'' = f(x, y, y')$, *Amer. J. Math.* **73** (1951), 390–404.

[271] S. Heikkila and V. Lakshmikantham, *Monotone Iterative Techniques for Discontinuous Nonlinear Differential Equations*, Marcel Dekker, New York, 1994.

[272] M. Helal and A. Ouahab, Existence and solution sets of impulsive functional differential inclusions with multiple delay, *Opuscula Math.* **32** (2012), 249–283.

[273] J. Henderson, *Boundary Value Problems for Functional Differential Equations*, World Scientific, Singapore, 1995.

[274] J. Henderson and A. Ouahab, On solution sets for first order impulsive neutral functional differential inclusions in Banach spaces, *Dyn. Contin. Discrete Impuls. Syst. Ser. A Math. Anal.* **18** (2011), 53–114.

[275] J. Henderson and A. Ouahab, Impulsive version of Filippov's theorem and the Filippov-Wazewski theorem for second order impulsive semilinear functional differential inclusions, *Int. J. Mod. Math.* **3** (2008), 111–133.

[276] J. Henderson and A. Ouahab, Controllability of nondensely defined impulsive functional semilinear differential inclusions in Fréchet spaces, *Int. J. Appl. Math. Stat.* **9** (2007), 35–54.

[277] J. Henderson and A. Ouahab, Impulsive hyperbolic differential inclusions with infinite delay and variable moments, *Comm. Appl. Nonlinear Anal.* **13** (2006), 61–78.

[278] J. Henderson and A. Ouahab, Extrapolation spaces and controllability of impulsive semilinear functional differential inclusions with infinite delay in Fréchet spaces, *Appl. Anal.* **85** (2006), 1255–1270.

[279] J. Henderson and A. Ouahab, Local and global existence and uniqueness results for second and higher Order impulsive functional differential equations with infinite delay, *Australian J. Appl. Math.* **4** (2007), 1–26.

[280] J. Henderson and A. Ouahab, Global existence results for impulsive functional differential inclusions with multiple delay in Fréchet spaces, *PanAmerican Math. J.* **15** (2005), 73–89.

[281] J. Henderson and A. Ouahab, Impulsive hyperbolic differential inclusions with infinite delay, *Commun. Appl. Anal.* **13** (2006), 49–67.

[282] J. Henderson and A. Ouahab, Impulsive hyperbolic differential inclusions with infinite delay and variable moments, *Commun. Appl. Anal.* **13** (2006).

[283] J. Henderson and A. Ouahab, A Filippov's theorem, some existence results and the compactness of solution sets of impulsive fractional order differential inclusions, *Mediterr. J. Math.* **9** (2012), 453–485.

[284] J. Henderson and A. Ouahab, Impulsive differential inclusions with fractional order, *Comput. Math. Appl.* **59** (2010), 1191–1226.

[285] J. Henderson, A. Ouahab, S. Youcefi, Existence and topological structure of solution sets for phi-Laplacian impulsive differential equations, *Electron. J. Differ. Equ.* (2012), Paper No. 56, 16 p.

[286] H. R. Henríquez and C. Lizama, Compact almost automorphic solutions to integral equations with infinite delay, *Nonlinear Anal.* **71** (2009), 6029–6037.

[287] E. Hernández, A comment on the papers: "Controllability results for functional semilinear differential inclusions in Fréchet spaces," [*Nonlinear Anal.* **61** (2005), 405–423] by M. Benchohra and A. Ouahab, and "Controllability of impulsive neutral functional differential inclusions with infinite delay" [*Nonlinear Anal.* **60**(2005), 1533–1552] by B. Liu, Nonlinear Anal. **66** (2007), 2243–2245.

[288] F. Hiai and H. Umegaki, Integrals, conditional expectations, and martingales of multivalued functions, *J. Multivariate Anal.* **7**(1977), 149–182.

[289] T. Hida, *Analysis of Brownian Functional*, Carleton Mathematical Lecture Notes, 13, Carleton Univ., Ottawa, Ont., 1975.

[290] T. Hida, H. H. Kuo, J. Potthoff, and L. Streit, *Wite Noise : An Infinite Dimensional Calculus*, Kluwer Academic Publishers, 1993.

[291] C. Himmelberg and F. Van Vleck, One the topological triviality of solution sets, *Rocky Mountain J. Math.* **10** (1980), 621–252.

[292] Y. Hino, S. Murakani, and T. Naito, *Functional Differential Equations with Infinite Delay*, in: Lecture Notes in Mathematics, **1473**, Springer-Verlag, Berlin, 1991.

[293] A. L. Hodgkin and A. F. Huxley, A quantitative description of membrane current and its application to conduction and excitation in nerve, *J. Physiol.* **117** (1952), 467–472.

[294] S. Hong, The method of upper and lower solutions for nth order nonlinear impulsive differential inclusions, *Dyn. Contin. Discrete Impuls. Syst. Ser. A Math. Anal.* **14** (2007), 739–753.

[295] S. Hong, Existence of solutions to initial value problems for the second order mixed monotone type of impulsive differential inclusions, *J. Math. Kyoto Univ.* **45** (2005), 329–341.

[296] S. Hong and M. Zhang, Comparison results for initial value problems of second-order impulsive integro-differential inclusions, *Nonlinear Anal.* **74** (2011), 67–80.

[297] J. Hu and X. Liu, Existence results of second-order impulsive neutral functional integrodifferential inclusions with unbounded delay in Banach spaces, *Math. Comput. Modelling* **49** (2009), 516–526.

[298] M. Hukuhara, Sur les systémes des équations differentielles ordinaires, *Japan J. Math.* **5** (1928), 345–350.

[299] J. Hu and Y. Shen, Existence theory of impulsive evolution neutral functional integrodifferential inclusions with unbounded delay in Banach spaces, *Math. Appl. (Wuhan)* **20** (2007), 568–573.

[300] J. Hu, Y. Shen and X. Liao, Nonresonance nth-order impulsive functional differential inclusions, (Chinese) *J. Huazhong Norm. Univ. Nat. Sci.* **42** (2008), 19–24.

[301] Sh. Hu and N. Papageorgiou, *Handbook of Multivalued Analysis, Volume I: Theory*, Kluwer, Dordrecht, 1997.

[302] Sh. Hu and N. Papageorgiou, Delay differential inclusions with constraints, *Proc. Amer. Math. Soc.* **123** (1995), 2141–2150.

[303] Sh. Hu and N. Papageorgiou, On the topological regularity of the solution set of differential inclusions with constraints, *J. Differential Equations* **107** (1994), 280–289.

[304] S. Hu and V. Lakshmiknthm, PBVP for second order impulsive differential systems, *Nonlinear Anal.* **13** (1989), 75–85.

[305] S. T. Hu, *Theory of Retracts, Detroit, Wayne State University Press*, 1965.

[306] S. T. Hu, *Cohomology Theory, Markham, Chicago*, 1968.

[307] D. M. Hyman, On decreasing sequeness of compact absoulte retracts, *Fund. Math.* **64** (1969), 91–97.

[308] K. Itô and H. P. McKean, Jr., *Diffusion Processes and their Sample Paths*, reprint, of the 1975 edition, Classics in Mathematics Series, Springer, 1996.

[309] J. Jarnik and J. Kurzweil, On conditions on right hand sides of differential relations, *Casopis Pest. Math.* **102** (1977), 334–349.

[310] S. Ji and G. Li, Existence results for impulsive differential inclusions with nonlocal condition, *Comput. Math. Appl.* **62** (2011), 1908–1915.

[311] M. Kamenskii, V. Obukhovskii, and P. Zecca, *Condensing Multivalued Maps and Semilinear Differential Inclusions in Banach Spaces*, Walter de Gruyter & Co., Berlin, 2001.

[312] K. Karthikeyan and A. Anguraj, Solvability of impulsive neutral functional integro-differential inclusions with state dependent delay, *J. Appl. Math. Inform.* **30** (2012), 57–69.

[313] Z. Kamont and S. Koziel, Mixed problems for hyperbolic functional differential equations with unbounded delay, *Nonlin. Anal.* **58** (2004), 489–515.

[314] D. Kandilakis and N. Papageorgiou, Existence theormes for nonlinear boundary value problems for second order differential inclusions, *J. Differential Equations* **132** (1992), 107–125.

[315] R. Kannan and D. O'Regan, A note on the solutions set integral inclusions, *J. Integral Equations Appl.* **12** (2000), 85–94.

[316] F. Kappel and W. Schappacher, Some Considerations to the fundamental theory of infinite delay equations, *J. Differential Equations* **37** (1980), 141–183.

[317] S. K. Kaul and X. Z. Liu, Vector Lyapunov functions for impulsive differential systems with variable times, *Dynam. Contin. Discrete Impuls. Systems* **6** (1999), 25–38.

[318] S. K. Kaul and X. Z. Liu, Impulsive integro-differential equations with variable times, *Nonlinear Studies* **8** (2001), 21–32.

[319] H. Kellermann and M. Hieber, Integrated semigroup, *J. Funct. Anal.* **84** (1989), 160–180.

[320] I. T. Kiguradze and B. L. Shekhter, Boundary value problems for the second-order ordinary differential equations, *Itogi Nauki i Tekh., Ser. Sovrem. Probl. Mat.* **30**, VINITI, Moscow, (1987), 105–201.; English transl. *J. Soviet Math.*, **43** (1988), 2340–2417.

[321] J. F. C. Kingman, *Poisson Processes*, Oxford, 1993.

[322] M. Kisielewicz, *Differential Inclusions and Optimal Control*, Kluwer, Dordrecht, The Netherlands, 1991.

[323] H. Kneser, Uber die Lösungen eine system gewöhnlicher differential Gleichungen, das der Lipschitzchen Bedingung nicht genügt, *S. B. Preuss. Akad. Wiss. Phys. Math. Kl.*, (1923), 171–174.

[324] A. Kneser, Unteruchung und asymptotische darstellung der integrale gewisser differentialgleichungen bei grossen werthen des arguments, *J. Reine Angen. Math.* **116** (1896), 178–212.

[325] M. A. Krasnosel'skii, *Topological Methods in the Theory of Nonlinear Integral Equations*, MacMillan Company, New York, 1964.

[326] M. A. Krasnosell'ski, *Translation Operator Along the Trajectories of Differential Equations*, Nauka, Moscow, 1966. (Russian).

[327] E. Kruger-Thiemr, Formal theory of drug dosage regiments. I. *J. Theoret. Biol.* **13** (1966), 212–235.

[328] E. Kruger-Thiemr, Formal theory of drug dosage regiments. II. *J. Theoret. Biol.* **23** (1969), 169–190.

[329] N. V. Krylov, *Introduction to the Theory of Random Processes*, Amer. Math. Soc. Vol 43, 2002.

[330] W. Kryszewski and S. Plaskacz, Periodic solutions to impulsive differential inclusions with constraints, *Nonlinear Anal.* **65** (2006), 1794–1804.

[331] S. Kyritsi, N. Matzakos and N. S. Papageorgiou, Nonlinear boundary value problems for second order differential inclusions, *Czechoslovak Math. J.* **55** (2005), 545–579.

[332] V. Kolmanovskii and A. Myshkis, *Introduction to the Theory and Applications of Functional Differential Equations*, Kluwer Academic Publishers, Dordrecht, 1999.

[333] V. B. Kolmanovskii and A. R. Nosov, *Stability of Functional Dfferential Equations*, Mathematics in Science and Engineering, 180, Academic Press, Inc. [Harcourt Brace Jovanovich, Publishers], London, 1986.

[334] S. K. Kaul, V. Lakshmikantham and S. Leela, Extremal solutions, comparison principle and stability criteria for impulsive differential equations with variable times, *Nonlinear Anal.* **22** (1994), 1263–1270.

[335] V. Lakshmikantham, D. D. Bainov and P. S. Simeonov, *Theory of Impulsive Differential Equations*, World Scientific, Singapore, 1989.

[336] V. Lakshmikantham, S. Leela, and S. K. Kaul, Comparison principle for impulsive differential equations with variable times and stability theory, *Nonlinear Anal.* **22** (1994), 499–503.

[337] V. Lakshmikantham, S. Leela, and J. Vasundhara Devi, *Theory of Fractional Dynamic Systems*, Cambridge Scientific Publishers, 2009.

[338] V. Lakshmikantham and S. G. Pandit, The Method of upper, lower solutions and hyperbolic partial differential equations, *J. Math. Anal. Appl.* **105** (1985), 466–477.

[339] V. Lakshmikantham, N. S. Papageorgiou, and J. Vasundhara, The method of upper and lower solutions and monotone technique for impulsive differential equations with variable moments, *Appl. Anal.* **15** (1993), 41–58.

[340] V. Lakshmikantham, L. Wen, and B. Zhang, *Theory of Differential Equations with Unbounded Delay*, Mathematics and its Applications, Kluwer Academic Publishers, Dordrecht, 1994.

[341] J. M. Larsry and R. Robert, *Analyse Non-Lineaire Multivoque*, Publication No. 7611, Centre de Recherche de Mathémtique de la Decision, Universites de Pairs, Dauphine, pp 1–190.

[342] A. Lasota and Z. Opial, An application of the Kakutani-Ky Fan theorem in the theory of ordinary differential equations, *Bull. Acad. Pol. Sci. Ser. Sci. Math. Astronom. Phys.* **13** (1965), 781–786.

[343] A. Lasota and J. A. Yorke, The generic property of existence o,f solutions of differential equations in Banach spaces, *J. Diff. Equs.*, **13**, (1973), 1–12.

[344] G. Li and X. Xue, Controllability of evolution inclusions with nonlocal conditions, *Appl. Math. Comput.* **141** (2003), 375–384.

[345] W. Li, Y. Chang and J. Nieto, Solvability of impulsive neutral evolution differential inclusions with state-dependent delay, *Math. Comput. Modelling* **49** (2009), 1920–1927.

[346] A. Lin and L. Hu, Existence results for impulsive neutral stochastic functional integro-differential inclusions with nonlocal initial conditions, *Comput. Math. Appl.* **59** (2010), 64–73.

[347] B. Liu, Controllability of impulsive neutral functional differential inclusions with infinite delay, *Nonlinear Anal.* **60** (2005), 1533–1552.

[348] K. Liu, Carathéodory approximate solutions for a class of semilinear stochastic integral equations, *J. Math. Anal. Appl.* **220** (1998), 349–364.

[349] X. Liu, and S. Zhang, A cell population model described by impulsive PDEs-existence and numerical approximation, *Comput. Math. Applic.* **36** (8) (1998), 1–11.

[350] Y. Liu, Boundary value problems for second order differential equations on unbounded domain in a Banach space, *Appl. Math. Comput.* **135** (2003), 569–583.

[351] Y. Liu, Boundary value problems on half-line for functional differential equations with infinite delay in a Banach space, *Nonlinear Anal.* **52** (2003), 1695–1708.

[352] Y. Liu and Z. Li, Second order impulsive neutral functional differential inclusions, *Kyungpook Math. J.* **48** (2008), 1–14.

[353] Y. Liu, J. Wu, and Z. Li, Impulsive boundary value problems for dynamical inclusions on time scales, *Dyn. Contin. Discrete Impuls. Syst. Ser. A Math. Anal.* **15** (2008), 381–389.

[354] Y. Liu, J. Wu, Z. Li, Impulsive boundary value problems for Sturm-Liouville type differential inclusions, *J. Syst. Sci. Complex.* **20** (2007), 370–380.

[355] Y. Liu, J. Wu, and Z. Li, Impulsive boundary value problems for first-order ordinary differential inclusions, *Acta Math. Appl. Sin. Engl. Ser.* **23** (2007), 411–420.

[356] E. Liz and J.,J. Nieto, Positive solutions of linear impulsive differential equations, *Commun. Appl. Anal.* **2** (1998), 565–571.

[357] C. Lizama, Regularized solutions for abstract Volterra equations, *J. Math. Anal. Appl.* **243**, (2000), 278–292.

[358] C. Lizama, On an extension of the Trotter-Kato theorem for resolvent families of operators, *J. Integral Equations Appl.* **2** (1990), 269–280.

[359] C. Lizama, A characterization of periodic resolvent operators, *Results Math.* **18** (1990), 93–105.

[360] C. Lizama, Uniform stability of resolvent families, *Amer. Math. Soc.*, **132**, (2003), 175–181.

[361] C. Lizama, On Volterra equations associated with a linear operator, *Amer. Math. Soc.*, **118** (1993), 1159–1166.

[362] C. Lizama, On multiplicative perturbation of integral resolvent families, *J. Math. Anal. Appl.*, **327** (2007), 1335–1359.

[363] C. Lizama and J. Sánchez, On perturbation of k-regularized resolvent families, *Taiwan. J. Math.*, **7**, (2003), 217–227.

[364] A. Mambriani, Su un teoreme relativo alle equazioni differenziali ordinarie del 2^0 ordine, *Atti Accad. Naz Lincei, Cl. Sci. Fis. Mat. Nat.* **9** (1929), 620–622.

[365] W. S. Massey, *Homology and cohomology theory. An approach based on Alexander–Spanier cochains, Monographs and Textbooks in Pure and Applied Mathematics*, **46**, Marcel Dekker, New York–Basel, 1978.

[366] G. Marino, P. Pielramala, and L. Muglia, Impulsive neutral integrodifferential equations on unbounded intervals, *Mediterr. J. Math.* **1** (2004), 3–42.

[367] D. R. Marques and M. D. P. Monteiro Manuel, A sweeping process approach to inelastic contact problems with general inertia operators, *Eur. J. Mech., A, Solids* **26** (2007), 474–490.

[368] I. V. Melnikova and A. Filinkov, *Abstract Cauchy Problems: Three Approaches* , Chapman & Hall/Crc Boca Raton London New York Washington, D.C. 2001.

[369] M. Matos and D. Pereira, On a hyperbolic equation with strong damping, *Funkcial. Ekvac.* **34** (1991), 303–311.

[370] K. G. Mavridi and P. Ch. Tsamatos, Positive solutions for first order differential nonlinear functional boundary value problems on infinite intervals, *Electron. J. Qual. Theory Differ. Equ.*, (2004), No. **8**, 1–18.

[371] K. G. Mavridis and P. Ch. Tsamatos, Positive solutions for a Floquet functional boundary value problem. *J. Math. Anal. Appl.* **296** (2004), 165–182.

[372] M. A. McKibben, Second-order damped functional stochastic evolution equations in Hilbert space, *Dynam. Systems Appl.* **12** (2003), 467–487.

[373] R. E. Megginson, *An Introduction to Banach Space Theory*, Graduate Texts in Math., **183**, Springer Verlag, New York, 1998.

[374] S. Migórski and A. Ochal, Nonlinear impulsive evolution inclusions of second order, *Dynam. Systems Appl.* **16** (2007), 155–173.

[375] V. D. Milman and A. A. Myshkis, On the stability of motion in the presence of impulses, *Sib. Math. J.* **1** (1960), 233–237, (Russian).

[376] V. D. Milman and A. A. Myshkis, Randorn impulses in linear dynamical systems, in "Approximante Methods for Solving Differential Equations," Publishing house of the Academy of Sciences of Ukainian SSR, Kiev, in Russian, (1963) 64–81.

[377] M. D. P. Monteiro Marques, *Differential Inclusions in Nonsmooth Mechanical Problems-Shockes and Dry Friction*, Birkhäuser, Boston, 1993.

[378] J. J. Moreau, Rale par un convexe variable (premiére partie), exposé no 15. Séminaire d'analyse convexe, University of Montpellier, 43 pages, 1971.

[379] J. J. Moreau, Rale par un convexe variable (deuxiéme partie) exposé no 3. Séminaire d'analyse convexe, University of Montpellier, 36 pages, 1972.

[380] J. J. Moreau. Probléme d'évolution associé á un convexe mobile d'un espace hilbertien. *C. R. Acad. Sc. Paris*, Série A-B, (1973), 791–794.

[381] J. J. Moreau, Evolution problem associated with a moving convex set in a Hilbert space. *J. Differential Equations* **26** (1977), 347–374.

[382] J. J. Moreau, Liaisons unilatérales sans frottement et chocs inélastiques. *C. R. Acad. Sci. Paris, Serie II* **296**, (1983), 1473–1476.

[383] J. J. Moreau, Dynamique de systémes á liaisons unilatérales avec frottement sec éventuel; essais numériques. Note Technique n. 85-1, LMGC, USTL, Montpellier, 1985.

[384] J. Munkres, *Topology*, Second Edition, Prentice-Hall, 2000.

[385] J. Musielak, *Introduction to Functional Analysis*,(in Polish). PWN, Warszawa, 1976.

[386] M. Nagumo, Über die Lage der Integralkurven gewöhnlicher Differentialgleichungen, *Proc. Phys.-Math. Soc. Japan*, **III.** Ser. 24, (1942), 551–559.

[387] F. Neubrander, Integrated semigroups and their applications to the abstract Cauchy problems, *Pacific J. Math.* **135** (1988), 111–155.

[388] J. J. Nieto, Aronszajn's theorem for some nonlinear Dirichlet problem, *Proc. Edinburg Math. Soc.* **31** (1988), 345–351.

[389] J. J. Nieto, Impulsive resonance periodic problems of first order, *Appl. Math. Lett.* **15** (2002), 489–493.

[390] J. J. Nieto, Periodic boundary value problems for first-order impulsive ordinary differential equations, *Nonlinear Anal.* **51** (2002), 1223–1232.

[391] J. J. Nieto, Basic theory for nonresonance impulsive periodic problems of first order, *J. Math. Anal. Appl.* **205** (1997), 423–433.

[392] H. Ning and B. Liu, Existence results for impulsive neutral stochastic evolution inclusions in Hilbert space, *Acta Math. Sin. (Engl. Ser.)* **27** (2011), 1405–1418.

[393] S. Ntouyas, Neumann boundary value problems for impulsive differential inclusions, *Electron. J. Qual. Theory Differ. Equ.* **2009** (2009), 13 pp.

[394] S. Ntouyas, Boundary value problems for first order impulsive differential inclusions, *Commun. Math. Anal.* **7** (2009), 37–49.

[395] V. Obukhovskii and J. Yao, On impulsive functional differential inclusions with Hille-Yosida operators in Banach spaces, *Nonlinear Anal.* **73** (2010), 1715–1728.

[396] A. Ouahab, *Some Contributions in Impulsive differential equations and inclusions with fixed and variable times, PhD Dissertation,* University of Sidi-Bel-Abbés (Algeria), 2006.

[397] A. Ouahab, Local and global existence and uniqueness results for impulsive differential equtions with multiple delay, *J. Math. Anal. Appl.* **323** (2006), 456–472.

[398] A. Ouahab, Existence and uniqueness results for impulsive functional differential equations with scalar multiple Delay and infinite delay, *Nonlin. Anal.* **67** (2006), 1027–1041.

[399] A. Ouahab, Filippov's theorem for impulsive differential inclusions with fractional order, *Electron. J. Qual. Theory Differ. Equ.* **2009** (2009), 23 pp.

[400] H. Oka, Linear Volterra equations and integrated solution families, *Semigroup Forum* **53** (1996), 278–297.

[401] D. O'Regan, Integral inclusions of upper semicontinuous or lower semicontinuous type, *Proc. Amer. Math. Soc.* **24** (1996), 2391–2399.

[402] B. G. Pachpatte, On some integral Inequalities and applications, *J. Inequal. Pure Appl. Math.* **3** (2002), 1–7.

[403] N. S. Papageorgiou, Topological properties of the solution set of a class nonlinear evolutions inclusions, *Czechoslovak Math. J.* **47 (122)** (1997), 409–424.

[404] N. S. Papageorgiou, Existence of solutions for hyperbolic differential inclusions in Banach spaces, *Arch. Math.,* (Brno) **28** (1992), 205–213.

[405] N. S. Papageorgiou, Convergence theorems for Banach space valued integrable multifunctions, *Internat. J. Math. Math. Sci.* **12** (1987), 433–442.

[406] N. S. Papageorgiou, On the structure of the solution set of evolution inclusions with time-dependent subdifferentials, *Rend. Sem. Mat. Univol. Padova*, **97**, (1997), 163–186.

[407] N. S. Papageorgiou and S.Th. Kyritsi-Yiallourou, *Handbook of Applied Analysis, Adv. in Mech. and Math.*, **19**, Springer, New York, 2009.

[408] J. Park, S. Park, and Y. Kang, Controllability of second-order impulsive neutral functional differential inclusions in Banach spaces, *Math. Methods Appl. Sci.* **33** (2010), 249–262.

[409] S. K. Patcheu, On the glabal solution and asymptotic behaviour for the generalized damped extensible beam equation, *J. Differential Equations* **135** (1997), 299–314.

[410] A. Pazy, *Semigroups of Linear Operators and Applications to Partial Differential Equations*, Springer-Verlag, New York, 1983.

[411] G. Peano, Démonstration de l'integrabilite des équations differentielles ordinaires, Mat. Annalen **37** (1890), 182–238.

[412] A. Petruşel, *Operatorial Inclusions,* House of the Book of Science, Cluj-Napoca, 2002.

[413] N. A. Perestyuk, V. A. Plotnikov, A. M. Samoilenko and N. V. Skripnik, *Differential Equations with Impulse Effects Multivalued Right-hand Sides with Discontinuities,* Walter de Gruyter & Co., Berlin, 2011.

[414] C. Pierson-Gorez, *Problémes aux Limites Pour des Equations Différentielles avec Impulsions,* Ph.D. Thesis, Univ. Louvain-la-Neuve, 1993 (in French).

[415] R. Pietkun, Periodic solutions to impulsive functional differential inclusions with constraints, *Nonlinear Anal. Forum* **14** (2009), 175–190.

[416] S. Plaskacz, Periodic solutions of differential inclusions on compact subsets of $\mathbb{R}^n$, *J. Math. Anal. Appl.* **148** (1990), 202–212.

[417] S. Plaskacz, On the solutions of differential inclusions, *Boll. Un. Mat. Ital.* **7** (1992), 387–394.

[418] A. Pietsch, *Nuclear locally convex spaces*, Springer-Verlag, Berlin 1972.

[419] J. Prüss, *Evolutionary Integral Equations and Applications* Monographs Mayh. **87**, Bikhäuser Verlag, 1993.

[420] I. Podlubny, *Fractional Differential Equation*, Academic Press, San Diego, 1999.

[421] B. Przeradzki, On a two-point boundary value problem for differential equations on the half-line, *Ann. Polon. Math.* **50** (1989), 53–61.

[422] T. Pruszko, *Some applications of the topological degree theory to multi-valued boundary value problems, Dissertationes Math.* **230** (1984), 1–50.

[423] N. Plotnikova, Averaging of impulsive differential inclusions, *(Russian) Mat. Stud.* **23** (2005), 52–56.

[424] M. D. Quinn and N. Carmichael, An approach to nonlinear control problem using fixed point methods, degree theory, pseudo-inverses, *Numer. Funct. Anal. Optim.* **7** (1984–85), 197–219.

[425] I. Rachunková, On a Kneser problem for a systems of nonlinear ordinary differential equations, *Czech. Math. J.* **31** (1981), 114–126.

[426] I. Rachunková, On Kneser's problem for differential equations of the 3rd order, *Čas. Pěst. Mat.* **115** (1990), 18–27.

[427] I. Rachunkov'a, Nonnegative nonincreasing solutions of differential equations of the 3rd order, *Czech. Math. J.* **40** (1990), 213–221.

[428] I. Rachunková, Singular Dirichlet second-order BVPs with impulses, *J. Differential Equations* **193** (2003), 435–459.

[429] Y. Ren, L. Hu, and R. Sakthivel, Controllability of impulsive neutral stochastic functional differential inclusions with infinite delay, *J. Comput. Appl. Math.* **235** (2011), 2603–2614.

[430] D. Revuz and M. Yor, *Continuos Martingales and Brownian Motion*, third edition, Springer, New York, 1999.

[431] B. Ricceri, Une propriété topologique de l'ensemble des points fixe d'une contraction moultivoque á valeurs convexes, *Atti Accad. Naz. Linei Cl. Sci. Fis. Mat. Natur. Rend. Lincei (9) Mat. Appl.* (1987) 283–286.

[432] S. P. Rogovchenko and Yu. V. Rogovchenko, Periodic solutions of a weakly nonlinear hyperbolic impulse system. (Russian) Asymptotic integration of nonlinear equations, *Akad. Nauk Ukrainy, Inst. Mat., Kiev*, (Russian) (1992), 97–103.

[433] B. Rzepecki, Scorza-Dragoni theorems for upper semicontinous multivalued functions, *Bull. Acad. Polon. Sci* (1980), 61–67.

[434] R. Sakthivel, E. Anandhi and S. Lee, Approximate controllability of impulsive differential inclusions with nonlocal conditions, *Dynam. Systems Appl.* **18** (2009), 637–653.

[435] A. M. Samoilenko and N. A. Perestyuk, *Impulsive Differential Equations*, World Scientific, Singapore, 1995.

[436] B. Satco, Solution tube method for impulsive periodic differential inclusions of first order, *Nonlinear Anal.* **75** (2012), 260–269.

[437] K. Schumacher, Existence and continuous dependence for differential equations with unbounded delay, *Arch. Rational Mech. Anal.* **64** (1978), 315–335.

[438] V. Šeda, On a generalization of the Thomas-Fermi equation, *Acta Math. Univ. Comen.* **39** (1980), 97–114.

[439] A. Sghir, On the solution set of second order delay differential inclusions in Banach space, *Ann. Math. Blaise Pascal* **7** (2000), 65–79.

[440] T. Shimokawa, K. Pakdaman, S. Satos, Time-scale matching in the response, of a leaky integale-and-fire neuron model to periodic stimulation with additive noise, *Physical Review E* **59** (1999), 3427–3443.

[441] J. S. Shin, An existence of functional differential equations, *Arch. Rational Mech. Anal.* **30** (1987), 19–29.

[442] N. Skripnik, Full averaging of fuzzy impulsive differential inclusions, *Surv. Math. Appl.* **5** (2010), 247–263.

[443] N. V. Skripnik, Averaging of impulsive differential inclusions with the Hukuhara derivative, *(Russian) Neliniini Koliv* **10** (2007), 416–432; *translation in Nonlinear Oscil.* **10** (2007), 422–438.

[444] N. Skripnik, The partial averaging of fuzzy impulsive differential inclusions, *Differential Integral Equations* **24** (2011), 743–758.

[445] G. V. Smirnov, *Introduction to the Theory of Differential Inclusions,* Graduate Studies in Mathematics 41, American Mathematical Society, Providence, 2002.

[446] E. H. Spanier, *Algebraic Topology, McGraw-Hill New York,* 1966.

[447] H. Sobczyk, *Stochastic Differential Equations with Applications to Physics and Engineering*, Klüwer Academic Publisher, London, 1991.

[448] R. Srzednicki, Periodic and bounded solutions in blocks for time-periodic nonautonomus ordinary differential equations, *Nonlinear Anal.* **6** (1994), 707–737.

[449] V. A. Staikos, Differential equations with deviating arguments-oscillation theory, *Unpublished manuscripts.*

[450] G. T. Stamov and I. M. Stamova, Second method of Lyapunov and existence of integral manifolds for impulsive differential equations, *SUT J. Math.* **32** (1996), 101–107.

[451] S. Szufla, Sets of fixed points of nonlinear mappings, *Funkcial. Ekvac.* **22** (1979), 121–126.

[452] S. Szufla, Some remarks on ordinary differential equations in a Banach spaces, *Bull. Acad. Polon. Math.* **16** (1968), 795–800.

[453] S. Szufla, Structure of the solutions set of ordinary differential equations, *Bull. Acad. Polon. Math. Sci.* **21** (1973), 141–144.

[454] S. Szufla, Solutions sets of nonlinear equations, *Bull. Acad. Polon. Math. Sci.* **21** (1973), 971–976.

[455] S. Szufla, Some properties of the solutions set of ordinary differential equations in a Banach spaces, *Bull. Acad. Polon. Math. Sci* **22** (1974), 675–678.

[456] S. Szufla, On the structure of solutions of differential and integral equations in Banach spaces, *Ann. Polon. Math.* **34** (1977), 165–177.

[457] S. Szufla, On the equation $x' = f(t,x)$ in Banach spaces, *Bull. Acad. Polon. Math. Sci.* **22** (1978), 407–413.

[458] S. Szufla, On the existence of solutions of differential equations in Banach spaces, *Bull. Acad. Polon. Math. Sci.* **26** (1982), 507–515.

[459] S. Szufla, On the equation $x' = f(t,x)$ in locally convex spaces, *Math. Nachr.* **118** (1985), 175–185.

[460] S. Tang and R. A. Cheke, State-dependent impulsive models of integrated pest management(IPM) stratgies and their dynamic consequences, *J. Math. Biol.* **50** (2005), 257–292.

[461] T. Taniguchi, Successive approximations to solutions of stochastic differential equations, *J. Differential Equations* **96** (1992), 152–169.

[462] L. Thibault, Sweeping process with regular and nonregular sets, *J. Differential Equations* **193** (2003), 1–26.

[463] H. Thieme, Integrated semigroups and integrated solutions to abstract Cauchy problems, *J. Math. Anal. Appl.* **152** (1990), 416–447.

[464] Y. Tian and J. Henderson, Three anti-periodic solutions for second-order impulsive differential inclusions via nonsmooth critical point theory, *Nonlinear Anal.* **75** (2012), 6496–6505.

[465] A. A. Tolstonogov, *Differential Inclusions in Banach Space,* Kluwer Academic Publishers, Dordrecht, 2000.

[466] A. A. Tolstonogov, On differential inclusions in a Banach space, Dokl. Akad. Nauk SSSR **244** (1988), 461–488.

[467] A. A. Tolstonogov, On properties of solutions of differential inclusions in a Banach spaces, *Math. Sbornik* **46** (1983), 1–15 (in Russian).

[468] C. C. Travis and G. F. Webb, Second order differential equations in Banach spaces, Proc. Int. Symp. on Nonlinear Equations in Abstract Spaces, Academic Press, New York (1978), 331–361.

[469] C. C. Travis and G. F. Webb, Cosine families and abstract nonlinear second order differential equations, *Acta Math. Hung.* **32** (1978), 75–96.

[470] C. P. Tsokos and W. J. Padgett, *Random Integral Equations with Applications to Life Sciences and Engineering*, Academic Press, New York, 1974.

[471] N. N. Vakhania, *Probability Distribution in Linear Spaces.* (Russian) Met-sniereba, Tbilisi 1971; English translation: North Holland, 1981.

[472] M. Valadier, Quelques problèmes déntrainement unilatéral en dimension finie, *Sém. Anal. Convexe*, Montpellier Exposé No. 8, (1988).

[473] A. S. Vatsala, and Y. Sun, Periodic boundary value problems of impulsive differential equations, *Appl. Anal.* **44** (1992), 145–158.

[474] G. Vidossich, On Peano-phenomenon, *Bull. Un. Math. Ital.* **3** (1970), 33–42.

[475] G. Vidossich, On the structure of the set of solutions of nonlinear equations, *J. Math. Anal. Appl.* **34** (1971), 602–617.

[476] A. Vinodkumar and A. Anguraj, Existence of random impulsive abstract neutral non-autonomous differential inclusions with delays, *Nonlinear Anal. Hybrid Syst.* **5** (2011), 413–426.

[477] I. I. Vrabie, *C_0-Semigroups and Applications*, Elsiver, 2003.

[478] H. Wang, Existence results for fractional functional differential equations and impulses, *J. Appl. Math. Comp.* **38** (2012), 85–101.

[479] L. Wei and J. Zhu, A second order multivalued boundary-value problem and impulsive neutral functional differential inclusions in Banach spaces *Neliniini Koliv.* **11** (2008), 191–207; *translation in Nonlinear Oscil.* **11** (2008), 200–218

[480] P. Weng, Global existence in integrable space for impulsive FDE with P-Delay, *Dynam. Contin. Discrete Impulse. Systems* **9** (2002), 321–337.

[481] Pui-Kei Wong, Existence and asymptotic behavior of proper solutions of a class of second order nonlinear differential equations, *Pacific J. Math.* **13** (1963), 737–760.

[482] X. Xiang, Y. Peng, and W. Wei, Solution map of strongly nonlinear impulsive evolution inclusions. Control theory and related topics, 338–353, World Sci. Publ., Hackensack, NJ, 2007.

[483] X. Xiang and W. Wei, Mild solution for a class of nonlinear impulsive evolution inclusions on Banach space, *Southeast Asian Bull. Math.* **30** (2006), 367–376.

[484] J. Xiao, X. Zhu, and R. Cheng, The solution sets for second order semilinear impulsive boundary value problems, *Comp. Math. Appl.* **64** (2012), 147–160.

[485] J. Xie, G. Chen and X. He, Integral boundary value problems for first order impulsive differential inclusions, *Int. J. Math. Anal.(Ruse)* **3** (2009), 581–590.

[486] B. Yan, On L^p_{loc}-solutions of nonlinear impulsive Volterra integral equations in Banach spaces, *SUT J. Math.* **33** (1997), 121–137.

[487] B. Yan, The existence of positive solutions of nonlinear impulsive Fredholm integral equations in Banach spaces, *Dynam. Contin. Discrete Impuls. Systems* **6** (1999), 289–300.

[488] B. Yan and X. Liu, Multiple solutions of impulsive boundary value problems on the half-line in Banach spaces, *SUT J. Math.*, **36** (2000), 167–183.

[489] G. Ye, J. Shen, and J. Li, Existence results for mth-order impulsive functional differential inclusions, *Indag. Math. (N. S.)* **22** (2011), 1–11.

[490] J. Yorke, *Spaces of Solution*, Lect. Notes Op. Res. Math. Econ., **12**, Springer-Verlag, Berlin, 1969, 383–403.

[491] T. Yoshizawa, *Stability Theorey by Liapunov's Second Method*, The Mathematical Society of Japan, Tokyo, 1966.

[492] K. Yosida, *Functional Analysis*, 6th ed., Springer-Verlag, Berlin, 1980.

[493] X. Yu, X. Xiang, and W. Wei, Solution bundle for a class of impulsive differential inclusions on Banach spaces, *J. Math. Anal. Appl.* **327** (2007), 220–232.

[494] D. Yujun, Periodic boundary value problems for functional differential equations with impulses, *J. Math. Anal. Appl.* **210** (1997), 170–181.

[495] D. Yujun and Z. Erxin, An application of coincidence degree continuation theorem in existence of solutions of impulsive differential equations, *J. Math. Anal. Appl.* **197** (1996), 875–889.

[496] F. Zanolin, Continuation theorems for the periodic problem via the translation operator, Univ. of Udine, 1994, preprint.

[497] J. Zhang and J. Hu, Existence of mild solutions of impulsive evolution differential inclusions on half-line, (Chinese) *Math. Appl. (Wuhan)* **23** (2010), 185–193.

[498] Q. J. Zhu, On the solution set of diferential inclusions in Banach space, *J. Diferential Equations* **93** (1991), 213–237.

Index

absolute neighborhood retract, 200
absolute retract, 200
absolutely continuous, 361, 362
acyclic map, 200
acyclic set, 358
acyclic space, 200
acyclically contractible, 200
admissible contraction, 37
almost separably, 360
almost separately valued, 360
antiderivative, 362
approximate selection, 29
Aronszajan results, 201
Arzela–Ascoli theorem, 363
asymptotically stable, 67

(BC), 35
β-condensing, 39
Bielecki-type norm, 147
Bochner integrable, 360
Bochner integral, 359
Bouligand cone, 328
bounded variation, 362

C_0-semigroup, 40
Carathédory map, 11, 35
Castaing representation, 33
Castaing's theorem, 33
Čech homology, 357
characteristic function, 34
closed multimap, 14, 26
co$F$, 29
cofinite topology, 19
cohomological acyclicity, 358
compact carrier, 357
compact map, 12, 27
contractable space, 199
contraction, 12, 31
controllable, 268, 276

decomposable selection, 35
directional topology, 137
dominated convergence theorem, 361

ϵ-net, 38
Eberlein-Kakutani theorem, 367
equicontinuous, 363
equi-integrable, 365
ET, 9
exponentially bounded, 42, 333

Filippov's, 99
Filippov regularization, 2
Filippov–Wazewski theorm, 303
finite measure space, 365
fixed moments, 58
Fréchet space, 12
Fubini's Theorem, 361

Γ-continuous, 137
generator, 42
generator of an integral resolvent, 332
graph measurable, 32

H-continuous, 30
H-lower semicontinuous, 30
H-upper semicontinuous, 30
Härmondar's formula, 18
Haudorff metric, 15
Hausdorff distance, 15
Hausdorff measure of noncompactness, 38
Hausdorff pseudometric, 30
Hille–Yosida operator, 43, 45
Hodgkin–Huxley, 10
homotopy, 199, 227

impulsive stochastic differential equations, 321
infinitely controllable, 268

infinitesimal generator, 41
integrably bounded, 365
integral resolvent family, 39, 332
integral solution, 256, 268, 276
integrated semigroup, 42
IPM, 9
Itô integral, 307

K-measurable, 32
Kruger–Thiemer model, 8
Kuratowski–Ryll–Nardzewski selection, 33

L^1-Carathédory, 11, 35
l.s.c., 20
Laplace transformation, 332
Lasota–Yorke Approximation, 201
Lebesgue point, 362
Lipschitz, 31
locally compact, 27
locally compact map, 27
Lotka–Volterra model, 8
lower semicontinuous (l.s.c.), 20, 21

Mazur's Lemma, 32, 367
measurable-locally-Lipschitz (mLL), 208
measurable multifunction, 32
measure of noncompactness, 38
measures of noncompactness, types, 38
metric retraction, 219
Michael selection, 28
Michael's selection theorem, 29
mild solution, 232, 233, 292, 310, 335
modulus of equicontinuity, 38
modulus of fiber noncompactness, 38
multifunction, 21
multivalued function, 13

neighborhood retract, 199
neutral functional differential equations, 155
Niemytzki operator, 36
nonatomic probability measure, 35
nondegenerate, 43
nondensely defined operators, 254
Nonlinear Alternative, 12, 37
$N^p(x, K)$ the normal cone, 327
nuclear operator, 307

Poincaré operators, 226
projection, 327
proper map, 201, 367
proximal normal, 328
proximate retract, 219
pulse-vaccination model, 9

quasicompact, 27

R_δ-contractible, 201
R_δ-set, 199
relaxation impulsive differential inclusion, 303
retract, 199
Riemann–Liouville kernel, 332

selection function, 28
selection theorem of Browder, 30
semicompact, 39, 366
semilinear functional differential inclusions, 254
seminorms, 12
σ-Ca-selectionable, 208
simple functions, 359
stable, 67
state-dependent impulses, 154
stochastic differential equation, 306
strong Carathédory map, 222
strongly continuous cosine family, 41
Strongly dissipative, 131
Strongly measurable, 32
subdifferential, 126
superposition operator, 364
support function, 18
sweeping process, 327

terminal (or target) problem, 245

u.s.c., 20
uniformly asymptotically stable, 67
uniformly continuous, 333
uniformly equicontinuous, 363
uniformly integrable, 365
uniformly stable, 67
unstable, 67
upper-Scorza–Dragoni, 209
upper semicontinuous (u.s.c.), 20, 21, 131

Vietoris continuous, 20, 21
Vietoris topology, 18, 20

weakly relatively compact, 366